Herwig O. Gutzeit and
Jutta Ludwig-Müller

Plant Natural Products

Related Titles

Dewick, P.M.

Medicinal Natural Products - A Biosynthetic Approach, 3rd Edition

2009

ISBN: 978-0-470-74167-2, also available in digital formats

Wink, M.J. (ed.)

Annual Plant Reviews Volume 40 - Biochemistry of Plant Secondary Metabolism, 2nd Edition

2010

ISBN: 978-1-405-18397-0, also available in digital formats

Wink, M. (ed.)

Annual Plant Reviews Volume 39 - Functions and Biotechnology of Plant Secondary Metabolites, 2nd Edition

2010

ISBN: 978-1-405-18528-8, also available in digital formats

Ashihara, H., Crozier, A., Clifford, M. (eds.)

Plant Secondary Metabolites – Occurrence, Structure and Role in the Human Diet

2006

ISBN: 978-1-405-12509-3, also available in digital formats

Crozier, A., Ashihara, H., Tomás-Barbéran, F. (eds.)

Teas, Cocoa and Coffee - Plant Secondary Metabolites and Health

2012

ISBN: 978-1-444-33441-8, also available in digital formats

Krauß, G., Nies, D.H. (eds.)

Ecological Biochemistry
Environmental and Interspecies Interactions

2014

ISBN: 978-3-527-31650-2

Herwig O. Gutzeit and Jutta Ludwig-Müller

Plant Natural Products

Synthesis, Biological Functions and Practical Applications

WILEY Blackwell

Authors

Prof. Dr. Herwig O. Gutzeit
Technische Universität Dresden
Institute of Zoology
Helmholtzstr. 10
01069 Dresden
Germany

Prof. Dr. Jutta Ludwig-Müller
Technische Universität Dresden
Institute of Botany
Helmholtzstr. 10
01069 Dresden
Germany

Library of Congress Card No.: applied for

British Library Cataloguing-in-Publication Data
A catalogue record for this book is available from the British Library.

Bibliographic information published by the Deutsche Nationalbibliothek
The Deutsche Nationalbibliothek lists this publication in the Deutsche Nationalbibliografie; detailed bibliographic data are available on the Internet at <http://dnb.d-nb.de>.

Wiley-Blackwell is an imprint of John Wiley & Sons, formed by the merger of Wiley's global Scientific, Technical, and Medical business with Blackwell Publishing.

Print ISBN: 978-3-527-33230-4 (Softcover)
978-3-527-68198-3 (Hardcover)
ePDF ISBN: 978-3-527-68197-6
ePub ISBN: 978-3-527-68200-3
Mobi ISBN: 978-3-527-68199-0

Cover Design Formgeber, Mannheim

Typesetting Thomson Digital, Noida, India

Printing and Binding betz-druck GmbH, Darmstadt

Printed on acid-free paper.

Contents

Preface

All organisms synthesize biologically active organic compounds that confer a selective advantage for the respective species. Natural substances play a central role in the evolutionary struggle for survival and hence Nature has invented a plethora of organic substances with different activities that, for example, help to cope with stressful situations, permit communication with other organisms, or deter enemies. We feel that this important interdisciplinary topic should be represented in student courses of biochemistry, pharmacology, and, of course, biology. The book is based on a lecture course on the biology of natural substances, which is part of the curriculum for biology students at the TU Dresden.

The book covers a wide range of different aspects concerning the importance of secondary metabolites for the plants in their respective ecological niche. We discuss, for example, the biosynthesis of biologically active substances, their role in the development of plants, and their interaction with beneficial organisms as well as pathogens, and present examples of the communication between plants and animals feeding on the plants. The toxicity of some substances in vegetarian food is of concern and a toxicological assessment is essential. For this reason, we outline basic principles of toxicological tests with relevant natural compounds. The public discussion about health-promoting substances addresses important scientific questions that are not easy to answer. In the chosen examples of natural substances affecting human health, we summarize the results of intensive research and illustrate the complexity of the elicited reactions. Estrogenic compounds and hallucinogenic drugs exemplify the perils and promises of natural compounds. Are some compounds in our food disease-preventing? In the book, we discuss current attempts to answer this important question. Finally, the inspiration of Nature for medicinal chemistry will be briefly addressed.

The focus of the book is on plant secondary metabolites but the reader is occasionally reminded that animals and bacteria also contain most interesting compounds. In view of the vast published literature on the subject, we had to resist the temptation to address every interesting aspect concerning the biology of natural substances. Rather, we present current concepts of general biological mechanisms and illustrate the essential interdisciplinary approach in the functional analysis of natural substances. We hope that the reader will share our enthusiasm for the subject and will find this book stimulating and an incentive for further studies.

Dresden
February 2014

Herwig O. Gutzeit
Jutta Ludwig-Müller

1 Biosynthesis and Chemical Properties of Natural Substances in Plants

The number of known so-called "secondary metabolites" (also referred to as "natural products") that have been discovered to date is increasing at a constant rate. Yet, it is not only plants (as described in this book) that produce these bioactive compounds; rather, other organisms such as bacteria, fungi, sponges, as well as animals, are also capable of synthesizing a plethora of these metabolites.

Whilst some of these metabolites are discussed in Chapters 4 and 5, a large number remain undiscovered. Moreover, secondary metabolites often possess interesting pharmacological properties, and therefore their characterization is very important. It should not be forgotten that plants synthesize these compounds as part of their own survival strategies, typically as defense compounds or as signals for pollinators or symbionts. In addition, recent evidence has pointed to additional roles for secondary metabolites in plant development. Although the term "secondary metabolites" perhaps infers a less important role for these compounds than those involved in primary metabolism, this is not the case. In fact, many essential and nonessential compounds in this group are found in plants, and even so-called "nonessential materials" can play a role in a plant's responses against abiotic and biotic stress. In this situation, the deletion of a biosynthetic pathway would cause damage to the plant, even if the pathway was not needed under favorable conditions. Interest in the secondary metabolites of plants was further increased when more sensitive analytical instruments became available, as well as genome sequence data for many plant species. Together, these tools allow the details of biosynthetic pathways to be investigated, an example being biotechnological manipulation. Notably, attention also began to be focused on the evolutionary aspects of secondary metabolite synthesis.

In general, secondary metabolites occur as complex mixtures (Figure 1.1). Their biosynthesis can be influenced by a variety of factors during development, in addition to stress, which makes the determination of their complete pattern essentially impossible. Whilst secondary metabolites can occur in the tissues as active compounds, they can also be synthesized as inactive compounds that must be transformed into active products; such metabolites that pre-exist are known as "phytoanticipins." Compounds that are biosynthesized under stress conditions are typically not detectable in unstressed tissues; when they are synthesized after the invasion of plants by various pests these metabolites are termed "phytoalexins."

Plant Natural Products: Synthesis, Biological Functions and Practical Applications, First Edition.
Herwig O. Gutzeit and Jutta Ludwig-Müller.

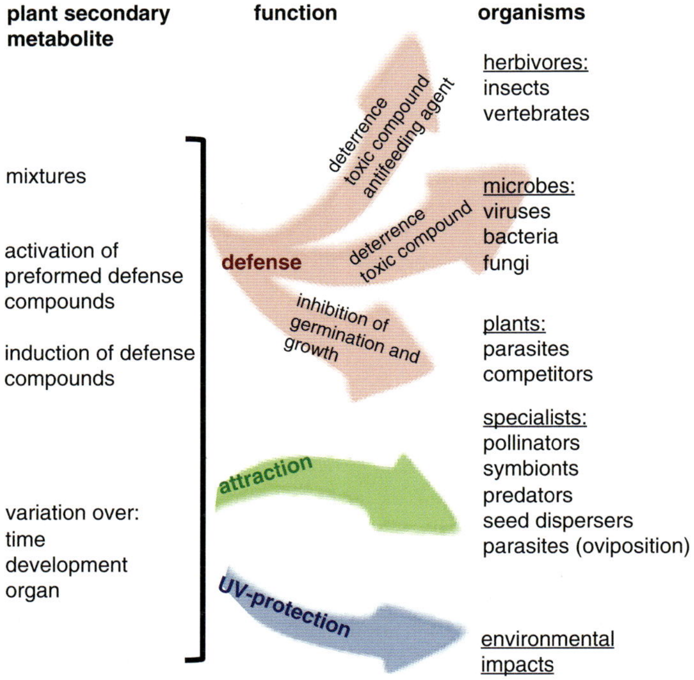

Figure 1.1 Correlation between secondary metabolites, their occurrence, and various functions. Secondary metabolites occur always as complex mixtures, and each plant has a specific set of compounds. The composition varies over time, development, and in different tissues and organs. In addition, the metabolites can be inactive and must be metabolized to the active form. Others will be produced only in response to biotic stress factors and other stress situations. The latter have a function in deterring a variety of herbivores, but can also attract insects for pollination.

Secondary metabolites play a variety of roles in the interaction of plants with their environment, but they may also be involved in plant development. They can function in defense mechanisms against herbivores such as nematodes, insects or vertebrates, as well as against microbes and competing plants. Metabolites may also play important roles in the attraction of symbionts, such as insects for pollination or zoophagous insects in multitrophic (when more than two organisms are involved) defense mechanisms. Animals that are required for seed-dispersal purposes are attracted by color and aroma compounds; however, some specialist insects and vertebrates have developed mechanisms to cope with the toxic effects of these compounds, and may generate an ecological niche in recognizing a specific host plant for either feeding or oviposition. Abiotic signals also contribute to the specific patterns of secondary metabolites; for example, some compounds are good protectors against ultraviolet (UV) light and photo-oxidation (i.e., oxidation reactions induced by light).

In order to understand the complexity of these compounds it is essential to have a good knowledge of their biosynthesis and the regulation of genes that encode the enzymes involved in syntheses, under changing conditions. A brief description of the biosynthesis of selected compounds is provided in the following sections of this chapter, but for additional information relating to the enzymes and their biochemistry, the reader is referred to respective textbooks on plant physiology and plant biochemistry.

1.1 Selected Classes of Secondary Metabolites

1.1.1 Occurrence and Compartmentation

The patterns of secondary metabolites will differ depending on the species, the developmental stages, and the stress situations occurring in that species. The differences in pattern are the result of the metabolite biosynthesis, and the regulation of genes encoding the biosynthetic enzymes, transcription factors and transport proteins (see Sections 1.1.3 and 1.1.4). It is therefore difficult to generalize the metabolic patterns for a given plant species. The occurrence of some metabolites can be confined to a certain plant species, such as betalains in the Caryophyllales, glucosinolates in the Brassicales (with few exceptions), and polyacetylenes mainly in the Asteraceae and Apiaceae families. In the latter case, the compounds are important as bioactive substances in food crops (see Chapter 6) such as carrot (*Daucus carota*), celery (*Apium graveolens*) and fennel (*Foeniculum vulgare*). A compound from the Asteraceae *Carlina acaulis* has a strong antimicrobial activity and is also toxic for trypanosomes. Based on their chemical structures, the secondary metabolites can be placed into different classes (Table 1.1).

The synthesis of secondary metabolites can occur in all plant organs, including the roots, shoots, leaves, flowers, fruit, and seeds. Some metabolites are stored in specific compartments, which may be either whole organs or specialized cell types. Within these compartments the concentration of toxic secondary metabolites may be very high, so that they can exert an efficient defense against herbivores. For example, the glandular trichomes are often rich in toxic compounds, and some pathways – an example being the synthesis of the labdane diterpene Z-abienol in tobacco – occur exclusively in these trichomes. The glandular trichomes of peppermint leaves also contain large amounts of aromatic oils. The latex of *Euphorbia* is a good example of a toxic mixture of compounds being stored in specialized cells, the laticifers (Figure 1.2).

Lipophilic compounds can be found in the membranes of organelles or in the endoplasmic reticulum (ER), although specific storage vesicles have been reported for some compounds, such as the alkaloid berberin. Hydrophilic compounds are stored preferentially in vacuoles in the inner parts of the organelles and in the cytosol, as well as in the extracellular space (apoplast). Biosynthesis can occur in the cytosol, organelles and the ER, whereas transcriptional control takes place in the nucleus.

Table 1.1 The major classes of secondary metabolites based on their chemical structures. Chemical structures are from The PubChem Project.

Class of compound	Example	Structure
N-containing		
Alkaloids	Nicotine	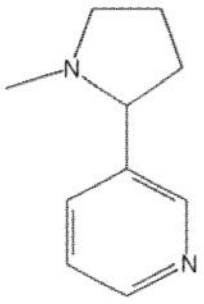
Nonproteinogenic amino acid	Canavanine	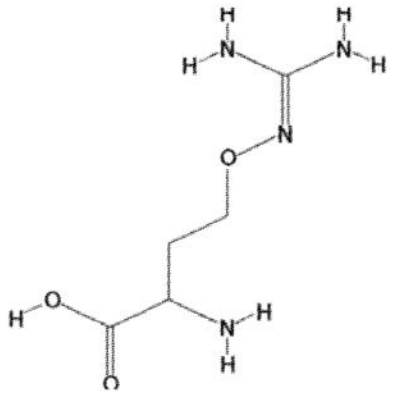
Amines	Butylamine	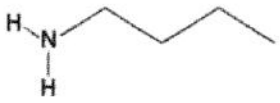
Cyanogenic glycosides	Dhurrin	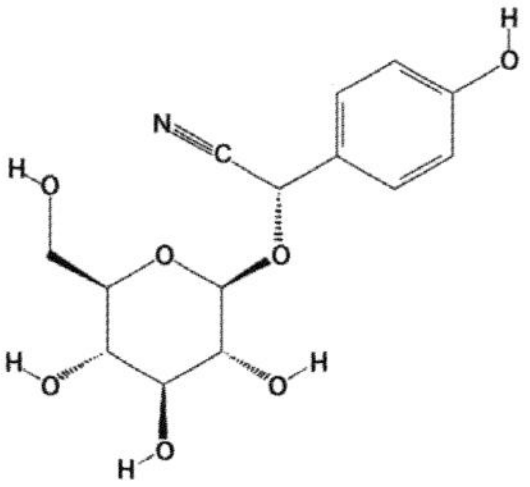
N- and S-containing		
Glucosinolates	Glucobrassicin	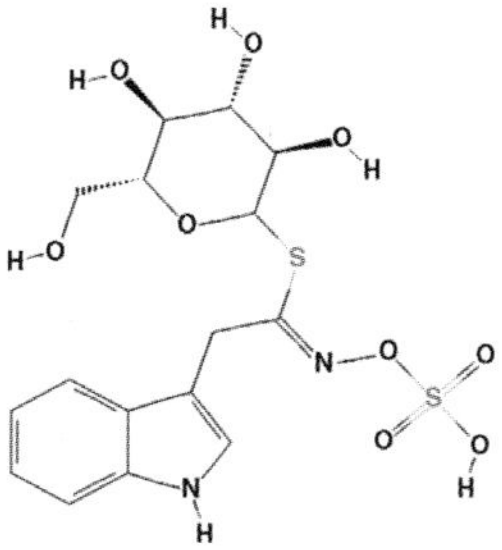
Without N		
Anthraquinones	Emodin	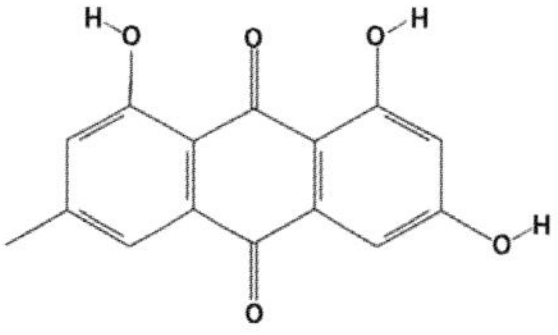

Flavonoids	Quercetin	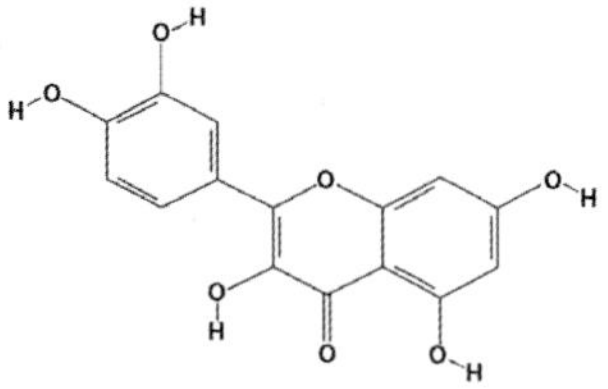
Polyacetylenes	Falcarinol	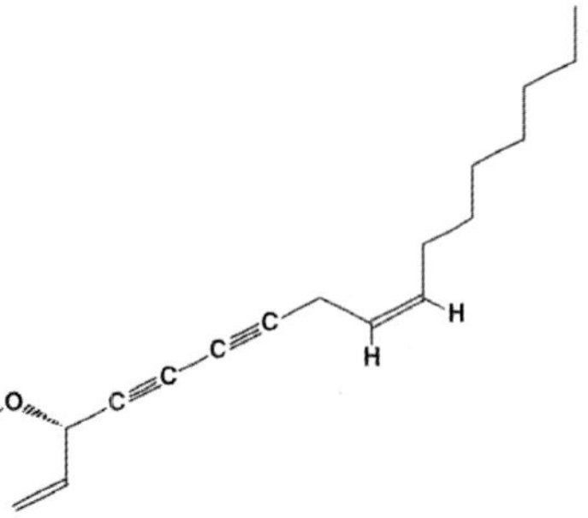
Polyketides	Aloeresin	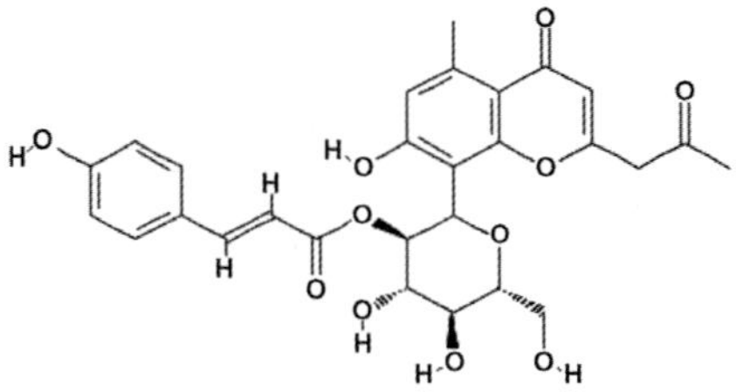
Phenylpropanoids	Rosmarinic acid	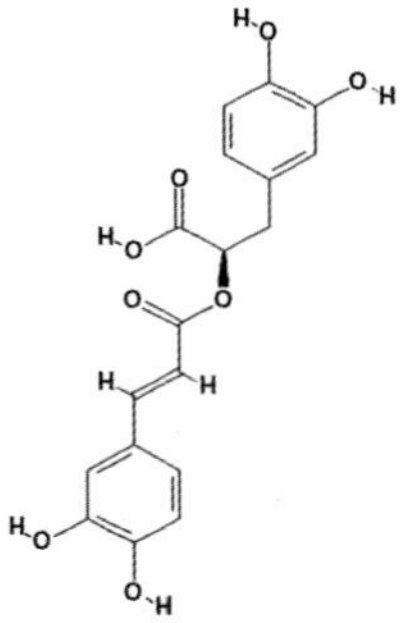
Terpenes		
Mono-	Thymol	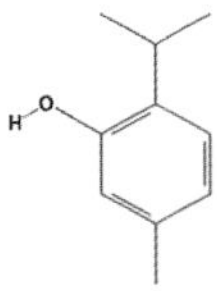
Sesqui-	Helenaline	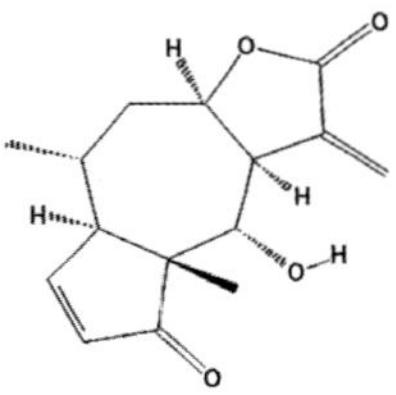

(continued)

Table 1.1 (*Continued*)

Class of compound	Example	Structure
Di-	Gingkolide	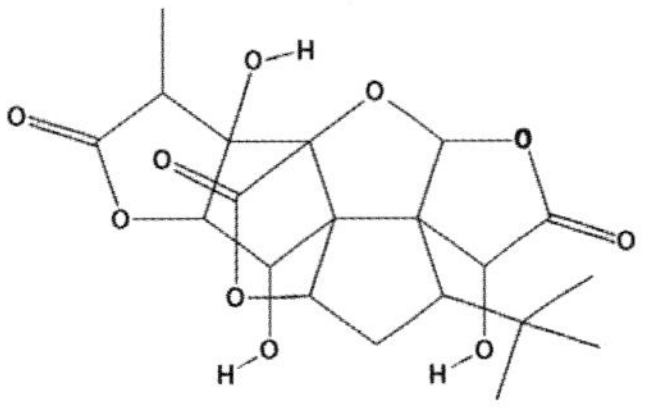
Tri-	Oleanolic acid	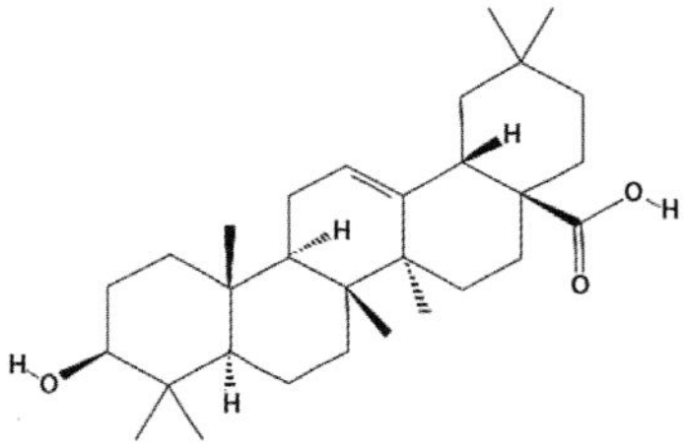
Tetra-	ß-Carotin	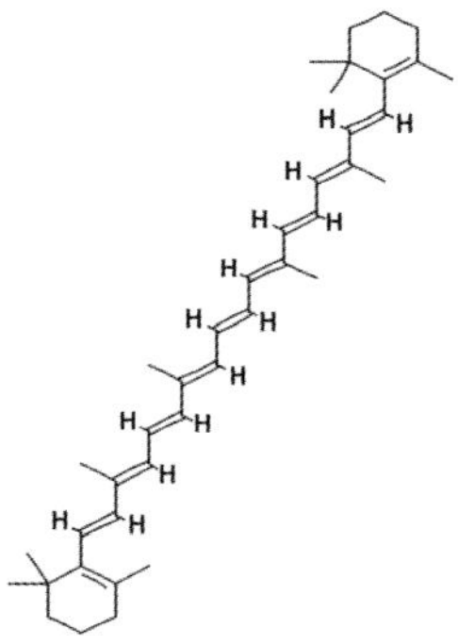
Poly-	Rubber	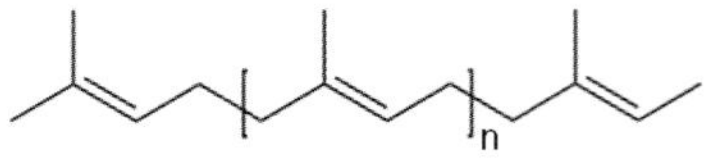

Some metabolites occur only in specialized subcellular compartments. In such cases, the tissue- and cell-specific localizations will depend on the solubility of the compound, notably whether it is lipophilic or hydrophilic (Table 1.2). Most hydrophilic compounds accumulate in vacuoles, and are typically present as inactive precursor substances that must be activated either chemically and/or enzymatically; this is the case for cyanogenic glycosides and glucosinolates. Antimicrobial tannins, which exert their protective effects in the extracellular space, are located in the apoplast, while the laticifers contain both hydrophilic and lipophilic compounds. The metabolite patterns are species-specific, but all are toxic; if human skin comes into contact with a latex extract, a severe irritation results. Lipophilic storage compartments also contain antimicrobial and/or defense compounds. Metabolites located

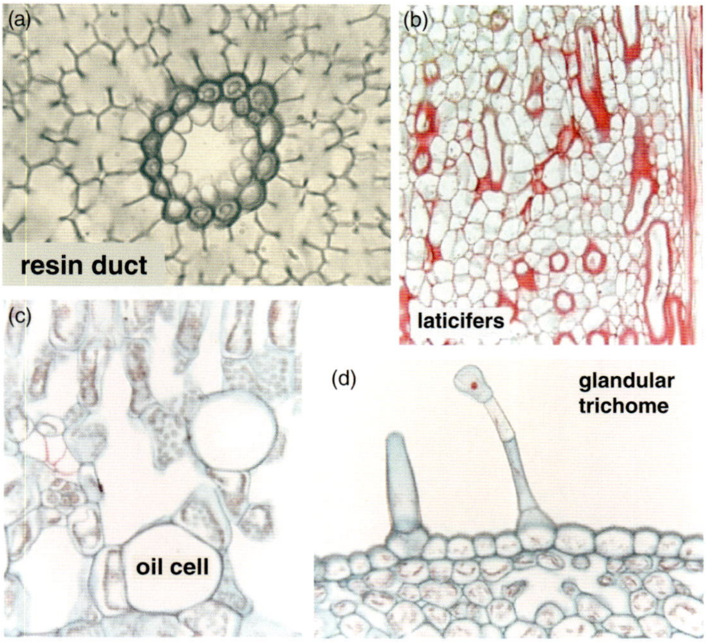

Figure 1.2 Some examples of secondary metabolite-containing plant compartments or cells. (a) Resin duct; cross-section through the needle of a *Pinus* species; (b) Laticifers; longitudinal section through the shoot of an *Euphorbia* species; (c) Oil cells; cross-section through a magnolia (*Magnolia* sp.) leaf; (d) Trichome; section through a foxglove (*Digitalis* sp.) leaf with glandular trichome. Image (a) from Bernd Schulz and Jutta Ludwig-Müller, Technische Universität Dresden, Germany; images (b–d) from James D. Mauseth, University of Texas, USA.

in the membranes of the plastids and mitochondria are necessary for primary metabolism, an example being electron-transport chains and light absorbance. Plants secrete volatile compounds either to attract insects for pollination purposes, or zoophagous insects to defend themselves against herbivorous insects (Figure 1.1; see also Chapter 2). Plants may also deliver nonvolatile compounds into the soil, thereby altering the rhizosphere (termed allelopathy; see Section 2.6) that leads to an inhibition of the growth of other, competing, plants.

The process of compartmentation has three important implications:

1) The plant itself may not be in contact with the toxic compounds, because it is only after wounding or attack by other organisms that the metabolites will be released from the storage compartments.
2) The biosynthetic pathways are also compartmentalized, but this necessitates the involvement of transport processes for metabolite movement.
3) The target molecules for the secondary metabolites may be present in all compartments; however, the target is not necessarily colocalized with an active compound.

Table 1.2 Tissue-specific and subcellular compartmentation of hydrophilic and lipophilic compounds.

Characteristics	Storage compartment		Class of compounds
	Tissue	Cell	
Hydrophilic	Laticifer		Some alkaloids
			Nonproteinogenic amino acids
			Digitalis glycosides
			Cyanogenic glycosides
		Vacuole	Many alkaloids
			Nonproteinogenic amino acids
			Saponins
			Glycosides
			Flavonoids
			Anthocyanidins
			Betalains
			Tannins
			Cyanogenic glycosides
			Glucosinolates
			Amines
		Apoplast	Tannins
		Cytoplasm	Most hydrophilic compounds during their biosynthesis
		Nucleus	Flavonoids
Lipophilic	Trichome		Terpenoids
	Resin		Flavonoids
	duct		Terpenoids
	Laticifer		Diterpenes
			Quinones
			Flavonoids
			Polyterpenes
	Oil cell		Anthraquinones
			Terpenoids
		Cuticula	Wax
			Flavonoids
			Terpenoids
		Mitochondrion	Alkaloids (*Conium*)
		Plastid	Terpenes
			Alkaloids (coniin, quinolizidine, coffein)
		Specialized vesicles	Alkaloids (protoberberin)
		Plastid membrane	Ubiquinones
Tetraterpenes			
		Endoplasmic reticulum	Lipophilic substances during enzymatic hydroxylation steps

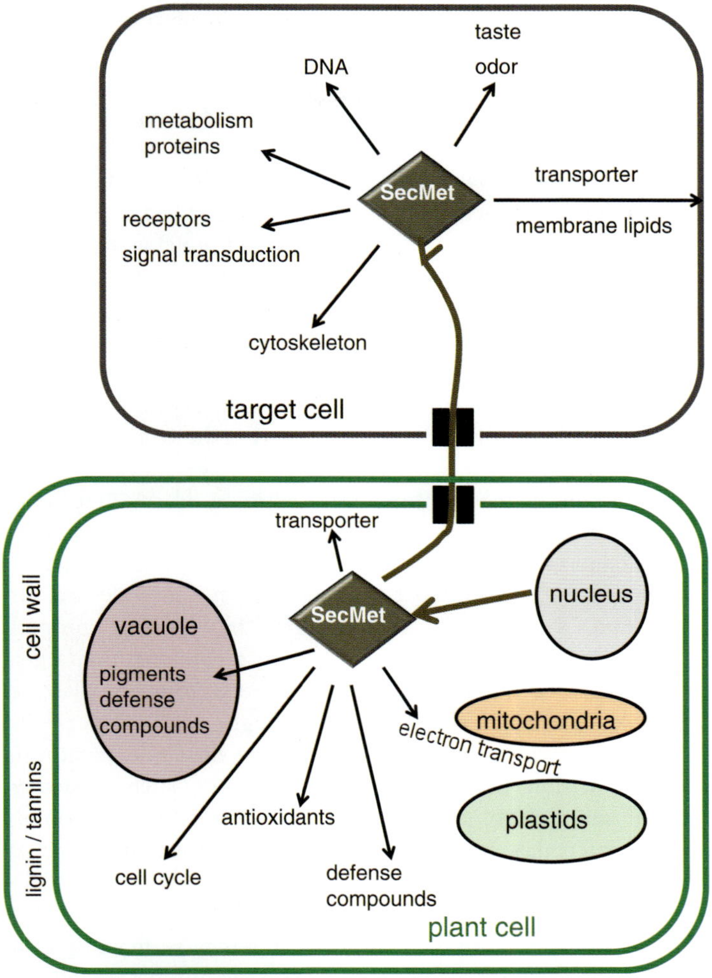

Figure 1.3 Cellular and molecular targets for secondary metabolites in plants and other organisms. In the plant cell (green), secondary metabolites (SecMet) are synthesized and can exert their effects in different compartments, such as the electron-transport chain in organelles, transporters at the plasma membrane, as antioxidants, and as fortification elements in the cell walls. They are stored in vacuoles or cell wall compartments as coloring or defense compounds. In the cells of the target organisms (brown), into which the metabolites must be transported, the targets can be in all compartments. Target molecules encompass transporters and channels, enzymes and proteins, nucleic acids, the cytoskeleton, receptor molecules and signal transduction chains, as well as membrane lipids.

A few cellular targets will be discussed in the following section (see Figure 1.3), and further descriptions are provided in Chapters 5 and 6, with details of the compounds' biological activities. Many compounds that are used to protect plants against microbes and herbivores exert their biological activities outside the plant tissues; however, if these compounds were to exert any toxic effect on the plant, their compartmentation would provide protection against self-poisoning.

Membrane targets can be found at the plasma membrane, ER or Golgi apparatus, where the functions of molecules such as transporter and channels, and also of membrane receptors, can be altered. The electron-transport chains of both plastids and mitochondria may also serve as targets, and plant compounds may inhibit their function. In addition, the structural proteins of the cell wall and of enzymes may be affected, while the cytoskeleton, with its component proteins myosin and tubulin, may be a target for secondary metabolites and result in an altered cell cycle (see Figure 2.5). Within the nucleus and the organelles, desoxyribonucleic acid (DNA) is a target for interaction with secondary metabolites; it is well known that some molecules alkylate DNA, which in turn affects its replication, transcription, and repair mechanisms. It should be noted that, in order to be active in these compartments, the metabolites must be transported across the respective membranes (see Section 1.1.4).

Although the major role of these compounds seems to be the defense of plants against other organisms, humans are also affected by secondary metabolites (see Chapters 3.2 and 6). When defending plants against microbes, the toxic effect of the secondary metabolites on one or several target molecules constitutes a major role, although their "bitter-tasting" properties will generally deter attacks by herbivores. In these organisms, the targets may be located as indicated in Figure 1.3, with the exception of plastids, which only play a role as target in allelopathy. This might not be the case for parasitic plants, however, which are dependent on the host photosynthesis for their nutrient acquisition. Many secondary metabolites, such as sterols, can influence membrane fluidity (see Figure 2.28), while other compounds may interact with or intercalate in DNA, and are thus cytotoxic; mitosis may also be inhibited (see Figure 2.5).

Secondary metabolites can also inhibit the activity of various enzymes. Alkaloids and amines are often similar to the neurotransmitters of vertebrates, or to hormones involved in the development of insects, and thus mimic their effects. Cyanogenic glycosides release toxic cyanide, which inhibits cytochrome c oxidase in the mitochondrion and, as a consequence, energy production in the form of adenosine triphosphate (ATP). Isothiocyanates, which are released from glucosinolates, can influence membrane fluidity. Terpenes also interact with membranes, though some demonstrate similarities to biologically important sterols in fungi and animals. For example, the Na^+-, K^+-ATPase, which is important in animals, can be inhibited by several groups of metabolites, including the pyrrolizidine alkaloid monocrotalin, the digitalis glycosides, and/or polyketides such as anthraquinones. The latter also interfere with the formation of cyclic AMP (cAMP) by inhibiting adenylate cyclase, which itself influences many signal transduction pathways. Interestingly, plants contain cyclic guanosine monophosphate (cGMP) rather than cAMP. Flavonoids also have the ability to inhibit enzymes, but may cause additional mutagenic and toxic effects on DNA via alkylation.

1.1.2
Biosynthesis

The biosynthetic pathways derive from various precursors of primary metabolism (Figure 1.4). The precursor is defined as a molecule used by a biosynthetic enzyme as a substrate and converted to a product. The product can be an intermediate in the pathway, and in this case it is used as precursor for the next biosynthetic enzyme, or it is the final product of the reaction chain.

$$\begin{array}{l} \text{substrate (precursor)} \rightarrow \text{intermediate} \rightarrow \text{intermediate} \rightarrow \text{product} \\ \qquad\qquad\qquad\qquad\qquad\qquad\quad \Updownarrow \\ \qquad\qquad\qquad\qquad\qquad\quad \text{precursor} \rightarrow \text{intermediate} \rightarrow \text{product} \end{array}$$

In a complex reaction scheme, which has many junctions, an intermediate is simultaneously also a precursor for another part of the pathway. In Figure 1.4, shikimic acid would be an intermediate for the amino acid metabolism, and also

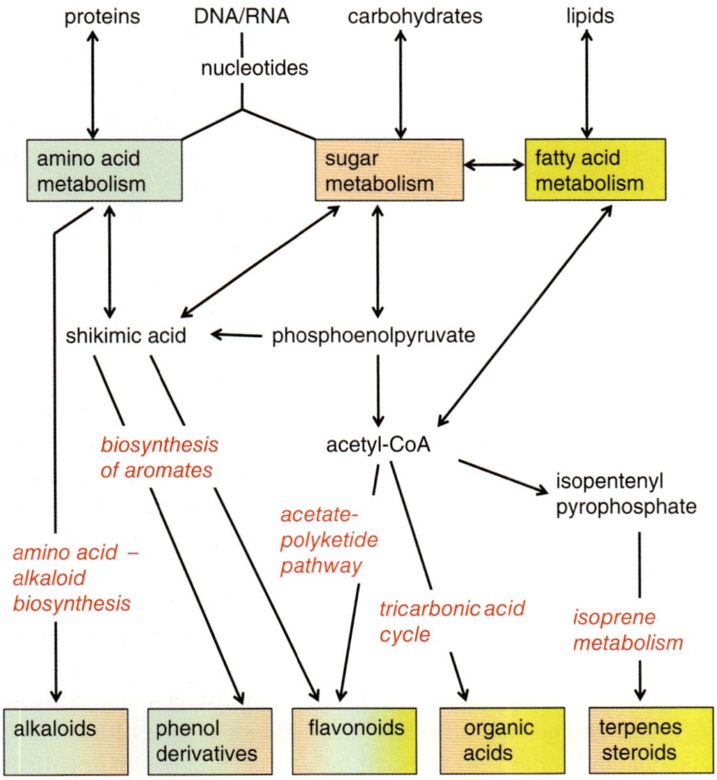

Figure 1.4 General scheme of biosynthetic pathways and precursors for the major classes of secondary metabolites. Pathways deriving from aromatic amino acids are shown in green; pathways deriving from carbohydrates are shown in brown; pathways deriving from lipids are shown in yellow. Modified from Larcher, W. (1995) *Physiological Plant Ecology*, 3rd edn, Springer-Verlag.

a precursor for the biosynthesis of aromatic secondary metabolites. Similar precursors are used within one class of compounds for the biosynthesis, but the same precursors can be also used for a range of different metabolites.

The final products are derived from three major classes of compounds and marked in the same color in Figure 1.4. The major precursors are derived from protein (amino acids), carbohydrate (sugars) and lipid (fatty acid) metabolism. The biosynthetic pathway for aromatic amino acids is one of the major sources of aromatic compounds such as phenols, flavonoids and some alkaloids. Acetyl-CoA is a central metabolite formed by glycolysis and also via the β-oxidation of fatty acids, and is used in the tricarboxylic acid cycle in the synthesis of organic acids, which are also precursors for secondary metabolites. In addition, acetyl-CoA is involved in the synthesis of terpenes, which form a distinct class of metabolites. In the following sections some selected biosynthetic pathways will be discussed, but for more detail the reader is referred to textbooks on plant biochemistry. Knowledge of the biosynthetic pathways of natural compounds is essential for the targeted manipulation of these pathways in biotechnology. In addition to knowledge on the enzymes and genes involved in these biosyntheses, it is important to understand the transcriptional control of the genes. Therefore, a few examples will be provided where the transcriptional control of pathways has been well elucidated, notably of the glucosinolates and flavonoids (Section 1.1.3). Finally, details on biosynthetic pathways and corresponding genes will help to provide an understanding of how these secondary metabolites have evolved (Section 1.2).

When the basic skeleton of a secondary metabolite has been synthesized, it is the many enzymatic modifications (which are added later to the structure) that form the main reason for the large variations among these molecules. The chemical reactions involved include oxidations, hydroxylations, reductions, acylations, methylations, prenylations and glycosylations (see also Table 2.5).

1.1.2.1 **Alkaloids**

Within the group of alkaloids many compounds with stimulating, hallucinogenic and analgesic properties can be found (see Chapter 5). Many of these compounds are toxic, or they can be converted from nontoxic to toxic compounds. Depending on the chemical structure of an alkaloid, different precursors are required for the biosynthesis. Examples of the major groups of alkaloids, together with a typical chemical structure, a plant in which the compound is found and the main precursor(s) involved in the biosynthesis, are shown in Table 1.3.

The biosynthetic pathway for an individual compound is very complex. Many different precursors are involved for the various pathways, including not only aromatic amino acids such as tryptophan, tyrosine and phenylalanine but also aspartate, glutamine, lysine, glycine and valine for other alkaloids (Figures 1.5 and 1.6; Table 1.3). In addition, the nonproteinogenic amino acid ornithine is an important precursor for various alkaloids. Anthranilic acid, from which tryptophan is synthesized, is the precursor for acridine alkaloids, and thus the shikimate pathway (which is responsible for the biosynthesis of all aromatic amino acids) is involved. For several alkaloids two different precursors are needed for the biosynthetic

Table 1.3 The major classes of alkaloids and their precursors, according to their chemical structure. One example, with chemical structure, and one plant species where it occurs, is given for each class. Chemical structures are from The PubChem Project.

Basic structure	Precursor	Example of structure (Plant)	Formula basic structure
Acridine	Anthranilate 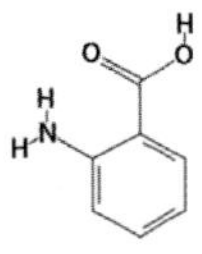	OH-N-Methylacridone (*Ruta graveolens*)	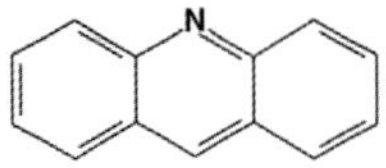
Benzophenanthrinidine	Tyrosine 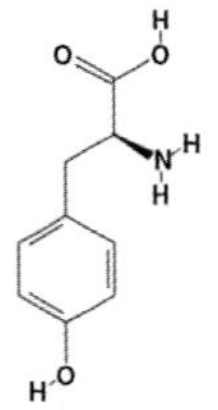	Sanguinarine (*Eschscholzia californica*)	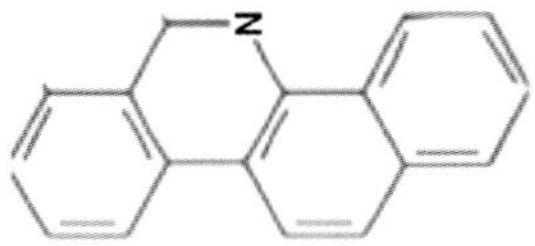
Quinazoline	Ornithine 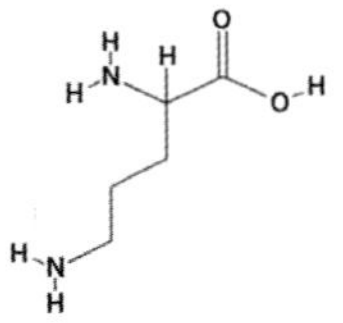	Peganine (*Peganum harmala*)	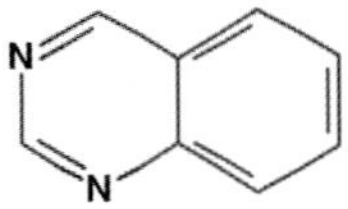
Quinoline	Anthranilate Tryptophan 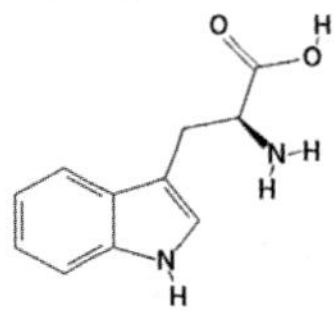	Quinine (*Cinchona officinalis*)	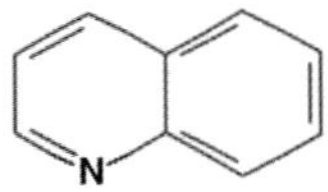
Quinolizidine	Lysine 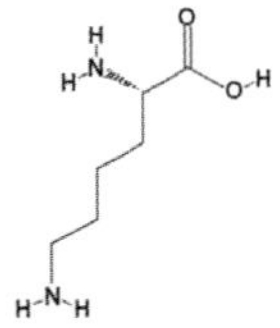	Lupanine (*Lupinus polyphyllus*)	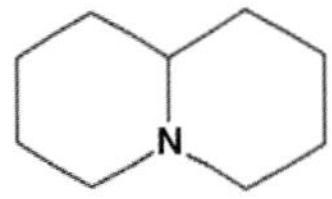
Imidazole	Histidine 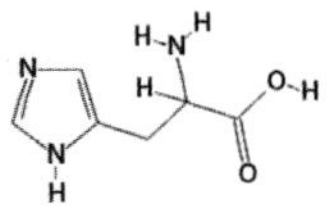	Pilocarpine (*Pilocarpus jaborandi*)	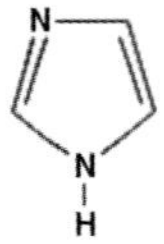

(*continued*)

Table 1.3 (*Continued*)

Basic structure	**Precursor**	**Example of structure (Plant)**	**Formula basic structure**
Indole	Tryptophan	Serotonin (*Musa x paradisiaca*)	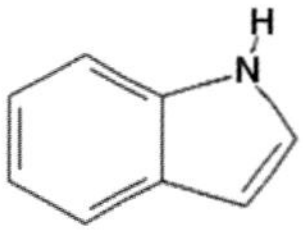
Isoquinoline	Tyrosine	Morphine (*Papaver somniferum*)	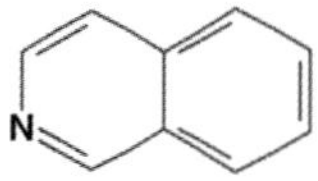
Phenylalkylamine	Phenylalanine 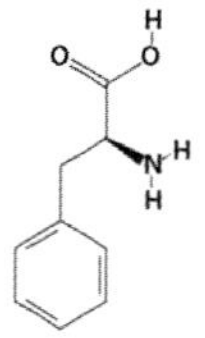	Cathinone (*Catha edulis*)	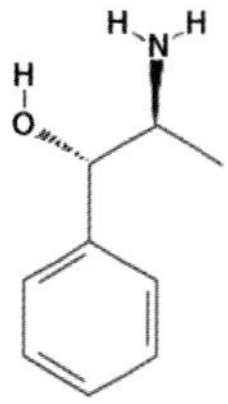
Piperidine	Lysine	Coniine (*Conium maculatum*)	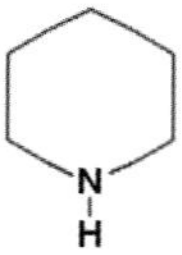
Purine	Aspartate 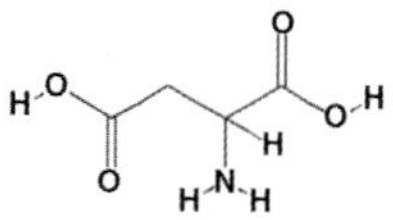 Glycine 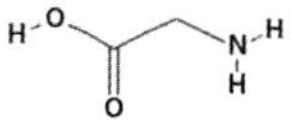 Glutamate 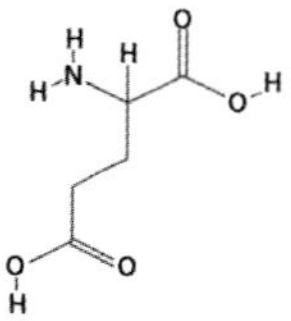	Caffeine (*Coffea arabica*)	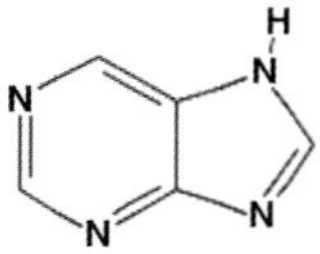
Pyridine	Aspartate Ornithine	Nicotine (*Nicotiana tabacum*)	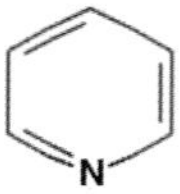
Pyrrolidine	Aspartate Ornithine	Muscarine (*Amanita muscaria*)	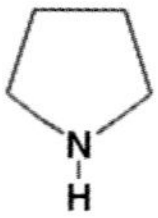

Pyrrolizidine	Ornithine Valine 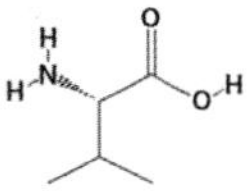	Senecionine (*Senecio jacobaea*)	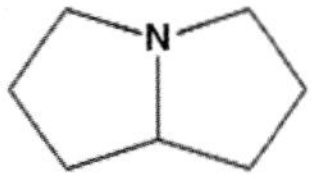
Terpene indole	Tryptophan Isoprene 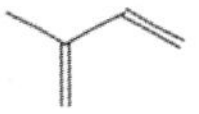	Physostigmine (*Physostigma venenosum*)	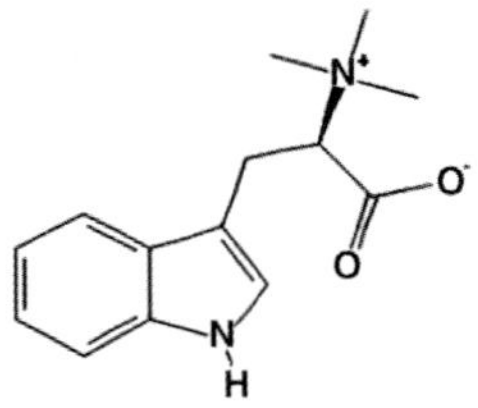
Tropane	Ornithine Arginine 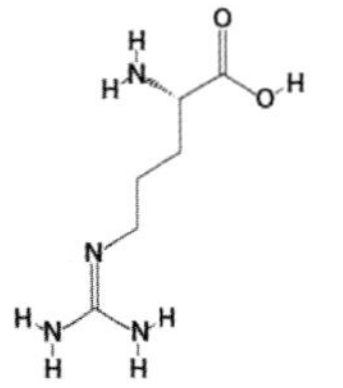	Cocaine (*Erythroxylon coca*)	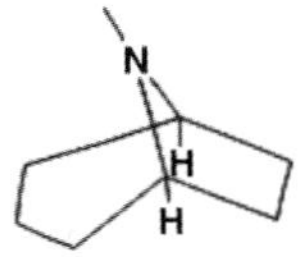
Tropolone	Tyrosine Phenylalanine	Colchicine (*Colchicum autumnale*)	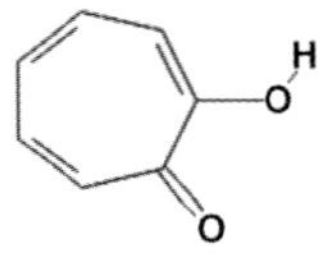

pathways. In the case of terpene indole alkaloids (Figures 1.5 and 1.6; see also Figure 1.34), it is not only tryptophan that is involved as a precursor for the indole moiety, but also monoterpenes for the synthesis of side chains. Another example is the biosynthesis of the tropane alkaloids hyoscyamine and scopolamine, where ornithine and phenylalanine are required for the different parts of the molecule (Figure 1.5; Table 1.3).

Although many alkaloids are of importance to humans (Chapter 5), their biosynthetic pathways are often not fully understood. However, it is important to know as much as possible about the precursors, enzymes and limiting steps in the pathway, if these secondary metabolites should be changed in abundance by breeding or biotechnological methods employed in plants (Section 1.3). It is important that changes in one pathway do not lead to limitation of a precursor for a second pathway, in case the two pathways compete for the same precursor. This can be observed especially for aromatic amino acids, which play many roles in different pathways in the plant; examples include their requirement in proteins, a role in the synthesis of cell wall components, and as precursors for the plant hormone indole-3-acetic acid (IAA). In secondary metabolite synthesis, amino acids

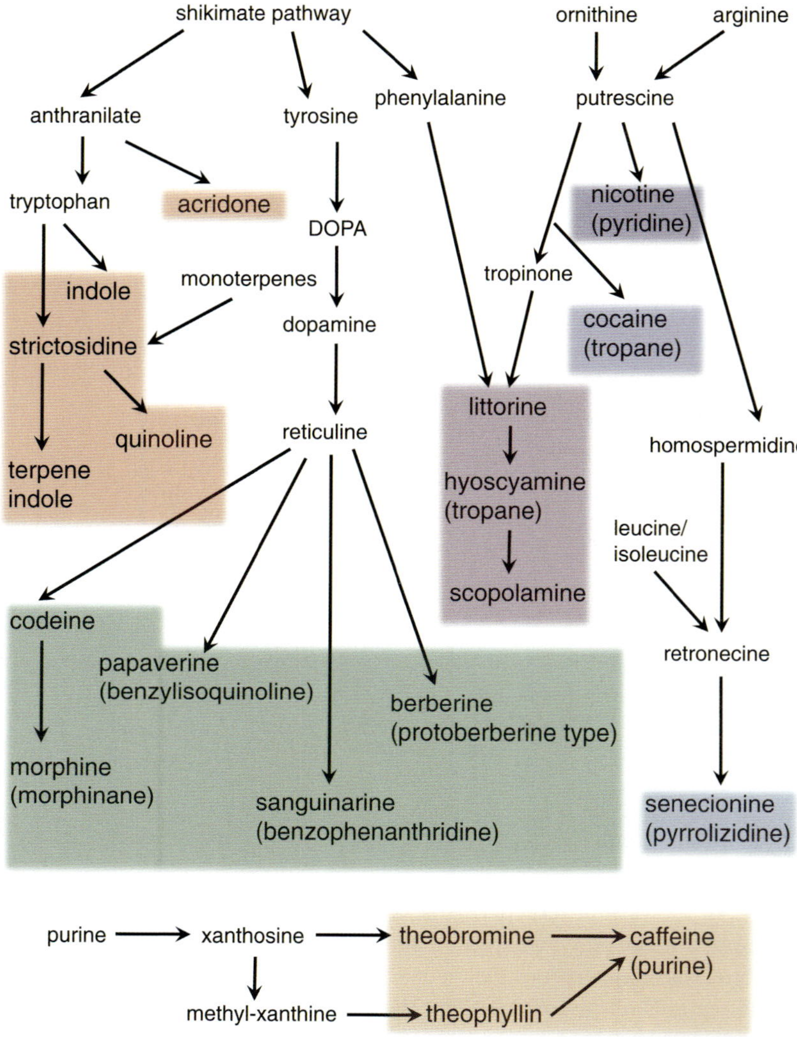

Figure 1.5 Overview on the biosynthesis of selected alkaloids. The shikimate pathway leads to the synthesis of many different alkaloids given in colors depending on their origin (red = anthranilate/tryptophan, green = tyrosine). In addition, ornithine and arginine via putrescine result in the synthesis of pyridine (dark blue), tropane (blue) and pyrrolizidine alkaloids (light blue). Phenylalanine together with ornithine is needed for the synthesis of a second group of tropane alkaloids (violet). Caffeine and related substances is derived from purine (brown). The class of compounds is given in brackets.

are involved in the synthesis of alkaloids, of phenolic compounds, and also of pigments such as anthocyanins and betalains. This indicates that, even under natural conditions, alkaloid biosynthesis competes with many other pathways, because the other amino acids are also needed for protein synthesis.

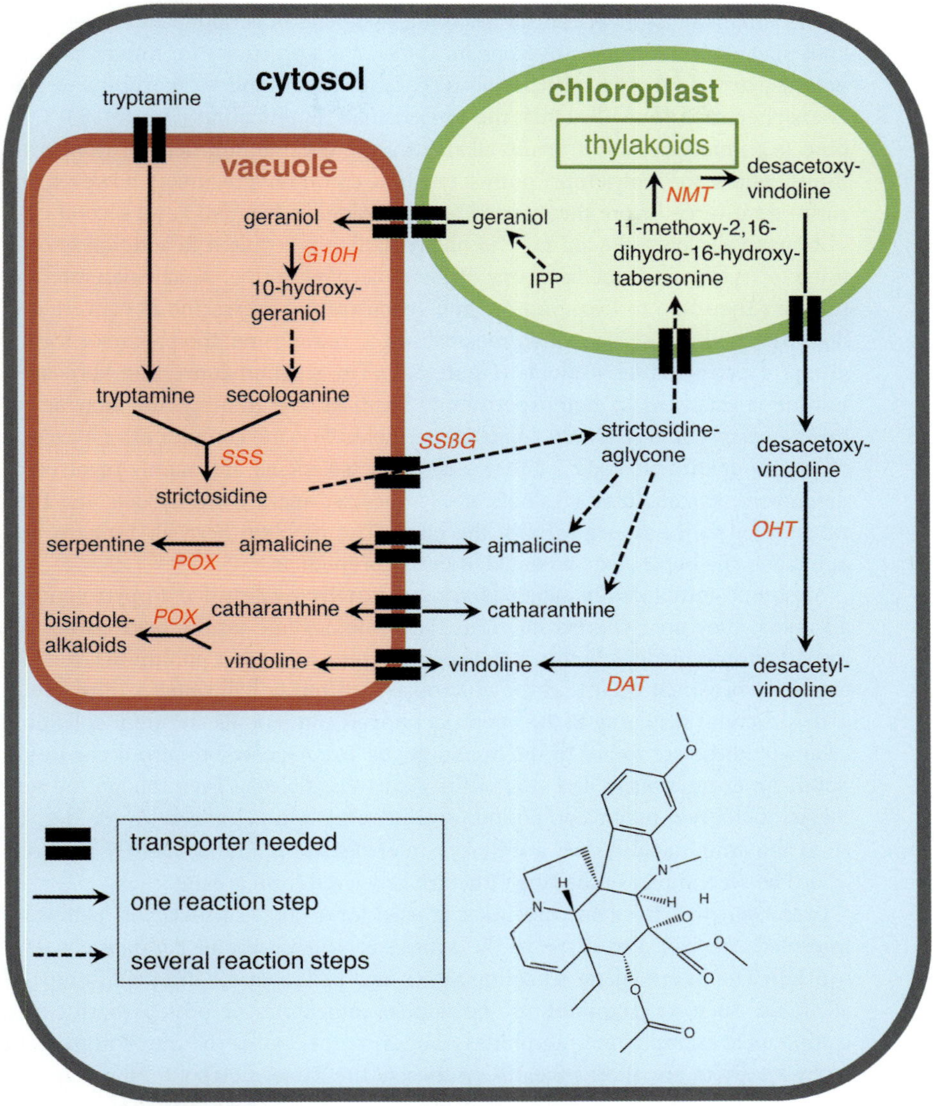

Figure 1.6 Compartmentation of terpene indole alkaloid biosynthesis in *Catharanthus roseus*. The cellular compartments (plastids, cytosol and vacuole) participate in the biosynthetic steps. Enzymatic reactions with dashed arrows are hypothetical. Circles symbolize membrane-associated enzymes. Putative transporters are also indicated. The structure for the terpene indole alkaloid vindoline is shown. IPP: isopentenylpyrophosphate; G10H: geraniol-10-hydroxylase; NMT: *S*-adenosylmethionine hydroxytabersonine-*N*-methyltransferase; DAT: acetyl-coenzyme-A deacetylvindoline 17-*O*-acetyltransferase; OHT: 2-oxoglutarate-dependent dioxygenase; POX: peroxidase; SSβG: strictosidine-β-glycosidase; SSS: strictosidine synthase. Modified from Verpoorte, R. (2000) *Metabolic engineering of plant secondary metabolism*. Kluwer Academic Publisher, pp. 1–29. Chemical structure is from The PubChem Project.

The shikimate pathway contributes to the synthesis of acridone and indole alkaloids via anthranilic acid or tryptophan. Tyrosine is a precursor for a series of compounds, namely the protoberberine type, benzylisoquinoline, morphinane, and benzophenanthridine alkaloids; the intermediate is reticuline (Figure 1.5). Pyridine (e.g., nicotine) and tropane alkaloids (e.g., cocaine) are derived from ornithine, where the polyamine putrescine is a common precursor; however, the pathways diverge before the intermediate tropolone is formed. For a second class of tropane alkaloids, the amino acid phenylalanine is required in addition to ornithine for synthesis of the backbone of the molecule littorine, which is an intermediate on the route to hyoscyamine and scopolamine. Putrescine is converted to homospermidine, which is then metabolized to retronecine, the precursor for the group of pyrrolizidine alkaloids (Figure 1.5). The reaction converting homospermidine is catalyzed by homospermidine synthase, the first specific enzyme for this pathway. For an additional part of the molecule – the complex ester structure of mono- or dibasic acids – either leucine or isoleucine are mainly involved as precursors. Pyrrolizidine alkaloids are synthesized mainly in the roots and then transported to the upper parts of the plant. The problem here for humans and animals is the hepatotoxic potential of these compounds.

Although quinolizidine alkaloids are structurally very similar to pyrrolizidine alkaloids, they are synthesized from the proteinogenic amino acid lysine. An important intermediate in this pathway is cadaverine (another polyamine), the biosynthesis of which occurs in the chloroplasts of mesophyll cells. A third group with structural similarity to the above-mentioned compounds are the indolizidine alkaloids; these are found in the orchid genus *Dendrobium*, but also occur in the South American poison dart frog of the genus *Dendrobates*. Even though the synthesis of the two major compounds swainsonine and castanospermine derives from the amino acid lysine, a different intermediate is found, namely pipecolic acid. The second part of the ring structure is derived from acetate.

In the way that the same precursor is used for a range of metabolic pathways, intermediates can also serve in the synthesis of a variety of compounds (Figure 1.5). The intermediate reticuline serves as a precursor for different groups of alkaloids, such as isoquinoline-, benzophenanthridine- or protoberberine-type compounds; examples are morphines, sangurinarine, and berberine. Within these pathways there are alternatives to synthesize the same metabolite; for example, nicotine can be produced from either ornithine alone or by a fusion of ornithine with putrescine. Morphine (for structure, see Figure 5.30) can either be synthesized via thebaine, oripavin and morphinone (the latter is the direct precursor for morphine), or by the conversion of thebaine to neopinone, which is then converted via codeinone to the bioactive codeine, a known remedy against cough (see Section 4.1); ultimately, codeine is converted to morphine.

Amphetamine-like compounds such as cathinone from *Catha edulis* are also derived from an aromatic amino acid (phenylalanine in this case). The mild stimulating activity of the plant can be achieved by chewing leaves, whereby cathinone is absorbed via the mucosa of the mouth. The genus *Ephedra* produces ephedrine and pseudoephedrine, which are used in medicine to reduce nasal congestions

(Section 5.3.7). The slightly stimulating activity is, again, due to the similarity with amphetamine. The biosynthesis might proceed via a route similar to phenylpropanoid synthesis, but with benzoic acid or benzoyl-CoA as intermediates.

Simple indole alkaloids such as serotonin most likely derive from tryptophan via tryptamine as intermediate. The complex structure of terpene indole alkaloids is also dependent on the terpene pathway for the side-chain moiety. The biosynthesis of terpene indole alkaloids is also an example of a highly compartmented pathway (Figure 1.6). On the one hand, the localization of enzymes is critical, but on the other hand the solubility of the compounds plays a role and these factors necessitate the compartmentalization of the compounds (Figure 1.3; Table 1.2), and therefore transporters are essential for the respective intermediates. Tryptophan is synthesized in the plastids, whereas tryptamine is formed in the cytosol. The latter must be transported between the two compartments, and the same is true for the monoterpene moiety, which is again synthesized in the plastids (see Figure 1.14). Tryptamine is further converted in the cytosol, whereas the first step to the terpene indole alkaloid takes place in the vacuole. Although further enzymatic modification steps are observed again in the cytosol, the later steps also occur in the vacuole (Figure 1.6).

1.1.2.2 **Phenol Derivatives, Especially Flavonoids**

As a representative group for bioactive phenols, the biosynthesis of flavonoids is described in more detail. The biosynthetic pathways derive from the shikimate pathway (Figure 1.7), which is shared by indoles, and by several alkaloids and betalains. The precursor for the cinnamic acid derivatives and flavonoids is phenylalanine, which is converted by the key enzyme for this part of the pathway, phenylalanine ammonia lyase (PAL), to cinnamic acid. Simple molecules (e.g., cinnamic acids) as well as very complicated molecules (e.g., rosmarinic acid) are synthesized via this pathway. Rosmarinic acid has a high antioxidative potential and also good aromatic qualities. Coumarin occurs in woodruff as an aroma and an intoxicating (poisonous) substance. The cinnamic acid derivatives also serve as precursors for polymers such as lignin, which is synthesized via cinnamaldehydes and the respective alcohols, termed monolignols. Flavonoids and stilbenes, such as resveratrol from grapes, are also formed via this pathway. Activation by coenzyme-A plays an important role in this biosynthesis network; the cinnamic acid is activated and acts then, as cinnamoyl-CoA, as a precursor for other pathways in the synthesis of flavonoids and coumarins.

Flavonoid biosynthesis has been quite well elucidated in the model plant *Arabidopsis thaliana*, in which only a small set of secondary metabolites are synthesized. Much information is also available from maize and legumes, the latter also contain isoflavonoids. The scheme for the biosynthetic steps leading to the major flavonoid compounds includes information derived from several plant species (Figure 1.8). The key enzyme for the flavonoid pathway is chalcone synthase, which catalyzes the reaction to naringenin chalcone, the first flavonoid in the pathway. Chalcone synthase mutants of *Arabidopsis* have a transparent seed coat (testa), and are therefore referred to as *transparent testa* (*tt*) mutants. However, other mutants in the pathway of, for example, the next enzyme encoding chalcone

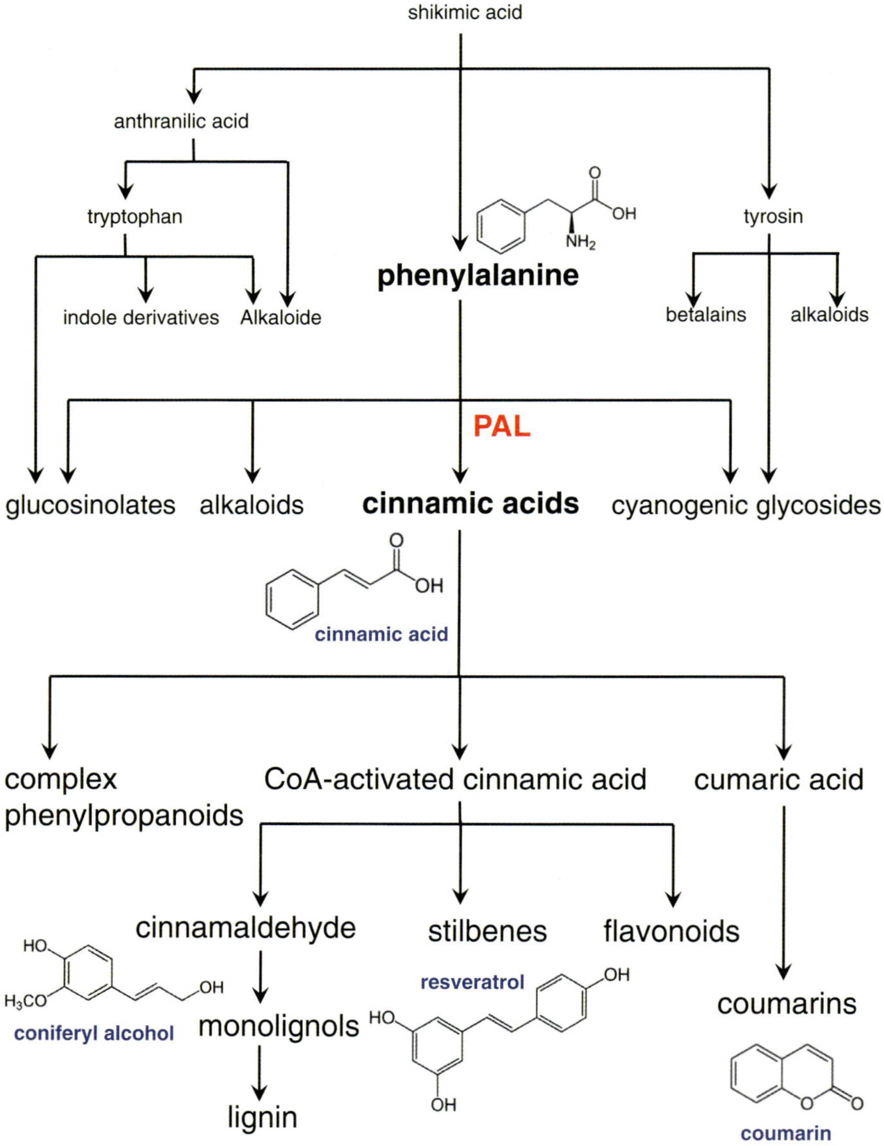

Figure 1.7 Schematic overview for the biosynthetic pathways of selected phenols from phenylalanine as precursor (bold). A key enzyme, phenylalanine ammonia lyase (PAL), is shown (red). Some example structures are depicted (blue). Chemical structures are from The PubChem Project.

isomerase (which is responsible for the synthesis of naringenin), also show this phenotype, and consequently the mutations were numbered consecutively, starting with "1." Mutations in the transcription factors (see Section 1.1.3) that control the synthesis of flavonoids have similar phenotypes.

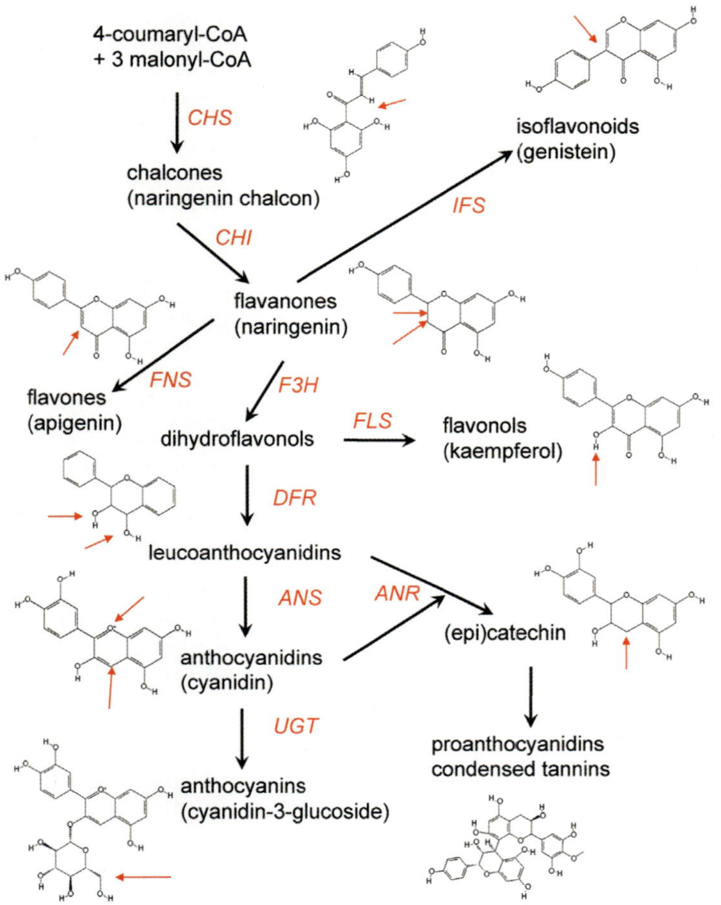

Figure 1.8 The main pathways for flavonoid synthesis derived from different plant species. Because of the complexity of the different biosynthetic steps, only the basic features of the pathways are illustrated. The names in brackets are examples for each class of compounds, which is then also presented in a chemical structure. The red arrows indicate changes in the individual flavonoid groups compared to naringenin. CHS: chalcone synthase; CHI: chalcone isomerase; IFS: isoflavonoid synthase; FNS: flavone synthase; F3H: flavanone-3-hydroxylase; FLS: flavonol synthase; DFR: dihdroflavonol reductase; ANS: anthocyanidin synthase; UGT: glycosyltransferase; ANR: anthocyanidin reductase. Chemical structures are from The PubChem Project.

Several biosynthetic pathways originate from naringenin, which lead to (i) the isoflavonoids of legumes, (ii) flavones, and (iii) flavonols, which constitute the largest group among the flavonoids (Table 1.4). The flavonols in the model organism for molecular biology, *Arabidopsis thaliana*, are kaemfperol and quercetin, and their metabolites (glycosides). Dihydroflavonols are intermediates for the formation of flavonols, but are also converted to the anthocyanin pigments. In addition, they give rise to the proanthocyanidins (also known as "condensed tannins"),

Table 1.4 The major classes of flavonoids with the estimated number of compounds and some functions (see also Chapter 2).

Class	No. of known compounds	Biological function in plants
Anthocyanins	250	Red and blue pigments[a]
Aurones	20	Yellow pigments
Biflavonoids	65	Not known yet
Catechins	40	Antimicrobial
Chalcones	60	Yellow pigments
Dihydrochalcones	10	Feeding deterrents
Flavones	350	Light yellow pigments
Flavonols	350	Feeding deterrent
Isoflavonoids	15	Antimicrobial, attraction
Proanthocyanidins	50	Antimicrobial

a) Pigmentation can have the function in flowers of pollination, but also in fruit of seed dispersal or the deterrence of herbivores in combination with bitter-tasting compounds. In leaves, the pigmentation can serve as a protection against high levels of light.

which are found in the seed coats of many plants. The catechins are derived from intermediates of the anthocyanin pathway, the (leuco)anthocyanidins. The highly diverse structures are achieved by glycosylation reactions with various sugars, where mostly monoglycosides or diglycosides are formed (see Figure 2.13). The sugar moiety is added either as a monosaccharide or disaccharide to one position, though the sugars can be also added at different positions of the molecule. Glycosylations also render the molecules more hydrophilic, and this may change the storage compartment involved. Further details on the water-soluble anthocyan(in)s are provided in Chapter 2.

1.1.2.3 Betalains

Betalains fulfill the function of anthocyanins in the Caryophyllales as pigments and UV-protecting compounds. They have never been detected in combination with anthocyanins in any plant, and the reason for that remains a mystery. In addition to plants, betalains occur in some mushrooms, such as fly agaric (*Amanita muscaria*). Betalain-containing plant species cannot convert flavane-3-dioles to the respective anthocyanin, whereas flavonoids up to the step of leucoanthocyanidins occur in betalain-containing plants.

The group of N-containing betalains are derived from tyrosine, though an alternative pathway based on tyramine has been reported for some plant species. In contrast to anthocyanins and carotenoids, the biosynthetic pathway of betalains is only partially understood. Like anthocyanins, they also occur in a glycosylated form, but the color does not depend on the pH. Betalains are comprised of two major groups: the yellow betaxanthins and the red-pink betacyanins (Figure 1.9). The more than 50 betacyanins known to date are water-soluble and stored in vacuoles. The most important intermediates are DOPA (L-3,4-dihydroxyphenylalanine; see Figure 1.9) and betalamic acid. Next to betanidin, amararanthin is a prominent compound in

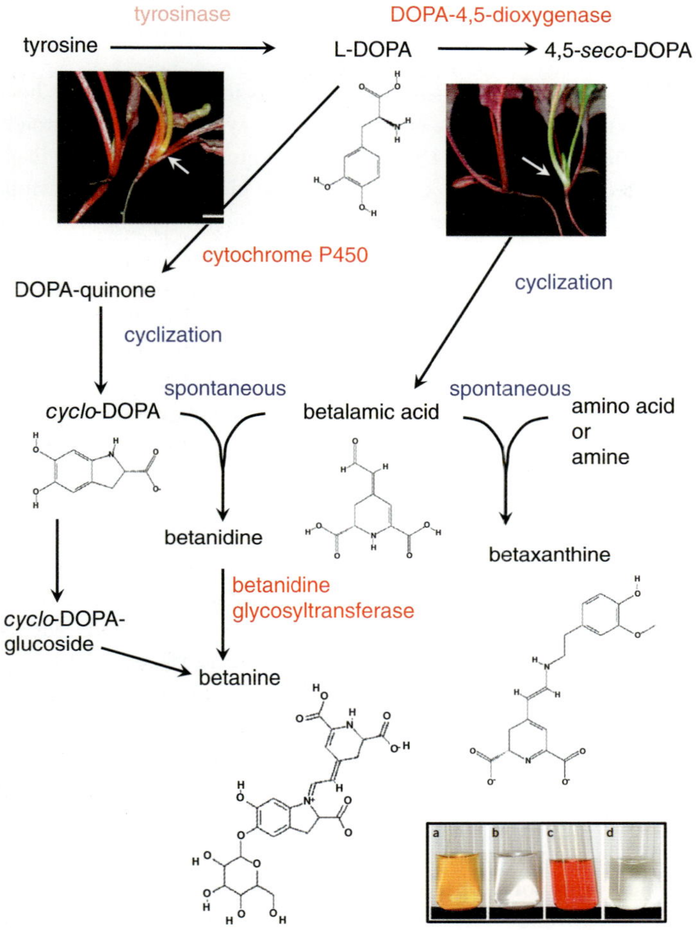

Figure 1.9 Betalain biosynthesis from the aromatic amino acid tyrosine. Enzymatic steps are shown in red, spontaneous reactions in blue. The inferred enzyme tyrosinase is shown in light red. DOPA = (2*S*)-2-amino-3-(3,4-dihydroxyphenyl)propanoic acid; DODA = DOPA 4,5-dioxygenase. Insert: Expression of the cytochrome P450 gene and the *DODA* gene in yeast (a. *DODA* expression + DOPA as substrate; b. *DODA* expression – DOPA; c. both genes + DOPA as substrate; d. both genes – DOPA). Photographs from Hatlestad, G.J. *et al.* (2012) *Nat. Genet.*, **44**, 81. Reprinted with permission from Nature Publishing Group. Chemical structures are from The PubChem Project.

Amaranthus species, while betalamic acid serves as intermediate for the betaxanthins and betanidins. The reaction is nonenzymatic in both cases; rather, a spontaneous addition occurs of either cyclo-DOPA to yield betanidin, or of an amino acid or amine, which results in the formation of a betaxanthin. The glycosylated betanin is formed by a glycosyltransferase reaction with either cyclo-DOPA or betanidin. The enzyme tyrosinase is known from the synthesis of melanin in animals, and in this case two reactions are being catalyzed by the one enzyme; but this enzyme has not been characterized in plants. In contrast, evidence has been advanced that for these two steps a different enzyme is required.

A novel gene involved in the biosynthesis of betacyanins in *Beta vulgaris* has been isolated with the help of yellow mutants encoding a novel cytochrome P450 protein that is involved in the conversion of DOPA to DOPA-quinone, and most likely again by the spontaneous cyclization to cyclo-DOPA. The biosynthesis of betalamic acid is dependent on a DOPA 4,5-dioxygenase (DODA), which results in the formation of both betacyanins and betaxanthins. Functional proof of this was derived from the silencing of red wild-type *B. vulgaris* plants using both genes alternatively. Whereas loss of the cytochrome P450 gene resulted in yellow-pigmented plants, loss of the *DODA* gene yielded white (unpigmented) plants; this showed that in beet, betalamic acid is indeed the precursor for both the yellow and red pigments (Figure 1.9). Complementation of the yellow beet mutant with the cytochrome P450 gene led to red tissues. The expression of these genes in yeast, with subsequent addition of the substrate DOPA, resulted in a red color in the case of both enzymes, in a yellow color when only the *DODA* gene was expressed, and in no color when the DOPA substrate was omitted. The results of this experiment indicated that both genes are essential for the red betacyanin formation, whereas the *DODA* gene alone is sufficient to yield the yellow betaxanthins.

1.1.2.4 **Cyanogenic Glycosides, Glucosinolates, and other Indole Derivatives**

While alkaloids, phenols, flavonoids and terpenes are present in most plant families – albeit with their plant-specific patterns – glucosinolates occur mainly within the crucifer family (Brassicaceae). This family includes many crop plants, vegetables and spices are found, such as cabbages, radish, mustard, or oilseed rape (canola) (Figure 1.10); the model plant *A. thaliana* is also a member of the Brassicaceae. Although glucosinolates and cyanogenic glycosides share similar intermediates in their biosynthetic pathways, it has been hypothesized that both have evolved separately (Section 1.2), and both classes of metabolites have been identified only in a few plant species, among them in papaya (*Carica papaya*).

The cyanogenic glycosides and glucosinolates are both present as glycosides in the vacuoles; however, they are inactive and must first be metabolized to form the active compounds, either via an enzymatic reaction or spontaneously. This process occurs only following damage of the tissue due to colonization by microbes, or by the feeding of herbivores, and consequently the glycosides and glucosinolates are considered as preformed defense compounds (see Section 2.5). Although the glucosinolates are localized in the vacuole, the degrading enzyme – β-thioglycosidase (myrosinase) – is compartmentalized in the cytosol. Following tissue disruption, however, both components of the system come together, whereupon the glucosinolates are metabolized to their toxic active components that are referred to as "mustard oils"; hence, the whole system is known as "The Mustard Oil Bomb." As these compounds might also be toxic towards the plant cell, compartmentation is essential in order to maintain healthy tissues. The application of immunohistochemical techniques has shown that myrosinases also occur in the vacuoles, albeit in specialized "myrosin cells," named

Figure 1.10 Some examples of crop, vegetable and spice plants belonging to the Brassicaceae. Only the species, not cultivars, are given. Source: Fotolia.com (Chinese cabbage, © Bernd, S.; oilseed rape, © Schliemer; radish, © photocrew; cauliflower, © dimakp; Brussels sprout, © narumol pug; cress, © eyetronic; mustard, © Andrey Starostin; rucola, © hvoya; horseradish, © Hetizia; broccoli, kohlrabi, © Natika).

after the enzyme myrosinase. In contrast, the glucosinolates are localized in so-called S-cells (Figure 1.11), so-named due to their high sulfur (S) content. This specific cell-type localization of the glucosinolate–myrosinase system has been identified in the inflorenscences of *A. thaliana* and *Brassica napus* (oilseed rape).

Among the *Brassica* species used as food or spices, compartmentation is especially important because it gives rise to the special taste of these crops. Some of the metabolites also have anticancerogenic potential, so that their formation is desirable (Chapter 6). Other cell types contain the so-called "specifier proteins"; these are important when determining the metabolic fate of the glucosinolate as either volatile (iso)thiocyanates, or as less-toxic nitriles that can be further converted to carboxylic acids (Figure 1.12).

The precursors required for the biosynthesis of all these compounds are amino acids (Figure 1.12). The glucosinolates can be divided into three large groups, depending on the respective precursor amino acid: (i) aliphatic glucosinolates deriving from methionine; (ii) aromatic glucosinolates deriving from phenylalanine and tyrosine; and (iii) indolic glucosinolates originating from tryptophan. Not all brassicaceous plants contain all groups of glucosinolates. For example, the

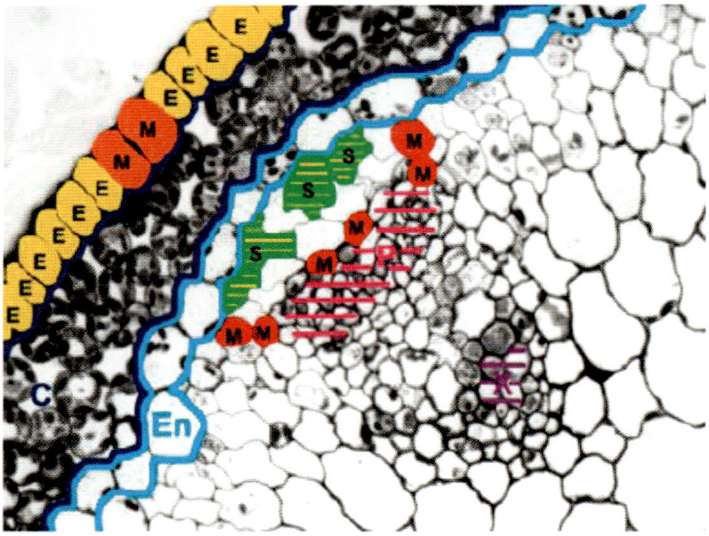

Figure 1.11 Cellular compartmentation of the glucosinolate–myrosinase system in inflorescences of *Arabidopsis thaliana*. A light microscopy cross-section shows the different cell types labeled in different colors: glucosinolate-containing S-cells (green), the myrosin cells (M) containing the β-thioglycosidase myrosinase (red and orange), the epidermal cells (E), which contain epithiospecifier proteins (yellow), the cortex (C) (dark blue), the endodermis (En; cyan), phloem (P; pink) and xylem (X; violet). From Kissen, R. *et al.* (2009) *Phytochem. Rev.*, **8**, 69 (see Further Reading). Reprinted with permission from Springer.

model plant *Arabidopsis* contains aliphatic and indole glucosinolates, but no aromatic representatives. On the other hand, this is the major glucosinolate group in the garden cress (*Lepidium sativum*) and in nasturtium (*Tropaeolum majus*) which, in addition, contains indole glucosinolates.

The major variation in the aliphatic glucosinolates is derived from enzymatic chain elongation reactions of the amino acid methionine. These chain-elongated precursors are then used for the formation of the next intermediate aldoxime. This enzymatic step, catalyzed by cytochrome P450-dependent monooxygenases, is quite specific for the amino acid substrate. In contrast, the later steps – that is, the addition of a sulfur group and glycosylation – are not very specific, and the respective enzymes accept a variety of different glucosinolate structures. This implies that only a modification of the specific step could lead to novel compounds in other plant species, because they can transform the new substance by their set of enzymes to the final glucosinolate. This has been achieved with *Arabidopsis*, where the biosynthesis of benzylglucosinolate was engineered by transformation with the gene encoding the protein for the aldoxime synthetic reaction. The late modification steps, such as hydroxylation or methylation, are again specific for the individual glucosinolates. The range of possible enzymatic reactions can differ greatly between the aliphatics on the one hand and indole glucosinolates on the other hand (Figure 1.12). Finally, unusual modification reactions have been reported for glucosinolates, such

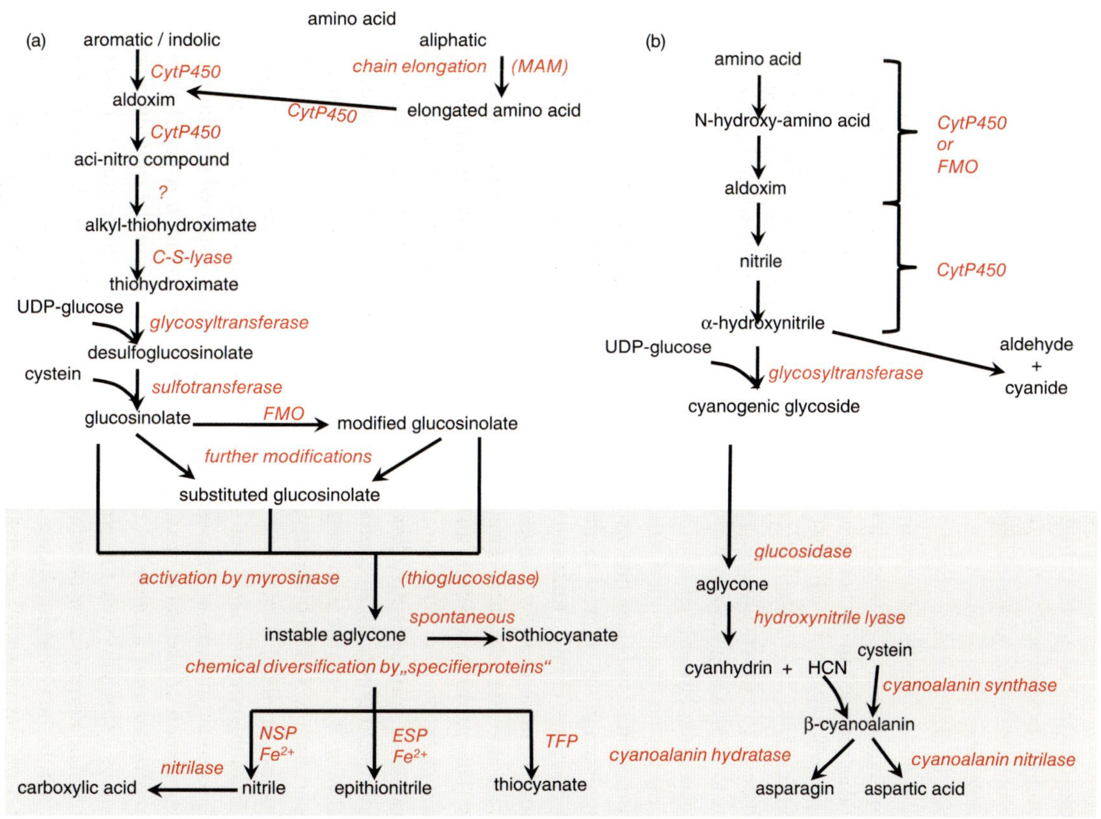

Figure 1.12 Biosynthetic pathways (light gray) and metabolism (gray) of glucosinolates and cyanogenic glycosides in comparison. For the glucosinolates the general scheme is shown for the different amino acids as precursors (a) and for cyanogenic glycosides the main pathway is shown (b). In both cases the metabolism leads to toxic compounds. For cyanogenic glycosides the deglycosylation yields cyanide, whereas for glucosinolates the metabolism is more complicated involving so-called specifier proteins. MAM = methylthioalkylmalate synthase; FMO = flavin-dependent monooxygenase; CytP450 = cytochrome P450-dependent monooxygenase; "?" indicates a reaction step for which no enzyme has yet been identified; ESP = epithiospecifier protein; NSP = nitrile-specifier protein; TFP = thiocyanate-forming protein.

as benzoylation and sinapoylation. In these cases, a benzyl or sinapoyl moiety is attached following the esterification of hydroxylated glucosinolates to the core structure.

Closely related to the indole glucosinolate pathway is the biosynthesis of other indolic compounds, for example the indole-type phytoalexins in *Brassica* species. In *Arabidopsis* there is one phytoalexin (camalexin) with an indole moiety present, whereas other *Brassica* species contain several indole phytoalexins, such as cyclobrassinin, brassilexin, rutalexin, or spirobrassinin (see Section 2.5). The biosynthetic pathway of camalexin is a branch of the indole-3-acetaldoxime pathway leading to indole glucosinolates (see Figure 1.19), whereas the biosynthesis of the *Brassica* indole phytoalexins derives from either tryptophan, indole or indole-3-glycerolphosphate, thus representing an earlier branch point in the metabolic network of indole metabolites.

Apart from the indole glucosinolates and indole phytoalexins of the Brassicaceae, other indole derivatives – which are also considered as defense compounds – are synthesized in different plant species. Among these are the terpene indole alkaloids (Figure 1.6) and a volatile compound in maize (DIMBOA; 2,4-dihydroxy-7-methoxy-1,4-benzoxazin-3-one). A detailed description of the biosynthetic pathway of this compound is provided in Section 1.2 (see Figure 1.25). Briefly, the precursor is also indole-3-glycerol phosphate, which is converted in two enzymatic steps to DIBOA (2,4-dihydroxy-2*H*-1,4-benzoxazin-3(4*H*)-one). DIBOA is glycosylated by a uridine diphosphate (UDP)-dependent glycosyltransferase; the glycosylated DIBOA is then converted to DIMBOA-glycoside, which is stored in the vacuoles. As in the case of glucosinolates, the glycosylated form is not toxic and must be deglycosylated to yield the toxic metabolite (the chemical structure is shown in Figures 2.30 and 3.12).

The cyanogenic glycosides are also synthesized from an amino acid (see Figure 1.12) although, in contrast to the glucosinolates, they do not occur predominantly in one plant family. The best-understood pathway is that yielding dhurrin, the cyanogenic glycoside of millet (*Sorghum bicolor*). In general, the amino acid is converted to an aldoxime, which is reminiscent of a biosynthetic step in glucosinolate synthesis. The aldoxime is further converted via a nitrile as an intermediate to the cyanogenic glycoside. The nitrile can also be converted to an aldehyde and cyanide. Detoxification of cyanide is therefore an essential part of the metabolic pathways leading to these compounds. In contrast to the formation of aldoxime in the glucosinolate pathway, the enzyme catalyzing the first step in cyanogenic glycoside synthesis is dependent on NADPH. Metabolic channeling was also shown to occur for the plasma membrane-bound enzymes, which means that the metabolites are not released from the enzymes after each step of the reaction. Nonetheless, this is a very efficient procedure in terms of the availability of precursors in low concentrations.

The degradation of glucosinolates and cyanogenic glycosides to toxic metabolites closely resembles one another in the first steps (see Figure 1.12). Metabolic conversion is especially vital for glucosinolates to become bioactive, and their fate is mainly determined by protein factors, which influence the outcome of the reaction. While the β-thioglucosidase myrosinase catalyzes the deglycosylation, the so-called "specifier proteins" determine the next reaction product. Isothiocyanates may be formed spontaneously. An epithiospecifier protein is involved in the formation of an epithionitrile, a nitrile specifier protein in the formation of a nitrile, which is less toxic. These reaction products can be further metabolized to the corresponding carboxylic acid by nitrilases (see Figures 1.12 and 1.24). Finally, a thiocyanate-forming protein gives rise to the toxic thiocyanate. If none of these specifier proteins is present, then the isothiocyanate is mostly released, because of its volatility. Cyanogenic glycosides are also deglycosylated, which results in an aglycone that is further converted to cyanohydrin and the toxic cyanide (HCN) by the enzyme hydroxynitrile lyase. The cyanide group can be transferred enzymatically to the amino acid cysteine, such that β-cyanoalanine is the final product. The latter may then be converted by a specialized nitrilase to aspartate, or by a nitrile hydratase to asparagine (see Figure 1.24). These reactions are important to the plant, because of the extreme toxicity of the cyanide.

1.1.2.5 **Terpenes**

Among the terpenes some intoxicating or hallucinogenic compounds can be found; examples are the cannabinoids (see Section 5.3.2), which are synthesized predominantly via the methylerythritol–phosphate (MEP) pathway in the plastids. Terpenes can be classified as either essential or nonessential compounds (Figure 1.13). Essential terpenes include the carotenoids, which play an important role in photosynthesis as components of light-harvesting complexes, and also as protective compounds against high light intensities, but also antioxidative compounds such as tocopherol. Many plant hormones also belong to the terpenoid family, but as these are essential signaling molecules for growth and development and occur in rather small amounts, they are not usually viewed as secondary metabolites. Among nonessential terpenes antimicrobial substances such as the monoterpenes can be found. The term nonessential means that the plant is at an advantage if it can produce these substances under stress conditions; however, under normal conditions nonessential terpenes will not result in a phenotype or prove to be fatal if the biosynthetic pathway is mutated.

The biosynthesis of terpenes can proceed via two different pathways: (i) the mevalonate (MVA) pathway; or (ii) the methylerythritol-phosphate (MEP) pathway (Figure 1.14) (these names derive from the major metabolites involved). The MVA pathway is located in the cytosol, while the MEP pathway is in the plastids. Although both reactions can occur in the plant simultaneously, some organisms use only one pathway for the synthesis of terpenes. The sesquiterpenes and triterpenes are synthesized in the cytosol, whereas the plastid is responsible for synthesis of the mono-, di-, and tetraterpene structures. When coupled to the ER, the oxidation steps are catalyzed by cytochrome P450-dependent monooxygenases. Terpenes can also be produced in non-green plastids, for example in the roots of plants inoculated with arbuscular mycorrhizal fungi (see Figure 2.23). It has been assumed that these two pathways interact with each other, as both are essential for synthesis of the so-called "building blocks" isopentenyl-diphosphate (IPP) and dimethylallyl-diphosphate (DMAP), which are used in the synthesis of higher-molecular-weight terpenes via head-to-tail additions. This reaction is catalyzed by prenyltransferases.

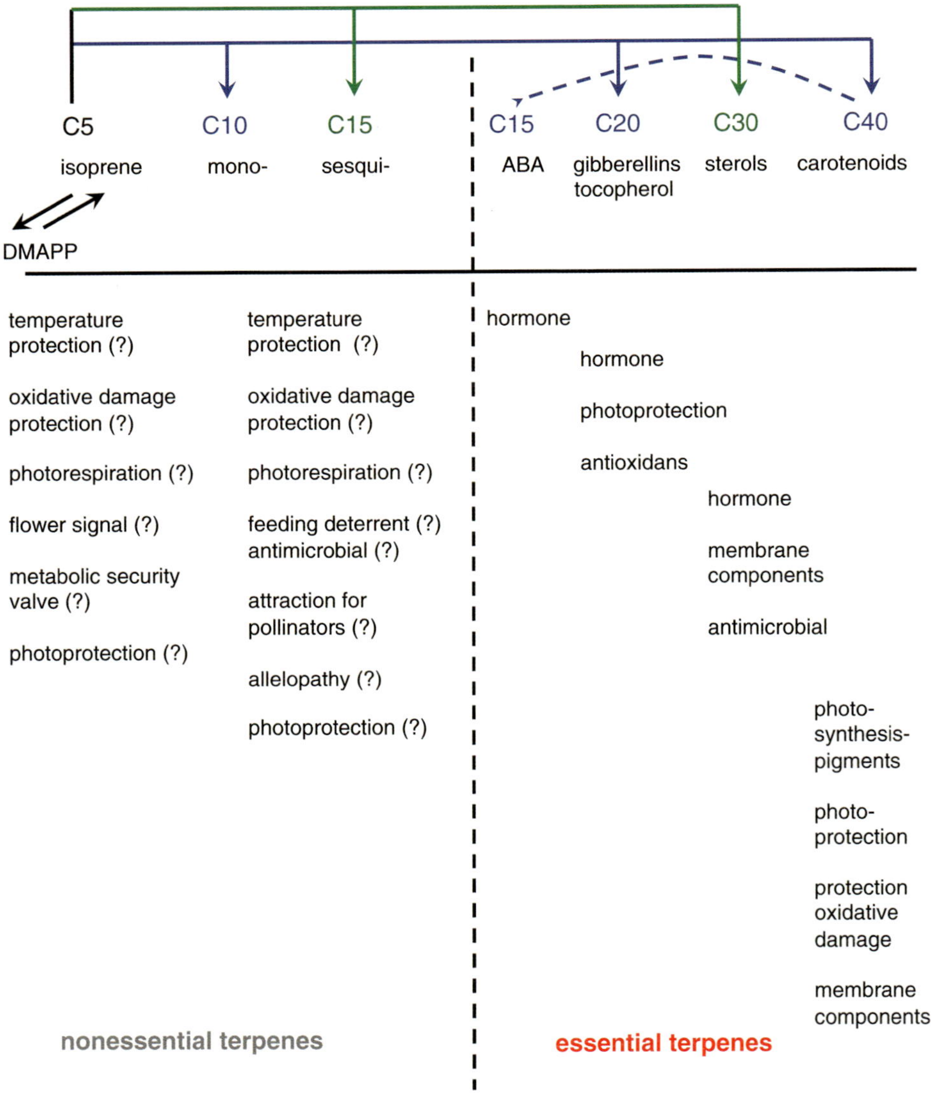

Figure 1.13 Classification of terpenes into essential and nonessential compounds for the plant. Examples of compounds and their biological function are given. Within the respective chemical scaffold, substances deriving from odd numbers of carbon atoms are shown in green; those deriving from even numbers of carbon atoms are shown in blue.

Longer-chain terpenoids are synthesized from two diphosphate molecules, where one phosphate group is lost. This reaction is typical for the synthesis of chlorophylls and ubiquinones. The key enzymes for the MVA pathway are hydroxymethylglutaryl-CoA synthase and -reductase, which catalyze the formation of mevalonic acid. Similarly, there are also key enzymatic steps for the MEP-pathway, involving 1-deoxy-D-

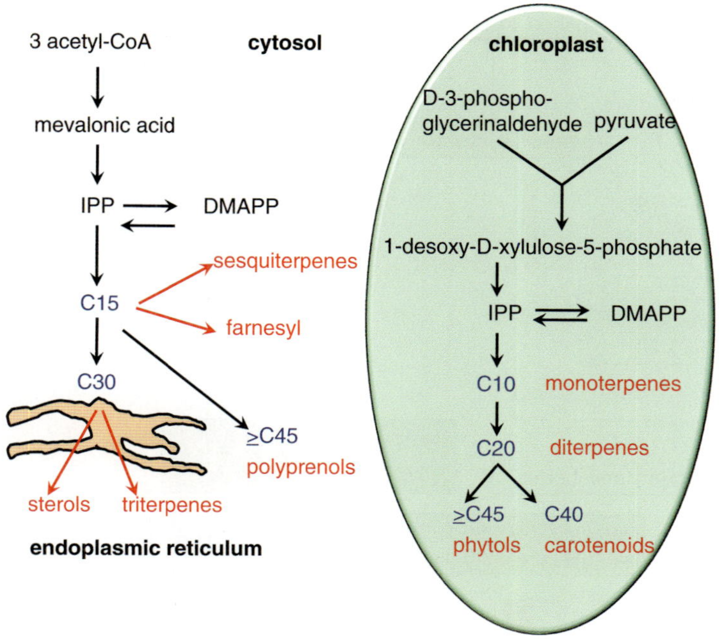

Figure 1.14 General scheme for terpene biosynthesis. The biosynthesis of terpenes can take place in two different compartments, the cytosol and chloroplast, using different pathways. The mevalonate (MVA) pathway in the cytosol delivers sesqui- and triterpenes, whereas the plastidic MEP (methylerythritol phosphate) pathway results in the synthesis of mono-, di-, and tetraterpenes. The endoplasmic reticulum hosts cytochrome P450-dependent monooxygenases, which catalyze various oxidation reactions. Both pathways are thought to interact with each other. IPP = isopentenylpyrophosphate; DMAPP = dimethylallylpyrophosphate.

xylulose-5-phosphate synthase and -reductoisomerase, which form 2-*C*-methyl-D-erythritol-4-phosphate. The later steps for both pathways are then catalyzed by terpene synthases (see Section 1.2) that are encoded by multigene families and are able to synthesize various longer-chain terpenes from diphosphate molecules. The number of C-atoms in the molecule is only dependent on the respective precursor; for example, geranyl-geranyl-diphosphate is converted to diterpenes by terpene synthases. Terpenes are also important precursors for the terpene indole alkaloids (Figure 1.6).

1.1.3 Transcriptional Control

The biosynthesis of secondary metabolites must be regulated in a timely and spatial manner, mainly because of the large energy costs associated with the process (Figure 1.15). As a plant cannot provide a complete range of secondary metabolites needed to survive stress situations at all times, regulation of the process is especially important for compounds that contain several elements, such as nitrogen-

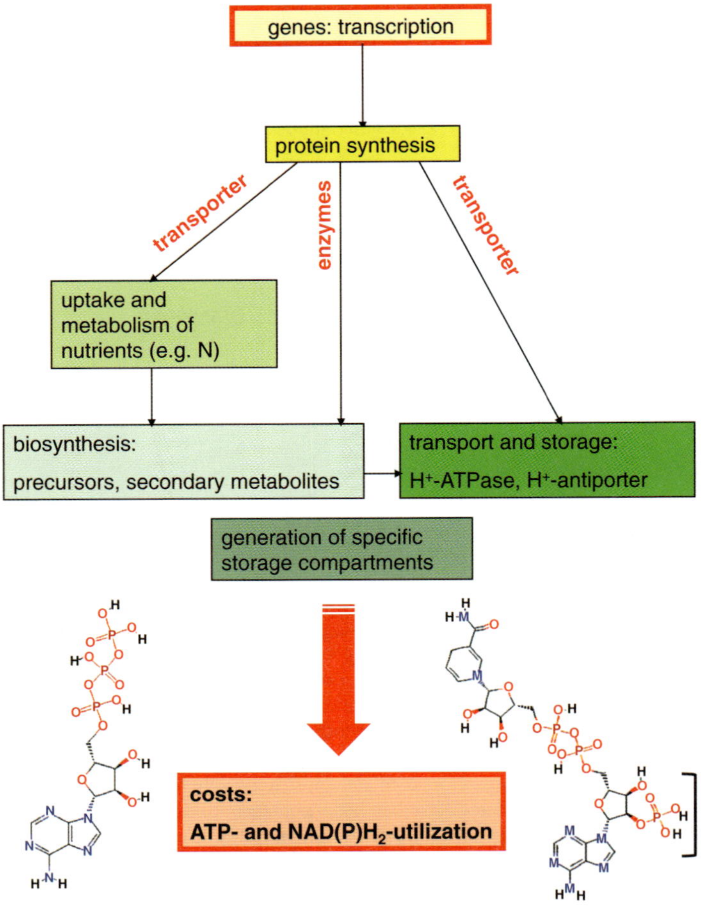

Figure 1.15 The biosynthesis of secondary metabolites is energy-consuming; therefore not all genes are transcribed at a given time point, and subsequently not all proteins present in the cells. Proteins are needed for biosynthesis, uptake and transport within the plant. For all these steps ATP and reduction equivalents, that is, $NAD(P)H_2$, are needed. The structures of adenosine triphosphate (ATP; left side) and nicotinamide-adenine dinucleotide phosphate (NADPH; right side) are shown. The bracket indicates the phosphate group not present in nicotinamide-adenine dinucleotide (NADH). Chemical structures are from The PubChem Project.

containing alkaloids and nitrogen- and sulfur-containing glucosinolates. Previously, a modeling approach has shown that the equivalent of about 15% of the total energy needed to synthesize all components of a leaf cell must be invested in the synthesis of all the glucosinolates that occur in *A. thaliana*. This results in a drain of oxygen, carbon dioxide, water, sulfur and nitrogen to the amino acid precursors needed to synthesize a secondary metabolite moiety. Clearly, the production of only one compound in large amounts under stress conditions might lead to disadvantages for the plant if a second stressor were to occur because, in the case of this second stress, the "wrong" metabolite had been generated (see Sections 2.3 and 2.5). This of course implies that many metabolites must be generated simultaneously, while others would not be synthesized if not specifically required.

The energy to drive the chemical reactions must be provided by adenosine triphosphate (ATP) and reduction equivalents ($NAD(P)H_2$), which are needed not only for the enzymatic reactions and transport processes but also for the assimilation of nutrients. The tissue also requires specific storage compartments within the cell, or within an organ. Finally, messenger ribonucleic acid (mRNA) and proteins need to be synthesized. Consequently, regulation of the biosynthesis of enzymes and transporters on a transcriptional or post-transcriptional level is extremely important. For some compounds, the metabolites must be enzymatically activated (Figure 1.12), at both the right time and location.

Several options can be envisioned to regulate the synthesis of secondary metabolites, the most important being perhaps transcriptional control (see Figures 1.17 and 1.19). Transcription factors are DNA-binding proteins that interact with the regulatory promoter regions of the target genes and modulate the rate of transcriptional initiation by RNA polymerase. Several families of transcription factors have been shown to participate in controlling the biosynthesis and accumulation of secondary metabolites (Table 1.5). To date, some have only been found in one plant species, whilst others seem to occur more globally within the higher plants. Transcription factors belong mainly to seven distinct families, and their regulators integrate both internal (e.g., developmental) and external signals. These regulators bind to the corresponding control elements in the promoter (*cis*-elements) to either activate or repress the expression of enzyme-coding genes, while some will interact with other transcription factors to form a complex (see Figure 1.17).

In addition, both post-transcriptional (via microRNAs; Figure 1.16) and post-translational mechanisms (such as the activation of proteins by modifications; i.e., glycosylation, phosphorylation) can regulate the synthesis of secondary metabolites. Phosphorylation is often involved in the activation of transcription factors by protein kinases (Figure 1.19), while deactivation can be achieved by dephosphorylations catalyzed by protein phosphatases. Enzymes can also be activated or inhibited by substrates or reaction products.

MicroRNAs (miRNAs) are short, highly conserved noncoding RNAs that are capable of mediating the silencing of target mRNA molecules. This ability was first detected as a suppressive strategy of plants against viral infections. In addition, miRNAs are involved in the regulation of developmental processes, and can control the interaction of a plant with both abiotic and biotic environments. The same ability is also used to control post-transcriptional gene regulation *in vitro*, by transforming plants with miRNA precursors. Post-transcriptional regulation could also be involved in regulating secondary metabolism. The miRNAs are encoded by MIR genes, which are transcribed from the genomic DNA (Figure 1.16), and the resultant primary miRNA transcript then folds back into a double-stranded structure that is processed by a protein named "dicer" to produce a 21–30 nucleotide-long miRNA. The gene encoding a miRNA precursor has two features: one feature forms part of the complementary DNA, while the second feature forms a

Table 1.5 Families of transcription factors involved in the regulation of plant secondary metabolite genes. These belong mainly to seven classes: MYB, bHLH, AP2/ERF, WRKY, Zinc finger, DOF and NAC. For the listed transcription factors a plant species is given as an example for the presence of the respective transcription factor. The relevant metabolic pathway is also given. Compiled from Yang, C.Q. *et al.* (2012) *J. Integr. Plant Biol.*, **54**, 703–712.

Transcription factor	Metabolic pathway	Plant species
Superfamily MYB		
MYB	Anthocyanins	*Arabidopsis thaliana*
MYB	Anthocyanins	Asiatic hybrid lily
MYB	Anthocyanins	*Malus × domestica*
MYB	Anthocyanins	*Vitis vinifera*
MYB	Glucosinolates	*Arabidopsis thaliana*
MYB	Phenylpropanoids	*Nicotiana tabacum*
Ruby	Anthocyanins	*Camellia sinensis*
TT2	Proanthocyanidins	*Arabidopsis thaliana*
PAR	Proanthocyanidins	*Medicago truncatula*
Superfamily bHLH		
GL3	Anthocyanins	*Arabidopsis thaliana*
TT8	Anthocyanins	*Arabidopsis thaliana*
MYC	Terpene indole alkaloids	*Catharanthus roseus*
MYC	Terpene	*Arabidopsis thaliana*
MYC	Nicotine	*Nicotiana tabacum*
bHLH	Nicotine	*Nicotiana benthamiana*
Superfamily AP2/ERF		
ORCA	Terpene indole alkaloids	*Catharanthus roseus*
ERF	Artemisinin	*Artemisia annua*
Superfamily WRKY		
WRKY	Gossypol	*Gossypium arboreum*
WRKY	Artemisinin	*Artemisia annua*
WRKY	Camalexin	*Arabidopsis thaliana*
WRKY	Terpene indole alkaloids	*Catharanthus roseus*
Superfamily Zinc finger		
ZCT	Terpene indole alkaloids	*Catharanthus roseus*
Superfamily DOF		
OBP	Glucosinolates	*Arabidopsis thaliana*
DOF	Flavonoids	*Arabidopsis thaliana*
Superfamily NAC		
NAC	Camalexin	*Arabidopsis thaliana*

loop that has no complementary sequences but is structurally important for folding of the RNA. The miRNA∗ strand (see Figure 1.16) is later degraded, while the miRNA strand binds to its complementary mRNA molecule. The cell recognizes the partially double-stranded RNA, which is then degraded in a ribonucleoprotein complex, the so-called RNA-induced silencing complex (RISC), of which an essential component is the protein argonaute. Finally, the initially transcribed mRNA is degraded into short fragments. In the case of a virus infection, the viral nucleic

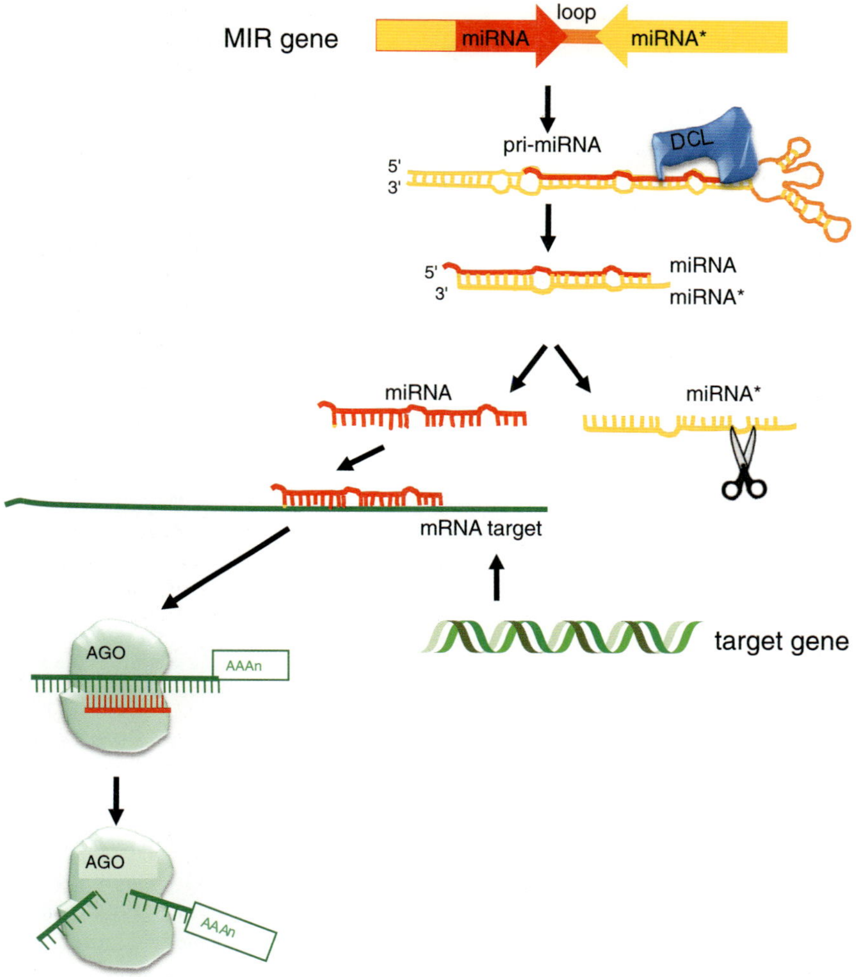

Figure 1.16 MicroRNA-mediated silencing of mRNA and translational repression. miRNAs are encoded by MIR genes. The primary miRNA (pri-miRNA) transcript folds back into a double-stranded structure, which is processed by dicer (DCL1). The miRNA* strand is degraded and the miRNA strands binds to the target gene (in green). Modified from *Teaching Tools in Plant Biology*, The American Society of Plant Biologists.

acids dimerize spontaneously when a certain number of molecules is present. This RNA duplex is also recognized by the plant, and degraded by dicer. The resultant short RNAs, which are referred to as short interfering RNAs (siRNAs), can bind further RNA molecules so that their numbers are rapidly decreased.

Many of the transcription factors involved in controlling secondary metabolite synthesis are also controlled by signaling molecules, such as jasmonic acid (Figures 1.17–1.19). Jasmonic acid and salicylic acid are both important signaling molecules in plant defense reactions against microbes and herbivores. Jasmonates, for

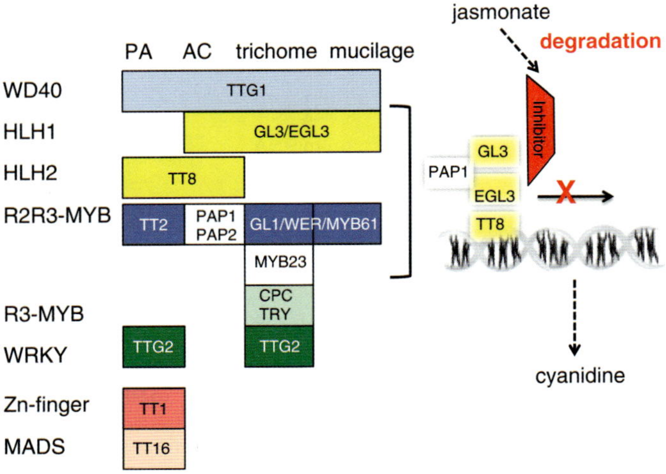

Figure 1.17 Transcriptional control of pigment and proanthocyanidin synthesis in *Arabidopsis thaliana*. The different families of transcription factors involved in the different biosynthetic pathways, proanthocyanidin (PA) and anthocyanidin (AC) synthesis are shown on the left. The developmental processes such as pigmentation, trichome and mucilage formation in *Arabidopsis* are shown at the top. The transcription factors sometimes need to form complexes to gain activity. They can also be regulated by inhibitor molecules (red). The inhibitor is degraded in a jasmonate-dependent manner (see also Figure 1.18). The gene names of some transcription factors are inspired by the phenotype of the mutant. TT = transparent testa; GL = glabrous (no trichomes); PAP = production anthocyanin pigment; WER = werewolf; EGL = enhancer of glabrous; TRY = triptychon; AN = anthocyanin.

example, can regulate many transcription factors such as the MYB family, AP/ERF family, and WRKY family, which implies that jasmonic acid can regulate the synthesis of a wide spectrum of secondary metabolites. Among these are quite different structures such as glucosinolates and camalexin in *Arabidopsis* (Figure 1.19), flavonoids and anthocyanidins in many plant species (Figure 1.17), terpene indole alkaloids in *Catharanthus roseus* (Figure 1.34), nicotine in tobacco, and artemisinin in *Artemisia*. Consequently, jasmonic acid is often used to elicit the synthesis of secondary metabolites in biotechnology (Table 1.9). The transcriptional control of secondary metabolite biosynthesis will be explained by using two classes of compounds as examples, namely the flavonoids and the glucosinolates.

The biosynthesis of flavonoids and anthocyanins is regulated in a compound-specific manner, and the induction is triggered either tissue- and/or stress-specifically (Figure 1.17). Often, several transcription factors need to interact with each other to exert their correct function, although in some cases one transcription factor may be active alone. Pigmentation is important not only for flowers and fruit coloring, but also for seed coatings to provide protection against high levels of irradiation. Developmental processes, such as the formation of trichomes or the production of mucilage, are also regulated by this group of transcription factors and flavonoids. The transcription factor complexes are able to regulate either one

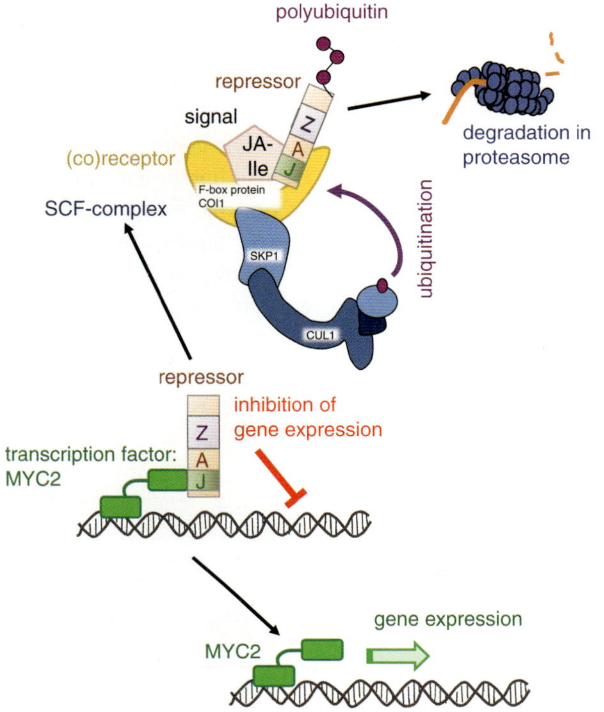

Figure 1.18 The COI signaling pathway of jasmonate-mediated gene expression regulation. Jasmonate-isoleucine (JA-Ile) binding by the COI1-JAZ coreceptor is shown. Ubiquitination and degradation of the JAZ inhibitor in the 26 S proteasome is the prerequisite for transcriptional activation. The repressor is targeted for degradation by addition of a poly-ubiquitin tail in the SCF-complex (SCF = SKP, CUL, F-box, where the F-box protein is the receptor COI1). Transcriptional activation is then exerted by MYC2 transcription factors and others. Modified from *Teaching Tools in Plant Biology*, American Society of Plant Biologists.

specific or several processes. For example, at least six transcription factors are involved in the synthesis of proanthocyanidins, whereas for mucilage formation only three are involved (Figure 1.17). However, the regulatory network is much more complex than depicted here, mainly because no environmental factors such as light, pathogens, or other abiotic stresses have been taken into account.

There are proteinaceous inhibitors of jasmonic acid-induced transcription, which have to be degraded before gene activation is possible (Figure 1.18). In this case, it is not jasmonic acid itself but rather its conjugate with the amino acid isoleucine that triggers this process. Under noninducing conditions a transcriptional repressor (JAZ) binds to the promoter of jasmonate-inducible genes, and this prevents transcriptional activation by the MYC transcription factors. In order for the repressor to be degraded, the amino acid conjugate of jasmonic acid with isoleucine is needed as signal. Binding of the jasmonate conjugate together with the repressor protein to a receptor (COI1), which is an F-box protein, targets the repressor to proteolytic degradation in the 26S proteasome. For this, the target protein must be ubiquitinated as only proteins with a polyubiquitin tail are recognized by the proteasome. Ubiquitination is achieved in the SCF (SKP, CUL, F-box) complex, which is composed of the receptor itself (the F-box protein), the ubiquitin ligase, and other adapter proteins. Once the repressor is degraded, the MYC-type transcription factor can activate the transcription of the jasmonate-inducible gene.

The biosynthesis of glucosinolates is regulated by many different environmental factors (see Figure 1.19), and consequently many transcriptional events need to be induced under these conditions. Specific transcription factor families were isolated from *Arabidopsis,* which can control preferentially either the indole or aliphatic glucosinolate pathway (for details on the biosynthetic enzymes, see Section 1.1.2).

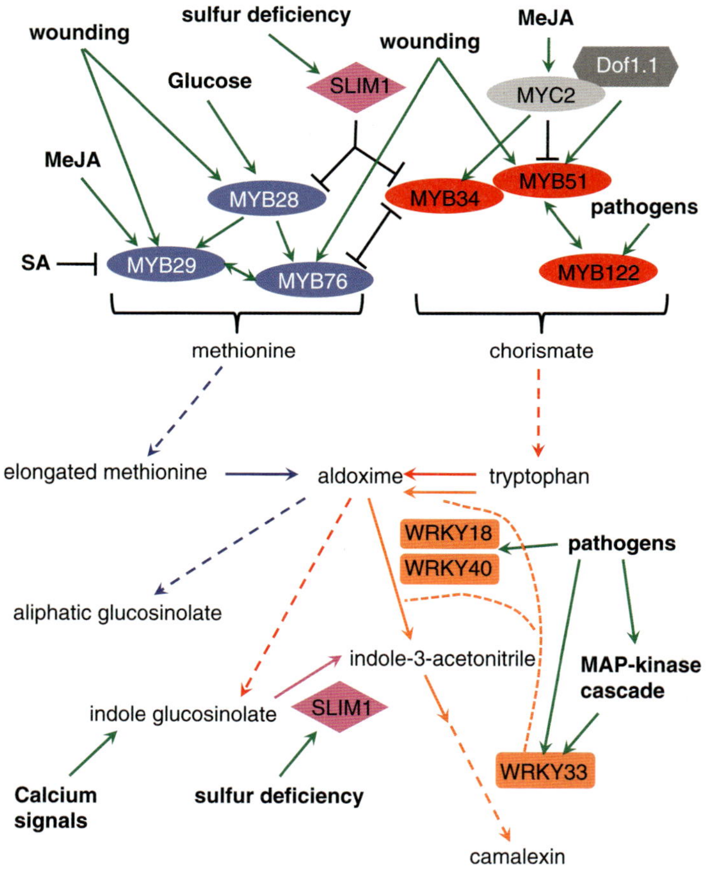

Figure 1.19 Transcriptional control of glucosinolate biosynthesis. Transcription factors involved in the control of aliphatic glucosinolate synthesis are shown in blue; those controlling indole glucosinolate synthesis are shown in red. Camalexin synthesis (orange) is regulated by WRKY transcription factors. WRKY33 itself is regulated by activation of a MAP kinase cascade. Green arrows point to an induction by environmental factors. A solid arrow symbolizes one reaction step, and a dashed arrow several reaction steps. Dof1.1/SLIM1/MYB/MYC/WRKY = transcription factors; MAP = mitogen-associated protein; Me-JA = methyl-jasmonate.

Environmental factors such as pathogens, sulfur deficiency and wounding, or nutrients such as glucose, can each induce transcription factors which, in turn, control genes involved in glucosinolate synthesis. This transcriptional control also involves the synthesis of precursors such as tryptophan. In addition, pathways connected to indole glucosinolates, such as the synthesis of the phytoalexin camalexin and the plant growth hormone IAA (Section 2.2.1), are controlled by members of these transcription factor families.

Some genes are regulated only by one transcription factor, and others by two or even three different factors, depending on the signal(s) to which they respond. Plant signals involved in biotic stress responses, such as salicylic acid or jasmonic acid, can also induce these transcription factors and in turn increase the levels of the glucosinolate defense compounds. Since glucosinolates are sulfur-containing molecules, it is feasible that their synthesis is transcriptionally regulated by sulfur. Sulfate deficiency induces the transcription factor SLIM1, which acts in fact as a repressor of the MYB-transcription factor family. The latter are positive regulators of aliphatic and indole glucosinolate biosynthesis, so that the upregulation of SLIM1 represses the biosynthesis of glucosinolates under unfavorable conditions. On the other hand, the metabolism of glucosinolates is induced by SLIM1, so that the sulfur is released for reuse in other metabolic reactions. Thus, the glucosinolates can be also viewed as S-storage compounds.

The biosynthetic pathway to the phytoalexin camalexin is closely connected to indole glucosinolate and IAA biosynthesis (Figure 2.4). However, the genes encoding the enzymes for the respective biosynthetic pathways are controlled by different transcription factors. While the glucosinolates are controlled by the MYB transcription factor family, WRKY transcription factors regulate camalexin synthesis (Figure 1.19). WRKY transcription factors are not only important for camalexin synthesis but also play a general role in the plant defense activation. One of these, WRKY33, is post-translationally regulated by a MAP kinase cascade, which results in the phosphorylation of the transcription factor. A negative control is achieved by dephosphorylation via a protein phosphatase.

Recently discovered proteins with signatures that associate them with the calcium response of the cell are also involved in the control of indole glucosinolate accumulation (Figure 1.19). These possess a calmodulin-binding motif. Calmodulin then interacts with calcium to activate signaling cascades. The calmodulin-binding protein from *Arabidopsis* has a nuclear-targeting sequence, which implies an effect on gene activation. This protein is especially induced after pathogen colonization. In addition, plants that overexpress the gene for this protein are more resistant to herbivore attack. This is in accordance with the high level of gene expression in the vasculature, especially in the phloem.

1.1.4 Transport of Secondary Metabolites

The molecular transport mechanisms of secondary metabolites (Figure 1.20) are less well understood than the biosynthesis, and very few details concerning transport

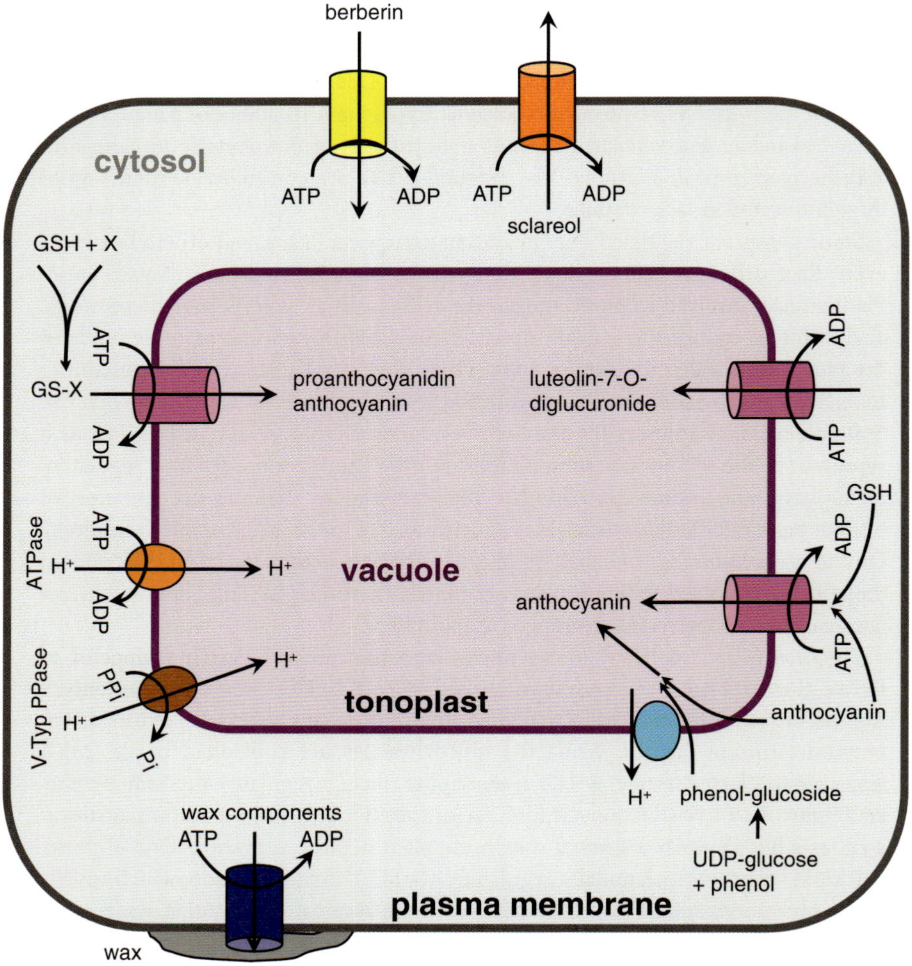

Figure 1.20 Transport of secondary metabolites in different cellular compartments. The ABC-transporters occur in high numbers and transport many different molecules. Representative secondary metabolites, which are transported by one class of these transporter types are shown: ABC-transporter of the MDR-family (yellow), of the PDR-type (orange), of the WBC-family (dark blue) and of the MRP-family (magenta) (for abbreviations see text). The ATP/H^+-dependent transporters (cyan) use a different mechanism. These are mainly dependent on the proton gradient and only indirectly on ATP. ATPase and PPase proteins (brown and dark orange) are involved in the formation of the proton gradient.

proteins and mechanisms are available. The compounds or precursors need to be transported into either the vacuole or other cellular compartments, where they are stored or further metabolized. One major transporter family which is thought to be involved in the transport of secondary metabolites is the ABC-transporters (ABC = ATP binding cassette). These are subdivided into three classes, which occur at the plasma membrane and are ATP-dependent: (i) the MDR (multidrug resistance)-type family; (ii) the PDR (pleiotropic drug resistance)-type; and (iii) the WBC (white-brown complex)-type. The MRP-type-related proteins represent another form of ABC-transporter which is located at the tonoplast. These transporters differ, for example, in the number of transmembrane domains and the position of the nucleotide-binding site(s). The transport of some compounds is coupled to a proton antiport, and in this case the proton gradient must be maintained by either H^+-dependent ATPases or pyrophosphatases (PPases). Some flavonoids can be transported as glutathione derivatives while other compounds, such as simple phenols, are transported as glycosides.

Flavonoid glycosides, including anthocyanins, are usually transported into the vacuole. The transport mechanisms can depend on the plant species and organs, but the most likely mechanisms involve the transport of anthocyanins via a glutathione *S*-transferase (GST)-like protein or a MDR-like protein (a type of ABC transporter). The involvement of the former has been shown in maize, petunia and *Arabidopsis*, while the latter has been identified in maize. The molecular mechanism of how these protein conjugates (especially GSTs) are transported has not yet been clarified.

A second mechanism involves the vesicle-mediated mass transport of anthocyanins to vacuoles, as has been observed in lisianthus (*Eustoma grandiflorum*). Anthocyanins are also targeted directly to the protein storage vacuole via ER-derived vesicles in *Arabidopsis* seedlings, and this process does not depend on either GST activity or an ATP-dependent transport mechanism. For flavonoids, it has been suggested that their transport can be coupled to their synthesis by an enzyme complex located at the ER, where the final product is directly included in the ER vesicles and transported to the vacuole, where the vesicle is fused with the tonoplast. This type of transport is suitable only for compounds which cannot diffuse through membranes.

A third mechanism may involve an *Arabidopsis* multidrug and toxic compound extrusion (MATE) transporter (TT12). This vacuolar flavonoid/H^+-antiporter is necessary for the vacuolar accumulation of proanthocyanidins, and has been shown to mediate anthocyanin transport *in vitro*. The application of flavonoids to *Arabidopsis* roots can be used to determine their fate after uptake due to their fluorescence after a specific staining procedure *in planta*, using fluorescence microscopy (see also Figure 1.29). *Arabidopsis* roots grown in complete darkness do not accumulate flavonoids, as the expression of genes encoding enzymes of flavonoid biosynthesis is light-dependent. However, flavonoids do accumulate in the root tips of plants with light-grown shoots and light-shielded roots, consistent with shoot-to-root flavonoid movement. In addition, flavonoids were accumulated in the tissues at some distance from the site of application, which is indicative of uptake and movement systems. Flavonoid applications to the root tips resulted in a basipetal movement in the epidermal layers, with their detection 1 cm from the application sites after some incubation time. In addition, the movement of flavonoids towards the root tip occurred mainly in vascular tissues. The results of experiments with inhibitors of ABC-transporters mediating flavonoid transport suggested that a MDR-associated protein facilitates long-distance movement of the flavonoids.

Digitalis glycosides do not require a transporter, and have been observed to pass the plasma membrane simply by diffusion along a concentration gradient. In the cell, they are modified so that they cannot pass the plasma membrane by diffusion and so are trapped within the cell. At this point, a transporter is most likely needed to relocate these compounds to the vacuoles, where they are stored.

If secondary metabolites have to be transported from the origin of synthesis to the tissue where they are accumulating, then long-distance transport is essential. Very little is known about these processes, as most information is based on correlative experimental data. For example, if the site of biosynthesis and the accumulation patterns of an individual compound differ, then the existence of long-distance transport must be assumed. In tobacco (*Nicotiana tabacum*), pyridine alkaloids are synthesized in the roots and transported to the leaves, where their bioactivity is needed (see Sections 2.5.3 and 3.1). The transport within cells and from cell to cell is most likely possible for nicotine via a nonspecific transporter. Although long-distance transport can be achieved in the xylem vessels and the phloem, this type of transport is dependent on the direction and velocity of water and assimilated transport, and so is not very specific for individual compounds. Differences in nicotine distribution have been observed in two closely related tobacco species, *Nicotiana alata* and *N. langsdorffii*; whereas *N. langsdorffii* can translocate nicotine into the shoot, *N. alata* is not capable of this transport and consequently stores nicotine in the roots. Genetic analyses of these two species have shown that the characteristic of "no transport" is dominant over the ability to transport nicotine. The underlying genetic mechanisms for this property will eventually allow the genes responsible for long-distance nicotine transport in *Nicotiana* species to be identified. These findings also indicate that the trait of "transport" or "no transport" has evolved independently, and can be attributed to different environmental conditions where the accumulation of nicotine in leaves provides the plant with a defensive advantage against herbivores.

In contrast, more information is available for the glucosinolates, which exist as organic anions under the pH conditions of the cell and are therefore unable to cross membranes by simple diffusion. Similar to other ionic compounds, once the glucosinolates are inside the cell they become trapped and must be actively transported. As glucosinolates are found in high concentrations in S-cells, and in vacuoles within those cells, there is a need for an active transport mechanism for these compounds (Figure 1.21). The same has to be anticipated for the long-distance transport of glucosinolates, as they are transported through either the phloem or xylem; however, in order for them to be loaded into the vascular system they must also be able to pass membranes. It is possible that their synthesis in companion cells simplifies such uptake, without transporters, into the phloem system, from where they are further distributed (Figure 1.21).

Currently, biochemical evidence exists for the active transport of aliphatic glucosinolates. The application of radioactively labeled glucosinolates to rosette leaves of *Brassica napus* resulted in an accumulation of radioactivity within the maturing seeds, especially the embryo. Moreover, the phloem was also labeled, which points to a long-distance transport mechanism. Two closely related members of the nitrate/peptide transporter family were identified as high-affinity, proton-dependent,

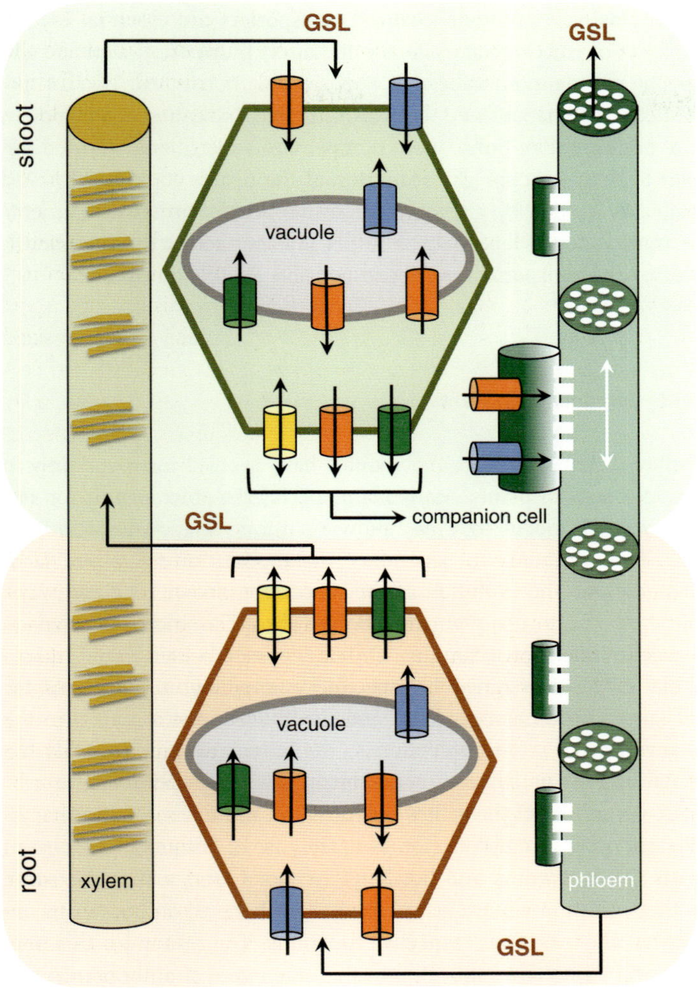

Figure 1.21 Long distance transport and loading/unloading at source and sink tissues of glucosinolates. Glucosinolates are synthesized close to the vasculature and possibly also in companion cells. Part of the unloading for glucosinolates is symplastic, so no transporter is needed, but some part is also apoplastic, and for this a transporter is needed. Glucosinolates are also transported into the vacuole by similar mechanisms, but maybe only by three different transport molecules, which mediate import and export. Blue = proton symporters PUP, NRT/PTR; orange = primary transporters ABC type, green = proton antiporters MATE type, yellow = not further specified permeases (no genes known). For simplification, only the direction of the transport of glucosinolates is shown and not dependence on H^+ or ATP. Redrawn from Nour-Eldin, H.H. and Halkier, B.A. (2013) *Curr. Opin. Biotechnol.*, **24**, 263.

glucosinolate-specific transporters that were most likely needed for the translocation of glucosinolates from the apoplast into the phloem. A double-mutant failed to accumulate glucosinolates in the seeds, but showed a more than tenfold higher accumulation of glucosinolates in source tissues such as leaves and silique walls. This indicated that both plasma membrane-localized transporters are essential for the long-distance transport of glucosinolates. Finally, the direct transport of aliphatic glucosinolates across the plasma membrane of *B. napus* was demonstrated. Two components were identified for transport: one that was saturable, indicating a transporter; and one that was nonsaturable. Subsequent competition experiments showed the saturable transport to be specific for glucosinolates, as the uptake could be inhibited by other glucosinolates but not by glycosides or amino acids. Further experiments revealed that the transport was dependent on a pH gradient across the membrane, which indicates that ABC-transporters are not responsible for this transport.

1.2
Evolution of Natural Compounds in Plants

It is generally believed that secondary metabolites have evolved independently in different systematic groups of plants. Many plant species are polymorphic for secondary metabolites, which means that they show a natural variation between species or cultivars. This variation in secondary metabolites offers a wealth of information for studies on the evolution of the respective biosynthetic pathways. It is not possible, however, to use plant metabolites for a classification based on "chemical evolution" or chemical taxonomy. Only a few metabolites would fit into such a category that could be assigned to a specific systematic group of plants; for example, the betalains in the Caryophyllales and the glucosinolates mainly in the Brassicales. However, there is one exception to the latter grouping, namely the glucosinolate-containing genus *Drypetes*, which belongs to the Euphorbiaceae.

The betalains are composed of the yellow betaxanthins and the magenta betacyanins (Figure 1.9), and occur as aglycones or, as in the case of anthocyanins, also in glycosylated forms. Anthocyanins and betalains are not found within the same plant species as they are mutually exclusive; however, where no anthocyanins are present the betalains take over their respective functions. The reason for this phenomenon is unknown. Although, interestingly, the precursors of anthocyanin biosynthesis can be converted to leucoanthocyanins in species containing betalains, the later steps in the pathway are absent, which suggests that this part of anthocyanin synthesis has been lost in the Caryophyllales.

Tannins, lignins and lignans are among the "oldest" secondary metabolites (Figure 1.22), and for their synthesis the development of the shikimate pathway was necessary. In this case, aromatic amino acids serve as the precursors and are converted to cinnamic acid derivatives. These secondary metabolites are important for the structure and robustness of land plants, and are found in cell wall components; they are also vital for the development of complex plant architecture. Tannins have antimicrobial activities.

Benzylisoquinones evolved at a rather early stage, whereas terpenes and indoles developed only later. Isoflavonoids have evolved on several occasions, independently of one another. For many of these compounds, two different pathways for the delivery of precursor molecules have evolved in parallel; examples of this are

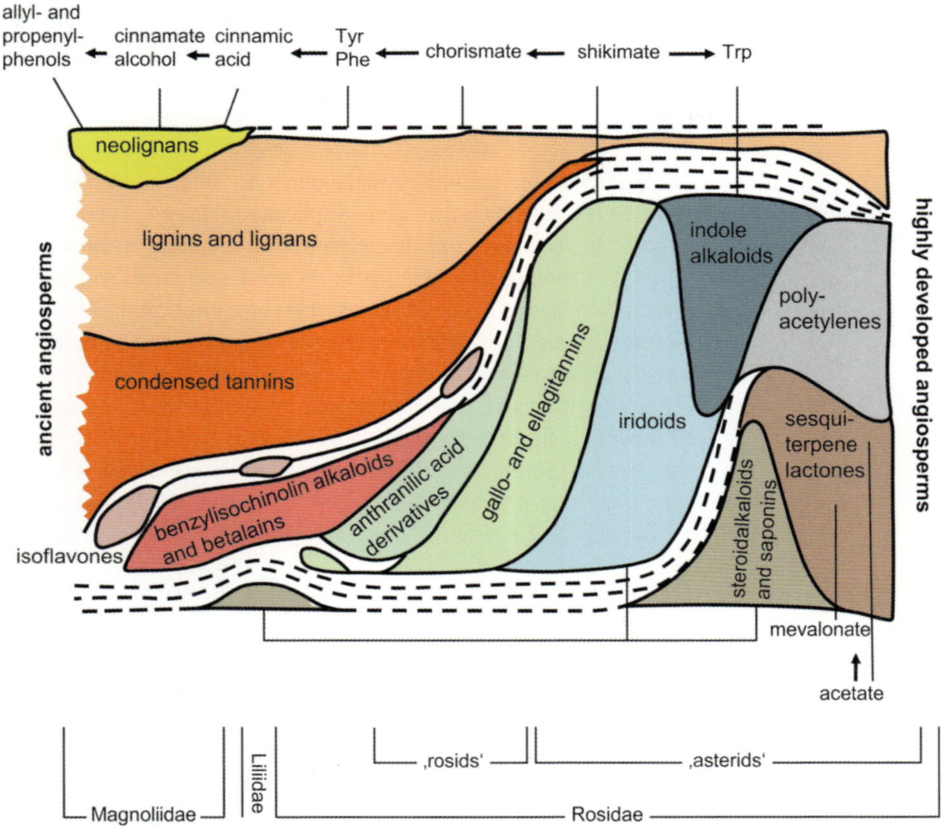

Figure 1.22 Development of different classes of natural compounds in the evolution of plants. The different colors symbolize large groups of secondary metabolites. The horizontal axis as a "time scale in evolution" shows the different plant groups. Above the figure the respective pathways for individual classes of compounds are indicated, which are necessary for their production. The vertical axis shows the compounds, which have been evolved simultaneously.

the shikimate and MVA pathways for the synthesis of terpene and indole alkaloids. Some aspects of the evolution of selected examples are presented in the following subsections. The genes involved in major pathways are known to have been present in the early land plants (e.g., mosses), which had a simpler body plan and where the number of homologous genes within one species would be much lower than in higher plants (Table 1.6). The greatest numbers of genes within a given gene family are found in angiosperms.

The size of the gene families can be affected by several molecular mechanisms:

1) Gene duplication or whole-genome duplication is an ongoing contributor to genome evolution, and occurs within the same order of magnitude as the mutation rate per nucleotide site.

2) The acquisition of new functions for existing genes, followed by gene duplication, is the cause of new gene families.
3) A reduction in gene families can also occur by the deletion of a single gene, or even of several connected genes.

The biosynthetic pathways for secondary metabolites are often modified pathways of the primary metabolism by way of gene duplication. However, gene duplications may also occur in secondary metabolite pathways, resulting in two "new" pathways (Figure 1.23). Although the "old" pathways can be retained, some parts might be lost or altered by new enzyme activities, thereby complicating investigations into the evolutionary descendance of metabolic pathways. In the case of a common evolutionary ancestor protein, the enzymes are recruited from a single precursor. However, closely related compounds can be synthesized by pathways that have evolved separately in different plant families, such that a protein, which catalyzes the same reactions in the two different species may have developed from completely separate precursor proteins. In this case, the whole biochemical pathway has evolved separately in the different plant groups (Figure 1.23).

1.2.1
Parallel Evolution of Similar Pathways

The biosynthetic pathways of two large groups of glycosides – the glucosinolates and cyanogenic glycosides (see Section 1.1.2) – possess obvious similarities in their biosynthetic pathway precursors (amino acids) and intermediates, and metabolites such as aldoximes and nitriles. In addition, nitriles are intermediates in the biosynthesis of the indole phytoalexin camalexin and the plant hormone IAA. While cyanogenic glycosides have been identified in many plant species within the pteridophytes, gymnosperms and angiosperms, the glucosinolates are found predominantly in the order Brassicales. However, a few species exist that contain both classes of compounds, such as *Carica papaya* (Caricaceae) and *Alliaria petiolata* (Brassicaceae).

A current hypothesis implies that the glucosinolate pathway has recruited the genes from the biosynthesis of cyanogenic glycosides, because glucosinolates are evolutionarily younger. The acquisition of such a pathway could be important for detoxification reactions. A mutated cyanogenic glycoside biosynthesis could have produced toxic intermediates that were then metabolized by enzymes; this later evolved into the glucosinolate pathway. Other genes of the glucosinolate pathway may have been taken from basic biosynthetic reactions, for example the C–S lyase and glucosyltransferases (Figure 1.12). Evidence for a common evolution of both metabolic pathways has been found in *A. petiola*, where both the hydroxynitrile glucoside alliarinoside and the glucosinolate 2-propenylglucosinolate are synthesized from the precursor homomethionine. The first step to alliarinoside is catalyzed by a cytochrome P450-dependent monooxygenase, which was recruited for this pathway to metabolize the oxime intermediate. The glucosinolate and cyanogenic glycoside biosynthesis also involves cytochrome P450-dependent reactions.

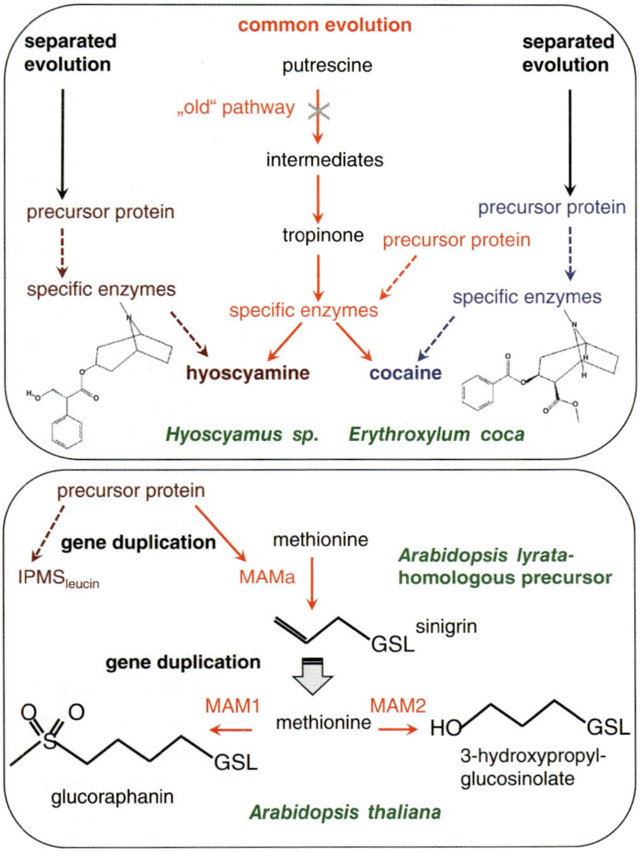

Figure 1.23 Several hypotheses for the evolution of natural substances biosynthetic pathways, using tropane alkaloids (upper panel) and aliphatic glucosinolates (lower panel) as examples. For tropane alkaloids two different possibilities for the evolution of the biosynthetic pathways are shown. The common pathway descends from one precursor, which is converted by homologous proteins to the next intermediate. The enzymes were recruited from one precursor protein. Pathway marked with a light gray cross symbolizes that the "old" pathway could have been lost. For later steps, separate evolutionary events recruit specific enzymes from different precursor proteins. For glucosinolates a hypothesis for the evolution of the biosynthesis of aliphatic glucosinolates in *Arabidopsis* species is shown. The MAM-enzymes catalyze the chain elongation reaction of methionine, whereas the other enzymes are needed for the modification of side chains. They could be derived by gene duplication of *MAMa*. In the glucosinolate structures only the respective side chains are shown. GSL = glucosinolate moiety; $IPMS_{Leucin}$ = alpha-isopropylmalate synthase gene; MAMa = methylthioalkylmalate synthase gene. Chemical structures in the upper panel are from The PubChem Project.

The further metabolic steps to yield alliarinoside share enzymatic activities with the 2-propenylglucosinolate synthesis. The hydroxynitrile may also function as a precursor for the cyanogenic glycosides in *A. petiola*, which are responsible for the liberation of cyanide. By changing the sequences of different genes encoding for cytochrome P450-dependent monooxygenases during evolution, this complex network has been altered.

The synthetic pathways for alkaloids have some evolutionarily conserved features, and some special modifications in certain groups of plants. In those pathways with a common evolution, the "old" parts may be lost whereas the "new" pathways are present and eventually develop further (Figure 1.23). In contrast to the examples explained above, a separate evolution for tropane biosynthesis in Solanaceae and Erythroxylaceae has been established, based on an analysis of the reduction of tropinone. Importantly, the enzyme occurs in the leaves of *Erythroxylum coca*, but not in the roots, whereas the root system is the major location of tropane alkaloid biosynthesis in Solanaceae. Moreover, the enzymes have evolved from different protein precursors, even though they catalyze the same reaction in both plant families. This is most likely also true for the other biosynthetic steps leading to tropane alkaloids. In consequence, this biosynthetic pathway has been invented at least twice independently during evolution, but its evolution has occurred most likely more often because other, unrelated, plant families also contain these compounds.

1.2.2
Gene Duplication

Gene duplication is one of the most important hallmarks for the evolution of secondary metabolite pathways. In many cases, a gene from primary metabolism has been duplicated and then evolved into an enzyme which is involved in secondary metabolism (see Figure 1.25). In the case of gene duplication events, the two genes are quite often found in a tandem array on the chromosome, indicating rather recent events. However, the gene can also be translocated to different parts of the genome.

Pyrrolizidine alkaloids are mainly found in plant families such as Asteraceae, Boraginaceae and Fabaceae, where they occur in only a few genera. The biosynthesis of pyrrolizidine alkaloids has most likely evolved from the polyamine biosynthetic pathway. The enzyme homospermidine synthase, which catalyzes the first committed step to pyrrolizidine alkaloids, has most likely been developed by gene duplication from deoxyhypusin synthase, a protein that is essential to the plant. This enzyme is involved in the post-translational activation of the eukaryotic initiation factor 5A. This activation is, in turn, essential for cell growth and division, and was also related to seed germination, senescence, and apoptosis.

The duplication of a gene from primary metabolism is most likely also the basis of the biosynthesis of the indole derivative DIMBOA in maize (see Section 1.1.2). The alpha-subunit of tryptophan synthase was recruited into this other pathway to catalyze the synthesis of a different indole derivative (see Figure 1.25). Similarly, genes described for the synthesis of secondary metabolites from *Arabidopsis* and rice (Section 1.2.3) have been recruited by gene duplication from primary metabolism; in one case this was essential sterols, and in the other case the growth-promoting plant hormones, gibberellins. Genes for chain elongation reactions in aliphatic glucosinolate synthesis are possibly derived from the precursor protein methylthioalkylmalate synthase (MAMa) by gene duplication, after the gene had been recruited from primary metabolism. The *MAMa* gene evolved from primary metabolism itself, by the duplication of a gene that was important for biosynthesis of the amino acid leucine, alpha-isopropylmalate synthase (Figure 1.23). The genes derived from *MAMa* are now termed *MAM1* and *MAM2*. Originally, these might have evolved in *Arabidopsis lyrata*, which is a close relative of *Arabidopsis thaliana*, and where a *MAMa* homologous sequence is still present.

The metabolism of glucosinolates and cyanogenic glycosides depends on the deglycosylation by glucosidases and subsequently, in some cases, further detoxification by nitrilases. In the case of glucosinolates, the product of the nitrilase reaction is a less-toxic carboxylic acid. In addition, indole glucosinolates may serve as precursors for biosynthesis of the growth-promoting hormone IAA (see Section 2.2.1). Other carboxylic acids could be further metabolized to yield different products. In *Arabidopsis*, a nitrilase activity is also involved in the synthesis of the phytoalexin camalexin (Section 1.1.2).

The metabolism of cyanogenic glycosides requires the presence of nitrilases for two distinct steps. A hydroxynitrile lyase is involved in the formation of cyanide, whereas a cyanoalanine nitrilase is necessary for the detoxification (Figure 1.12). In addition, a third group of enzymes – the nitrile hydratases – forms an amide group from the nitrile, catalyzing an alternative reaction.

In Brassicaceae, a large group of nitrilases can be found with different substrate preferences, which are dependent on the number and characteristics of the glucosinolates present in a particular species. In *Arabidopsis*, four nitrilase genes are known (Figure 1.24); three of these genes (nitrilases 1–3) are found in a tandem array on one chromosome, indicating a close relationship among these sequences and most likely gene duplication events, whereas the fourth gene (nitrilase 4) is located on a different chromosome. The latter is characterized by significant sequence differences, and the enzyme is more distributed ubiquitously among plant species, whereas the nitrilases 1 to 3 are specific for specific *Brassica* species. Nitrilases 1 to 3 are involved in the metabolism of glucosinolates, whereas nitrilase 4 catalyzes the detoxification of cyanide by converting the intermediate β-cyanoalanine to asparagine, aspartic acid, and ammonium.

Another type of nitrilase 4 was identified in *Sorghum bicolor*, and shown to catalyze the metabolism of the cyanogenic glycoside dhurrin, without the synthesis of toxic cyanide. However, other reactions exist in plants that also produce cyanide as a toxic byproduct; an example of this reaction is the biosynthesis of the plant hormone ethylene. Therefore, the detoxification of cyanide via nitrilase 4 is widespread among different systematically but unrelated plant groups such as mosses, moss ferns, gymnosperms and, of course, angiosperms, this being proof of their importance in metabolism (Figure 1.24). In grasses, gene duplication events have yielded another group of nitrilase genes (nitrilase 4B), which have then further evolved in species producing cyanogenic glycoside to groups B1 and B2. Interestingly, some of the encoded proteins can only show enzymatic activities as heterodimers (B group), whereas others are also active as homomers (nitrilase A). This results in a larger number of combinations for the resulting enzymatic complexes, which can in turn yield different substrate specificities.

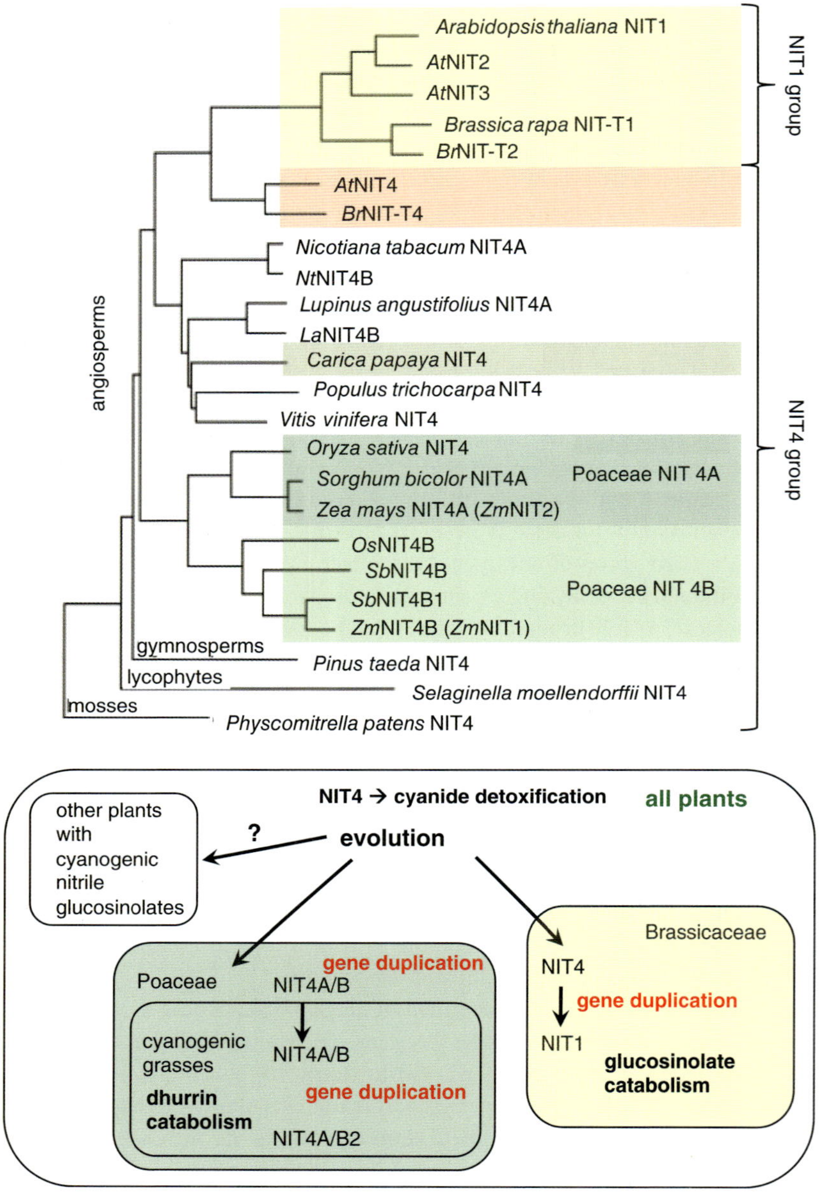

Figure 1.24 Evolution of nitrilase genes. Top: Part of a phylogenetic tree for nitrilase sequences. The nitrilase 1 family is typical for Brassicaceae and related plant species, whereas nitrilase 4 sequences can be also found in lower land plants. In Brassicaceae the nitrilase 1 family is most likely a descendant from nitrilase 4. The nitrilase 4 genes in grasses have also evolved from other nitrilase 4 genes. Yellow = nitrilase 1–3 from Brassicaceae; orange = nitrilase 4 from Brassicaceae; green = nitrilase 4 from grasses and from *Carica papaya*, a glucosinolate and cyanogenic glycoside containing plant species. Bottom: Evolutionary gene duplication events yielding the different nitrilase groups. Modified from Piotrowski, M. (2008) *Phytochemistry*, **69**, 2655–2667.

Gene duplications can also be detected in other phylogenetic trees for other synthetic pathways. An example of this is terpene synthase (Table 1.6), which catalyzes the synthesis of terpenes from different precursor moieties (see Figure 1.25). As each plant species can only synthesize a typical range of compounds, a specialization of the enzymes must be postulated. The sequence information of whole genomes allows the comparison of synthetic pathways in selected species. Whereas only one gene for a terpene synthase was identified in the moss *Physcomitrella patens* (this is actually a bifunctional copalyl synthase/kaurene synthase capable of forming *ent*-kaurene), the moss fern *Selaginella moellendorffii* has already 14 genes, which most likely encode full-length proteins based on their length. In other genomes the number is much higher, with between 40 and 152 terpene synthase genes having been identified, though not all are functional (Table 1.6). Many proteins have only one function – either copalyl synthase or kaurene synthase – and so can catalyze only one reaction step. In gymnosperms, bifunctional enzymes can still be found among the diterpene synthases, whereas this trait is lost in angiosperms. In contrast, monoterpene synthases contain only one functional domain in all of the plant species analyzed to date. Although terpene synthase genes can be classified into seven subgroups according to their sequence homologies, most plant species contain only one or two subgroups. An evolutionary plasticity within the terpene synthase families is shown by closely related enzymes, though these differ in their substrate specificities and also their intracellular localization. Differences in substrate preferences are also due to gene duplication and further evolution of the respective enzymes.

Table 1.6 Number of terpene synthase genes in different organisms and assignment to seven subgroups according to their sequence homologies. The number of individual genes represent those which are most likely full length and thus functional.

Plant species	Systematic classification	No. of terpene synthase genes	No. of related terpene synthase families in one organism
Physcomitrella patens	Moss	1	1
Selaginella moellendorffiii	Moss fern	14	3
Vitis vinifera	Dicotyledonous angiosperms	69	5
Populus trichocarpa	Dicotyledonous angiosperms	32	5
Arabidopsis thaliana	Dicotyledonous angiosperms	32	5
Oryza sativa	Monocotyledonous angiosperms	34	4
Sorghum bicolor	Monocotyledonous angiosperms	24	5

1.2.3
Gene Clusters

While gene clusters are common among bacteria (operons) and fungi for the genes encoding secondary metabolism enzymes, they have recently also been discovered in plants. Normally, many genes that mediate metabolic pathways in plants do not cluster (i.e., they occur in the form of continuous stretches on the DNA), but several examples have now been identified with either full or partial clusters for a specific metabolic pathway (Table 1.7). In the latter case, a cluster consists only of a few genes which are sometimes disrupted by genes not belonging to the cluster. These partial gene clusters can provide information on their formation during evolution, because they could have been formed by the transposition of large DNA pieces. Most likely, however, these events are preceded by gene duplication.

Several examples of gene clusters can be found for genes coding for enzymes in the synthetic pathways of secondary metabolites, which are active against phytopathogens (Figure 1.25). Among these is the gene cluster directing the synthesis of avenacin, a bioactive compound that is produced by some grasses against a phytopathogenic fungus (Section 2.5). For this gene cluster the rearrangement of genes has been postulated, because each gene in the cluster still has its own promoter and hence is not of the operon-type structure. To date, exactly how genes can rearrange is not clear, but possibilities include gene duplication, the relocation of individual genes, or the rearrangement of larger parts of the genome by defragmentation. These mechanisms could possibly lead to novel functions of relocated genes in the new cluster. Gene clusters need also to be simultaneously transcriptionally controlled, so that genes within one pathway are coexpressed.

Other examples of gene clusters in secondary metabolite synthesis include the clusters in *Arabidopsis* for thalianol, in maize for DIMBOA, and in rice for phytocassane and momilactone (Figure 1.25). A gene cluster normally contains genes that encode enzymes for synthesis of the core structure as well as side-chain modifications, which introduce specificity to the molecule (Table 1.7). The genes in a cluster are not necessarily in the same order in which they function in the pathway. For example, the cluster for thalianol in *A. thaliana* starts with a gene for a modifying enzyme, while the terpene synthases are located at the end. In the close relative *A. lyrata*, the gene cluster already has a different structure, such that the acyltransferase is strongly modified and a 10 kilobase insertion can be found between the acyltransferase gene and the *THAD* gene. The reading frames can also be changed for individual genes within the cluster.

Gene clusters can be rather large, as for example the cluster for phytocassane in rice, which consists of ten genes. While most gene clusters contain arrays of genes without any interruption, the cluster for DIMBOA is discontinuous and contains insertions of many other genes. A gene for DIMBOA synthesis is even located on a different chromosome, which means that not all genes for a given pathway are present in the cluster. The gene clusters for the metabolites DIMBOA, thalianol and avenacin are expressed during development of the healthy plant, while synthesis of the diterpenes momilactone and phytocassane can be induced by elicitors

Table 1.7 Gene clusters in the biosynthesis of secondary metabolites in plants. Six secondary metabolite gene clusters, which have been described in detail for plants, are presented. The size of the cluster (in kilobases), the number and function of genes and the role within core structure or side chain modifications are given. The genes are shown in their order (indicated by the small numbers) on the chromosome. The cluster sizes are not to scale. The asterisks mark genes that are not transcriptionally regulated. Modified from: Chu, H.Y. *et al.* (2011) *Plant J.*, **66**, 66–79; and Winzer, T. *et al.* (2012) *Science*, **336**, 1704–1708.

Plant	Substance	Reading frame and size of clusters (kb)	Enzymes for core structure	Enzymes for modifications
Arabidopsis thaliana	Thalianol	 35	Triterpene synthase[4]	BAHD Acyltransferase[1] CYP76[2] CYP85[3]
Avena strigosa	Avenacin	 202	Triterpene synthase[4]	Methyl-transferase[1] Sugar transferase[2] Serine carboxy-peptidase like acyltransferase[3] CYP51[5]
Oryza sativa	Momilactone	 170	Class II Diterpene synthase[1] Class I Diterpene synthase[4]	CYP99[2] Dehydrogenase[3] CYP99[5]
Oryza sativa	Phytocassane	 245 * * * *	Class I Diterpene synthase[4] Class II Diterpene synthase[7] Class I Diterpene synthase[8] Class I Diterpene synthase[9]	CYP76[1] CYP76[2] CYP76[3] CYP71[5] CYP71[6] CYP76[10]

(continued)

Table 1.7 (*Continued*)

Plant	Substance	Reading frame and size of clusters (kb)	Enzymes for core structure	Enzymes for modifications
Papaver somniferum	Noscapine	 214		Dehydrogenase[1] Carboxyl-esterase[2] CYP82[3] CYP82[4] Acetyl-transferase[5] Methyl-transferase[6] CYP82[7] Methyl-transferase[8] CYP719[9] Methyl-transferase[10]
Zea mays	DIMBOA	 15 618	Tryptophan-synthase alpha homolog[4]	CYP71[1] CYP71[2] Sugar transferase[3] CYP71[5] CYP71[6] Methyl-transferase[7]

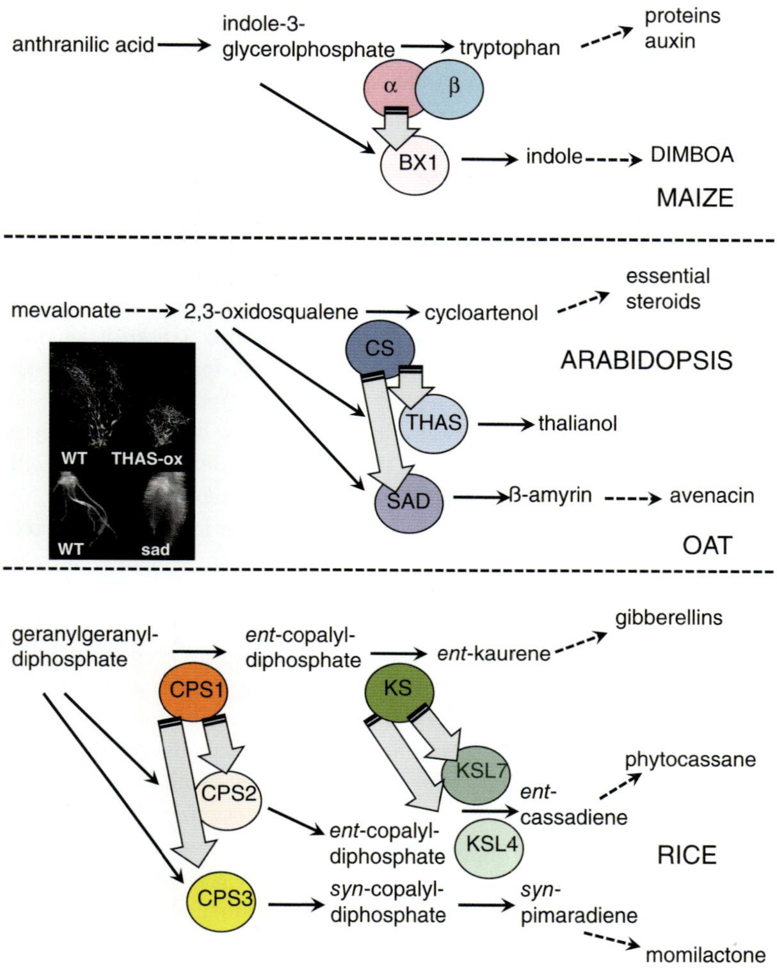

Figure 1.25 A hypothesis for the evolution of core genes within gene clusters involved in the formation of some secondary metabolites. The genes were recruited into the cluster from genes involved in primary metabolism or plant hormone synthesis genes. Top: Evolution of DIMBOA (= 2,4-dihydroxy-7-methoxy-1,4-benzoxazin-3-one) biosynthesis from the shikimate pathway in maize (*Zea mays*). Middle: Evolution of thalianol biosynthesis in *Arabidopsis thaliana* and of avenacin in oat (*Avena sativa*) from sterol biosynthesis. Bottom: Evolution of diterpene biosynthesis from the pathway to gibberellins in rice (*Oryza sativa*). BX = benzoxazin synthase; THAS = thalianol synthase; SAD = β-amyrin synthase; CS = cycloartenol synthase; CPS = class II- and KSL = class I-diterpene synthase. The pictures show *Arabidopsis thaliana* overexpressing the *THAD* gene (THAS-ox) and a mutant of oat for avenacin synthesis (*sad*), which accumulates a toxic intermediate, compared to the respective wild-type tissues. Figure based on Chu, H.Y. *et al.* (2011) *Plant J.*, **66**, 66–79. Photographs from: Osbourn, A.E. (2010) *Plant Physiol.*, **154**, 531–535; see Further Reading). Reprinted with permission from American Society of Plant Biologists.

and pathogens. Although many genes within the phytocassane cluster are coregulated, four of them (indicated by asterisks below the arrow representing the respective gene) are not regulated together with the others (Table 1.7).

A cluster containing only genes for modifying enzymes has been isolated from the opium poppy (*Papaver somniferum*), whereas the genes that encode the proteins involved in the core structure of benzylisoquinolines have been found outside the cluster. The genes in the cluster are involved in synthesis of the alkaloid noscapin; this cluster was identified by transcriptome analyses of noscapin-producing and nonproducing plant lines with ten genes that showed a high genetic coupling in a crossing population and were present only in the noscapin-containing plants. The gene cluster was verified by using bacterial artificial chromosomes containing the whole cluster. The compound noscapin is of great medical interest due to its anticancer potential that results most likely from its interaction with tubulin. Noscapin can also arrest metaphase and induce apoptosis, and also has a cough-suppressant effect (similar to codeine). However, as with thalianol, the biological function of noscapin in plants is currently unknown.

Partial gene clusters can be found encoding for proteins involved in the biosynthesis of the cyanogenic glycosides linamarin and dhurrin. For this pathway, the phenomenon of metabolic channeling has been described, where products are not released but rather are directly reused as substrates for the next reaction step. This phenomenon might also be associated with clustering of the respective genes.

What, then, could be the advantage of gene clusters? One possible explanation is that within gene clusters the chances of joint inheritance of the trait are higher. Also, when gene clusters result in metabolic channeling, toxic intermediates can be avoided. In addition, coordinated gene expression could avoid the creation of toxic metabolites, because the genes encoding metabolic enzymes for detoxification are expressed at the same time. Finally, during plant defense gene clustering may be advantageous when the genes are transcriptionally regulated at the same time, and by the same signals. An example of such a coordinated gene regulation is the synthesis of momilactone, where the genes in the cluster can all be induced by a chitosan elicitor. Only one transcription factor, from the family of leucine zippers, is responsible for this transcriptional activation. However, it is not known whether this transcription factor is responsible only for this gene cluster, or whether it is also involved in the regulation of other genes. In oat (*Avena sativa*), the gene cluster for avenacin synthesis was shown to be localized specifically in the outer epidermal cells of the root tips by using high-resolution DNA fluorescence *in situ* hybridization (FISH). A tissue-specific chromatin decondensation was also shown to be associated with the expression of the gene cluster.

Clustered genes which encode proteins involved in the synthesis of the core structure of a substance are often derived from genes in primary metabolism, and which have evolved after gene duplication (Figure 1.25). The protein BX1, which is involved in the synthesis of DIMBOA, was recruited from tryptophan synthesis, by duplication of the gene for the alpha-subunit of tryptophan synthase. Tryptophan is a proteinogenic amino acid but is also a precursor for many other secondary metabolites, such as terpene indole alkaloids and the plant hormone IAA.

Genes for thalianol synthase in *Arabidopsis* and β-amyrin synthase in oat evolved independently from the pathways to essential steroids in both plant species by recruiting cycloartenol synthase. The proteins for the synthesis of momilactone and phytocassane in rice have evolved from genes encoding diterpene synthases involved in the synthesis of the plant hormone gibberellin. In *Arabidopsis*, overexpression of the *THAD* gene, which encodes thalianol synthase THAS, results in a dwarfed growth and demonstrates that thalianol itself is also toxic to the plant producing it. Mutation of the *SAD* gene in oat results in a phenotype with fewer lateral roots, due to the accumulation of a toxic intermediate.

Whilst the question regarding the evolution of gene clusters in plants remains only partially solved, novel information can be acquired by using fungi as a model. Following a gene duplication event, the gene can relocate in the genome into a cluster. At the point of origin, and if the gene in the cluster is a single copy, the gene can be either retained or lost. Otherwise, it should be possible to find copies elsewhere in the genome. The relocation of a gene into a cluster is an advantage in meiosis, where closely linked genes are most likely inherited together in the progeny. This implies that these events should occur more often in regions with low recombination rates on the chromosome, which would stabilize such a cluster. These events may be advantageous for an organism during changing environmental conditions, when a greater evolutionary pressure would be exerted on the development of novel pathways. Such conditions might include nutrition as well as abiotic and biotic stress factors. This hypothesis is supported by the observation that some of the metabolites synthesized by enzymes encoded in gene clusters (i.e., DIMBOA, avenacin, momilactone and phytocassane) are involved in plant defense reactions against pathogens and insects.

Among the different plant species, gene clusters can either develop further or be retained in the original stages. For example, the biosynthesis cluster for DIMBOA in maize is quite well developed in the sense that it contains all of the necessary genes for the complete biosynthesis of DIMBOA, whereas the cluster in wheat (*Triticum*) does not contain all genes necessary for a complete biosynthesis. These observations can also help to provide an understanding of the evolution of a single pathway in different organisms within one order, for example, grasses. In maize, a noncomplete cluster could have been the precursor, and this was then further developed to the complete cluster. Alternatively, a complete cluster could have been the precursor, so that secondary event genes could have been lost in one species. In this latter case, the biosynthetic pathway would need to recruit other gene loci for the full biosynthesis of the respective metabolite or, as in rye (*Secale*), the biosynthetic pathway will be completely lost, as indicated by the synthesis of an alternative indole compound instead of DIMBOA.

As complete genome sequences for different organisms become available, bioinformatic methods can be applied to locate further putative gene clusters for secondary metabolite synthesis. In addition to the known gene clusters, five further putative clusters were predicted in *Arabidopsis* by computer-based methods. These have been postulated on the basis of homologous genes encoding cytochrome P450-dependent enzymes in *Brassica*, even though their specific function in *Arabidopsis* is not yet known.

1.2.4
Natural Variation

Many examples of natural variations in secondary metabolites have been described, an example being nicotine in tobacco (see Section 3.1 and Figure 3.12). At this point, the concept will be explained by using the examples of flavonoids and the glucosinolates.

The glucosinolates are a group of compounds which demonstrate a high variability among different plant species within the Brassicaceae, and also within cultivars and ecotypes (these are also termed "accessions" for *A. thaliana*). They can be used as examples to study the genetic variation, the aim being to understand their evolution. *Arabidopsis thaliana* is used as a model plant to investigate natural variations, because of the availability of hundreds of ecotypes. Here, the differences are used to identify quantitative trait loci (QTLs) to locate novel traits associated with differences in glucosinolate patterns, and which can encode either genes for either biosynthetic and metabolic enzymes or regulatory proteins – that is, transcription factors. Evolutionary pathways can also be investigated using this approach. The MAM enzymes (see Figure 1.23) are good examples as they have become diversified in evolution after gene duplication, and could be allocated to specific glucosinolate patterns because they can now use different substrates. However, there are some ecotypes of *Arabidopsis* which do not contain all three known *MAM* genes, and these will be deficient in the synthesis of several aliphatic glucosinolates. These different patterns are termed "natural variations" between accessions. For indole glucosinolates, the variation has been detected in relation to the side-chain modifications rather than to the core structures. This is a result of the differences between genes for modifying enzymes such as hydroxylases and methyltransferases.

One important metabolic reaction is removal of the sugar moiety of glycosinolates by myrosinase (see Figure 1.12), while further metabolic reactions depend on the specifier proteins that determine the outcome of the reaction. Both, epithionitrile and nitrile specifier proteins, as well as thiocyanate-forming proteins, direct the formation to respective products. In the case of nitrile the reaction product is less toxic than the two others, the epithionitriles and thiocyanates. The so-called myrosinase-binding proteins from *Brassica* also belong to this protein group, based on their sequence homology. A lectin-binding motif is necessary in order to bind to myrosinase, and this is common to all of these proteins. Lectins bind to specific sequence motifs via covalently bound sugar moieties of glycosylated proteins, of which myrosinase is a member. Epithionitrile- and nitrile-specifying proteins are found in *Arabidopsis,* while garden cress (*Lepidium sativum*) possesses a thiocyanate-forming protein. Consequently, these different species also exhibit different patterns of glucosinolate metabolites. As these compounds play a major role in defense for these plants, an accumulation of toxic metabolites is essential to provide their resistance against pathogens and insects. Thus, it is also important which of the specifier proteins has been developed in a given plant species. If only nitrile specifier proteins are present, then fewer toxic compounds will accumulate and this could have overall a negative effect on the defensive performance against pathogens and herbivores. Consequently, the evolution of these proteins has an important implication for defense responses. Among the different *Arabidopsis* ecotypes, the metabolites of the glucosinolates occur in different concentrations and also in different proportions to one another, which indicates that different specifier proteins are present. Interestingly, the evolution of the specifying proteins – and especially of the nitrile specifiers – has also been demonstrated in specialized insects that feed on *Brassica* species, which are essential for their nutrition (Section 3.1).

A structural variation of flavonoids has also been found in different *Arabidopsis* ecotypes, and a comparison of these natural variations led to the identification of novel compounds and regulatory proteins for the synthesis of these flavonoids. Whilst regulatory elements are most likely involved if only quantitative differences are present in the patterns, a novel compound implies a new biosynthetic reaction. Flavonoids can be grouped differently based on their occurrence in different ecotypes that occur all over the world. One flavonoid group was seen to occur only in ecotypes from America and Asia while European ecotypes did not show any common pattern, although the numbers of samples taken was most likely insufficient to corroborate such a classification for flavonoids. Nevertheless, the results of this geographic distribution could point to evolutionary pathways of groups of flavonoids on a worldwide basis.

1.2.5
Modulation of Different Molecules with Similar Functions

During the evolution of secondary metabolites, the functional aspect of molecules with which they interact must also be taken into account. One such example is the inhibition of polar auxin transport by flavonoids (see Section 2.2.1; Figure 2.3). Polar auxin transport from cell to cell is regulated by efflux transporters of two different structural families, the PIN and PGP proteins (Figure 1.26), both of which can be interfered with by flavonoids. The PGP-type transporters belong to

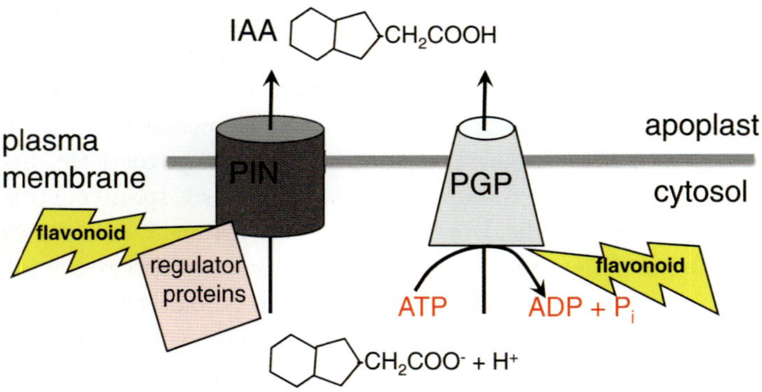

Figure 1.26 Flavonoids modulate two different types of auxin transporters, the evolutionary younger PIN-type and the older ABC-transporter PGP-type, which is dependent on ATP. The flash symbolizes possible interaction sites with flavonoids and transporters.

the ABC-transporter family, where flavonoids inhibit their ATPase activity. The same situation occurs in mammals (Section 4.1.1), where flavonoids have an inhibitory effect on the transporters, but in plants the effect is less pronounced, perhaps because the molecules have been able to adapt to the flavonoids over a longer time period. In contrast, the flavonoids can indirectly influence the PINs by affecting regulatory proteins such as kinases/phosphatases that regulate PIN activity. The evolutionarily younger PIN-transporters are less strongly inhibited by flavonoids than the PGP-family, which is indicative of their better adaptation to a plant environment during the evolution of their specific transport activity towards auxin. The modulation of auxin transport is used by bacterial symbionts (rhizobia) to establish auxin maxima at the site of nodule formation (see Section 2.4). Plants also use flavonoids to modulate their auxin transport capacities during developmental processes. Thus, auxin transport proteins and their regulators may be regarded as cellular components that have been adapted to a certain environment during evolution, rather than to act as specific target molecules for flavonoids.

In the model plant *Arabidopsis*, as a member of the higher plants, the PIN-family consists of eight different genes that are expressed in different tissues and localized intracellularly at the plasma membrane, with one exception. Recently, a PIN protein in the ER was described which was also a site of flavonoid synthesis. These proteins appear not to be involved in intercellular transport, but rather to control intracellular auxin homeostasis. Since in the moss *Physcomitrella patens* this specific PIN member is the only one present, it was assumed that the endoplasmic PIN proteins are evolutionarily older than those found at the plasma membrane. It has been assumed that the earliest function of the PIN proteins was to regulate intracellular IAA levels, and that cell-to-cell transport was evolutionarily younger. Thus, an attractive hypothesis might be that polar auxin transport is a hallmark of the development of higher land plants with more complicated body plans, whereas the control of cellular auxin levels was already necessary in lower land plants.

1.2.6 Evolution by Stress Factors

1.2.6.1 Abiotic Stress

Another possible approach to investigate the evolution of secondary metabolites uses flavonoids as examples. Based on their UV-absorbing properties, flavonoids were described as the "sunprotection screen" of plants. Their absorption maxima are within the wavelengths of hazardous UV light, by which the expression of genes encoding enzymes for biosynthetic steps are also induced (see Section 2.3). Although, in *Arabidopsis*, UV-B light induces quercetin synthesis, there are indications that flavonoids are not the primary protection compounds, despite their high induction under light stress. Rather, they might have an indirect effect as antioxidants against reactive oxygen species (ROS) produced during high light stress. Other candidates providing protection against hazardous light include sinapine derivatives.

In the lower land plants that occurred at the early stages of evolution, a mycosporin-like amino acid metabolism was lost in favor of flavonol metabolism, despite the mycosporins having better effects on leaf surfaces in terms of protection against light stress. Thus, it was hypothesized that flavonoids could have evolved their UV-protective function as a secondary trait from other roles in the plant. Interestingly, most genes encoding flavonoid biosynthetic enzymes are already present in lower land plants, such as liverworts and mosses. One characteristic of these "old" genes is a strong and rapid induction after light stress and other abiotic stress factors, which typically leads to the formation of ROS. In addition, transcription factors of the R2R3-MYB type (Figure 1.17) are present in mosses and can be induced by a high light irradiation. Observations that the flavonoid pathway had been relatively well conserved over many million years indicated a central role for the evolution of these secondary metabolites, which play diverse roles in the stress adaptation of plants. Such a general function may be very important when considering unpredictable changes in environmental conditions, and organisms with a wide range of metabolic compounds might be better protected against such variations in the environment. Indeed, flavonoids may be considered a class of plant metabolites that have conferred the ability to adapt to changing environments over long periods of time.

The role of flavonoids is in accordance with the hypothesis concerning the regulation of auxin transport. Quercetin is a better inhibitor of polar auxin transport than kaempferol, and therefore UV-B induced quercetin synthesis should have an indirect effect on plant growth and development, as the distribution of a growth-promoting hormone would be changed. Likewise, stressed plants could react with a different habitus to changing environmental conditions.

1.2.6.2 **Biotic Factors**

The interaction of plants with their environment, especially with changing pathogen and herbivore populations, necessitates the synthesis of different secondary metabolites (see Section 3.1). The development of novel defense compounds during evolution is dependent on changes in genes, gene function, or a combination of different proteins to result in novel pathways. One important mechanism in this respect is gene duplication (see Section 1.2.2). It is disadvantageous for a plant to focus on only a few metabolite classes, although this would result in a highly adaptive defense behavior in relation to any parasites invading the plant. The energy costs to synthesize only a few metabolites would also be much lower, and this again would be advantageous. During evolution, however, the parasites may also adapt to the chemical defense of the plants, and consequently a plant with only a few different metabolites would be less well protected against pests that were capable of adapting. Clearly, a wider variation of secondary metabolites would be advantageous within the natural environment, and the evolution of a wider spectrum of compounds by one plant species would be necessary if a plant were to be attacked by a range of parasites at different stages of its development. Thus, a broad series of defense reactions would hamper the adaptation of a parasite.

1.3
Biotechnological Applications

1.3.1
Identification and Quantification of Natural Plant Products

In order to identify or quantify a compound of interest, the metabolite must first be extracted from the plant tissues. However, the chemical properties of a material under investigation is of great importance in the development of a relevant purification scheme (Figure 1.27). The most important issues to be taken into account include:

1) It must be defined whether a compound that is already known should be extracted and quantified (targeted approach), or whether a broad range of unknown (bioactive) compounds should be identified (untargeted approach). The latter experimental approach is also coined "metabolomics" (Figure 1.28).
2) For individual compounds, it must be determined which properties are already known, and which solvents can be used for their extraction.
3) The purity of the compound might be important for identification and also for bioactivity assays; in this situation the metabolite must be further purified using chromatographic methods (Figure 1.27).
4) If identification and quantification is the only purpose then the sample will not need to be recycled. If it should be subjected to further assays for bioactivity, a larger amount of the compound(s) will be needed after sample purification.

The first decisions to be made involve the extraction solvent. It is important to know whether polar or unpolar compounds are to be extracted, as different pH values can be used to further differentiate compounds according to their molecular properties. Following the evaporation of organic solvents, the remaining aqueous phases can be further extracted using organic solvents with different polarities. At this point the extract will still contain many molecules, and these must be further enriched by applying chromatographic methods. The complex extract can be subjected to different chromatographies, depending on the molecule(s) to be isolated. Such methods include size fractionation, separation according to charge/polarity, or affinity chromatography. The chromatographic steps can be analytical, using small volumes, or preparative for the purification of larger samples. If the chromatographic methods are used together with a suitable detector (i.e., for high-performance liquid chromatography; HPLC), then a certain amount of information, perhaps about the absorption properties of the molecules, can be obtained at the early purification steps. For affinity chromatography, additional information must be gathered about the compound; in this case, ligands that bind to the molecule in question can be attached to a resin and will retain a small group of compounds.

The analytical methods applied to secondary metabolites include paper chromatography (this is barely used any more), thin-layer chromatography (TLC), HPLC, gas

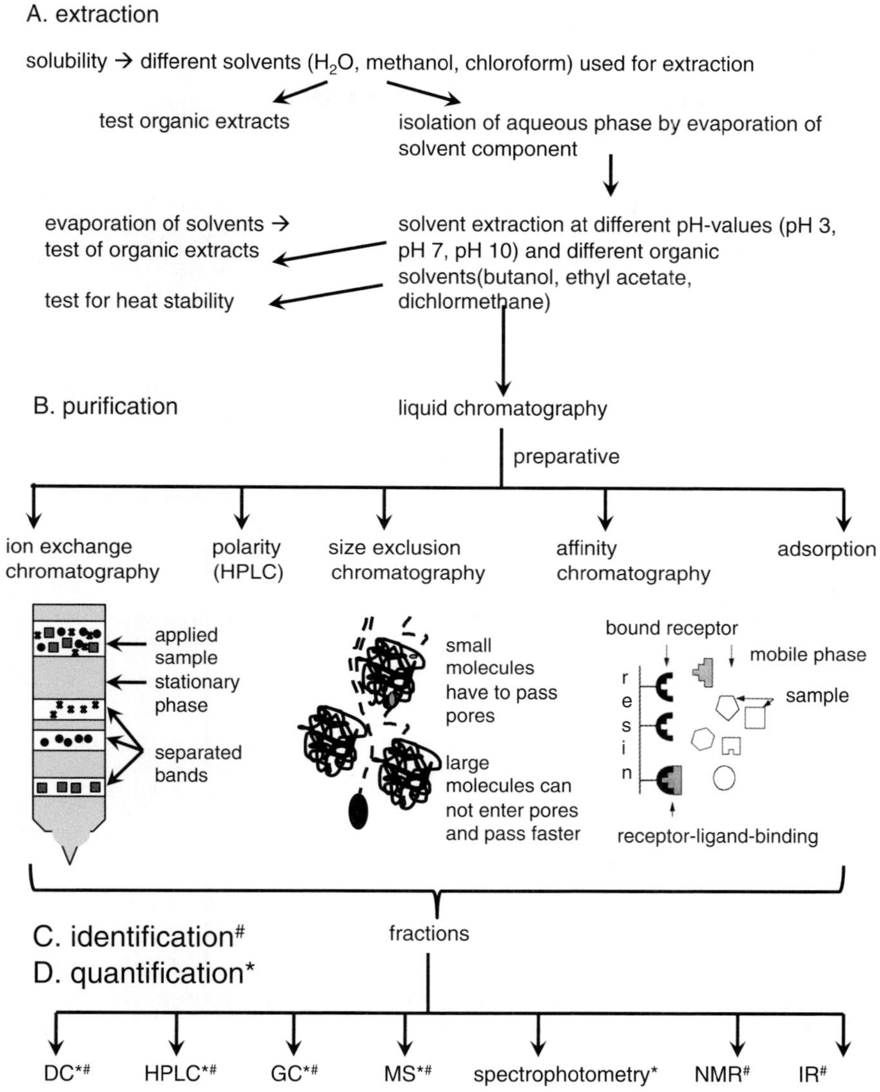

Figure 1.27 Example of a scheme for the extraction, purification and possible identification of complex metabolite mixtures from plants. TLC = thin-layer chromatography; HPLC = high-performance liquid chromatography; GC = gas chromatography; MS = mass spectrometry; NMR = nuclear magnetic resonance; IR = infrared.

chromatography (GC), mass spectrometry (MS), nuclear magnetic resonance (NMR) spectroscopy, infrared (IR) spectroscopy, and spectrophotometry (Figure 1.27). Some of these methods can be combined, such as GC or liquid chromatography (LC) with MS as detector. Some methods also allow the quantification of metabolites as well as

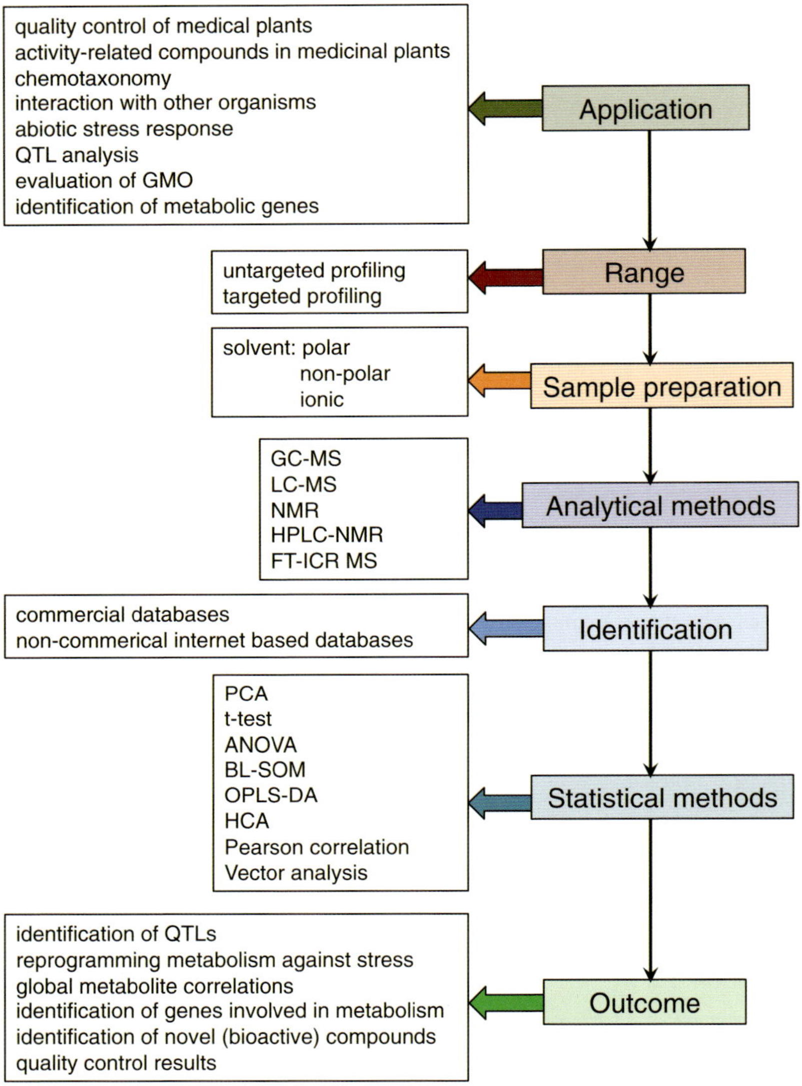

Figure 1.28 Metabolomics approaches in plant research, especially in biotechnology. Plant-specific components for the expectations and outcome are shown in green; possible experimental conditions are shown in brown; analytical conditions are shown in blue.

their identification. In particular, photometric assays with specific reagents can yield rapid and reliable results, and can also be used occasionally in combination with TLC (Figure 1.30). By using reference compounds, a tentative identification is possible with this technique in some cases. The chromatograms obtained can be scanned, and specific spots subsequently quantified. Comparisons with standards when using

TLC involves using the R_f value; this is a factor that indicates the relative migration of a given compound (for HPLC and GC, the retention times are used in similar fashion). If a reference substance is available, then standard calibration curves can be used to quantify a compound. If MS is used as a detection system for either LC or GC, identification via the mass spectrum should be possible. A quantification can also be achieved if reference compounds are available that have been labeled with heavy isotopes; these behave exactly as the natural compounds but, if added prior to the extraction, they can be used to determine the efficiency of the extraction procedure for a specific compound. On the basis of the characteristic masses for standard and endogenous compound, concentrations of compounds in the cells or tissues can be calculated. This method is accurate even for small molecules, which occur in minimal quantities. The accurate identification of a compound is sometimes not possible using MS, but can often be achieved using NMR- or IR-spectroscopy.

Metabolomics – which is defined as the mostly unbiased parallel analysis of as many compounds as possible – has attracted attention relating not only to the identification of novel metabolites but also to biotechnological aspects (see Section 1.3.3). This approach can provide an insight into the changing metabolic pattern under various conditions during the development, as well as the changing environmental conditions under abiotic and biotic stress situations, and thus help to identify novel compounds (Figure 1.28). By using modern biotechnology approaches, plants with different biological properties can be compared to identify additional bioactive substances, for example in medicinal plants (Figure 1.31). Other applications include the quality control of medicinal plants, the identification of QTLs, chemotaxonomy, the evaluation of genetically modified organisms (GMOs), and the identification of metabolic genes.

The choice whether a targeted or an untargeted approach is chosen determines the suitable solvent(s) used for the extraction. It should be noted that the perfect universal solvent for all applications does not exist, and that the solvent could be polar, nonpolar, or ionic. The following step determines the intensity, reliability and potential for the identification of novel compounds, and the analytical techniques to be used for the analysis of the compounds. While the pioneering metabolomics studies were performed using GC-MS, this method has now been complemented or replaced by other techniques such as LC-MS, NMR-spectroscopy, or combinations of these (Figure 1.28).

The identification of individual compounds within chromatograms obtained using either GC or LC relies on deconvolution software, which is needed to separate the peaks from one another, as well as databases that may be either open source internet-based or commercially available. Finally, the data need to be evaluated (Figure 1.28). Various statistical methods used in conventional genetic studies are applicable to metabolomic data by considering the amount of each metabolite as a trait value. For this, principal component analysis (PCA), which is a multivariate analysis method, is commonly used in metabolomic studies. The PCA model can provide an overview of all observations or samples in a data table by projecting and clustering each sample and highlighting any holistic differences

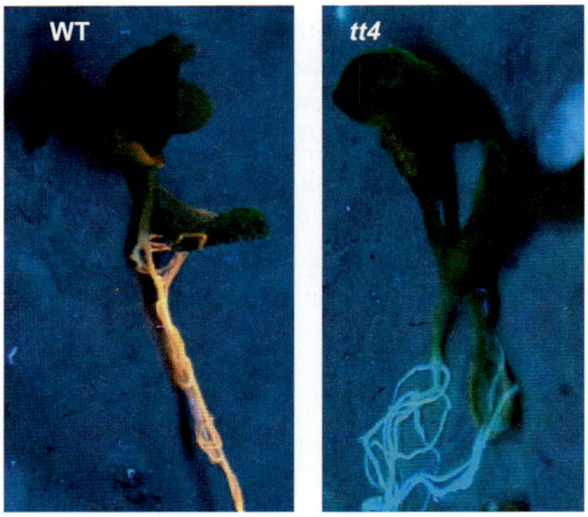

Figure 1.29 ***In situ*** **staining of flavonoids in** ***Arabidopsis thaliana*** **with diphenylboric acid-β-aminoethylester.** Seedlings on the left show orange fluorescence typical of quercetin and its derivatives, while the seedling on the right, mutated in chalcone synthase, *transparent testa 4* (*tt4*) is flavonoid-free and thus shows the cyan fluorescence indicating sinapates. WT = wild-type.

in a given sample set. Many reports have described the application of PCA to metabolomic data. In addition, several statistical analytical methods have been used to analyze metabolomic datasets, including hierarchical cluster analysis (HCA), partial least-squares-discriminant analysis (PLS-DA), and batch-learning self-organizing map analysis (BL-SOM).

Depending on the objective of each study, the most appropriate statistical analytical method should be exploited to evaluate the available metabolomic data. The outcome of such an experiment could then be the identification of QTLs, the reprogramming of metabolism against stress, global metabolite correlations, the identification of genes involved in metabolism, the identification of novel (bioactive) compounds, and quality control of the results.

In order to localize secondary metabolites within a plant tissue, a so-called *in situ* staining method can be used for some compounds (Figure 1.29). Many natural compounds, such as the flavonoids shown here, can be rendered visible under UV-light because they form fluorescing complexes with certain staining reagents. In the case of treatment with diphenylboric acid-β-aminoethylester (Figure 1.29; see also Figure 1.30), an orange fluorescence is typical for quercetin and its glycoside derivatives, while the cyan fluorescence in a flavonoid-free mutant of *Arabidopsis* reveals the presence of sinapates. In leaves, it is difficult to visualize the flavonoid complexes because of the high concentration of chlorophyll.

Laser microdissection can be used to obtain an insight into the more detailed distribution of different secondary metabolites. For this, small tissue sections can be generated or even single cells isolated; these are then collected and further

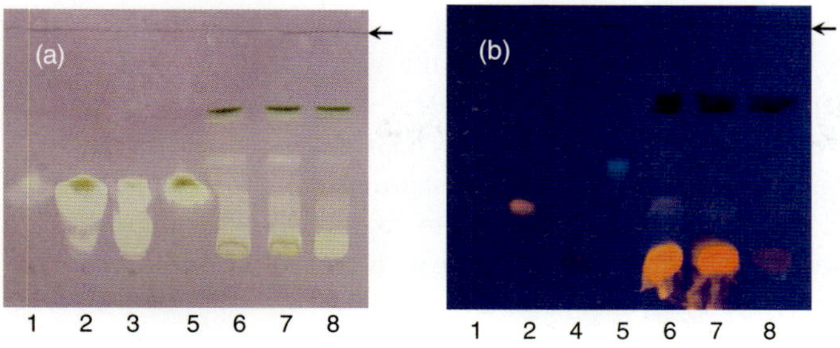

Figure 1.30 An example of "biochromatography" in comparison with fluorescence staining. Two silica gel TLC plates are compared. The upper plate was stained with a reagent to show antioxidative activity (2,2-diphenyl-1-picrylhydrazyl); the lower plate was stained using a reagent which forms fluorescing complexes with flavonoids and sinapates (diphenylboric acid-β-aminoethylester). The lower plate was photographed under UV-light. The solvent front has been marked with an arrow. The individual lanes show 1–5: standards of naringenin (1), quercetin (2), rosmarinic acid (3), rutin (4), and kaempferol (5); lanes 6–8: metabolites from a carnation plant, which were extracted using a methanol–water mixture. After evaporation of the organic solvent, the remaining aqueous phase was adjusted to different pH values and extracted using an organic solvent (lane 6: pH 3; lane 7: pH 7; lane 8: pH 10).

analyzed. Using this method, four different parts of mature seeds of oilseed rape (*Brassica napus*) were sampled to investigate the distribution of major secondary metabolites such as glucosinolates and sinapine, and of two flavonoids in the hypocotyl and radicle, the inner cotyledon, outer cotyledon, seed coat, and endosperm. Subsequently, these compounds were identified and quantified in dissected samples by using HPLC with diode array detection and MS. The differential localization of the secondary metabolites can also provide clues regarding their putative functions. MS can also be used on fresh tissue sections, and this results in an even higher spatial resolution of secondary metabolites. This technique has been used to demonstrate the distribution of glucosinolates in various *Arabidopsis* organs. This high-resolution "imaging" of secondary metabolites can also reveal information about the specific roles of certain tissues. For example, a high concentration of secondary metabolites at the leaf margin, which is among the first contact points between a leaf and an insect, could indicate a possible role against herbivores (see also Figure 2.35).

1.3.2 Biological Activity

Various test systems used for the determination of biological activities, together with the important targets, are described in Chapters 4 to 6. However, the possibility of conducting bioassays either together or immediately after a chromatographic separation should be mentioned at this point. Such test systems should be simple to apply and cheap to operate, especially if many compounds are to be screened. Consequently, the systems are directed towards the simple properties of secondary metabolites, such as their antioxidative potential, which can easily be determined by using spetrophotometric tests with colored compounds. Alternatively, these staining methods can be used in combination with TLC as a so-called "biochromatography," revealing the chromatographic behavior of a compound together with an indication of its antioxidative activity and other analytical tests (Figure 1.30). In this case, the TLC plate has been stained (after development) with a colored radical, leaving white spots where compounds with antioxidative activities migrate (Figure 1.30a). TLC of the same extracts, after staining with a reagent that forms complexes with flavonoids and sinapates, is shown in Figure 1.30b. In combination, these two methods reveal important biochemical properties of the compounds in the analyzed extract.

1.3.3 Biotechnological Production of Natural Compounds

Bioactive compounds have long been used by humankind for many different purposes (Table 1.8), and the requirements for compounds with better or even novel properties is increasing continuously. The exploitation of natural compounds of plants can be traced back generally to their medicinal use, with hallucinogenic, mind-altering properties and analgesic effects having received much of attention (see also Chapters 5 and 6). Other compounds are used in foods as flavor or spice components; examples include the spice capsaicin from peppers (which has also antimicrobial activity), vanillin from vanilla, and also the isothiocyanates released from the glucosinolates upon tissue disruption. Finally, colored substances have received much attention in the food industry and as natural dyes in other applications.

Plants are considered to be the organisms of choice for the production of metabolites that rely for their synthesis on complex pathways. The realization of such complex biosynthetic pathways in microbes is very difficult, because of missing intermediates and potential problems with protein folding and modifications. Yeasts can be seen as an alternative, but in this case reactions that need different plant compartments cannot be imitated. Today, sterile cell or organ cultures in plants are among the methods of choice.

The compounds used by humans can be employed for dyes (e.g., betanidine, indigo, shikonine), as aroma volatiles and spices, as food additives (see Chapter 6), in the cosmetics industry as aroma compounds and antioxidants, or in pharmacology (e.g., pilocarpine, scopolamine) (Table 1.8; see also Chapter 5). In addition, their use in agriculture as biopesticides has received much attention (see Section 2.5.2).

Many plants with beneficial properties are either difficult to cultivate and/or are on the list of endangered species; therefore, their harvest is difficult or

Table 1.8 Applications of secondary metabolites in medicine, pharmacy, as food additives, and in cosmetics industry.

Compound class	Product	Effect
Aroma volatiles	Perfume	
Aroma compounds	Spice	Many
Natural dyes	Indigo	
	Betanidine	
	Shikonine	
Biopesticides[a)]	Glucosinolate	Fungicide
		Herbicide
		Antiviral
		Antimicrobial
		Insecticide
Stimulants		Stimulating, hallucinogen
Pharmaceutically active compounds	Phytotherapy[a)]	Anti-inflammatory
		Antibacterial
		Antidepressant
		Fungicide
		Antimigraine
		Antiparasitic
		Muscle relaxant
		Diuretic
		Cardiotonic
		Dermatologic
		Receptor agonist/antagonist

a) Only examples are given for the listed classes of compounds or for their effects.

sometimes even prohibited. Hence, novel methods for the cultivation of such species are required to ensure the production of beneficial compounds from medicinal plants. Laboratory cultures have also the advantage that gene alterations created by molecular biological methods are possible in order to change pathways in favor of a desired metabolite (see Figure 1.32). Consequently, during the past few decades plant cell or organ cultures have become established for the production of bioactive metabolites. These contribute to a high-quality production of these compounds, based on the controlled environment of the cultures. In addition, large-scale bioreactors for plant cultures are now available whereby, the induction of secondary metabolites by elicitors is a feasible and easy-to-perform method under controlled conditions. The elicitors may be either stress-signaling compounds or changes in culture conditions, such as oxygen depletion. In addition, genetically modified cultures can be used to raise the concentration of a desired compound per biomass. These genetic alterations can lead to an activation of the biosynthetic pathway in a targeted fashion. Some features for the selection of plant materials used in secondary metabolite production are shown in Figure 1.31.

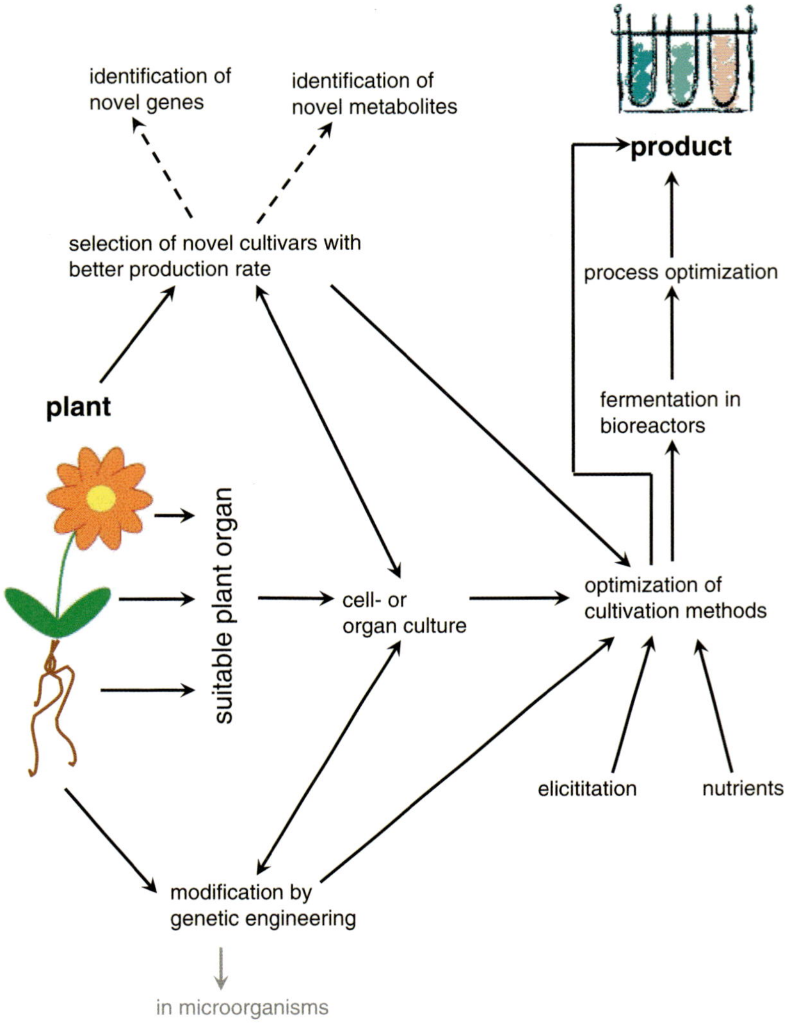

Figure 1.31 Production of secondary metabolites and optimization of plants for these procedures. It is indicated that microorganisms can also be used under certain conditions for the production of secondary metabolites of plants (gray arrows). Screening for novel genes and metabolites (broken arrows) can also be performed in cell or organ cultures and genetically modified organisms.

The first stage is to select the most suitable plant organ to generate the cell or organ culture (Figure 1.31), and to this end the metabolite pattern of the respective organs must be analyzed. The next step is to decide whether cell or organ cultures are to be established; the former are dependent on the continuous addition of plant hormones, while the latter are hormone-autonomous and therefore

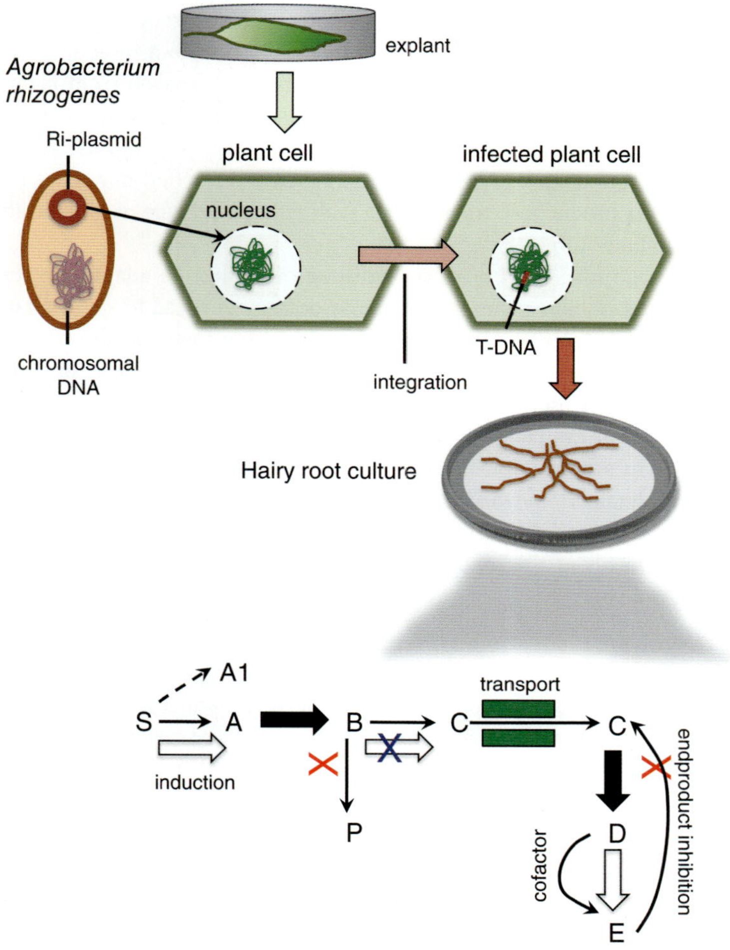

Figure 1.32 Generation of Hairy root cultures from leaf explants using *Agrobacterium rhizogenes*. These are inoculated with either wild-type or transgenic bacteria, which introduce their T-DNA from the Ti-plasmid (tumor-inducing) into the host genome, resulting in stable transformation events and a sterile Hairy root culture. Some factors are shown in the lower part of the figure which could lead to problems in secondary metabolite production after genetic modification. Closed arrows symbolize the situation in a wild-type culture or plant; open arrows and red-marked pathways situations in a transgenic culture or plant, resulting in better fluxes of metabolites. Small arrows show limiting steps. More factors are described in the text. Lower part based on Georgiev, M.I. *et al.* (2010) In: *Medicinal Plant Biotechnology* (Arora, R., ed.), CAB International, Oxon, UK, pp. 156–175.

cheaper to cultivate. The cultures can be induced from a variety of different plant tissues by the soil bacterium *Agrobacterium rhizogenes*, for example from the leaves, flowers, and also the roots (Figure 1.32). This phytopathogenic bacterium transforms plant roots with its tumor-inducing DNA (T-DNA), located on the

Ti-plasmid, that in Nature results in the symptoms of Hairy root disease. The genes transformed into the host genome are termed *rol* genes, but in addition to the *rol* genes the bacterium can be transformed by other genes taken from plants or other organisms, so as to alter the metabolic pathways or profiles in a desired manner. The resulting organ cultures are also termed Hairy roots, similar to the naturally occurring root disease. These cultures are used very often nowadays for secondary metabolite production, transformed with either wild-type or genetically modified bacteria (Table 1.9).

The genetically optimized cell or organ cultures must first be adapted to the cultivation conditions in bioreactors, and this may occasionally lead to stress situations due to the possible depletion of nutrients or shearing forces in the reactor. The cultivation can be also used to introduce stress factors deliberately, as this can increase the synthesis of the product, as well as elicitors (plant signaling molecules) to create biotic stress (Table 1.9). As the plants do not leave the laboratory during cultivation in a bioreactor, there is no (legal) conflict with these genetic modifications. A detailed knowledge of the biosynthetic pathways, transport between compartments, and their (transcriptional) regulation are necessary for targeted alterations. Details of the metabolites produced by cell and organ cultures

Table 1.9 Examples for the production of secondary metabolites in cell or Hairy root cultures. Often, a higher metabolite concentration was found in the cultures compared to the mother plant, even without genetic alterations. In many cultures higher metabolite contents was generated by elicitation using either biotic or abiotic stress factors or signals. It is also indicated by "yes" when the cultures were genetically modified by other genes, either from plants or bacteria (yes in brackets). In addition to the plant metabolites synthesized under normal conditions, after transformation with bacterial genes also other metabolites can be synthesized.

Plant species	Secondary metabolite	Cell culture	Hairy Root	Elicitor	Genetically modified
Ammi majus	Scopoletine		Yes	*Enterobacter*	
Argemone mexicana	Sanguinarine	Yes			
Atropa belladonna	Scopolamine Atropine [a]Hyoscyamine		Yes	Nitrate	Yes
Beta vulgaris	Betalaines		Yes	*Lactobacillus*	
Catharanthus roseus	Indole alkaloids (chlorinated indole alkaloids)	Yes	Yes	*Trichoderma*	Yes (yes)
Cinchona ledgeriana	Quinine (Strictosidine)		Yes		(Yes)
Datura innoxia	Hyoscyamine Scopolamine		Yes		
Datura quercifolia	Hyoscyamine Scopolamine		Yes		

Datura candida	Hyoscyamine Scopolamine		Yes		
Datura metel	Scopolamine		Yes	*Aspergillus* *Alternaria* *Fusarium* Yeast extract	Yes
Duboisia leichhardtii	Scopolamine		Yes		(Yes)
Fagopyrun esculentum	Catechin Epicatechin		Yes		
Hyoscyamus niger	Hyoscyamine Scopolamine		Yes		Yes
Lavandula officinalis	Rosmarinic acid Caffeic acid	Yes		Oxygen depletion	
Lithospermum erythrorhizon	Shikonine	Yes	Yes		(Yes)
Morinda citrifolia	Anthraquinone	Yes			
Mucuna pruriens	L-DOPA	Yes			
Panax ginseng	Ginsenoside		Yes	Chitosan Methyl-jasmonate	
Papaver somniferum	Alkaloids	Yes			
Rauwolfia serpentina	Reserpine	Yes			
Rubia tinctoria		Anthraquinone		yes	
Salvia miltiorrhiza	Rosmarinic acid	Yes		Yeast extract	
	Tanshinone		Yes	Yeast extract	
Solanum khasianum	Solasodine		Yes		
Solanum tuberosum	Sesquiterpenes		Yes	*Rhizoctonia*	
Tagetes patula	Thiophene		Yes	*Fusarium*	
Taxus brevifolia	Paclitaxel	Yes			
Valeriana officinalis	Valpotriate		Yes		

a) racemic mixture from hyoscyamine and scopolamine.

are summarized in Table 1.9, including several important compounds with medicinal uses (see Chapter 5).

The question of which factors can influence the outcome of the production of secondary metabolites by a given biosynthetic pathway remains unclear. The best results can be obtained by using simple changes in pathways, where only

Figure 1.33 Activation of a single MYB transcription factor in a mutant of cauliflower (*Brassica oleracea*) leads to the purple color compared to wild-type. Although this picture shows a cultivated form, similar results can be obtained with a transgenic approach, as described in the text. A green variety is shown, which has chlorophyll in its tissues, for comparison. Fotolia.com, © Lsantilli.

one enzyme or transcription factor is activated. For example, synthesis of the alkaloid scopolamine via hyoscyamine and an intermediate is catalyzed in a two-step reaction by the same enzyme. Another example is the overexpression of a gene encoding a maize transcription factor in other plants; this particular transcription factor controls many steps in anthocyanin biosynthesis, and the various tissues of the resulting transgenic plants are typically colored. A similar effect is shown for a natural mutation in Figure 1.33, where a single MYB-type transcription factor is activated, leading to the purple- headed cauliflower (*Brassica oleracea*).

In order to modify complex biosynthetic pathways, many factors must be taken into account to ensure that the desired metabolite can be produced, including:

- the identification of the step determining the velocity;
- the availability of precursors;
- the limitation of cofactors;
- competing pathways using the same precursor or intermediates;
- inhibition by the final product;
- transport over cellular compartments; and
- transcription factors (Figure 1.32).

First, an attempt is made to determine the critical step for the overall velocity of the pathway. This step is an ideal candidate for metabolic engineering, and changes can be achieved by overexpressing the gene(s) that encode the proteins responsible for the limiting steps. By increasing the flux from substrate S to metabolite A, the second metabolite A1 is not synthesized, and therefore production of the desired product C should be increased.

However, when homologous genes are used it must be guaranteed that there is no cosuppression effect that will reduce the expression of the transgene and of the original gene itself. The phenomenon of cosuppression was first observed after transforming petunia by a chalcone synthase gene; the effect led to white areas, which represented an indication of reduced gene expression and, consequently, to less pigment formation. In the example shown in Figure 1.32, intermediate C would no longer be generated, and this would lead to an accumulation of intermediate B. This in turn could lead to the undesired accumulation of side compound P.

The availability of precursors is also dependent on the number of biosynthetic pathways competing for a substrate. For example, tryptophan is a precursor for a wealth of different secondary metabolites, and also for protein synthesis and synthesis of the plant hormone IAA. Therefore, the overexpression of tryptophan decarboxylase could lead to a reduction in other indole metabolites, such as indole glucosinolates. In laboratory cultures this is not important, but in the field the plants would be less tolerant to various biotic stress factors. An alternative to the upregulation of the major pathway would be to suppress the side pathways, for example by applying antisense or RNA interference techniques. This would result in the availability of more precursors for the main pathway (Figure 1.32).

One major problem, however, would be the transport of intermediates between organelles, because the levels of transporters would also need to be increased; consequently, metabolic pathways that occur in only one compartment would be easier to manipulate. A modification of compartmented pathways might be possible by adding signal sequences to the heterologously expressed genes. Alternatively, a synthesis of metabolites in the culture medium could be attempted by adding secretion sequences to the protein(s). The secretion of a protein or reaction product ensures an easy purification of the product from the culture supernatant. Finally, the inhibition of an enzymatic reaction by either the product or an intermediate could also constitute a problem. In this case, even when the transgene is highly expressed, no product would accumulate. This is also the case when an essential cofactor is not present, although such a compound could be added to the culture, if not too expensive.

Catharanthus roseus produces the powerful anticancer drugs vinblastine and vincristine by dimerization of the terpenoid indole alkaloids vindoline and catharanthine (see Figure 1.6). Therefore, various modifications of these interesting metabolites have been genetically engineered in this species. The experimental strategy to modify terpene indole alkaloids by the Hairy roots technique is described here as an example. The biosynthesis of these metabolites starts with the generation of indole and terpene precursors via the shikimate and MEP pathways (see Section 1.1.2). The biosynthetic pathway of terpene indole alkaloids is especially highly compartmented, and therefore numerous transporters are also involved.

A key enzyme for the pathway is tryptophan decarboxylase, which has been used on many occasions as a target for genetic engineering, though several upstream enzymes can also be used (Figure 1.34). *C. roseus* Hairy root cultures expressing

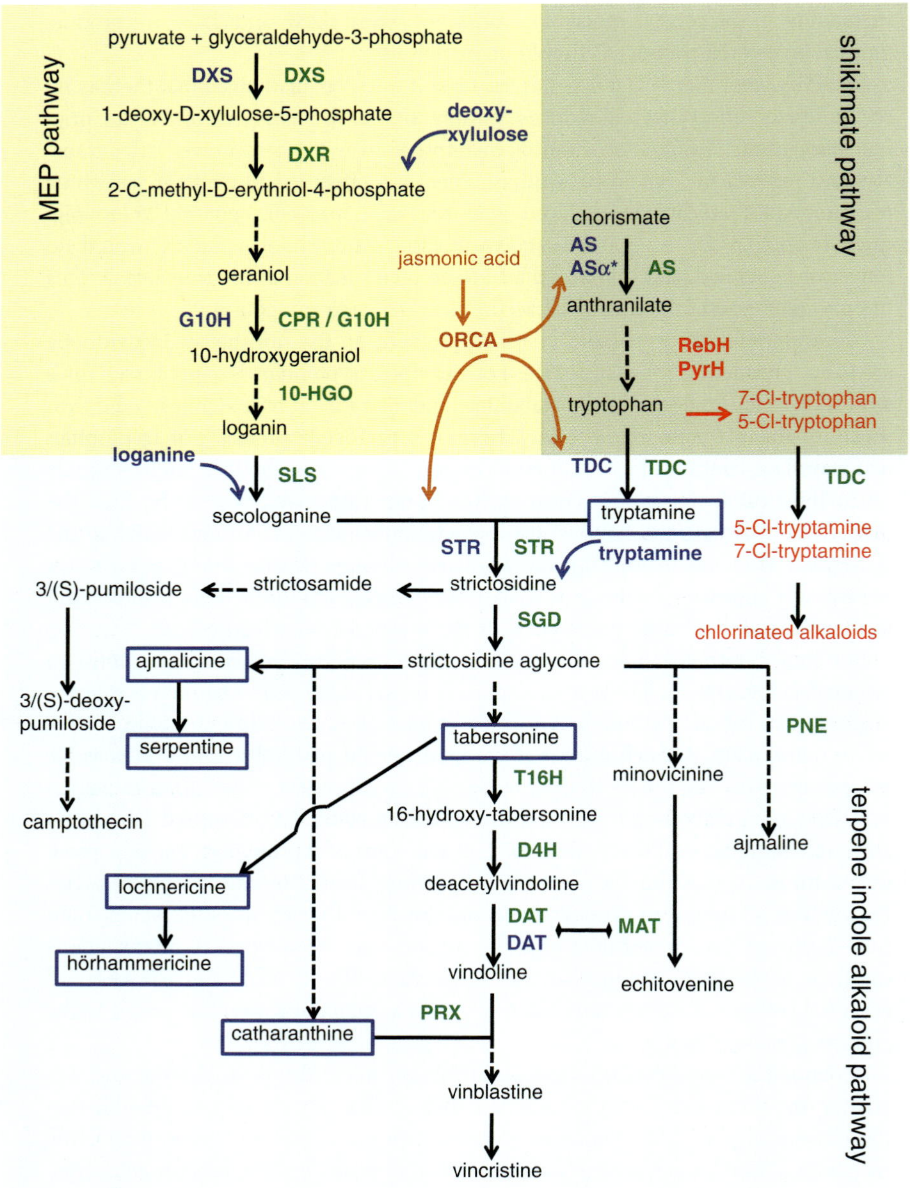

Figure 1.34 Genetically modified terpene indole alkaloid metabolism in *Catharanthus roseus*. Transgenes contain either homologous or heterologous genes or transcription factors. The MEP pathway (shaded in yellow) produces the terpene moiety; the part of the shikimate pathway delivering tryptophan is shaded in green; the core terpene indole alkaloid pathway is shaded in blue. Genes encoding native plant enzymes are shown in green; transgenic enzymes from the same plant in blue; heterologous enzymes from another plant in blue, and

various forms of anthranilate synthase and/or tryptophan decarboxylase have been generated with enhanced fluxes through the tryptophan branch of the terpene indole alkaloid pathway, such that enhanced levels of tryptamine and serpentine result. The additional treatment of anthranilate-overexpressing cultures with terpenoid precursors resulted in an elevation of various compounds, including catharanthine, ajmalicine, lochnericine, and tabersonine. Tabersonine is a direct precursor of the anticancer drugs vinblastine and vincristine, and therefore its increase could be beneficial for their production. In contrast, an increased production of side products such as lochnericine could result in a lower concentration of the precursors for the desired compounds. Increasing the expression of the gene encoding strictosidine synthase – an enzyme that catalyzes another rate-limiting step for terpene indole alkaloid biosynthesis – also enhances the synthesis of downstream metabolites, and their production is further enhanced if a precursor (loganin and/or tryptamine) is added. The overexpression of other genes from the pathway also alters metabolite composition.

In a different approach, the RNA-mediated suppression of tryptamine biosynthesis in *C. roseus* Hairy root culture can totally eliminate the production of terpene indole alkaloids, and be used to generate a spectrum of unscheduled products by introducing tryptamine analogs to the culture media. In addition, bacterial genes can be used to alter the metabolite spectra of Hairy roots. Previously, *C. roseus* has been successfully transformed using bacterial tryptophan halogenase genes, and this resulted in an accumulation of halogenated terpene indole alkaloids (Figure 1.34). Other possible strategies include the modulation of gene expression using transcription factors and/or elicitation by stresses, because the biosynthesis of terpene indole alkaloids is induced by many stress factors, including fungal elicitors, UV-B light, and jasmonate. The role of jasmonate has been confirmed, showing that it increases the levels of the transcription factor family ORCA 2 and 3 (octadecanoid-derivate responsive *Catharanthus* AP2-domain protein), thereby enhancing the transcription of several genes involved in the terpene indole alkaloid pathway.

marked with an asterisk; bacterial transgenes in red. Compounds that were increased in genetically modified lines are indicated by the corresponding genetic modification color (when not native to the plant) or blue boxes (if native). Elicitors such as jasmonic acid also increase flux through the pathway by inducing transcription factors (in orange). Arrows each indicate a single enzymatic step, dashed arrows indicate multiple steps. AS = anthranilate synthase; ASa = anthranilate synthase subunit a; CPR = cytochrome P450 reductase; D4 H = desacetoxyvindoline 4-hydroxylase; DAT = deacetylvindoline acetyltransferase; DXR = 1-deoxy-D-xylulose-5-phosphate reductoisomerase; DXS = 1- deoxy-D-xylulose-5-phosphate synthase; G10 H = geraniol-10-hydrolase; 10-HGO = 10- hydroxygeraniol oxidoreductase; MAT = minovincinine-19-hydroxy-*O*-acetyltransferase; MEP = methylerythriol-4-phosphate; ORCA = octadecanoid-derivate responsive Catharanthus AP2-domain protein; PNE = polyneuridine aldehyde esterase; PRX = peroxidase; SGD = strictosidine β-D-glucosidase; SLS = secologanin synthase; STR = strictosidine synthase; T16 H = tabersonine 16-hydroxylase; TDC = tryptophan decarboxylase. From Georgiev, M.I. *et al.* (2012) *Trends Biotechnol.*, **10**, 528–537; see Further Reading). Reprinted with permission from Elsevier.

Table 1.10 Examples for bioreactor configurations used for Hairy root cultivation in bioactive compound production. The operation mode, bioreactor type, volume of the bioreactor and an example for the plant, where a Hairy root culture was derived from, are shown. Modified from: Georgiev, M.I. *et al.* (2010) In: *Medicinal Plant Biotechnology* (Arora, R., ed.), CAB International, Oxon, UK, pp. 156–175.

Operation mode	Bioreactor type	Volume (liters)	Plant species
Mechanically driven	Stirred-tank reactor	5	*Beta vulgaris*
	Stirred-tank reactor with separate impeller	25	*Atropa belladonna*
	Wave reactor	0.5	*Panax ginseng*
Pneumatically driven	Bubble column reactor	2	*Harpagophytum procumbens*
	Bubble column reactor	2	*Beta vulgaris*
	Airlift reactor	2	*Nicotiana tabacum*
Bed reactors	Mist reactor	1.5	*Artemisia annua*
	Mist reactor	4	*Nicotiana tabacum*
	Trickle bed	14	*Hyoscyamus muticus*

In addition to changes in pathway fluxes, the molecular properties of plant enzymes can be altered to modulate pathways. In this case, the genes are mutagenized in such a way that is likely to modulate the activity of the encoded protein by, for example, altering the substrate-binding domain. This can only be achieved with proteins for which the structure is known in detail. The enzymatic properties can be first tested by gene expression in microorganisms and, if the desired mutant enzyme has been produced, then transformed into the plant. An alternative is "domain swapping," where complete domains with important properties are switched between enzymes, or between proteins. These proteins with novel structural elements could have completely new enzymatic properties; for example, the main biosynthetic pathway may be preferentially catalyzed compared to the side pathways. This has been shown to be effective for three genes encoding all of the dioxygenases of opium poppy (*Papaver somniferum*) morphine biosynthesis where novel mutant proteins were created after combining the various parts of the resulting protein, one of which had the desired property.

The type of bioreactor also influences the growth of the culture and the production of compounds (Table 1.10). A wide variety of bioreactor set-ups were used for the cultivation of transformed root culture, such as mechanically driven reactors (e.g., stirred-tank reactors, wave reactors, rotating drum reactors), pneumatically driven reactors (e.g., bubble column reactors and airlift reactors), and bed reactors (e.g., trickle bed reactors and mist reactors). Experimentally, it is challenging to select the type of bioreactor that will ensure the best media for the cultivation of transformed root cultures, although typical mechanically agitated reactors should not be used, if possible, because of the high stress-sensitivity of the transformed roots. This is indicated by the high level of so-called plant stress hormones, which could have either positive effects by elicitation (Table 1.9) or negative effects due to the highly variable levels of compounds produced. However, slight changes in bioreactor design could allow the successful cultivation of transformed root cultures. Until now, the volumes used for these Hairy root bioreactors have been rather small (Table 1.10), and generally insufficient for the large-scale production of metabolites, in contrast to what is possible when using cell cultures.

2
Function of Natural Substances in Plants

For many centuries, humans have exploited the potential of natural compounds not only as pharmaceuticals, hallucinogens and potent painkillers, but also as food additives, aroma compounds, and spices. Secondary metabolites are also used for technical applications (e.g., polyterpenes such as rubber) and as biopesticides. It is important to appreciate, however, that plants do not synthesize this wealth of compounds for the benefit of humans, but rather for their own use. This, in turn, implies the existence of high-energy-dependent biosynthetic pathways that must be induced at exactly the right moment to ensure that the metabolite is created for its specific function. Although not all such functions have been identified for the many compounds known to date, it is clear that the ability of plants to defend themselves against pathogens and herbivores is a major factor, though improved tolerances to abiotic stresses are also of great importance. These natural substances also serve as components of primary metabolism (e.g. in energy-supplying reactions; see Section 2.1), and play vital roles in plant development by attracting insects and animals for the purposes of both fertilization and/or seed dispersal. Details of these functions are summarized for selected compounds in Table 2.1.

2.1
Secondary Compounds in Primary Metabolism

Although, by definition, there is a clear difference between primary and secondary metabolism, it must be acknowledged that secondary metabolites can in fact play a variety of roles in primary metabolism. In this chapter, some examples will be provided where secondary metabolites – whether as components of membranes or organelles, or as cofactors of enzymes – play a leading role in primary metabolic reactions (Table 2.2). For example, terpenes such as carotenoids and xanthophylls are important components of light-harvesting complexes in photosynthesis, and also take part in antioxidative reactions within the xanthophyll cycle (Section 2.3). Quinones such as ubiquinone, plastoquinone or phylloquinone are essential components of electron-transport chains in plastids and mitochondria, where they are involved in ATP production. Heme components can be found in the chlorophylls and pheophytines which, again, are important for light absorption in photosynthesis,

Plant Natural Products: Synthesis, Biological Functions and Practical Applications, First Edition.
Herwig O. Gutzeit and Jutta Ludwig-Müller.

Table 2.1 Summary of functions of selected secondary metabolites according to structural classes. The shaded boxes indicate where a class of secondary metabolites has a function (as explained in the text).

Function	Alkaloids	Flavonoids	Antho-cyanins	Terpenes	Phenols	Gluco-sinolates	Indoles
Metabolism				■	■		
Development:							
Auxin transport		■					
Auxin effects	■					■	
Senescence			■	■		■	
Fertilization		■	■	■			
Male fertility			■				
Dissemination of seeds			■	■			
Abiotic stress:							
Light			■			■	
UV	■	■	■	■	■	■	
Heat			■	■	■		
Cold			■				
Drought		■	■		■		
Anoxia					■		
Heavy metals		■	■				
Air pollution				■			
Nutrient deficiency	■		■	■	■	■	
Antioxidants		■		■	■		
Symbioses:							
Rhizobia		■					
Arbuscular mycorrhiza		■		■			
Ectomycorrhiza	■						
Biotic Stress:							
Phytoanticipins				■		■	■
Phytoalexins	■	■	■	■	■	■	■
Deterrents	■	■			■	■	
Attraction of zoo-phagous insects				■			■
Allelopathy				■	■		

Table 2.2 A selection of secondary metabolites involved in various reactions within primary metabolism.

Metabolite	Metabolic pathway	Enzyme	Function
Carotenoids	Photosynthesis		Light-harvesting complex
Xanthophylls	Photosynthesis		Xanthophyll cycle
Chlorophylls	Photosynthesis		Light absorption
Quinones	Photosynthesis		Electron transport
	Respiratory chain		Electron transport
Cytochromes	Photosynthesis		Electron transport
	Respiratory chain	Cytochrome c oxidase	Electron transport
		Cytochrome P450-dependent enzymes	Metabolism
Other hemes		Peroxidases	Metabolism, detoxification
		Catalase	Detoxification
Biotin		Carboxylases	Metabolism
Flavin		Flavoproteins	Metabolism
Pyridoxalphosphate		Transaminases	Metabolism
		Decarboxylases	Metabolism

as well as in enzymes such as cytochrome c oxidase, peroxidases, catalase and cytochrome P450-dependent monooxygenases. Other enzymes not mentioned here may also incorporate secondary metabolites as cofactors or coenzymes; examples include biotin in carboxylases, flavins in flavoproteins, and pyridoxalphosphate in transaminases and dicarboxylases.

2.2 Development

Plant development includes seed germination, vegetative growth and the formation of reproductive and overseasoning organs, which are also involved in embryogenesis (Figure 2.1). Plant hormones play essential roles in all developmental stages. Some of these hormones, such as the auxin indole-3-acetic acid (IAA), resemble secondary metabolites such as serotonin in terms of structure. As plant hormones are synthesized in small amounts, and are specifically involved in signaling chains to regulate growth and development, they are not considered as secondary metabolites.

The complete absence of a secondary metabolic pathway might not lead directly to growth-related phenotypes, although under certain conditions (e.g., stress) it might become important for the plant and growth may be diminished. During the last past few decades, an increasing amount of evidence has been acquired on the roles of secondary metabolites in plant development, as follows:

1) The effect can be achieved directly by compounds which have a structural similarity to plant growth regulators and are synthesized, for example by microbes, to alter plant development (see Section 2.4).

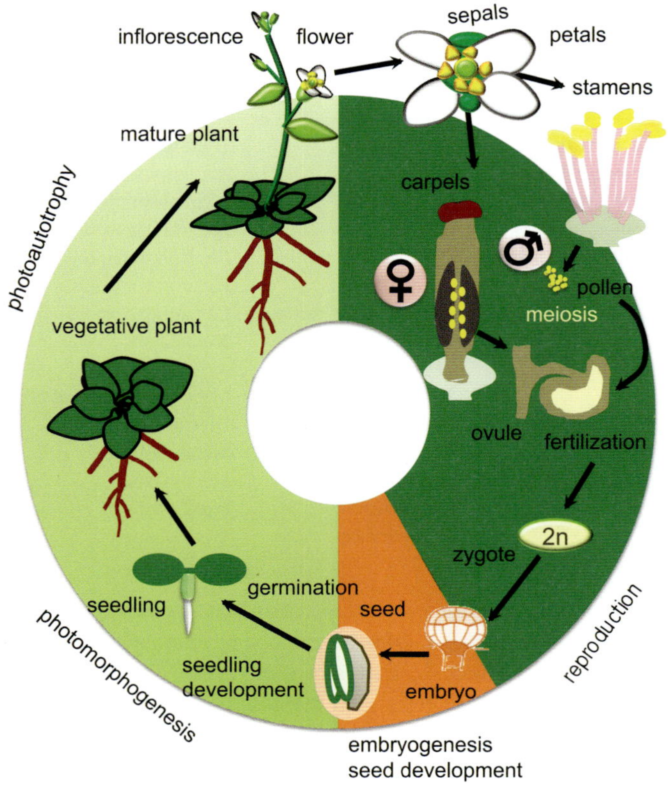

Figure 2.1 The development of higher plants, using *Arabidopsis thaliana* as an example. The developmental cycle starts with the embryogenic phase as dormant and overseasoning stage (light brown), followed by the vegetative growth period (light green), and the reproductive phase (dark green). In *Arabidopsis*, the inflorescence with the reproductive organs develops from the vegetative rosette.

2) Secondary compounds can influence other molecules that are essential for the signaling or transport of plant hormones.

Flavonoids are very good examples for such secondary metabolites, as they function not only as phytoalexins in plant defense and/or as antioxidants, but also as signals in symbioses and plant development. Because of their multiple roles in the life of a plant, flavonoids are involved in many aspects of growth and development, interaction with other organisms, and in stress responses (Figure 2.2). Flavonoids are able to modulate the plant growth hormone auxin by altering its rate of transport from cell to cell, and also have other functions as attractants for pollination. Many developmental processes regulated by secondary metabolites have been found in the model plant *Arabidopsis*, because of its rather simple morphology and

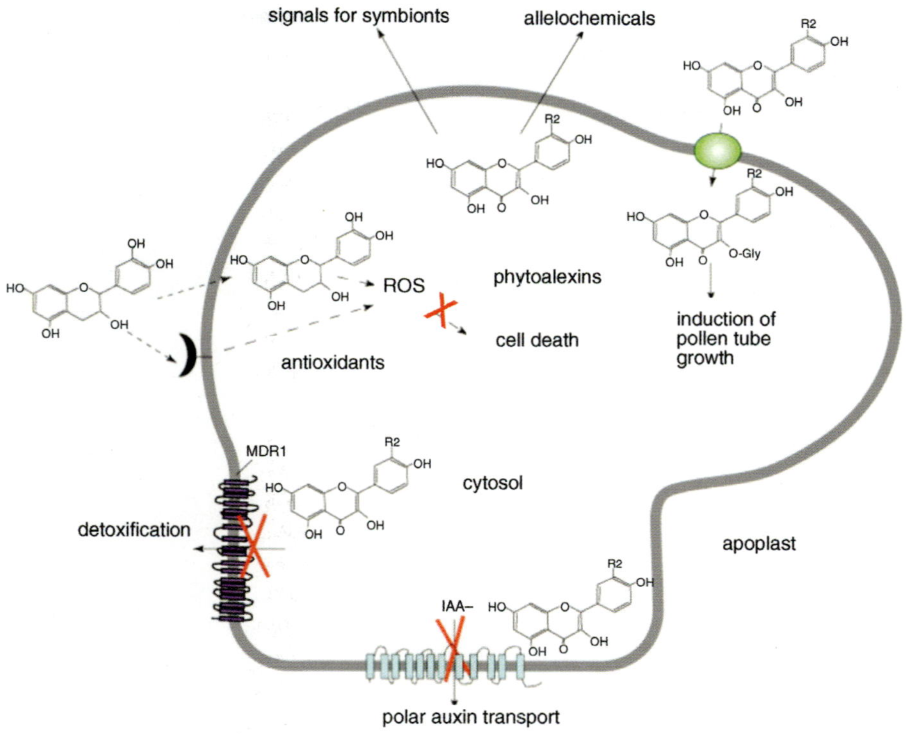

Figure 2.2 Summary of different developmental processes and interactions with the environment and other organisms as determined for flavonoids. From Taylor, L.P. and Grotewold, E. (2005) *Curr. Opin. Plant Biol.*, **8**, 317–323 (see Further Reading). Reprinted with permission from Elsevier.

the well-investigated molecular analysis of pathways and their regulation. Alkaloids can also alter growth, presumably by acting as structural mimics of growth hormones. However, whether this effect has a role in normal plant development, or only in the interaction of plants with symbionts, has still to be determined.

2.2.1
Vegetative Growth

Secondary metabolites can be involved in three different strategies in plant growth and development (Table 2.3):

1) They are directly involved in plant growth; an example are lignin monomers fortifying the cell walls and thus strengthening the plant tissues. The involvement of flavonoids and anthocyanins in pollen-mediated germination is also included in this category.

2) They can be involved by mimicking the structure of plant growth hormones, and so can act as either synergists or antagonists of plant growth.
3) They can change the processes necessary for correct hormone function, notably by modulating transport of the growth-promoting hormone auxin.

Table 2.3 Summary of possible effects of different secondary metabolites on plant growth and development. A few examples are given, which are further explained in the following subsections of the chapter. Leaf and flower pigments are not shown in this compilation, but these will be described in Section 2.2.2. The structure of IAA is shown as an example of an active auxin; the structure of gibberellic acid is shown as an example of an active gibberellin. Chemical structures are from The PubChem Project.

Secondary metabolite	Mechanism		
	Direct effect	Indirect effect	Agonist/Antagonist
Sinapoylalcohol 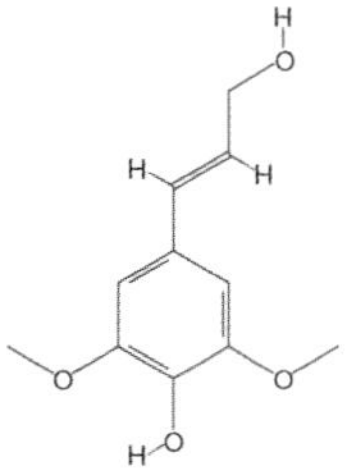	Fortification: Lignin		
Coniferylalcohol			
Cumaroylalcohol			
Naringenin		Auxin transport (binding to transport protein)	
Kaempferol 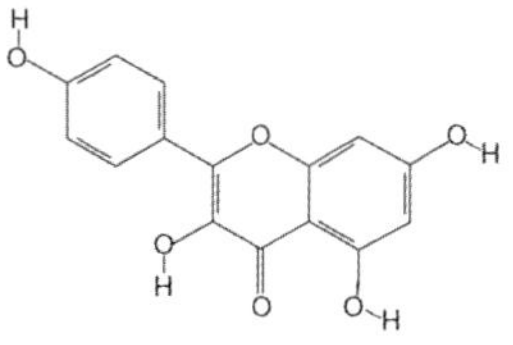			
Quercetin			
Flavonoids	Polar growth		
Colchicine	Mitosis (Mictrotubules)		
Vinblastine			
Paclitaxel 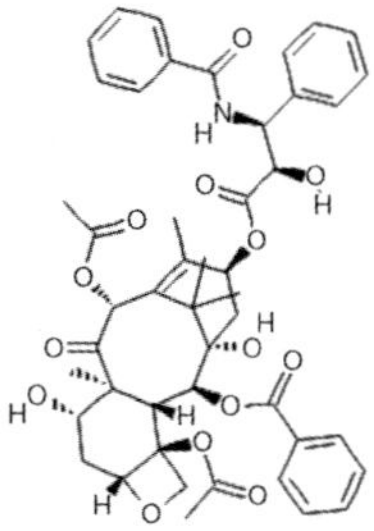			

Flavonoid-Glycosides	Cell cycle		
Delcosine 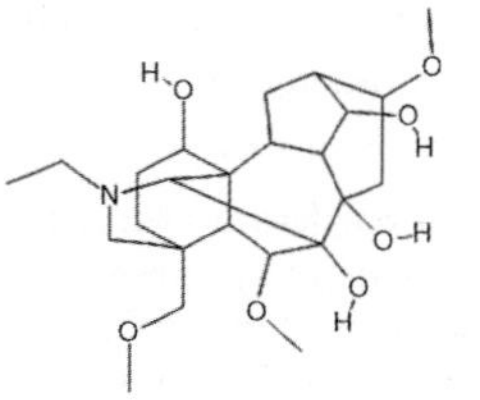			Growth: Gibberellin antagonist 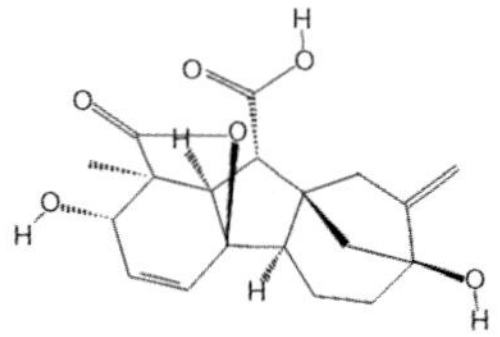
Melatonin 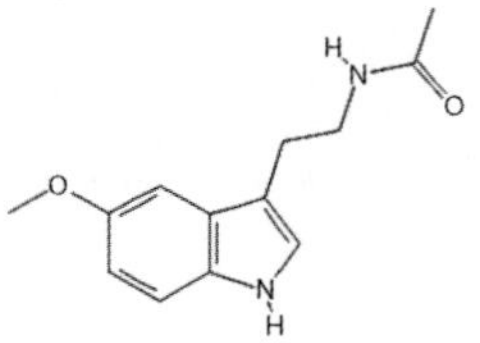			Growth: Auxin agonist 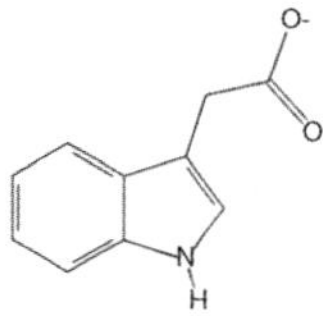
Serotonin 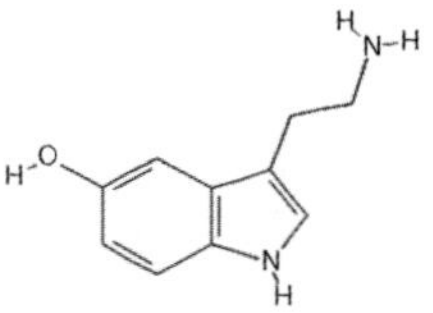			
Yohimbine 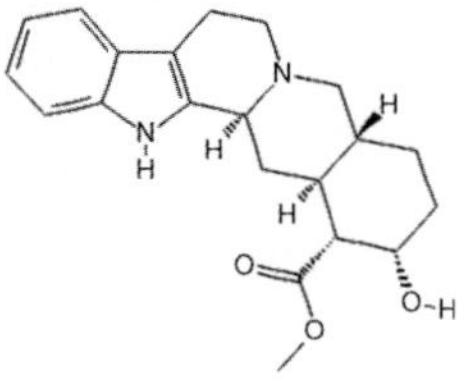	Root hair growth		Growth: Auxin antagonist Modulation of auxin response
Brucine			
Flavonoids Flavonol-3-O-glucoside	Pollen development		
Flavonoids	Symbiosis: Recognition	Auxin accumulation: Organ development	

2.2.1.1 **Structural Polymers**

Plant growth is dependent not only on the turgor pressure inside the cells but also on the rigid cell wall structure, both of which factors permits the growth of large organisms. In addition to the primary cell wall, further fortification is required for the synthesis of secondary and tertiary wall elements. As the vasculature – especially xylem vessels – is lignified, those molecules involved in the formation of lignin and other structural polymers (such as suberin and waxes) are important for the growth and development of land plants. Lignin components are one of the most abundant polymers in plants, and are derived via metabolism of the phenolics pathway (see Section 1.1.2). The differentiation of cells into the sclerenchyma, collenchyma and xylem vessels is dependent on the synthesis of lignin precursors from the phenylpropanoid pathway. Lignin monomers such as coniferyl-, cinnamoyl- and sinapoylalcohols are derived from cinnamic acid (Figure 1.7), and are oxidized by a lignin peroxidase, followed by spontaneous polymerization to form a three-dimensional network. This barrier also protects against abiotic factors, such as drought stress and pathogen invasion.

Suberin, another important polymer provides an efficient barrier against transpiration and water loss, and also affords protection against fungal pathogens. Like lignin, suberin also plays a role during wound-healing reactions. Suberin is composed of two different layers, one layer being polyaliphatic and the other polyaromatic. During its synthesis, the polyaromatic layer (which contains cinnamic acid derivatives) is first anchored in the cell wall, after which lipid moieties are added to the polymer. The detailed biosynthesis of suberin is not fully understood. Finally, epicuticular waxes consist of secondary metabolite monomers like, for example, various flavonoids.

2.2.1.2 **Flavonoids and Auxin Transport**

Hormone gradients are important for organ development, and also for responses to the environment – that is, photo- and gravitropism where asymmetric growth responses result in the bending of a shoot or root towards or away from a stimulus. The transport of hormones is an important level of regulation. Auxin is transported in a polar fashion in the shoots and roots, using a cell-to-cell mechanism. As a functional explanation of such transport, a chemiosmotic model was proposed which relied on the pK_a of IAA of 4.7, as this would result in a molecule that was mostly protonated in the acidic cell wall compartment. The non-charged molecule can be transported passively across the membrane, if a concentration gradient is present. In the more neutral pH of the cytosol, the IAA would be dissociated, with the result that the charged molecule could not diffuse across the membrane and would therefore be trapped within the cell. In order to leave this compartment, an active auxin efflux must take place (Figure 2.3), mediated by two efflux carriers (see Figure 1.26) – the family of so-called PIN proteins and the ABC-type transporters called PGP. In the chemiosmotic model only the PIN proteins are shown for clarity, and their role in relation to flavonoids is explained in more detail. Despite the observation that both classes of auxin efflux proteins can be modulated by flavonoids, the PIN family has been much better investigated in this respect. The influx is also mediated by a transport protein (AUX), in addition to the diffusion component. The polarity of auxin transport depends on the differential localization of the two types of transporters in the plasma membrane. The influx proteins are located on one side of the cell (in the case of shoots, on the apical side), whereas the PIN proteins of efflux carriers are located on the opposite side of the cell (in the case of shoots, on the basal side) (Figure 2.3).

The influence of flavonoids on transport of the auxin IAA has been investigated using flavonoid-free *Arabidopsis* mutants. The structural similarity of a flavonol

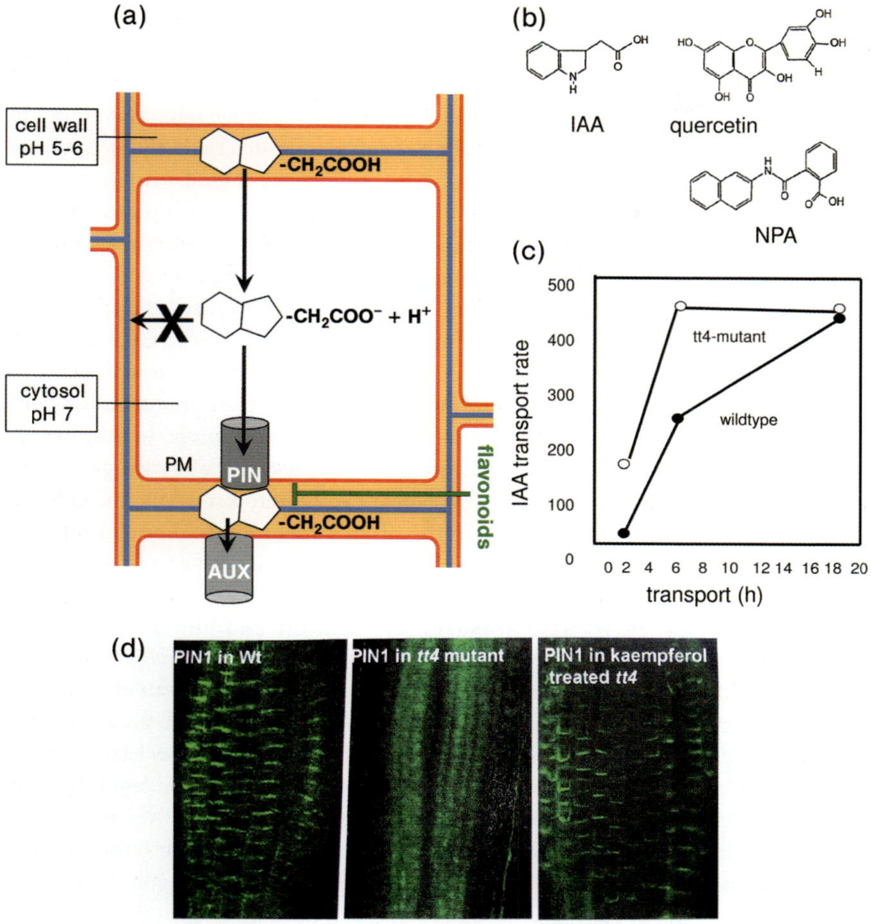

Figure 2.3 Flavonoids as inhibitors of polar auxin transport. (a) Chemiosmotic model for polar transport of the auxin indole-3-acetic acid (IAA). AUX = influx carrier; PIN = efflux carrier. IAA is undissociated (protonated, CH_2COOH) at low pH values in the cell wall, whereas in the cytosol at pH 7 the molecule is dissociated (deprotonated, $CH_2COO^- + H^+$) due to its acid constant of 4.7; (b) Flavonoids (as an example the structure of quercetin, central position) in comparison with the structure of the synthetic auxin transport inhibitor naphthylphthalamic acid (lower structure) is given. The structure of the auxin IAA is also shown; (c) Comparison of basipetal auxin transport in wild-type and flavonoid-free mutants of *Arabidopsis* (*tt*), where the faster distribution of radioactivity is shown for *tt* mutants; (d) Localization of PIN proteins is altered in flavonoid free mutants (*tt*), but the correct position can be restored by addition of different flavonoids, such as naringenin, kaempferol or quercetin. Photograph from Peer, W.A. *et al.* (2004) *Plant Cell*, **16**, 1898–1911 (see Further Reading).

(quercetin) and the artificial transport inhibitor for auxin efflux, naphthylphthalamic acid, is illustrated in Figure 2.3b. The possible similarity in structure might be one explanation for the effect of flavonoids on auxin efflux. Flavonoid-free mutants of *Arabidopsis* were able to transport the hormone faster, as shown by the distribution of radioactive auxin in the shoot. The mutation in flavonoid synthesis also altered the distribution of PIN proteins in the plasma membrane. The distribution was no longer asymmetric, but the correct orientation could be restored by the addition of different flavonoids, such as naringenin, kaempferol and quercetin. Therefore, the effect of flavonoids on auxin efflux might be twofold: one effect directly on the transporter itself, and the other effect on the correct localization of the transporter.

The correct localization of auxin efflux carriers is essential for many developmental processes. In the roots, the response to gravity is fundamental for the orientation of a plant within its environment, and especially in the case of seedlings after germination it is absolutely essential that the roots can reach their substrate. After graviperception, the asymmetric auxin distribution results in an asymmetric growth and thus a bending of the root towards the soil. Plants grown on media containing flavonoids do not demonstrate this behavior if they are turned on their axis away from gravity. This effect is most likely due to alterations in auxin efflux by flavonoids, and especially by quercetin, which has a strong effect on gravitropic curvature. The defect in gravitropic response can be interpreted as an incorrect orientation of the PIN transporters in the plasma membranes. Vesicle transport and the cytoskeleton play important roles in the reorientation of these proteins, with actin filaments directing the vesicles towards the plasma membrane. Some secondary metabolites may also influence the structure and function of the cytoskeleton (see Section 3.2).

Further differences in the phenotypes of flavonoid-free *Arabidopsis* mutants are the length and density of the root hairs, lateral root formation, growth of the hypocotyl and inflorescence, as well as the number of inflorescences per plant and the number of seed pods. However, as these parameters are not always regulated similarly in different *tt* mutants, there is still no unequivocal picture of the function of flavonoids in growth and development. As only some effects are in accordance with the modulation of auxin transport, other mechanisms relating to the involvement of flavonoids in plant development remain to be discovered.

2.2.1.3 **Glucosinolates and the Biosynthesis of IAA**

Glucosinolates, especially those derived from tryptophan, can have indirect effects on auxin production because their biosynthetic pathways share similar intermediates before they diverge (Figure 2.4). Modifications in the biosynthetic pathway to indole glucosinolates can affect auxin production, because this can lead to either an overaccumulation or a reduction of the hormone. One example of IAA overaccumulation in *Arabidopsis* is the *superroot* (*sur1*) mutation; in this case, the mutant synthesizes lesser amounts of indole glucosinolates, causing more precursor for auxin synthesis to be available, and this accumulates. Alterations in the homeostasis between indole glucosinolates and auxin in favor of the growth-promoting substance leads to changes in the phenotype. The mutant has an aberrant root growth, with especially many lateral roots (which explains its name). The phenotype demonstrates a reduced seedling growth caused by high concentrations of IAA having a growth-inhibiting effect; this is in contrast to moderate levels of IAA, which cause an increased plant growth.

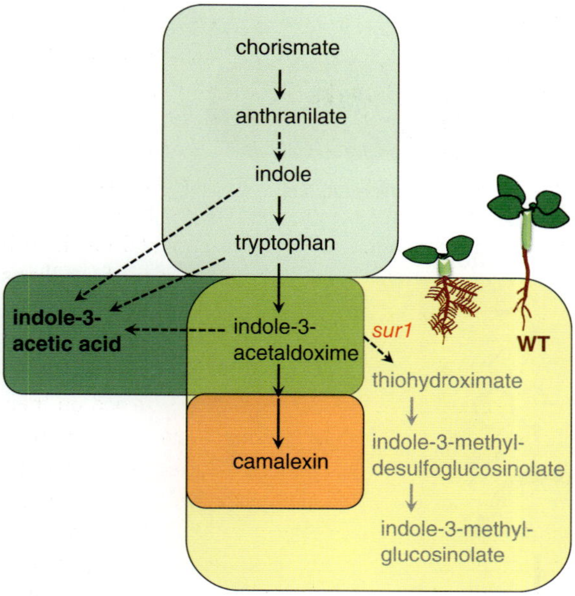

Figure 2.4 Connecting indole glucosinolate biosynthesis with the formation of the plant hormone IAA. The biosynthetic pathway to tryptophan is shown in light green, to IAA in dark green, and to indole glucosinolates in yellow. The camalexin pathway is also shown (orange). Solid arrows symbolize one enzymatic reaction, dashed arrows indicate several reaction steps. The position of the *sur1* mutation is shown in red, and the lacking compounds in this mutant in gray. The loss of this gene function would result in an accumulation of IAA (bold), which leads to the aberrant phenotype of the mutant *sur1* in comparison to the wild-type (WT).

2.2.1.4 **Polar Growth**

Root hairs are single cells that grow at the tip of the root, demonstrating a common growth pattern that is also found in pollen tubes (Section 2.2.2). Mutations in the flavonoid content of these tissues demonstrated an inhibition of this tip growth, with such plants producing fewer and shorter root hairs. This has important implications for water and nutrient uptake, because the root hairs contribute significantly to the large surface area of the root. In addition, polar growth is dependent on the cytoskeleton (see below), so that secondary metabolites – which bind to constituents of the cytoskeleton – might also change polar growth. The alkaloid hypaphorin (Section 2.4) has been shown to alter the number and growth rate of root hairs of *Eucalyptus* seedlings.

2.2.1.5 **Cytoskeleton and Cell Cycle**

The cytoskeleton consists mainly of actin filaments, microtubules and other components (see also Section 3.2). Many secondary metabolites can affect the cytoskeleton by binding to either actin or tubulin (Figure 2.5; Table 3.2); indeed, natural substances that interfere with the polymerization or depolymerization of

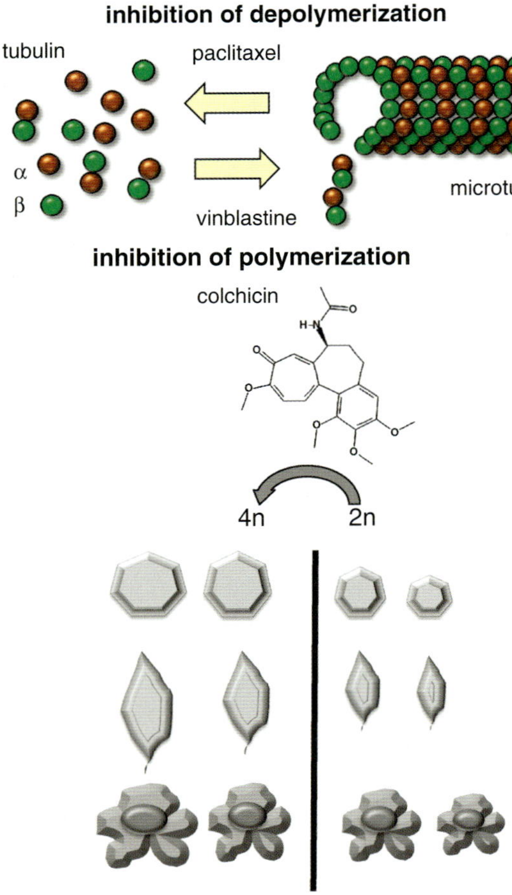

Figure 2.5 Effect of secondary metabolites on the polymerization of tubulin and subsequently on mitosis. Upper part: Paclitaxel inhibits the depolymerization of tubulin, while colchicine and vinblastine inhibit the polymerization. Lower part: Two tetraploid leaves, flowers and pollen grains of *Rhododendron* (left) and two diploid representative organs (right). Chemical structure is from The PubChem Project. Lower part based on Kehr, A.E. (1971) *J. Am. Rhododendron Society*, 25n1. Available at: http://scholar.lib.vt.edu/ejournals/JARS/

tubulin will inhibit microtubule function. The microtubules are the main constituents of the mitotic spindle, and therefore will greatly influence chromosome movement. As microtubules are essential for cell division, these secondary metabolites are often used as cytostatic drugs for the treatment of cancer; however, whether they also act on the microtubules in the plant itself is not yet clear. Processes inhibited by nonfunctional microtubules include the migration of organelles or chromosomes during mitosis. The tropolone alkaloid colchicine, in addition to other compounds already used in medicine (such as paclitaxel or the terpene indole alkaloid vinblastine; see Table 1.9), inhibits formation of the mitotic spindle and increases the ploidy of a given tissue (Figure 2.5).

Microtubules consist of 13 parallel chains of α-β-tubulin heterodimers, which form the filament. The dynamics of the microtubules needed to direct chromosome movement is dependent on a continuous polymerization and depolymerization. Vinblastine and colchicine each inhibit polymerization, whereas the taxanes inhibit depolymerization (Figure 2.5; see also Chapter 4).

Colchicine is used to generate polyploidic plants with larger, more attractive flowers, and is also applied to *in-vitro* cultures to produce meristematic tissues, especially for chromosome duplication in embryos from which microspores can be obtained to produce fertile plant lines. Cultures of pollen may also be stimulated to grow following the addition of colchicine. Unfortunately, as these *in-vitro* effects cannot be transferred to an intact plant system, a general function of the compounds that influence mitosis cannot be derived from these observations. Polyploidic plants often have larger organs, which is attractive in horticulture (Figure 2.5), although polyploidization in a plant in a stress situation can affect its growth and its responses to the environment. Subsequently, larger tissues (i.e., leaves, flowers or pollen grains) are formed which could, in turn, actually result in plants with a greater resistance to stress. In accordance with this hypothesis, seedlings in which polyploidy was induced during early development were more stress-resistant when they were grown in the presence of low concentrations of colchicine. This was especially the case for drought-mediated stress conditions, when the plants also showed higher chlorophyll levels.

Some flavonoids (e.g., quercetin) are known to influence the cell cycle, and this property has been studied extensively in human tissues as it forms the basis for the development of drugs with anticancer activities (see Chapters 4 and 5). Flavonoids show cytostatic effects on the cell cycle of plants (see also Chapter 4), and their levels are balanced by the activity of meristem-specific glycosyltransferases. However, if the expression of the respective glycosyl transferase gene were to be completely inhibited, this would prove lethal to the plant. A reduction of gene expression leads to a longer cell cycle, which in turn is associated with a reduced growth rate and confirms that flavonoids influence growth independently of the auxin-dependent mechanism. Conversely, overexpression of the glycosyl transferase gene results in a shorter cell cycle.

The indole alkaloid hypaphorin, which is synthesized by an ectomycorrhizal fungus (Section 2.4), can also alter the cytoskeleton by interacting with actin filaments. In roots treated with hypaphorin, the small filaments in the subapical region of the root hairs are replaced by larger actin filaments that are directed towards the root hair tip. As a consequence, the total number of filaments is reduced, most likely because they form aggregates that can stabilize the cytoskeleton and in turn would inhibit the vesicle-controlled tip growth of the root hairs. In consequence, the root hairs would remain shorter and thicker. Interestingly, auxin can reverse this development.

2.2.1.6 **Alkaloids and Growth**

Since *Arabidopsis* most likely does not contain complex alkaloids, even though homologous gene sequences were found in the genome, their possible role in plant development has to be investigated in other plant species. The growth-regulating activity of alkaloids was hypothesized to be due to their structural similarity with plant hormones. Alkaloids of the simple indole type have structural similarities to the indole ring of the auxin IAA, and could thus mimic its growth-promoting effects (Table 2.3). As shown for hypaphorin produced by a fungus, this might indeed be the case, so that indole derivatives induce certain growth patterns (Section 2.4). Whether this effect is used exclusively by the microbial partner, or whether plants also use this compound to regulate growth and development, is not clear. Other indole alkaloids for which a growth-regulating effect has been demonstrated, besides hypaphorin, are brucine and yohimbine. Investigations on Eucalyptus seedlings showed that the growth-inhibitory effect on hypocotyl segments of the ethylene (gaseous plant hormone) precursor aminocyclopropane carboxylic acid could be partially reduced by brucine and yohimbine. On the other hand, these two compounds reversed the growth-inhibitory effect of auxin on root elongation of the Eucalyptus seedlings. However, for other auxin-regulated processes they could act both as a positive modulator or attenuator, or have no effect at all. Thus, it was postulated that indole alkaloids could interact with some – but not all – components of a given signal transduction chain, based on their structural similarity to the plant hormone.

The indole alkaloid melatonin, which is synthesized in animals from serotonin, could be identified in some plant species. The putative biosynthetic pathway from tryptamine and its structure very much resembles the synthetic pathway of IAA (Table 2.3). Therefore, it was suggested that melatonin might possess growth-promoting activities as an auxin mimic. In particular, a role in the circadian regulation of the flowering time was attributed to this metabolite. Melatonin was also found to stimulate seed germination, as well as the root and hypocotyl growth of seedlings. As melatonin is also a radical scavenger it might also play a role in the antioxidative system of plants after exposure to stress. The role of melatonin as a growth regulator in plants is controversial.

A similar growth-modulating function has been proposed for the structurally related serotonin (5-hydroxy-tryptophan), which is a candidate precursor for the synthesis of melatonin also in plants, even though the respective pathway has not been described yet. Serotonin itself is synthesized from tryptamine, as are all indole alkaloids, and has been found in a conjugated form with phenol derivatives, especially cinnamic acids, and also caffeic acid. The structural similarity of serotonin to IAA has led to the hypothesis that it might be involved in vegetative growth and reproductive development. A separate role from that of an IAA mimic was postulated in the detoxification of ammonia. High proteolytic activities liberate ammonia from amino acids, which is inhibitory for germination; the ammonia is then transferred to glutamic acid and reduced in concentration in the germinating tissue. Further transfer to other amino acids such as tryptophan could also result in higher levels of serotonin, which would accumulate in the respective tissues. However, this is only correlative evidence. Like melatonin, serotonin has a high antioxidative potential *in vitro*.

A growth-inhibitory activity has been described for the diterpenoid alkaloid delcosine, isolated from *Delphinium ajacis*, which interacts negatively with the biosynthesis of another growth-promoting hormone, gibberellin.

2.2.1.7 **Senescence**

While flower pigments are usually attributed to the attraction of insects for pollination (Section 2.2.3), red leaf pigments play a role during protection against UV irradiation and other stress factors (Section 2.3). UV protection is important in young tissues, where photooxidation must be prevented, because other mechanisms such as thick walls and the cuticle are not yet fully developed. Photooxidation reactions are oxidations induced by light, leading to damage of the respective molecules, and thus functional losses. Often, radicals are formed during such reactions, causing further damage. Leaf pigments are also synthesized after pathogen invasion (Section 2.5).

Senescence is a developmental program that indicates the end of a life cycle (for annuals) or vegetation period (for perennials). Senescence also occurs after stress situations when a plant sheds off damaged or even dead organs, mostly leaves. This is termed "sequential senescence," and implies that older leaves are shed in favor of younger leaves, which stay attached to the plant. In the autumn, leaves turn from green to orange, yellow, and even purple, as the result of a coordinated degradation of chlorophyll at a time point when carotenoids are still present. In addition, some plants accumulate anthocyanins in the vacuole, which can protect against UV irradiation. Even in the state of senescence a plant should not lose too many metabolites, such as chlorophylls, by photooxidation. In perennials, a delay in cell death by irradiation damage and other stress factors can lead to the "recycling" of valuable compounds as nutrient storage for the plant. The photoinihibition of photosynthesis occurs mainly during a combination of strong light and cold temperatures. In the scenario of dropping temperatures while there are still high light intensities, anthocyanins can protect the leaves against the damage caused by radicals.

2.2.2
Reproduction: Male Fertility

Investigations using various plant mutants have shown that flavonoids are involved in the formation of pollen grains and their fertility (Figure 2.7). A mutation of *Arabidopsis* in a gene for an anther-specific enzyme with homology to a dihydroflavonol-4-reductase (DRL) attenuated pollen formation in comparison to the wild-type, and this led to a reduced fertility. Expression of the *DRL* gene in anthers and pollen correlates with this function. The tissue-specific gene expression was visualized using the *DRL* promoter fused to a reporter gene, in this case the gene for β-glucuronidase (Figure 2.6). The promoter of a desired gene is cloned by removing the native gene, after which the native gene is replaced by a reporter gene; this construct is then transformed into the plant. When the promoter is active in a given tissue, then the enzyme β-glucuronidase is synthesized,

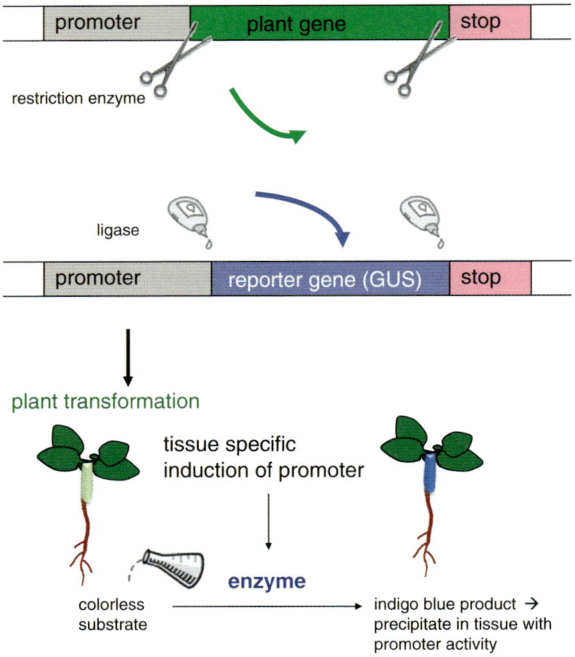

Figure 2.6 Model explaining the principle of the reporter technique. The promoter of a desired gene is cloned by removing the native gene, after which the native gene is replaced by a reporter gene (transcriptional fusion). Alternatively, the native gene is retained together with the reporter (translational fusion). The construct is transformed into the plant. This can be, as depicted here, the gene encoding the enzyme β-glucuronidase (GUS). After the promoter is activated, the enzymatic activity can be made visible by adding a substrate, which is converted to a blue product, precipitating in the tissue. After destaining of the plant tissue due to chlorophyll, the blue color can be seen, which is an indication of the site of the reporter activity. Scissors represent restriction enzymes to cut out DNA, and the glue symbolizes ligases, which are enzymes that can link two DNA pieces. Alternative methods to obtain such constructs are possible.

which can convert a colorless substrate into a blue product that precipitates in the tissue where the promoter activation had taken place. In the case of the *DRL* promoter, the blue staining is highly specific for the pollen grains (Figure 2.7).

The role of flavonoids in pollen formation has been demonstrated for chalcone synthase in some mutants, which are devoid of flavonoids. For example, in maize or petunia the flavonoid deficiency leads to male sterility. In petunia, the phenotype could be reverted by addition of the flavonols quercetin and kaempferol, which results also in the germination of pollen, demonstrating their role in fertility and growth of the pollen tube. Only flavonols with nonsubstituted hydroxyl groups were able to induce pollen germination, indicating that a structure specificity is needed. However, the flavonoid-free mutants of *Arabidopsis* do not show an aberrant phenotype with regards to pollen formation, which indicates that

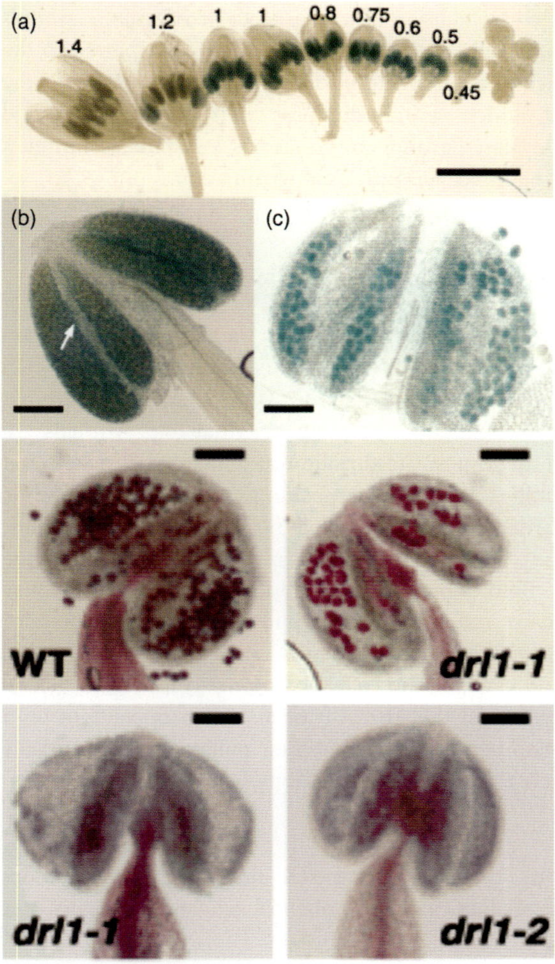

Figure 2.7 Pollen formation is dependent on flavonoids. The anther-specific expression of a gene encoding a dihydroflavonol-4-reductase (DRL) was visualized by using a promoter::reporter construct with β-glucuronidase (upper part). The blue staining represents the promoter::reporter activity in the tissue. Several mutants (*drl*) in the *DRL* gene attenuate pollen formation in comparison to wild-type (lower part). From Tang, L.K. *et al.* (2009) *New Phytol.*, **181**, 576–587 (see Further Reading). Reprinted with permission from John Wiley & Sons.

different plants need a species-specific flavonoid pattern in the anthers to regulate pollen formation.

Glycosylated flavonoids, especially flavonol-3-*O*-glycoside, also play a role in pollen development in petunia. This is corroborated by high levels of these specific compounds in pollen, which correlates with an enzyme activity capable of glycosylating flavonols. The pollen grains did not germinate without these compounds, and the phenotype was reverted by addition of the respective molecule. Flavonols are synthesized in the tapetum cells of the anthers and then transported to the developing pollen grain, which is able to take up these compounds. Only in the pollen they are glycosylated, which is important for their activity. The rapid response of the pollen grain to these flavonol glycosides has led to the assumption that they play a role as signaling molecules in pollen germination.

2.2.3 Flower Biology: Pollination

One of the most important developmental processes in plants is the fertilization which involves transfer of the pollen onto the carpel; this results in the formation of a zygote, which develops into the embryo (Figure 2.1). As many plants are not self-fertilizing, the attraction of pollinators plays an essential role. Animals are responsible for about 78% of angiosperm fertilization in temperate climates, and up to 94% in tropical communities. Bees are the most important pollinators, while pollinating animals are rewarded by nectar or protein-rich pollen. The signals, which are sent out by the plant, attract specific animals that are responsive to the respective signals (Table 2.4). Both, morphological and biochemical traits such as color, the form of flower organs and aroma compounds, may act as signals for recognition. The surface properties of petals and sepals can also play a role, as can accessibility to the anthers and carpels. The color and aroma compounds belong to the secondary metabolites, whilst the form of the flower is dependent on a genetic program during development. Flowers and their pollinators are mutually dependent on each other in terms of recognition patterns.

2.2.3.1 Nutrients

Besides carbohydrates, proteins, amino acids and other nutrients, the nectar contains also secondary plant metabolites, including phenols, alkaloids, terpenes and

Table 2.4 What is attracting pollinators? For the plant it is essential that a compatible pollinator is attracted; only then is the pollen transferred to a different plant for fertilization. Only few plant species are self-fertile; most are dependent on pollinators, which may be insects, birds, or other animals; however, bees are mainly responsible for the fertilization processes. The different recognition cues are summarized below.

Cue	Flower organ		
	Sepals/Petals	**Carpel**	**Stamens**
Visual cue	Color		
	Form		
	Pattern	Accessibility	Accessibility
Chemical cue		Nectar	Nutrients
Aromatic cue	Volatiles		
	Deterrents		
Touch	Surface properties		Pollen surface

flavonoids. The role of these compounds for the plant lies in a higher attractiveness of the nectar because of essential compounds for the pollinators, and also in their antimicrobial activities in some cases. It is essential to fight off unwanted insects by emitting deterrent compounds, or to use an antimicrobial activity to prevent microbes from entering the flowers via the nectarthodes as natural openings. Nectarthodes constitute openings at the base of a flower from which nectar exudes. In addition, the nutritious nectar itself is a very good environment for microbes to multiply, so that defense is necessary at this point.

Bees either drink the nectar from the plant, or they take up oils and pollen as nutrition. Social bees are less host plant-specific because they have to take care of their offspring during a longer period, and therefore they need more different plant species flowering at different time points. In consequence, during the summer they have contact with many different plant species. However, some bees live in solitude and in many cases have fewer offspring to feed; they are more specialized in terms of the plants they take nectar and pollen from, because the period during which they need the pollen is shorter. A specialized flower morphology ensures that only adapted bees can harvest the pollen from these plants. Investigations using European bees have resulted in numbers of ten to one thousand flowers being needed to nourish one larva in a beehive.

As well as the aroma compounds/volatiles, color and form of a preferred flower, the ingredients of the pollen which can be taken up as nutrients are also important; these include sugars, starch, amino acids and minerals, and also secondary metabolites. Sterols are an example of compounds that need to be taken up by bees, because the insects cannot synthesize sterols themselves. However, they do possess enzymes capable of converting plant sterols such as Δ5-avenasterol and β-sitosterol to cholesterol. The latter is integrated into the membranes for stabilization (see also Figure 2.27). Ecdysone, which is essential in development of the insects, can also be synthesized from the plant sterol precursors, this being a good example of coevolution between plants and insects (see Section 3.1).

2.2.3.2 **Aroma Volatiles**

Volatiles emitted by plants can be divided into four major classes according to their metabolic origin: terpenoids (e.g., mono- and sesquiterpenes and apocarotenoids); phenylpropanoids/benzenoids (e.g., eugenol, benzaldehyde); fatty acid derivatives (e.g., hexenal, hexenol); and amino acid derivatives (e.g., 2- and 3-methylbutanal). Among these, terpenoids and lipid derivatives are probably the most abundant and expensive to produce in terms of energy and nutrient consumption. The volatile compounds emitted by plants can play a dual function as deterrents and attractants (Sections 2.5 and 3.1). Therefore, the regulated synthesis of volatiles, depending on the respective situation, is essential. The specific mixtures are also recognized by the insects themselves as either attractive or repelling (Figures 3.10 and 3.11). In the case of diecious plants, it is important the aroma mixtures emitted have related compositions, so that the same insects are attracted to both male and female flowers, which keeps the pollination rates high. For the genus *Salix*, it was shown that male and female flowers exude a similar blend of volatiles; in some species the similarity was up to 90%, whereas in others it was only 50%. While the former are almost exclusively dependent on pollination by insects, the latter also rely on the distribution of their pollen by the wind.

Insects which do not feed on plants can also be attracted by a particular scent. For example, the flowers of so-called carrion flower plants (or stinking flower plants) emit a mixture of volatiles that is very unpleasant to humans but highly suitable to attract flies for pollination. Members of the Arum family (Araceae), such as the Titan arum (*Amorphophallus titanum*), imitate the scent of carrion emitted from their flowers by an increase of the temperature in the spadix (Figure 2.8). The spadix is a type of inflorescence with small flowers, which is surrounded by a large bract, also called the spathe. The scent is mediated mainly by sulfur-containing secondary metabolites, including dimethyldisulfide, dimethyltrisulfide, putrescine, and cadaverine. The signaling compound responsible for mediating the increase in temperature is called calorigen, and was identified as salicylic acid. An increase in salicylic acid precedes the temperature increase, and the temperature can be elevated by the addition of salicylic acid or an extract from a male flower (Figure 2.8). The elevation in temperature may be up to 10 °C, which results in the emission of compounds which are nonvolatile at lower temperatures. This implies that no enzymatic reactions are necessary to transfer nonvolatiles into volatile compounds (Figure 2.10; Table 2.5) and that the scent can be regulated in a timely manner. However, the production of heat is also energy-dependent (Figure 2.8). An activation of the alternative oxidase (AOX) in mitochondria leads to a bypass of the normal electron-transport process used to generate the heat, and this results in a reduced ATP production in these flowers; this event is called "thermogenesis." The heat can be made visible using thermal imaging, as shown in Figure 2.8 for a mature male flower of *Dracunculus vulgaris*, where bright red symbolizes the regions with highest temperatures.

Another example of the importance of a correct volatile mixture in fertilization was obtained by investigating a variety of *Protea* species, among them flowers pollinated by beetles or birds (Figure 2.9). All of these flowers contained complex mixtures of volatiles, which differed one from another. The amount of scent emitted from inflorescences of beetle-pollinated species was about 10-fold greater than that of bird-pollinated species. More individual compounds were found in the former mixture. The flowers could be grouped into two classes by principal component analysis (PCA): one group encompassed those flowers pollinated by beetles, and the other flowers pollinated by birds. The floral scent of the bird-pollinated species consisted mainly of "green leaf volatiles" (Section 3.1) and benzenoid compounds, whereas the floral scent of beetle-pollinated species contained monoterpenes, especially linalool, and methyl benzoate. Consequently, in a choice experiment one beetle species (*Atrichelaphinis tigrina*) recognized the cues sent by flowers normally pollinated by beetles (*Protea simplex*), but not of flowers normally pollinated by birds (*Protea roupelliae*).

The emission of volatile compounds is dependent not only on their biosynthetic rate but also on the ability to diffuse rapidly out of the plant tissues. Therefore, biosynthesis takes place in cells or tissues located at the edge of an organ, so that

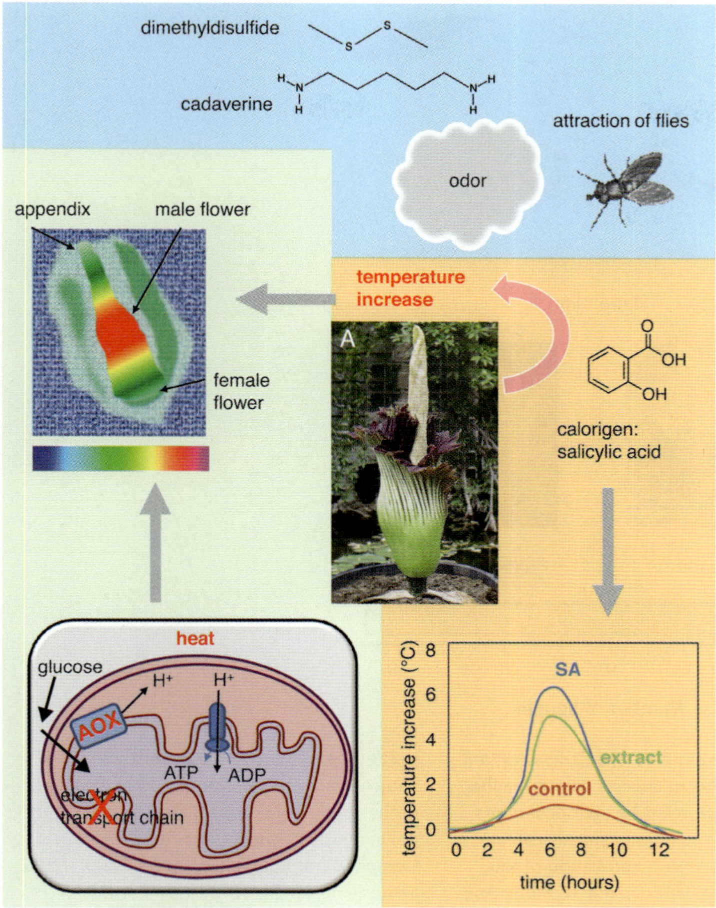

Figure 2.8 Emission of volatiles by induction with a calorigenic compound. The Titan arum (*Amorphophallus titanum*) is shown as an example of such a flower. The scent emitted by the flower attracts flies, which normally feed on carrion, for pollination. The volatile mixture consists mainly of sulfur-containing compounds, such as dimethyldisulfide and cadaverine. The increase in temperature is mediated by salicylic acid (SA), or by an extract from a mature male flower. At higher temperatures more volatiles can be emitted. The thermal image shows a mature flower of *Dracunculus vulgaris*, where bright red symbolizes the regions with highest temperatures. The alternative oxidase (AOX) in mitochondria is responsible for removing electrons from the normal electron transport chain, which results in heat development and reduction of ATP. Parts modified from *Teaching Tools in Plant Biology*, The American Society of Plant Biologists. Photo *Amorphophallus titanum* from Depositphotos.com © Alessandro Zocchi. Chemical structures are from The PubChem Project.

the compounds can be easily released into the environment. These are for example simple epidermal cells, and also trichomes and glands, but the main producers are flowers and fruits. Although many aroma compounds and their biosynthetic pathways are conserved, the mixture is unique so that it can be

Figure 2.9 The patterns of volatiles emitted from different *Protea* species determine the recognition of the scent by their pollinators. The non-metric multidimensional scaling (NMDS) (top) shows a correlation of total volatile emission patterns and 12 *Protea* species pollinated by beetles (white symbols) and birds (black symbols). In the center, one representative for a bird (*Protea roupelliae*) and beetle (*Protea simplex*) pollinated species is shown, whose whole flower scents were used in the choice experiments (bottom) using the pollinator of *P. simplex* (*Atrichelaphinis tigrina*). Compiled from Steenhuisen, S.-L. *et al.* (2012) *Phytochemistry*, **84**, 78–87 (see Further Reading). Photos reprinted with permission from Elsevier.

recognized by the respective specialized pollinator. This implies that many biosynthetic pathways are involved in the generation of a given blend of volatiles. The production can vary over tissues and time; for example, flowers can produce volatiles in a photoperiodic manner, guaranteeing the attraction of either daytime or night-time pollinators; these pathways are then controlled by light and the circadian clock of the plant. Promoter elements are also known that are specific for gene activation either during the day or in darkness. Although much is known about enzymes involved in the synthesis of aroma volatiles, their specific regulation by transcriptional, post-transcriptional or translational mechanisms is not well understood.

Aroma volatiles encompass many phenolic and terpenoid compounds (Figure 2.10). Simple modifications of the molecule that can result in a higher

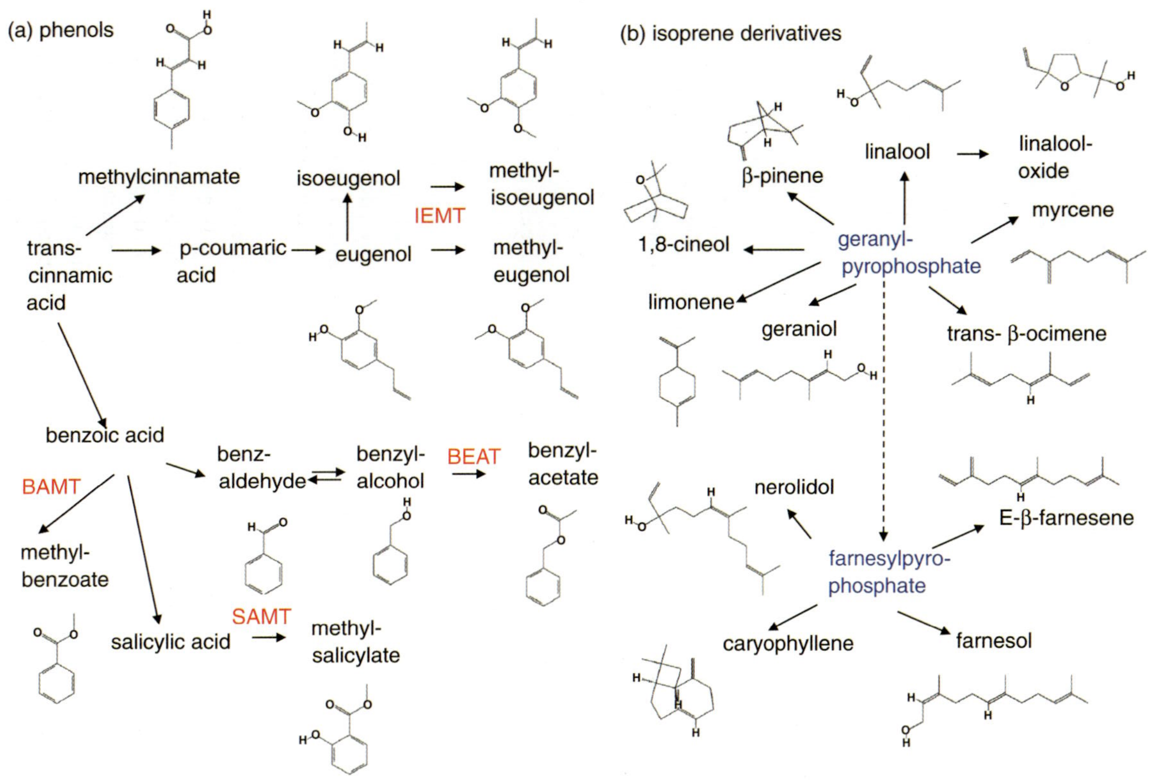

Figure 2.10 **Synthesis of volatile compounds in flowers.** Only the structures of the volatile compounds are shown. (a) Pathways and modifications of phenolic compounds. For some of the methylation reactions the SABATH-family enzymes are responsible. (SABATH stands for the initials of three members of this family (*s*alicylic *a*cid methyltransferase–*b*enzoic *a*cid methyltransferase–*th*eobromine synthase.) Acylations are catalyzed by the BAHD-family of proteins, also here the name stands for four members of this family (*B*EAT benzylalcohol acetyl transferase–*A*HCT anthocyanin-*O*-hydroxycinnamoyl transferase–*H*CBT anthranilate-*N*-hydroxycinnamoyl/benzoyl transferase–*D*AT deacetylvindoline-4-*O*-acetyl transferase). Abbreviations: BAMT = benzoic acid methyl transferase; BEAT = acetyl-Coenzyme A:benzylalcohol acetyltransferase; IEMT = isoeugenol methyltransferase; SAMT = salicylic acid methyltransferase. (b) Pathways leading to volatile monoterpenes and sesquiterpenes. Chemical structures are from The PubChem Project.

Table 2.5 Enzymatic modifications on basic structures, which can yield more volatile compounds. These modifications can be performed on many structures, not only those shown here.

Modification	Enzyme class	Example compound	Modified product
Hydroxylation	Cytochrome P450	Limonene (terpene)	*trans*-Isopiperitenole
Side-chain degradation	Cytochrome P450	Nerolidole (terpene)	Dimethylnonatriene
Side-chain degradation	β-Oxidation	C3-Phenylpropane	C1-Benzenoide
Oxidation	Dehydrogenase	Benzylalcohol (phenol)	Benzaldehyde
Methylation of hydroxyl groups	Methyltransferase	Chavicol (phenol)	Methylchavicole
Methylation of carboxyl groups	Methyltransferase	Salicylic acid (phenol)	Methylsalicylate
Acylation	Acyltransferase	Benzylalcohol (Phenol)	Benzylacetate

volatility (Table 2.5) include side-chain reactions of hydroxylation, oxidation, methylation, and acylation.

Phenols can be converted into the more volatile forms by methylation, either on hydroxyl or carboxyl groups. The former is realized in the methylation of eugenol or isoeugenol, whereas in the case of benzoic or salicylic acids the carboxyl groups are enzymatically methylated (Figure 2.10). Therefore, plant methyltransferases are highly important for the generation of flower scent. The methylation of salicylic acid is also important in pathogen defense, where the methylated form plays an important role in systemic resistance (Section 2.5.1). The major carboxy-methyltransferases belong to the so-called SABATH-family of methyltransferases; this name derives from the first three members of the family isolated (Figure 2.10). Generally speaking, all modifications resulting in a higher volatility of a compound are important for the specific blend of a plant tissue, and not only methylations; these may be hydroxylations, oxidations, or acylations (Table 2.5). The latter are catalyzed by acyltransferases of the BAHD-family, such as the conversion of benzylalcohol to benzylacetate. Here also, the name is derived from four important members of this family, which were isolated first (Figure 2.10). Oxidations are important for the conversion of benzylalcohol to benzaldehyde.

Many isoprene derivatives are known to be volatile, but they can also be converted to even more volatile compounds via modification (Figure 2.10 and Table 2.5); hydroxylations and oxidations by cytochrome P450-dependent enzymes play major roles in this case. The formation of ester and ether compounds is also important; some volatile compounds are synthesized by degradation of their side chain. The synthesis of some mono- and sesquiterpene volatiles employs geranylpyrophosphate and farnesylpyrophosphate as precursors, respectively (Figure 2.10).

moth	bee	kolibri
white color	violet color	red color
strong scent	little scent	little scent
much nectar	little nectar	much nectar
long tube	short tube	elevated sexual organs

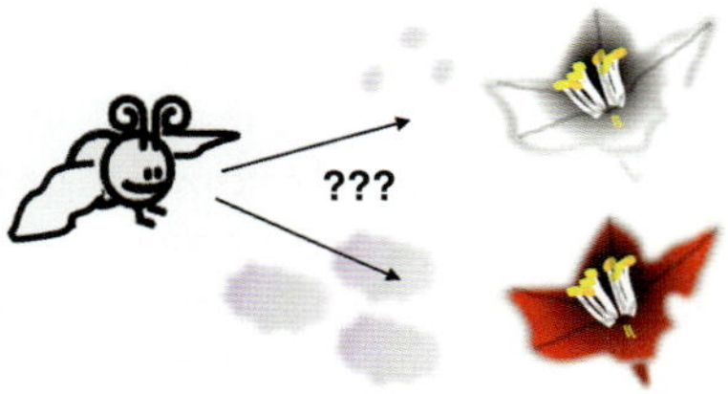

Figure 2.11 Combination of different cues are important for pollinators. The combination of color, together with scent, is more important than other cues such as the form of petals and accessibility of flower organs. If moths are exposed to the wrong combination of signals, they choose the flower randomly, as indicated by the question marks.

In other cases, it is the combination of color and scent that is vital (Figure 2.11). In an experiment using petunia flowers, if the cues were contradictory then the pollinator was irritated and could not find the correct flower. Three different petunia varieties are pollinated by three different organisms, moths, bees, and hummingbirds: moths prefer white flowers with a strong scent; bees prefer violet flowers with little scent, because the color red is more a gray shade in their vision; and the hummingbirds prefer red flowers with little scent. If moths were exposed to mixed cues – that is, red flowers with a strong scent and white flowers without scent – they could not find the correct flower in a choice experiment and so, without orientation, they chose the flowers randomly.

2.2.3.3 Color

The different colors of flowers also results in distinct visible patters on the petals. Bees, for example, can perceive different patterns than humans, because the visible wavelengths are shifted towards the UV light. Therefore, humans cannot recognize patterns on flower petals, which can be distinguished by bees (Figure 2.12).

Figure 2.12 Bees can see different color patterns than humans. The left-hand flower shows the yellow petals as can be seen by humans; the right-hand flower is a simulation of the UV absorbance pattern perceived by bees. Photograph of *Rudbeckia* from Fotolia.com; © M. Schuppich.

The glycosylated anthocyanins and their aglycones belong to the major group of water-soluble vacuolar pigments (Figure 2.13). These occur in many plant families, with the exception of the Caryophyllales, where betalains take over their function (Figure 1.9). The name anthocyanins is derived from Greek *anthos* (= flower) and *kyáneos* (= dark blue). All of these pigments occur in the vacuoles of leaves, the flowers and fruit (Figure 2.15). In the leaves they play important roles in the protection against high-light irradiation (Section 2.3.2), whereas in flowers they attract pollinators. In fruit, they play dual roles as attractants for seed dispersal, and also as a warning color to ward off herbivores.

The different colors of the anthocyanins are dependent on several parameters:

- molecular structure;
- substituents;
- pH-value;
- complexation of metal ions;
- glycosylation; and
- occurrence of mixtures (Figure 2.13).

Anthocyanins are present in the cell, mostly glycosylated in the 3- or 5-position as mono- or diglycosylated forms. Both positions can also be derivatized. Other substituents occur at different positions, mainly on the B-ring. The modifications are brought about by hydroxyl groups, but further modifications of these residues by methylation are also possible. Complex formation with metal ions also contributes to different colors (see also Figure 5.9); for example, increasing levels of aluminum result in the color change of petals of a hydrangea from pink, light red to violet and to dark blue. Therefore, it can be concluded that environmental factors are also important for the coloring of flowers. In addition, anthocyanins can play a role in abiotic stress tolerance (Section 2.3).

(a)

anthocyanidin	substituent	color
pelargonidin	----	orange-red
cyanidin	3'-OH	red
peonidin	3'-OCH_3	pink-red
delphinidin	3'-OH, 5'-OH	blue-violet
petunidin	3'-OCH_3, 5'-OH	light violet
malvidin	3'-OCH_3, 5'-OCH_3	red-violet

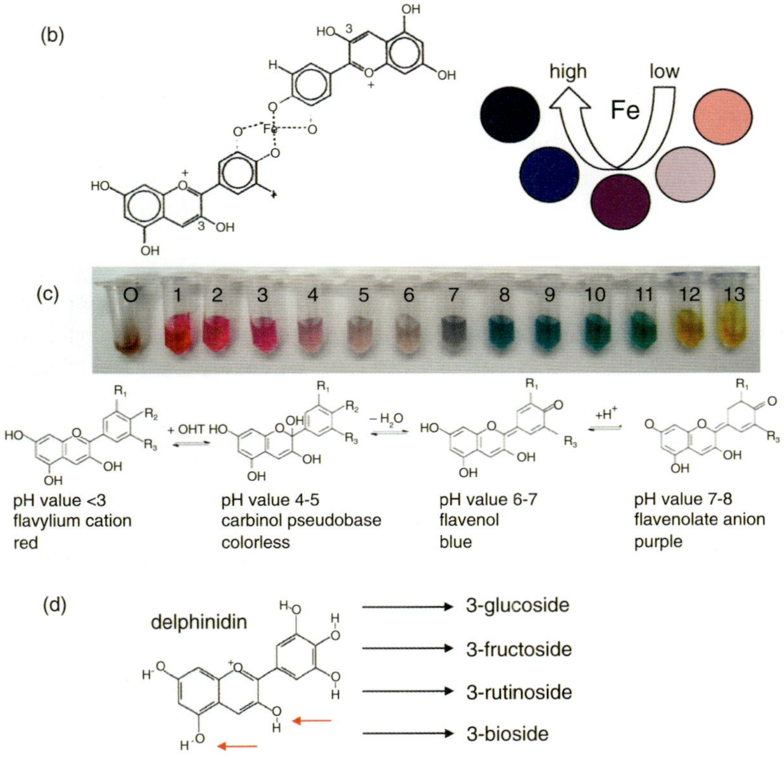

Figure 2.13 The color of anthocyanins is influenced by different parameters. (a) Substituents influence the color; (b) Complex formation of cyanidine with an iron atom changes the color of a hydrangea flower; (c) The color is dependent on the pH-value of the cells and also the environment. The biochemical reactions leading to different colors after changing the pH-values are shown, as well as an example extract from *Saintpaulia ionantha* at different pH-values. At left is the color of the original aqueous extract (O); (d) Different sugars can modify the structure, as shown for delphinidine. A bioside is modified at two different positions. Chemical structure from delphinidin is from The PubChem Project.

One additional environmental factor is the pH value of the soil, which also alters flower color. Anthocyanins can function as pH indicators. Plants of the same genotype can display either pink or blue flower petals, depending on the composition of the soil; hydrangea flowers are a very good example of this observation (Figure 2.13). In the laboratory, this effect can be demonstrated on aqueous extracts of flowers. For example, the petals of the violet *Saintpaulia ionantha* were used to demonstrate different colors at different pH values. The addition of protons (acidification) leads to the dissociation of oxygen and the simultaneous addition of a hydroxyl group (carbinol pseudobase). Under these conditions, the originally blue flavenol at pH 6–7 is converted into a colorless molecule at pH 4–5. Further acidification leads to a positive charge, and thus to the formation of a red flavylium cation at pH 3. In contrast, at pH 7–8 the purple flavenolate anion is formed. Further alkalinization opens the ring and results in the yellow chalcone.

The regulation of vacuolar pH, which greatly affects anthocyanin color, is only partly understood. The only known structural gene that regulates vacuolar pH directly with relevance to the resulting color is a Na^+/H^+-antiporter of the Japanese morning glory (*Ipomea nil*), that is specifically expressed before flower opening and increases the vacuolar pH to generate blue flowers. Petunia loci that regulate vacuolar pH have also been identified, but in this case the gene encodes a transcription factor of the R2R3 Myb family (see Section 1.1.3), which activates vacuolar acidification.

The color range from yellow to orange in flowers and fruits (Section 2.2.4) is produced by chalcones and flavonoids, and in particular by carotenoids (Figure 2.14). Chalcones are the yellow pigments of safflower, while aurones – a class of rare flavonoids – give brighter yellow flowers than chalcones and are found in a limited number of species, such as snapdragon (*Antirrhinum majus*). Flavonols and flavones are pale-yellow and mostly invisible to the human eye. However, as they absorb UV they produce colors and patterns to flowers to attract insects that can recognize these patterns in the UV light spectrum (Figure 2.12). A combination of different flavonoids results in the bright yellow color of pollen, which helps to attract insects and even larger animals. Although both animal and insects can detect the specific pollen color in visible or UV light for pollination, some plant species do not rely on the yellow pollen color but rather produce white pollen grains.

Carotenoids are tetraterpenes, and are synthesized via the isoprene pathway in the plastids. The precursors are colorless and are successively transformed into colored molecules (Figure 2.14). Lycopene has a light yellow color, β-carotene is already bright yellow, and zeaxanthin appears orange. Figure 2.14 shows the different structures of some carotenoid compounds; bacteria transformed with the respective biosynthesis gene, resulting in colored cultures, are also depicted. The xanthophyll cycle, which is important in abiotic stress resistance against strong irradiation (Section 2.3.2), is highlighted in gray. Dark red colors in the flowers of the pheasant's eye (*Adonis* sp.) are conferred by astaxanthin, whereas the spring pheasant's eye possesses yellow flowers that obtain their color from other carotenoids. Carotenoids as lipophilic pigments are not found in the vacuole; rather, they accumulate either in chromoplasts or special vesicles. Carotenoids occur also in large amounts in leaves, where they play important roles in the light-harvesting complex of photosynthesis (Section 2.1). Senescent leaves show predominantly the yellow color as a result of chlorophyll degradation (Section 2.1). The roots of some plants also contain carotenoids after colonization with beneficial fungi, such as arbuscular mycorrhiza (Section 2.4).

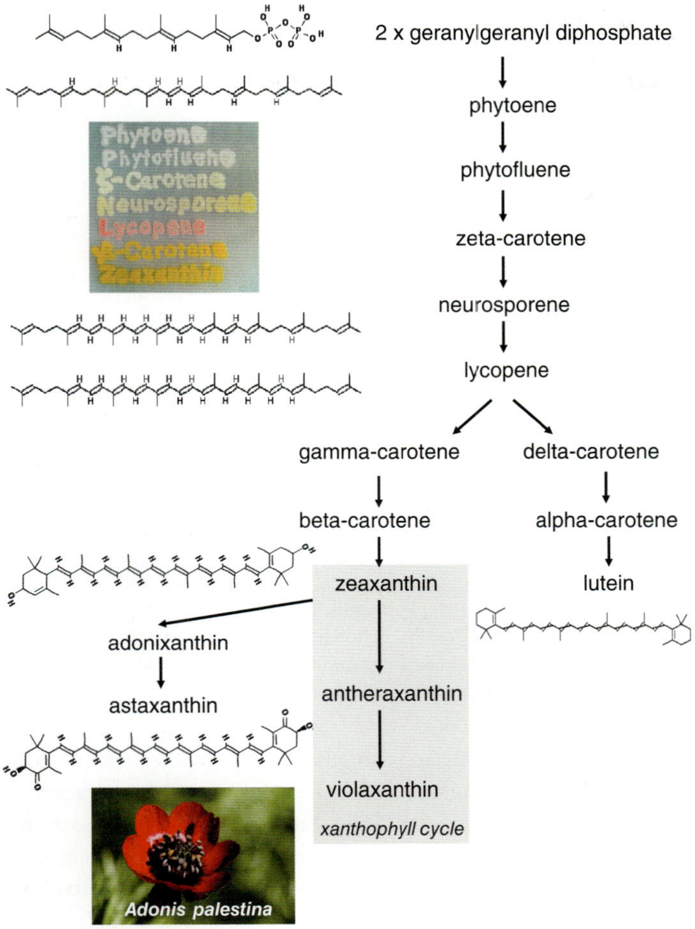

Figure 2.14 Carotenoids are plant pigments occurring in flowers. Some example structures from the biosynthetic pathway are shown. The upper image shows bacteria transformed with the respective biosynthesis gene yielding colored cultures. The compounds of the xanthophyll cycle are highlighted in gray. The photograph shows flowers of the pheasant's eye (*Adonis* sp.). Photograph of bacterial colonies from Cunningham, F.X. Jr and Gantt, E. (1998) *Annu. Rev. Plant Physiol. Plant Mol. Biol.*, **49**, 557–583 (see Further Reading). Reprinted with permission from Annual Reviews. Photograph of *Adonis* from: Fotolia.com; © Martina Berg. Chemical structures are from The PubChem Project.

2.2.4 Fruit and Seeds

2.2.4.1 Fruit Development

Many seeds are distributed by vertebrates, which are referred to as frugivores to highlight their preference for fruit. It is mainly birds and mammals that are responsible for the seed dispersal of fleshy fruit; most birds are attracted mainly

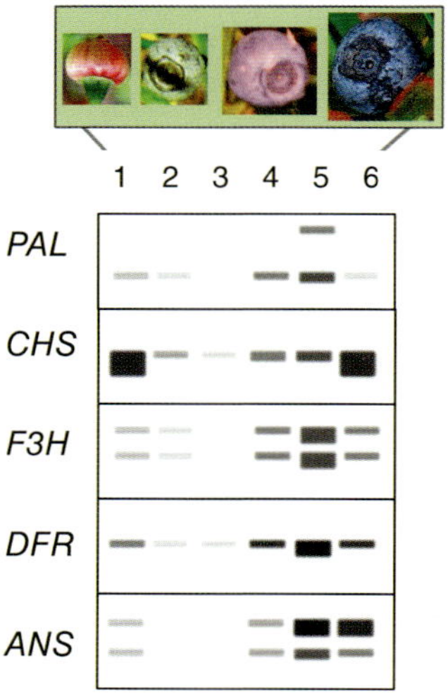

Figure 2.15 Anthocyanin accumulation during fruit ripening of blueberry. Expression analysis of selected genes encoding enzymes necessary for anthocyanin biosynthesis in different stages from flower (1) via non-ripe white-green (2, 3) and pink (4, 5) to ripe fruit (6), which are colored dark blue. PAL = phenylalanine ammonia lyase; CHS = chalcone synthase; F3 H = flavonole-3-hydroxylase; dihydroflavonole reductase; ANS = anthocyanidine synthase. For details of the flavonoid and anthocyanin biosynthetic pathway, see also Figure 1.8. Lower part of figure based on: Jaakola, L. *et al.* (2002) *Plant Physiol.*, **130**, 729–739.

by color, while frugivoral mammals are also attracted by volatiles. The fruit needs to be consumed by the seed dispersers which, therefore, must be attracted to the fruit first by either aroma and/or color. Later, fruit chemistry such as taste and digestibility also has an effect on fruit consumption. Following digestion of the flesh, the seeds are excreted, often rather far away from the mother plant, which promotes dispersal. In contrast, for protection some plants signal their toxicity by bright red colors of the fruit to the animals. The astringent properties of tannins, which can be found in seed coats, may also function as deterrents for animal herbivores. Millet seeds with high levels of proanthocyanidins proved to be less attractive to birds than seeds with low tannin levels (Section 2.5).

The anthocyanins are responsible for the red and dark blue colors (Figure 2.15), while the yellow/orange to red colors are provided by the carotenoids (Figure 2.16). Transcripts for genes encoding the proteins involved in anthocyanin biosynthesis accumulate during the ripening process of blueberry (*Vaccinium myrtillus*). Six different stages were analyzed, from flower over unripe to ripe fruit (Figure 2.15). Transcripts were also detected in the colored flowers, but this was not the case for white flowers. Transcription increases at the onset of fruit coloring, and peaks when the fruit is ripe and colored dark blue. The transcriptional control of anthocyanin accumulation is therefore strictly regulated during the ripening process. In the case of blueberry, the dark blue color is a signal for the ripe fruit to attract seed-dispersing animals.

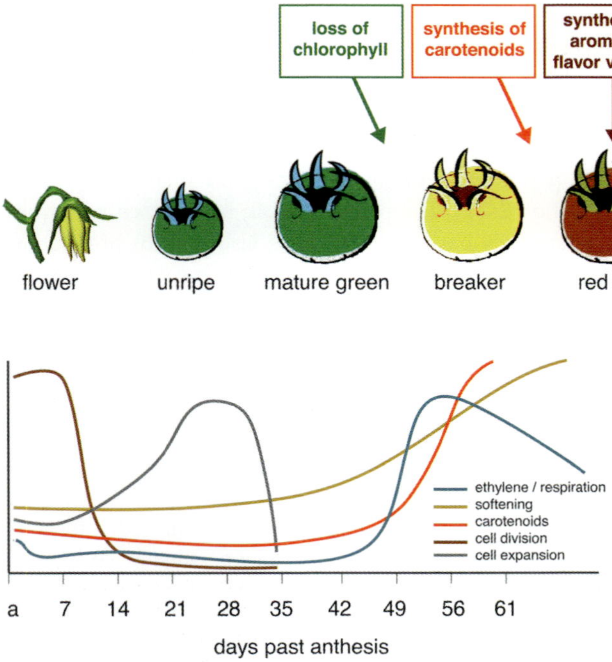

Figure 2.16 Tomato fruit development showing the ripening stages from flower over unripe to ripe fruit. There are four distinct developmental stages, the unripe and mature green fruit, the yellow breaker stage; and the red, fully ripe fruit. Green stages show an increased cell division and elongation rate (brown and green lines, respectively), while the later developmental stages from breaker to red ripe are correlated with the synthesis of carotenoids (red line), ethylene and respiration (blue line) and softening (yellow line). a = anthesis. Lower part of figure based on: Giovannoni, J.J. (2004) *Plant Cell*, **16**, S170–S180.

Another example is the fruit ripening of tomato (*Solanum lycopersicum*), where the synthesis of pigments strongly correlates with ripening of the fruit (Figure 2.16). The process can be divided into two phases: the first phase encompasses growth of the fruit to more or less its final size; the second phase is the actual ripening of the mature fruit. This implies that cell division and cell elongation processes occur during the first stage. When the mature green fruit has developed, synthesis of the fruit-ripening hormone ethylene begins concomitantly with the degradation of chlorophylls; in this case, the chloroplasts are degraded or develop into the carotenoid-containing chromoplasts. Carotenoid biosynthesis (mainly lycopene and β-carotene) is also induced by ethylene in the breaker stage, as well as softening of the fruit and the synthesis of aroma compounds during the ripening stage (which is also called “red ripe”). As a climacteric fruit, the increase in cellular respiration is a characteristic for tomato fruit in this late stage. The softening leads to a degradation of the fruit, which in turn liberates the ripe seeds.

2.2.4.2 **Seed Formation**

Flavonols play a role in seed formation, but not for fruit coloring. Silencing of the *chalcone synthase* gene results in parthenocarpy (seed production without fertilization) in tomato, but which flavonols are responsible for this has not yet been determined. The silencing of a flavonole synthase in tobacco causes the production of less-seeded fruits, and the pollen of these silenced lines was unable to produce functional pollen tubes (Section 2.2.2). The phenotype can be reversed with quercetin, which implies that flavonols have essential roles also in fruit production. Moreover, the results obtained indicated that silencing of the *flavonol synthase* gene could represent a new strategy to obtain plants with less-seeded or seedless fruits, though fertility could be chemically restored for seed production.

2.2.4.3 **Seed Dispersal**

Seed dispersal is dependent on the attractiveness of the fruit for animals but, on the other hand, the fruit must be protected from herbivorous organisms (Sections 2.5.3 and 3.1). This necessitates a strict control of the synthesis of aroma compounds in the fruit. Only a few structural changes can alter the taste of a compound from bitter to sweet, as shown for naringin from the peels of citrus fruit (Figure 2.17). The bitterness in citrus fruit is determined by the composition of

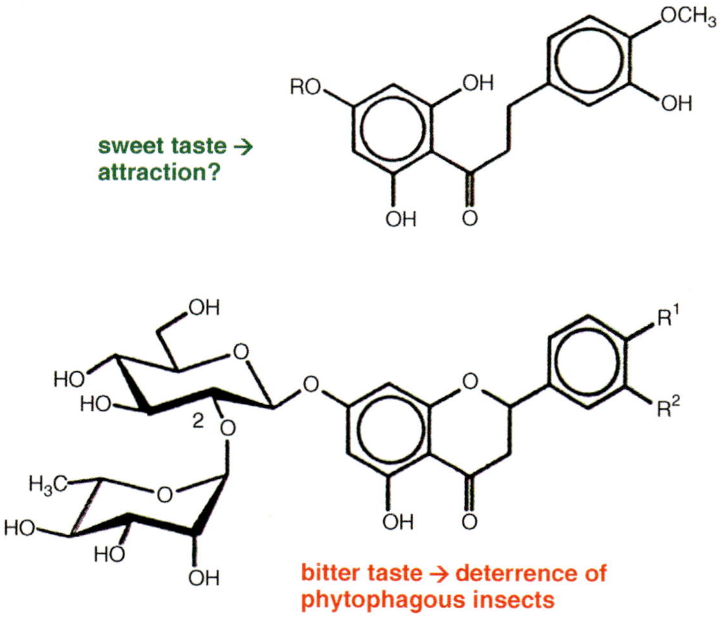

Figure 2.17 Flavonoids as aroma compounds. Flavanones as bitter- or sweet-tasting principle. Upper structure: Bitter-tasting naringin from citrus peels; rhamnose and glucose are linked together at the C2 atom of the glucose. Lower structure: The open chalcone ring changes the aroma to sweeter taste. Nonglycosylated forms are rather tasteless. Glycosylation does not necessarily result in a sweet-tasting compound!.

the branched-chain flavanone glycosides, which are the predominant flavonoids in citrus. The nonglycosylated substance with the open chalcone ring tastes rather sweet and could be involved in the attraction of insects for seed dispersal, whereas the glycosylated form with the closed ring system tastes bitter and could act as feeding deterrent. The flavor-determining biosynthetic step is catalyzed by two specific rhamnosyltransferases that utilize flavanone-7-*O*-glucose as substrate. These principles can also be used in breeding of citrus cultivars to make them more attractive to the consumers.

The evolution of attractive fruits for animals has led to the dispersal of seeds away from parent plants. The ripening stage of the fruit is very important for seed dispersal by animals. The secondary metabolites – plant volatile organic compounds, which are not essential compounds for plant survival – are involved in the interaction of fleshy fruits with seed dispersers and herbivores. From a chemical viewpoint, these volatiles can be classified as esters, alcohols, aldehydes, ketones, lactones, and terpenoids, and are usually accumulated at high levels specifically in ripe fruits. In general, volatile compound production increases between one- and more than one-thousand-fold during the maturation of most fruits, which is also accompanied by qualitative changes. For example, in many unripe green fruits the short-chained aldehydes, which confer a "green leaf odor," are abundant and their concentration decreases with ripening in fruits such as nectarines, guavas, apples, coriander, strawberries, and kiwis.

There is a switch in the production of the aroma blend during ripening, which is needed for seed dispersal. During the ripening of citrus fruits, for example, the monoterpene profile changes; in oranges (*Citrus* x *aurantium*) there is a reduction in the linalool content as maturation progresses, whereas the limonene content increases dramatically between the green and color-break stages (Figure 2.18). In tomato, next to sugars and acids, approximately 20 to 30 different aroma compounds are synthesized, among which β-ionone is the most prominent. The differences between green (ripe) and red-colored tomato fruit lies also within the composition of the volatiles emitted. Whereas green tomatoes emit high levels of volatile esters, red tomatoes do not. One major factor responsible for this is an esterase which is missing in the red fruit species, where an evolutionary ancient transposon insertion next to the esterase gene has occurred. Interestingly, the volatiles of red fruit are more attractive to humans, and maybe also to animals, so that the mutation in esterase activity might have conferred an advantage to the red fruit species. The species with red fruit forms a distinct clade that is quite different from that of the green fruit species.

2.2.4.4 **Seed Dormancy**

Finally, seed germination after seed dispersal is also regulated by secondary metabolites. Before a seed can germinate, it must be ensured that: (i) the embryo has matured; (ii) the environmental conditions are as optimal as possible for the growing seedling; and (iii) the seeds do not germinate on the mother plant (vivipary).

In order to ensure a controlled germination process, the timing of the initiation stage is important. The timing is first regulated by germination inhibitors, which

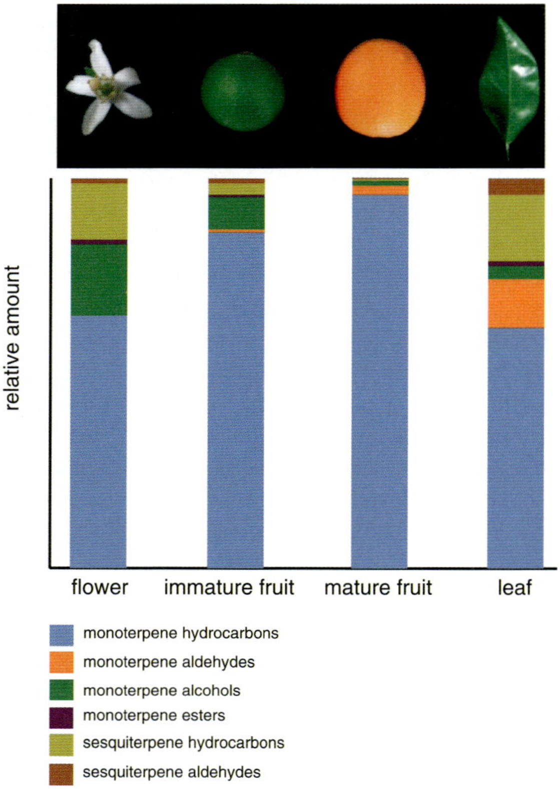

Figure 2.18 Volatile emission is different in ripe citrus fruit. An example of changes in volatile composition is shown here for four different organs of an orange plant. The ripe fruit emits predominantly monoterpene hydrocarbons, whereas other tissues show different distributions of volatiles. Diagram redrawn and photo from Rodriguez, A. *et al.* (2013) *New Phytol.*, **197**, 36–48 (see Further Reading). Reprinted with permission from John Wiley & Sons.

induce seed dormancy; however, if seed dormancy is not regulated, for example in mutant plants, then germination will take place on the parent plant (vivipary). Although, under normal conditions, vivipary is an unwanted phenomenon, some viviparous plants can produce seeds that germinate before they detach from the mother plant (e.g., some mangrove species). Later in the germination process stimulating signals are necessary, and this leads to an uptake of water, hydrolysis of storage products by the embryo, and finally growth of the root and leaf organs from the seeds. Seed dormancy is preserved by the plant hormone abscisic acid, and also by a range of other secondary metabolites such as cyanogenic glycosides, phenolic compounds (e.g., cinnamic acids), and monoterpenes. The germination inhibitors must be either diluted or degraded before germination can take place. Thus, secondary metabolites regulate an important developmental step in the plant's life cycle (see Figure 2.1), which occurs after seed dispersal.

2.3 Abiotic Stress

Many secondary metabolites are induced under abiotic stress conditions. Controlled stress treatments can be used to increase specific metabolites in cell or tissue cultures (Section 1.3.1). In particular, oxygen depletion and heavy metal treatment can increase the levels of certain compounds. Since a shortage of oxygen often occurs in bioreactors, these can be used for the induction of stress-related secondary metabolites (Tables 1.9 and 2.3). Some examples of the possible functions of secondary metabolites under abiotic stress conditions are summarized in Table 2.6; details of these are provided in the following subsections.

Table 2.6 The possible roles of selected secondary metabolites in response to abiotic stress factors. Stress tolerance can be induced directly or indirectly; the latter would be mediated by the possible antioxidative potential of a compound. As far as known, the possible mechanism is mentioned.

Stress	Substance	Mechanism
UV irradiation	Sinapine derivatives	Absorption
	Flavonoids	Absorption, antioxidant
	Terpene indolalkaloids	Absorption
	Tocopherols	Antioxidant
	Glucosinolates	?
Light/ Photooxidation	Anthocyanins	Absorption
	Tocopherols	Antioxidant
	Xanthophylls	Absorption of excess energy
Heat stress	Anthocyanins	?
	Phenols	?
	Tocopherols	Stabilization of membranes
	Monoterpenes	Protection of photosystem II from reactive oxygen species (?)
Cold stress	Anthocyanins	Osmotic adaptation (?) Protection from photooxidation (?)
Drought stress	Anthocyanins	Osmoticum, antioxidant
	Flavonoids	Antioxidant (?)
	Phenols, for example, rosmarinic acid	Antioxidant (?)
Salt stress	Theanine	Proteolysis of proteins and accumulation of osmoprotective compounds (?)
Oxygen depletion	Phenols, for example, rosmarinic acid, caffeic acid	Antioxidant (?)
Nutrient deficiency	Anthocyanins Others see Table 2.7	?
Heavy metal stress	Anthocyanins	Complex formation and detoxification (?)
	Flavonoids	Antioxidant, influence of phytochelatins (?)
Air pollution	Terpenes	?

2.3.1
Antioxidative Properties in Plants

Under both abiotic and biotic stress conditions, a wide range of compounds with antioxidative potential is generally synthesized within a plant. Abiotic stress conditions, which lead to the production of reactive oxygen species (ROS), include high irradiation, heat, cold and salt stress, and also high ozone levels. These ROS are also formed during many metabolic reactions (Table 2.7); in peroxisomes, they are produced during β-oxidation reactions and during photorespiration. Within the plant cell, the organelles are among the main producers of ROS; in both the chloroplasts and mitochondria they are produced in the electron transport chain. In plastids, the reduction of electron acceptors such as NADP can result in the reaction of ferredoxin with oxygen, with the generation of superoxide anions that are capable of oxidizing many other compounds and so must be detoxified very quickly. In the mitochondrion, oxygen radicals are also formed during electron transport, namely at complex III. UV-B irradiation, as well as high ozone levels, lead to the formation of ROS. Ozone is taken up via the stomata of the plant and can then react in the cell with oxygen species to form ROS. Osmotic stress and heavy metal toxification also lead to the formation of these oxygen radicals.

Biotic stress can induce a hypersensitive reaction, which is characterized by the production of ROS (Section 2.5.1). In this case, ROS are produced by a plasma membrane-bound enzyme called NADPH-oxidase, which is involved in the defense reactions (Figure 2.27). These radicals can also have an antimicrobial effect, and a similar reaction of ROS has been observed on mammal cell cultures (Section 5.1.2). However, since these ROS are also toxic to the healthy plant tissues, an antioxidative system to shut off this hypersensitive response must be present to prevent any damage in adjacent tissues.

Table 2.7 Production of reactive oxygen species (ROS) in plant cells via different mechanisms in metabolic reactions. The mechanism of ROS generation, the location in the cell, and the first oxygen radical formed are shown.

Mechanism	Cellular localization	Primary ROS formed
Photosynthesis: Electron transport photosystem I or II	Chloroplast	$O^{\bullet}{}_2^-$
Photorespiration	Chloroplast Mitochondrion Peroxisome	H_2O_2
Electron transport	Mitochondrion	$O^{\bullet}{}_2^-$
Glycolate oxidase	Peroxisome	H_2O_2
Excited chlorophyll	Chloroplast	$O^{\bullet}{}_2^1$
NADPH-oxidase	Plasma membrane	$O^{\bullet}{}_2^-$
β-Oxidation	Peroxisome	H_2O_2
Oxalate oxidase	Apoplast	H_2O_2
Xanthine oxidase	Peroxisome	$O^{\bullet}{}_2^-$
Peroxidases, Mn^{2+} and NADH	Cell wall	H_2O_2, $O^{\bullet}{}_2^-$
Amine oxidase	Apoplast	H_2O_2

The antioxidative system in plants encompasses secondary metabolites such as flavonoids and carotenoids, in addition to other molecules such as the vitamins C and E (tocopherol, a terpene), glutathione and antioxidative enzymes, such as peroxidase, superoxide dismutase, and catalase. Xanthophylls are involved in a cycle named after this class of compounds – the xanthophyll cycle – which is involved in protection against high irradiations in the chloroplasts (Figure 2.14). Terpenes are mainly localized in membranes due to their hydrophobicity, whereas flavonoids are found in many cellular compartments, which argues in favor of their general role as antioxidants in plants. The antioxidant properties of flavonoids represent a robust biochemical trait of organisms exposed to oxidative stress of different origin, and should be considered as a significant trait for plant – environment interactions. Flavonoids do not all possess the same antioxidative potential. For example, naringenin does not show an antioxidative activity, when assayed using the radical DPPH* as substrate, although in other test systems this might be different. Hence, the results obtained are dependent on the antioxidative assay used. It is not surprising that plants undergoing severe stress conditions preferentially accumulate dihydroxy B-ring-substituted flavonoids, such as quercetin, which are effective scavengers of ROS. Vacuolar dihydroxy B-ring-flavonoids found in the epidermis have long been suggested to possess antioxidant functions, as the epidermal cells themselves must be protected; they are not only aimed at preserving the underlying sensitive organs from photooxidative damage (Section 2.3.2). Whilst a beneficial role of secondary metabolites as antioxidants in plants is largely accepted, the biological effects of these substances in vertebrates are complex and cannot be described as simply antioxidative. Details of the role of secondary metabolites as antioxidants in humans are provided in Chapter 5.

Catechins were shown to have direct effects on the levels of transcripts and enzymatic activities of several antioxidative enzymes. The use of transgenic tobacco plants with elevated catechin levels, or the treatment of wild-type plants with catechin, revealed that catechin and epicatechin induced mRNA levels of glutathione reductase, ascorbate peroxidase and catalase, as well as their respective enzymatic activities. These flavonoid derivatives do not act directly as antioxidants in this system, but rather as inducers of the ROS-scavenging enzymes.

There is a close connection between antioxidants and the redox homeostasis in plants, with glutathione and ascorbate peroxidase being particularly important in the organelles. The redox state is also sometimes perceived as a signal (redox signal) in plant abiotic stress reactions, so that a deviation from the normal reaction can lead to the induction of stress resistance mechanisms. High concentrations of ROS would lead to changes in the redox homeostasis, and would therefore play a role in the induction of this signaling pathway. This is in line with the observation that ROS themselves play a role as signals in plants. In order to control the levels of ROS, it is important that scavenging molecules are synthesized, and therefore plant secondary metabolites exhibiting such a function would be indirectly involved in controlling the signaling cascades of ROS and, in particular, of hydrogen peroxide (H_2O_2). This molecule can pass easily between cellular compartments, and is therefore ideally suited as an inductor of stress responses. H_2O_2 could also interact with kinases, phosphatases, effectors of calcium signaling, ion channels or transcription factors. ROS scavengers would, in turn, control the level of H_2O_2 within the cellular compartments.

2.3.2 Light and UV Irradiation

Within the plant, photooxidation is one major source of ROS. Flavonoids – amongst which are the colored anthocyanins – play many roles in the protection of plant tissues against UV-B irradiation and strong light (Figure 2.19). Plants at different developmental stages can show altered distribution patterns of photoprotective compounds, such as high concentrations in the younger leaves, whereas the reddish coloring due to anthocyanins is reduced or completely vanishes in

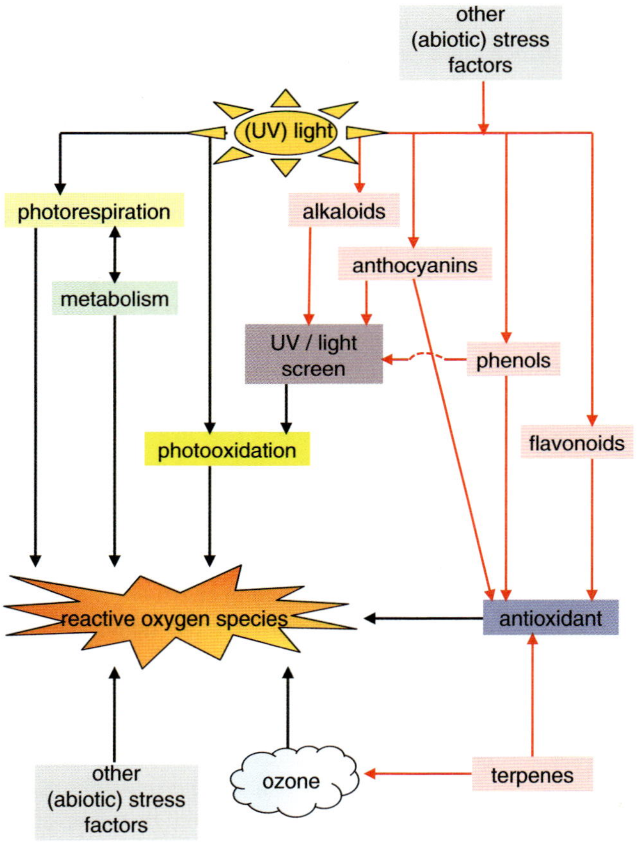

Figure 2.19 Light and UV irradiation, as well as ozone, induce many processes that result in the generation of reactive oxygen species (ROS). In addition, metabolic pathways leading to the production of ROS are indicated. Protective compounds are induced, which act as UV or light screen and/or antioxidants. Among these compounds, alkaloids, anthocyanins, flavonoids, some other phenolic compounds, and terpenes can be found. It is also indicated that other (abiotic) stresses can induce these mechanisms.

Table 2.8 Certain flavonoid glycosides are induced by high light. Leaves of *Ligustrum vulgare* grown at different intensities of sunlight over a 6-week period. Cinnamic acids, not shown here, were not changed.

Flavonoid glycoside			
Quercetin-3-O-glycosides	0.5	1.0	2.0
Luteolin-7-O-glycosides	1.7	3.0	4.8
Apigenin-7-O-glycosides	2.7	3.0	2.5

adult plants. Flavonoids may also accumulate in a cell-specific fashion to protect tissues; for example, the massive accumulation of flavonoids in external appendices (such as trichomes) is consistent with UV-protection functions, while leaf-surface flavonoids are effective UV-B absorbers. The ratios of the "effective antioxidants" quercetin and luteolin glycosides to the "poor antioxidants" kaempferol and apigenin glycosides are significantly increased upon strong light irradiation, irrespective of the relative proportions of different solar wavelengths that reach the leaf surface. The pigments have high protective properties, so that the photosystems can be shielded against high irradiation, and thus photooxidation. When anthocyanins accumulate in leaves they confer a reddish-brown color because they mix with the green leaf pigments. The accumulation of flavonoids and sinapate derivatives is due mainly to UV-B irradiation, although flavonoids can also accumulate under strong light conditions (Table 2.8). It is possible that this accumulation is also due to the UV component of the radiation. Whereas higher levels of flavonoids were found after exposure to 100% sunlight, the levels were decreased in the shade (20% or 40% sunlight), which indicated that the rise in flavonoid level may be attributed to their induction by light.

The induction of flavonoids by UV light can be shown by transcript analysis for genes encoding enzymes from the flavonoid pathway; for example, *chalcone synthase* gene expression is upregulated by UV irradiation. Chalcone synthase is one of the most important enzymes for flavonoid biosynthesis, and determines the first committed step to flavanones (Figure 1.8); hence, an increase in its activity would result in the induction of various flavonoids and anthocyanins. Flavonoids are noted as being protective metabolites against strong light, because of the correlation between the maximum extinction of quercetin in the UV wavelength. In flavonoid-free *Arabidopsis* mutants, sinapate derivatives protected against high light, and therefore these compounds have also been suggested as important players in the plant's "sunscreen." However, as no direct evidence for the protective role of flavonoids against intensive irradiation has yet been found, their effect might be more indirect due to their antioxidative potential when molecules have been destroyed by photooxidative processes (Section 2.3.1). Flavonoids are able to induce a systemic tolerance against UV irradiation via the signaling molecule, nitric oxide. In this case, the effect is a protection of the plant parts above the irradiated organ, so that even after partial exposure to light the whole plant can be protected. Shaded maize leaves do not synthesize the systemic response signal, whereas plants with illuminated leaves were found to produce flavonoids in systemic organs. Proanthocyanidins are also involved in stress protection, for example strong light and UV irradiation. Specifically, seed coats are decorated with these pigments so that the sensitive embryo is well protected.

Other secondary metabolites can also be induced by UV light. The total amount of glucosinolates was increased in broccoli sprouts, together with transcripts encoding biosynthetic enzymes and transcription factors. Interestingly, the induction of glucosinolates under this abiotic stress factor also had a positive effect on the plants under biotic stress, such as herbivore attack (Section 2.5.3). Terpene indole alkaloids can absorb UV light and may therefore serve as good compounds for protection against this irradiation. The biosynthesis is dependent on the conversion of tryptophan to tryptamine and, consequently, the transcript for tryptophan decarboxylase increases after UV irradiation. UV-responsive promoter elements were identified within the promoter region of the *tryptamine decarboxylase* gene, and were shown to be responsible for the upregulation of this gene.

Terpenes such as tocopherol and carotenoids belong to a plethora of metabolites that are capable of protecting plants against photooxidation, and play an important role in protecting cells from strong light. Carotenoids can revert to the so-called triplet state of chlorophyll, when they are unable to transfer electrons to the electron-transport chain of chloroplasts and are converted into triplet carotenoids. The surplus energy is then dissipated as heat. In addition, in order to reduce energy, reversible reactions occur between the xanthophylls (carotenoid derivatives) zeaxanthine and violaxanthine via antheraxanthine (also called xanthophyll cycle) (Figure 2.14). The synthesis of zeaxanthine is dependent on the reduction by ascorbate, whereas the reverse reaction requires $NADPH_2$ and oxygen. Zeaxanthine formation occurs on the inner side of the thylakoid membranes, at about pH 5, whereas the synthesis of violaxanthine occurs at the outer side of the membrane, at pH >7. This implies that the xanthophyll cycle is dependent on a large pH gradient between the thylakoids. From a biological point of view, this regulatory mechanism is reasonable because this situation – that is, a large difference in pH between the stroma and lumen – also occurs when photosystem II is in a high excitation state. This implies the existence of excess energy, which can then be transferred to the xanthophyll cycle.

2.3.3 High Temperatures

Under stresses such as high irradiation (Section 2.3.2) and high temperature, the xanthophylls are distributed between the light-harvesting complexes of the photosystems and the membrane lipids. One function of the xanthophylls, as radical scavenger molecules, is to protect the photosystems, while a second function is to stabilize the lipid part of the thylakoid membranes by reducing membrane fluidity and thus increasing thermostability. This leads to a lower lipid peroxidation under high-temperature stress.

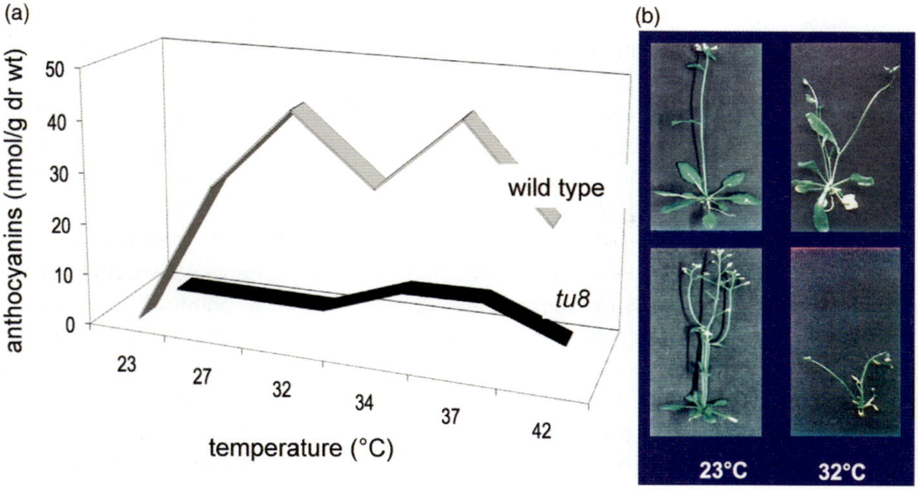

Figure 2.20 Anthocyanins are induced at high temperatures in *Arabidopsis*. (a) In the *Arabidopsis* ecotype Columbia (wild-type), anthocyanins are synthesized in high amounts already above 25 °C, whereas this is not the case for a thermosensitive mutant (*tu8*); (b) The wild-type tolerates temperatures up to 32 °C, but at this temperature the mutant is already completely inhibited in growth.

In addition to the direct protection against strong light, the synthesis of anthocyanins, flavonoids and other phenol derivatives is increased after other abiotic stresses, such as high temperatures (Figures 2.19 and 2.20). Strong light and high temperatures also often result in drought stress, and these stressors can be additive. The accumulation of phenolic compounds is accompanied by an induction of the transcript for phenylalanine ammonia lyase, the first committed step of phenylpropanoid biosynthesis (Figure 1.7). Using *Arabidopsis* as an example, elevated temperatures above 25–27 °C were shown to result in an accumulation of anthocyanins (Figure 2.20), whereas a thermosensitive mutant did not show this induction. Whereas the latter mutant showed complete growth inhibition at a temperature of 32 °C, the wild-type was still growing, albeit slowly. This implies a protective function of anthocyanins also for heat stress.

Some glucosinolates also follow the pattern of increased synthesis at higher temperatures, although their role under these conditions for the plant is not yet clear. Whereas, at 20 °C – which is close to the optimal growth temperature of broccoli – the accumulation of different glucosinolates is almost constant over a 24 h period, at 30 °C a strong circadian rhythm can be shown for their formation.

Volatile terpene compounds have also been suggested to play a role in the acquisition of tolerance against heat stress. They are thought to increase the resistance of the photosystems against higher temperature, and especially the protection of photosystem II against ROS. Oxygen is generated by the oxygenase function of ribulose-1,5-bisphosphate carboxylase, although under the experimental conditions employed barely any photosynthesis was detected. This experiment also demonstrated an induction of the antioxidative system of the plant, including xanthophylls and other metabolites (Section 2.3.1).

2.3.4
Cold, Salt, and Drought Stress

Some signaling pathways are similar for these three stress factors, even though there are also typical pathways induced for individual stressors. In particular, the so-called "osmolytes" – including the amino acid proline, glycerol, the polyamine spermidine and quaternary ammonia salts such as glycine betaine – can be induced by all three abiotic stress factors. Their role is to stabilize osmotic homeostasis in the cell so that, under stress conditions, the cell integrity is maintained. Organic osmolytes are typically termed "compatible solutes," based on the hypothesis that these solutes do not interact with macromolecules in detrimental fashion. Thus, they can be safely upregulated and downregulated, with little impact on cellular functions. Some osmolytes, such as glycerol or proline, may also function in maintaining redox states rather than have stabilizing properties. When these compounds accumulate during cold stress they are also termed "cryoprotectants." In contrast to the amino acids, sugars and glycerol, however, glycine betaine and related compounds are also considered to be secondary metabolites.

Leaf anthocyanins could also be involved in the osmotic adaptation and induce cold tolerance. However, there is only correlative evidence for this assumption, because cold- and drought-resistant plants have higher levels of anthocyanins in their leaves.

Additional secondary metabolites are induced under drought stress (Table 2.6). Many of these compounds, such as flavonoids or rosmarinic acid, have antioxidative properties, so that their protective effect is most likely by way of radical scavenging. This might be also true for these compounds when induced under oxygen depletion.

Finally, salt stress can increase those compounds also involved in the drought stress response; these are mainly phenolic compounds such as anthocyanins and flavonoids, based on their antioxidative properties. Transgenic tomato plants that accumulate zeaxanthine are better protected against salt stress induced by photoinhibition and photooxidation. The accumulation of zeaxanthine can induce tolerance to salt stress under these conditions, because of the antioxidative role of the metabolite in the scavenging of singlet oxygen or the prevention of lipid peroxidation. As a similar situation was observed after chilling stress at low irradiation, these findings place the xanthophyll cycle at the crossroads between several different stresses.

2.3.5
Nutrient Deficiency

In general, nutrient deficiency elicits nonuniform responses in terms of secondary metabolite synthesis (Table 2.9). While good N-nutrition is advantageous for alkaloid biosynthesis (because these metabolites contain nitrogen), N-deficiency has wider implications on different metabolite profiles. Phenol derivatives as well as glucosinolates accumulate after nitrogen depletion; phenols also accumulate under sulfur, phosphorus, and potassium shortages. Alkaloid production is also increased under low potassium conditions, but their synthesis is inhibited under nitrogen depletion; the latter effect is due again to the nitrogen content in the

Table 2.9 Effect of nutrition and water supply on the accumulation of secondary metabolites.

Secondary metabolite	Optimum supply	Deficiency				
	N	N	P	K^+	S	H_2O
Terpenoids:						
- Herbaceous plants		±	+	±		+
- Trees		--				--
Phenol derivatives		+	+	+	+	±
Alkaloids	+	--	±	+		+
Glucosinolates		+			--	+
Cyanogenic glycosides		--				+

N = nitrogen; P = phosphorus; S = sulfur; K^+ = potassium. "+" means induction compared to normal conditions, "--" reduction and "±" no specific trend found in different studies. Modified from: Larcher (1995) Physiological Plant Ecology, third Edition, Springer.

structures. Similarly, the concentration of N-containing cyanogenic glycosides is also decreased under nitrogen starvation. Finally, sulfur depletion negatively regulates the biosynthesis of S-containing glucosinolates (Figure 1.19).

Anthocyanins are induced under nutrient deficiency; this occurs especially when, under agricultural conditions, the nutrients in the soil are reduced and replaced by fertilizer. This is true for the depletion of nitrogen, phosphorus and potassium. Nitrogen starvation also leads to reductions in photosynthesis, and thus to the accumulation of ROS; under these circumstances, anthocyanins may function as antioxidants in leaves under nutrient deficiency. In contrast, sugars can enhance light signaling-induced anthocyanin accumulation in *Arabidopsis* seedlings by the differential regulation of several transcription factors. This would imply that low levels of sugars have a negative influence on anthocyanin accumulation under nutrient-stress conditions.

2.3.6 Environmental Pollution

2.3.6.1 Heavy Metal Stress

It should be noted that secondary metabolites also accumulate under heavy metal stress. It was hypothesized that flavonoids were involved in the heavy metal tolerance of *Arabidopsis*, because flavonoid-free *Arabidopsis* mutants were less tolerant to the heavy metals cadmium and zinc. This was demonstrated by a reduction of growth in comparison to the wild-type under these conditions. Likewise, transcript levels for the phytochelatin genes were less strongly induced in mutant leaves than in wild-type leaves. Phytochelatins are able to complex the heavy metals, and the complexes created are subsequently sequestered into the vacuole for detoxification. As anthocyanins can also form complexes with metal ions (see Figure 2.13), their accumulation under these stress conditions might be directly related to the detoxification of heavy metals. The latter complexes can also be stored in vacuoles.

2.3.6.2 Air Pollution

During a period of serious air pollution, plants were analyzed for their secondary metabolite patterns to identify clues regarding possible resistance mechanisms. In these studies, terpenes (notably monoterpenes) were especially in focus because they are emitted by gymnosperm trees, which were highly affected by air pollution. One example, which is typical for this type of analysis, showed for spruce (*Picea abies*) a dramatic increase in monoterpene hydrocarbon levels in injured needles when compared to healthy needles of the same internode from the same tree. A subsequent microanalysis of the distribution of volatile terpenes in different tissues of the needle showed clearly that all monoterpenes and sesquiterpene hydrocarbons, regardless of their chemical structure, are stored exclusively in the resin ducts, and most probably in the resin itself. As the damaged and undamaged parts of the same needle showed different monoterpene emission patterns, it was suggested each separate resin duct in the needle had reacted individually to the air pollution. A possible function for monoterpenes was deduced from the observation that they are highly reactive with $OH^{\bullet}$ and $NO_3^{\bullet}$ radicals and with ozone, and so it was assumed that they have a protective role against photooxidative stress (Figure 2.19). In particular, ozone causes more damage to plants than other pollutants, as it is taken up via the stomata of the plant and can then react in the cell with oxygen species to form ROS. If ozone is quenched before it passes into the intercellular space of a leaf, this should lead to a better resistance of the plant to this gaseous pollutant. In a process termed "terpene ozonolysis," α-pinene is for example converted to products such as pinonic acid, pinic acid, noropinonic acid, pinalic acid and other derivatives, by reacting directly with ozone. This reaction may lead to a reduction in ozone levels around the plant.

Isoprene, the simplest terpene structure, is emitted by plants in considerable amounts. It has been proposed that, under abiotic stress conditions, isoprene stabilizes the cell and particularly the chloroplast thylakoid membranes. The reduction in (chloroplast) damage showed that clones of plants sensitive to ozone but unable to emit isoprene would become resistant to acute and short exposure to ozone if they were fumigated with exogenous isoprene. Isoprene-induced ozone resistance is associated with the maintenance of photochemical efficiency as an indication for its role as a thylakoid membrane-stabilizing agent. In addition, when isoprene reacts with ozone within the leaves or in a humid atmosphere, it quenches the ozone concentration to levels that are less toxic or even nontoxic for plants, indicating a second mechanism by which isoprene can protect the plant. Irrespective of the mechanism employed, isoprene is one of the most effective antioxidants in plants (Section 2.3.1).

2.4 Symbioses

Many organisms are dependent on each other during their lifetime in various grades of interaction. The situation in which only one side gains an advantage is termed parasitism (Section 2.5), whereas when both sides gain advantages it is termed mutualism or symbiosis. The term symbiosis is used especially in cases where the two organisms form a close companionship. Symbioses are formed by plants with bacteria and fungi, where the interaction may even lead to the creation of new organs, and this is the case when bacteria from the genus *Rhizobia* form root nodules (Section 2.4.1). However, fungi can also interact with the plant and form distinct structures within the so-called mycorrhiza symbioses (Section 2.4.2). In these interactions, both partners can also gain advantages, but in these cases the homeostasis can be upset more easily by the environmental conditions. For example, a high nutrient availability will reduce mycorrhization, but this symbiosis is favored under reduced nutrient availability. Further advantages between endophytic fungi, such as mycorrhiza, and the host plant may lie in growth promotion, as with the endophytic basidiomycete *Piriformospora indica*. In addition, a higher stress tolerance is induced after colonization with selected *Acremonium* species, which are also defined as endophytes (Section 2.4.3).

Other symbioses include the interaction of fungi and cyanobacteria as photobionts in lichens (Section 2.4.4), and also the water fern *Azolla*, which has recruited nitrogen-fixing cyanobacteria of the genus *Anabaena* into its leaf cavity. Symbioses can be also formed between carnivorous plants (e.g., pitcher plants) and the bacterial communities that live inside the pitcher. The role of the bacteria is assumed to help in digesting prey that has fallen into the pitcher, by secreting digesting enzymes; these are in addition to the enzymes produced by the plant. In one case, mutualism between a zoobiont (an ant) and a photobiont (a plant) resulted in: (i) a better defense against herbivores; (ii) the pollination of flowers; (iii) seed dispersal; and (iv) the acquisition of nutrients such as potassium, phosphorus, and nitrogen.

2.4.1 Rhizobia

Flavonoids play a major role as signaling molecules in the recognition phase of the interaction between rhizobia and host plants of the Leguminosae family (Figure 2.21). These compounds not only contribute to the host specificity but also induce bacterial gene expression. It is interesting to note that these flavonoids can be toxic to other bacterial species, and that consequently the rhizobia must have developed mechanisms to either detoxify or exclude these plant metabolites (see Section 3.1).

With regards to specificity in the interaction with different microbes, the plant can synthesize different flavonoids in different tissues (Figure 2.21). For example, in the leaves of alfalfa (*Medicago sativa*) medicarpin is mainly found, which can act as a phytoalexin in plant defense, whereas in the roots the flavonoids luteolin and apigenin are synthesized. The latter are involved in the recognition between rhizobia and host root. Other flavonoids, which also determine specificity in the root nodule formation are naringenin, daidzein, and genistein; the latter two compounds are isoflavonoids and have a restricted occurrence in the plant kingdom. They are mainly found in legumes, which are the host plants of rhizobia. Isoflavonoids are colorless compounds and are synthesized, like all other flavonoids, from chalcone as a precursor (Figure 1.8). As flavanones are intermediates of this pathway, the enzyme-catalyzed migration of the B-ring must occur after the C15-skeleton has been synthesized. Interestingly, even

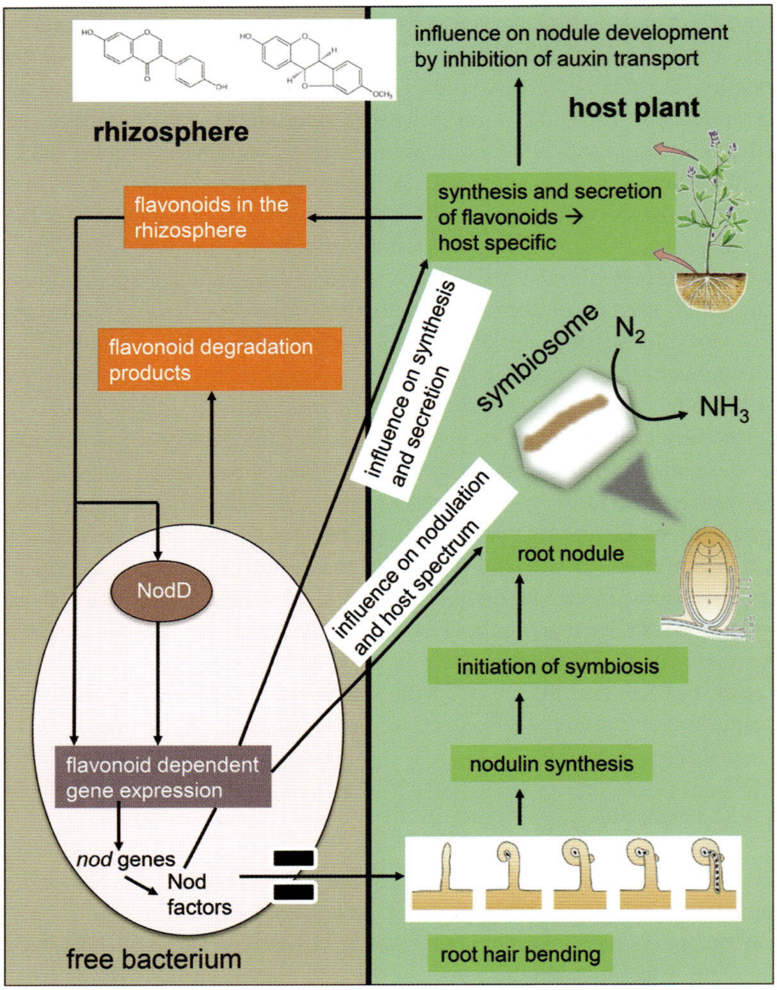

Figure 2.21 Involvement of flavonoids in the signal exchange and later in the establishment of the symbiosis of rhizobia with leguminous plants. The host plant attracts the free-living bacteria by secretion of flavonoids, which are then specifically recognized prior to colonization. Daidzeine and medicarpine are shown as examples of host plant-specific flavonoids. Upon signal molecule secretion by the plant and recognition by the bacteria, the latter react by the expression of specific genes, which leads ultimately to the establishment of the symbiosis with the formation of root nodules. The bacteria colonize via root hairs and then form a symbiosome, where molecular nitrogen is fixed. Chemical structures are from The PubChem Project.

plants such as *Arabidopsis* – which do not contain isoflavonoids – possess isoflavonoid synthase-like genes for which a function has not yet been assigned.

Flavonoids are secreted into the rhizosphere to attract the rhizobia. Here, they are recognized by specific bacterial receptors, which leads to a flavonoid-dependent gene expression in the free-living bacteria (Figure 2.21). The *nod* genes, which are induced by flavonoids, encode proteins important for colonization. The Nod factors are secreted into the plant by type III-secretion systems and cause an initiation of the symbiosis by induction of the plant *nodulin* genes. In the plant, the first response is a curvature of the root hairs, which allows colonization by the bacteria. Later, upon induction of the *nodulin* genes in the plant, a symbiosis is established during which the bacteria are no longer free-living organisms but occur in form of symbiosomes (specialized structures in the root). During this phase the fixation of molecular nitrogen by the symbionts takes place, leading to a better N-nutrition of the plant, while the bacteria obtain carbon, mostly as C4-carbonic acids, in return. The nitrogenase reaction is also sensitive to oxygen, so within the root nodules the plant provides a more or less oxygen-free environment. To achieve this, the plant-derived leghemoglobin plays an important role, which also renders the nodules red. Leghemoglobin, a protein with a similar structure to human hemoglobin, is able to bind O_2 very effectively in the nodules.

Flavonoids are not only recognition signals, but are also thought to be involved in the formation of the root nodule structure itself. Flavonoids accumulate at the entrance site of the rhizobia, and as they inhibit auxin efflux (Section 2.1.1; Figure 2.3) from the cells it is hypothesized that high levels of flavonoids can induce a local increase in the growth-promoting plant hormone auxin, which in turn results in a stimulation of cell proliferation. The latter effect is necessary for the formation of a novel structure at the root, the nodule. Root nodules are thought to be novel organs, much as lateral roots, and their initiation requires a plant-derived signal that could be auxin. Consequently, plants without flavonoid accumulation do not form nodule structures, even though their colonization by bacteria is permitted. Such flavonoid-deficient plants also show reduced auxin transport rates and less auxin accumulation at the infection sites with rhizobia.

2.4.2 Mycorrhiza

The interaction between different fungi and their host plants is known as Mycorrhiza (from the Greek "mykes" = fungus; "rhiza" = root), and it is assumed that more than 80% of all land plants have at least one type of mycorrhizal interaction. Mycorrhizal fungi can be grouped into the ectomycorrhiza and endomycorrhiza; the latter group is constituted mainly by arbuscular mycorrhizal fungi, which have a broader host spectrum and are evolutionarily older than the ectomycorrhiza. Ectomycorrhizal fungi evolved later and show a somewhat restricted host spectrum. However, both groups are in close contact with their hosts, and can exchange nutrients. The fungi facilitate the uptake of certain nutrients, such as phosphorus and nitrogen, in exchange for carbon, and also confer a better tolerance to abiotic and biotic stress situations.

Although, in general, it is advantageous for a plant to be colonized by mycorrhizal fungi, under certain environmental conditions – that is, if the nutritional supply is sufficient – the plant may prevent colonization. In view of these findings the question arises why some plant families, such as the Brassicaceae and Chenopodiaceae, generally do not form mycorrhizal associations. For the *Brassica* species, the glucosinolates as antimicrobial compounds were thought to trigger the inhibition of mycorrhizal colonization. Indeed, some glucosinolates and their degradation products were able to inhibit spore germination and hyphal growth. Investigations of several glucosinolate-containing species have shown that some of them can form arbuscular mycorrhizal associations. For example, nasturtium (*Tropaeolum majus*), a species with rather high benzylglucosinolate levels, can form functional arbuscular mycorrhiza with *Glomus* species. On the other hand, garden cress (*Lepidium sativum*) with a high phenylethylglucosinolate concentration, does not form arbuscular mycorrhizal associations. Thus, it was concluded that is not the presence of glucosinolates alone but rather their specific pattern that determines the outcome of a colonization of glucosinolate-containing plants with arbuscular mycorrhizal fungi. However, for Chenopodiaceae, no explanation has yet been found for the fact that they are not colonized by arbuscular mycorrhizal fungi.

Mycorrhizal fungi can influence the habitus of a host plant by changing their root structure. This is made possible by the synthesis of growth-promoting compounds such as the indole alkaloid hypaphorine by an ectomycorrhizal fungus (Figure 2.22), which is secreted into the rhizosphere. The indole alkaloid might mimic the endogenous auxin IAA by its structural similarity (Table 2.3). Larger root systems with more lateral roots provide more colonization sites for the hyphae.

As a symbiosis has great advantages for the plant, the recruitment of different partners is very important, and this implies a "first-come-first-serve" strategy for

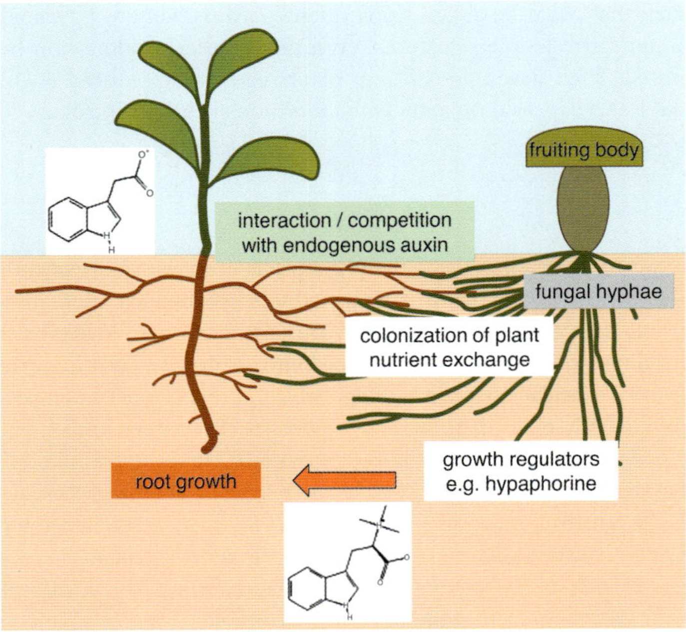

Figure 2.22 Ectomycorrhizal fungi synthesize and secrete growth-promoting indole alkaloids such as hypaphorine. This could lead to better colonization conditions for the fungus, and thus also to advantages for the host plant. The growth promotion could be achieved by imitation of the growth hormone auxin by structural similarity. Chemical structures are from The PubChem Project.

the microbes if the plant cannot regulate such species-specific preferences. Therefore, a type of regulation must exist that determines whether a plant can be colonized by one or by several species. Arbuscular mycorrhizal fungi react to flavonoids secreted by the roots of hosts by increasing both their spore germination rates and hyphal branching, both of which both either increase or facilitate colonization. After colonization, the host plant exudes more flavonoids which also accumulate within the roots, and this mechanism causes alterations in the composition of the plant-derived chemicals in the rhizosphere. If, among these chemicals, flavonoids are also secreted these can be sensed by rhizobia and this would result in colonization of the same plant (Figure 2.23). As arbuscular mycorrhiza and rhizobia can recognize the same group of plant-derived signals, they can influence each other during colonization. However, the accumulation of

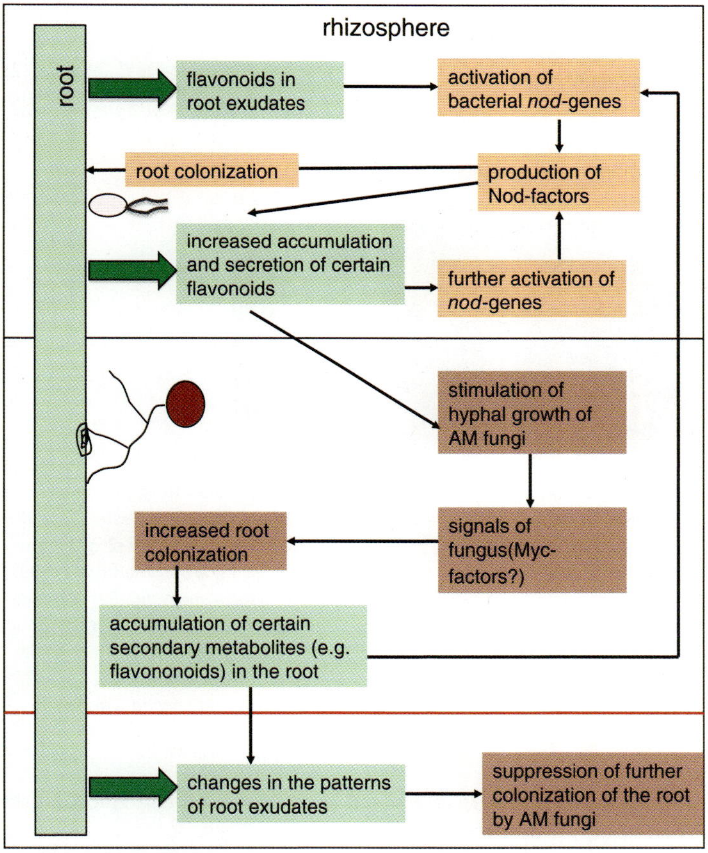

Figure 2.23 Flavonoids as common signals in the arbuscular mycorrhiza and rhizobia symbiosis. The secretion of secondary plant metabolites into the rhizosphere influences the behavior of microorganisms. Common signals prevent further colonization, which is termed autoinhibition. In another case, colonization by a second symbiont is not inhibited. An arbuscular mycorrhizal (AM) fungal signal (Myc-factor) is postulated in analogy to the Nod-factors of rhizobia.

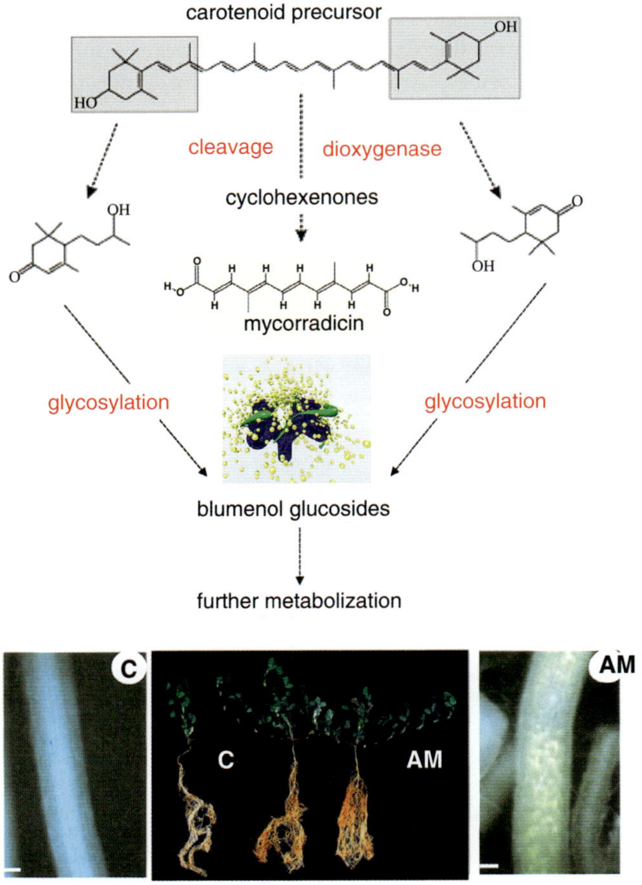

Figure 2.24 Synthesis and compartmentation of mycorradicin (highlighted in yellow) and derivatives in roots colonized by arbuscular mycorrhizal (AM) fungi. The pictures show in the center control (C) and inoculated (yellow) roots of medic barrel (*Medicago truncatula*), on the left and right microscopic views of control and infected roots, respectively. Microscopic image from Klingner, A. *et al.* (1995) *Protoplasma*, **185**, 50–57 (see Further reading). Reprinted with permission from Springer. Biosynthetic pathway based on Fester, T. *et al.* (2002) *Plant Cell Physiol.*, **43**, 256–265. Picture model of decomposing arbuscule from Fester, T. *et al.* (2007) *Phytochemistry*, **68**, 92–100 (see Further Reading). Reprinted with permission from Elsevier. Photo of plant roots from Walter, M.H. *et al.* (2007) *Phytochemistry*, **68**, 130–138 (see Further Reading). Reprinted with permission from Elsevier.

plant-derived compounds in the rhizosphere can prevent subsequent colonization by other mycorrhizal fungi, a process termed "autoregulation."

One prominent secondary metabolite which accumulates during arbuscular mycorrhizal symbiosis in the roots of some plant species is the so-called "yellow pigment," mycorradicin. This pigment, which is visible when the root shows high colonization rates (Figure 2.24), was identified as an apocarotenoid mixture occurring in colonized roots, where the individual cells appear yellow. Specifically, the apocarotenoids seem to accumulate in cells harboring arbuscules. Experiments to determine the biosynthetic pathway of C13-cyclohexenones and C14-mycorradicin have shown that these compounds are synthesized from a carotenoid precursor in the root plastids, but the compounds are also transported into the cytosol. Manipulation of the biosynthetic pathway to mycorradicin has shown that the metabolite is involved in the determination of the turnover of arbuscules. The reduction of apocarotenoids resulted in more apoptotic arbuscules, though the precise mechanism of this phenomenon is unclear. As these structures are responsible for the nutrient exchange between plant and fungi, their life-time is a crucial factor for the functionality of this symbiosis.

Other isoprenoids are also modulated by the colonization of plants by arbuscular mycorrhiza. In this case, a shift from nonessential to essential terpenes (see Figure 1.13) can be observed, especially under concomitant drought stress. Mainly, the synthesis of volatile terpenes is reduced under these conditions, but this might offer the possibility of carbon availability to the plant being altered under stress conditions.

2.4.3 Endophytic Fungi

The endophytic fungus *Piriformospora indica* has a broad host range and, in contrast to arbuscular mycorrhizal fungi, this organism can also colonize the roots of Brassicaceae. Therefore, the colonization of *Arabidopsis* is a good model to understand the interaction with glucosinolate-containing plant species. This fungus induces a short and transient defense reaction in the host plant after colonization, which has been also observed for arbuscular mycorrhizal fungi. During later stages of the interaction, the defense reactions are suppressed, most likely by the host plant itself; this suggests that the homeostasis between a fungal partner and a host plant is essential for a functional symbiosis. The fungus *P. indica* induces growth of the host plant and also resistance against abiotic and biotic stressors. Indole glucosinolates and the phytoalexin camalexin are important factors for the regulation of homeostasis between the plant and fungus. These compounds are induced after inoculation with *P. indica*, and mutants in the first step of the biosynthetic pathway to both metabolites show higher colonization rates. In addition, these plants were not promoted in growth of the above-ground tissues. In contrast, the mutant plants showed a reduction in their growth which was accompanied by a faster and stronger development of the fungus in the plant roots. The expression of defense genes was also reduced in these indole glucosinolate/camalexin mutants after colonization with *P. indica*, showing that the defense gene induction can be mediated by secondary metabolites.

2.4.4 Lichens

Lichens are defined by the association of a fungus (mycobiont) and an alga or cyanobacterium (photobiont), both of which form a specific structure, the thallus. The mycobiont is mainly an ascomycete, and only a few lichens are formed with basidiomycetes as partners. The new organism is able to synthesize a different metabolite spectrum than the individual parents. In contrast to rhizobial symbiosis, specific recognition mechanisms have not yet been observed, despite the synthesis of a variety of secondary metabolites. The typical secondary metabolites, termed "lichen acids" possess either antimicrobial potential to protect either the mycobiont, or the lichen, against herbivores (Section 2.5.3). They also assert an allelopathic activity, because mosses and other lichens can be inhibited in growth (see also Section 2.6). These compounds are rather specific for a given lichen species, but are derived from a few structural similar compounds such as two polysubstituted phenol or phenol-carboxylic acid residues bound to each other in a variety of combinations. They are created from acetate moieties and derive from depside/depsidone (depsides are condensation products of aromatic hydroxycarbonic acids), dibenzofurane or 1,4-diphenyl butadiene (e.g., vulpinic acid; Figure 2.39). Other secondary metabolites not related to lichen acids are induced under desiccation or presence of the mycobiont; they are also needed to protect the photobiont against UV irradiation, for example carotenoid derivatives (Section 2.3).

2.5
Defense

Plants are confronted with a plethora of parasites during their lifetime, among which are viruses, bacteria, protists, oomycetes, fungi, and many herbivorous organisms. Therefore, the plant defense mechanisms need to encompass many strategies against these different organisms. Such strategies range from structural means to chemical compounds and also antifungal proteins, but here attention will be focused on secondary metabolites as defense compounds. Plants which cope successfully with an invading organism are termed "resistant," while those that can be either colonized by pathogens or eaten by herbivores are termed "susceptible". A tolerant plant can deal with a pathogen over a longer time period, but is not able to completely fight off this parasitic organism. In plant pathology, the resistant interaction may be referred to as "incompatible," and the susceptible interaction as "compatible."

After pathogen invasion, many plant species synthesize a range of secondary metabolites that are involved either directly or indirectly in defense reactions. For example, *Arabidopsis* typically synthesizes anthocyanins after biotic attack, as already described under abiotic stress conditions (Section 2.3), although the role of these compounds in plant defense is not entirely clear. In contrast, many secondary metabolites are known to be induced that are directly involved in defense reactions. These can be grouped mainly into two groups: (i) preformed compounds or phytoanticipins, which are already present in the plant in large amounts before invasion; and (ii) inducible phytoalexins, which are present in detectable amounts only after a defense reaction against pathogens has been induced. Whilst both groups of compounds can be generated in response to a

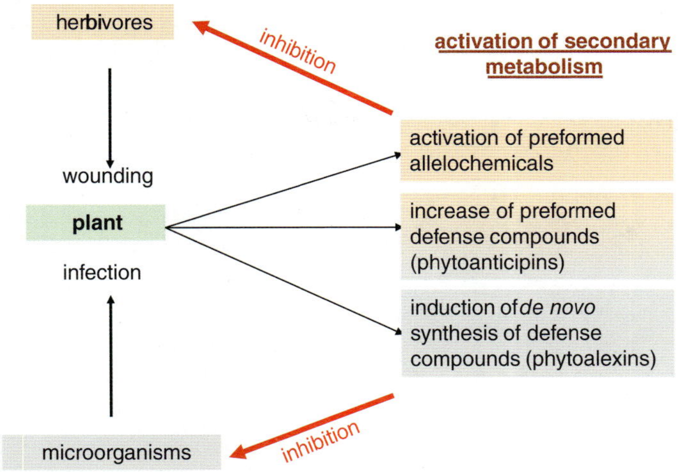

Figure 2.25 Examples of induced defense mechanisms in plants by secondary metabolites. The activation of secondary metabolite pathways leads to the induction and activation of phytoanticipins prior to infections in preparation to a potential threat, whereas damage done by bacterial and fungal pathogens induces the synthesis of phytoalexins.

variety of parasites, the phytoanticipins are quite often useful in defense against herbivores and the antimicrobial phytoalexins are highly effective against bacterial and fungal pathogens (Figure 2.25). Induced defense responses also involve the synthesis of polymers, which form barriers against invaders, such as callose against fungi or the systemic spread of viruses if inserted in sieve plates. Phytoanticipins are often found in specialized cells or subcellular compartments, such as terpene derivatives in trichomes or glucosinolates and cyanogenic glycosides in vacuoles; however, these compounds are also inducible after pathogen or insect attack. In addition, many have to be converted to their active principles in an enzymatic reaction (Figure 1.12).

All structural groups of secondary metabolites can be found in the defense response of plants (Figure 2.26). Photosynthesis generates all precursors required for the various pathways, notably carbohydrates and lipids, but the assimilation of nitrogen and sulfur is also important. The resulting substances encompass C-containing tannins, coumarins, quinones and flavonoids, and the N-containing glucosinolates, cyanogenic glycosides and alkaloids. The terpenes and polyketides are derived from lipids.

In addition, polymers that form physical barriers (e.g., lignins) derive from secondary metabolite pathways. Monophenols serve as precursors for lignin synthesis, but are also involved directly in plant defense as antimicrobial compounds. Lignin can be seen as a preformed barrier, but also as an induced defense reaction following contact of the plants with pathogens.

Some explanations of the function and targets of secondary metabolites in plant defense are provided in the following sections. Although specific molecular

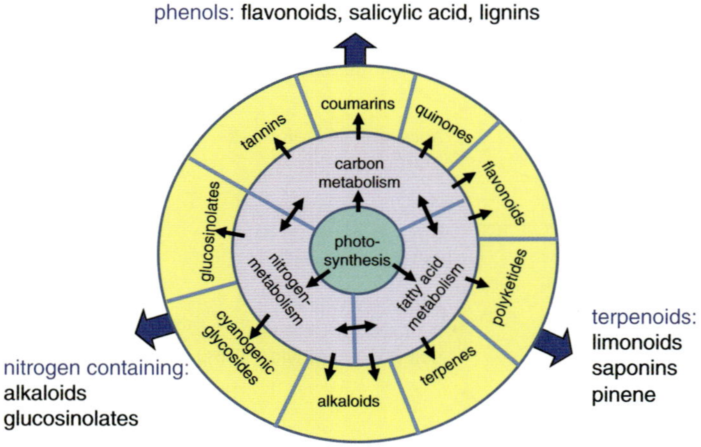

Figure 2.26 Many plant defense compounds are secondary metabolites. Some examples of defense compounds are shown here. The substances can be divided into three large groups: phenols; terpenes; and N-containing. From *Teaching Tools in Plant Biology* (TTPB18). Reprinted with permission from the American Society of Plant Biologists.

functions have not yet been elucidated for all metabolites, some mechanisms can be deduced from experiments using animal models (see Section 3.2) or non-plant pathogenic microbes.

2.5.1 Pathogens

In order to provide a better understanding of the role of secondary metabolites in plant defense, a short overview of the induction of general defense mechanisms will be provided here (Figure 2.27). If a pathogen is not recognized by the plant, it can colonize the host without problems and will lead to typical disease symptoms. If the pathogen is recognized by specific plant receptors (see also Figure 3.2), through either general cell wall or protein structures (elicitors or pathogen-associated microbial patterns; PAMPs) or by specific avirulence factors, then a plant defense reaction is initiated. Historically, the term "avirulence" is derived from studies to determine the interaction between a resistance gene from a plant and an avirulence protein from a pathogen (gene-for-gene interaction). In fact, the avirulence factors act as suppressors of plant defense (Figure 3.2), and the cascade often starts with the synthesis of ROS; this may lead to a hypersensitive reaction, including the apoptosis of only a few cells. This is followed or accompanied by the synthesis of secondary metabolites as antimicrobial compounds (phytoalexins), or parts of barriers against pathogens, such as cell wall polymers or callose. A preformed defense reaction is also involved in these responses against pathogens, often only after the activation of these compounds from inactive storage forms. Phytoalexins with different chemical structures (see Table 2.11) are synthesized in

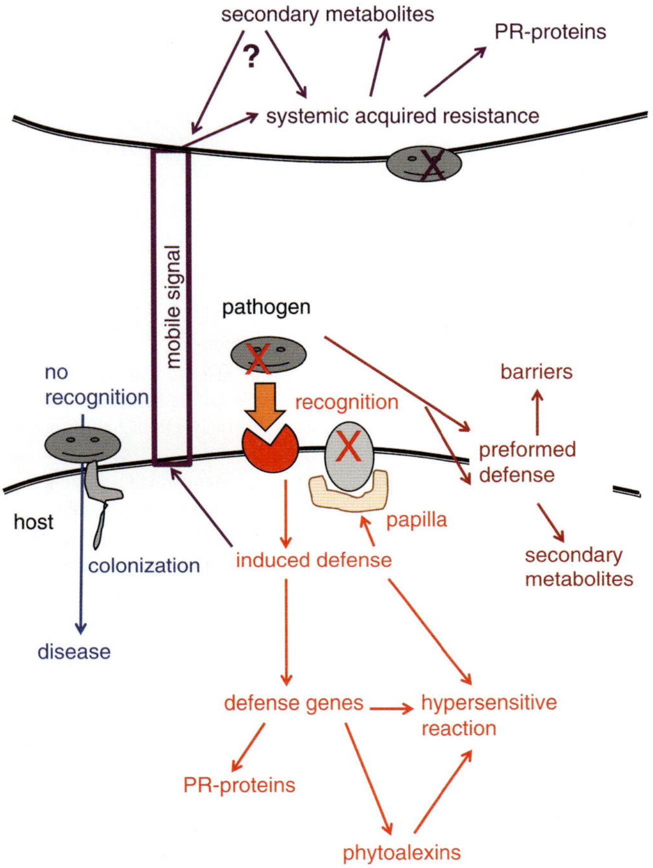

Figure 2.27 Events leading to a resistance response in plants after interaction with pathogens such as bacteria and fungi. The preformed response is indicated in brown, the induced defense in red. Systemic acquired resistance, the resistance response induced in plant parts above the infected tissues, is shown in violet. PR = pathogenesis-related. A red cross symbolizes the restriction of a pathogen at the indicated site. Based on Elstner, E.F. *et al.* (1996) *Phytopathologie*. Spektrum Akademischer Verlag.

detectable amounts only after pathogen recognition. Both groups of compounds can act together in plant defense responses, either at the same time or sequentially. The so-called pathogenesis-related (PR) proteins are directly involved in the plant defense responses; examples include the antimicrobial hydrolases chitinase and β-1,3-glucanase, which are able to digest fungal cell walls. Secondary metabolites also play a role in the establishment of the systemic responses against pathogens in neighboring tissues, for example systemic acquired resistance (SAR; see below). SAR is a resistance response induced in plant parts above the infected tissues by plant signals.

2.5.1.1 Preformed Defense Response

The outcome of a defense reaction is heavily dependent on the secondary metabolites present, either as preformed compounds or as an inducible defense (Figures 2.25 and 2.27). The phytoanticipins are already present in considerable amounts in the uninfected plant and can be liberated and exert their antimicrobial potential directly; alternatively, they may need to be converted to a toxic compound by an enzymatic or spontaneous reaction. The latter case is true for indole derivatives such as the benzoxazinoides, and also the glucosinolates and cyanogenic glycosides, which liberate cyanide (Table 2.10). Cyanide is an inhibitor of the mitochondrial respiration chain, and therefore is highly toxic to the plant itself.

Table 2.10 Examples of preformed defense compounds (phytoanticipins). The inactive and active compounds are shown; an example of a plant is given in which the substances can be found and a pathogen against which they play a role in defense. Chemical structures are from The PubChem Project.

Preformed compound	Active antimicrobial compound	Plant which possesses this compound	Pathogen against which this compound is active
Tuliposide B 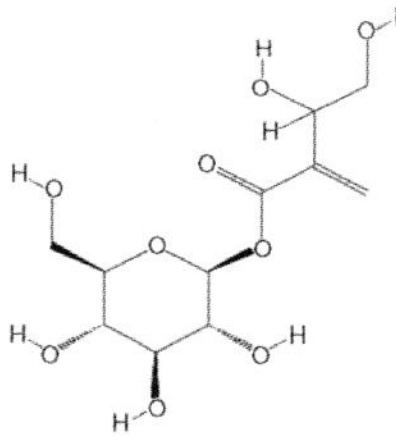	Tulipaline B 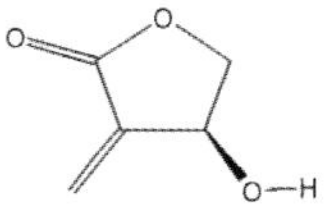	Tulip	*Botrytis cinerea*
Cyanogenic glycosides (example: amygdalin) 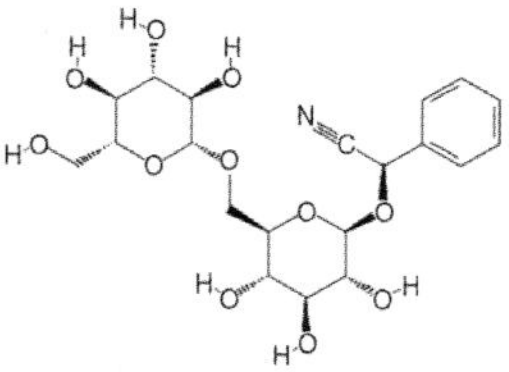	Cyanide N≡C-	Rosaceae	Generalists
Glucosinolates (example: sinigrin) 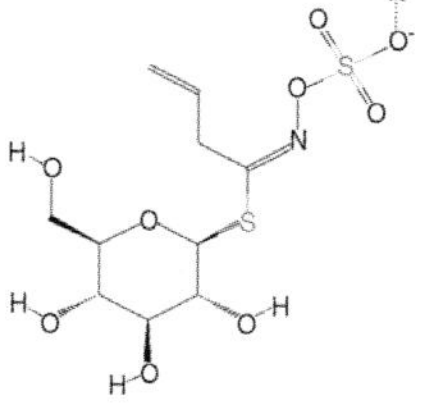	Isothiocyanate (example: allyl isothiocyanate) 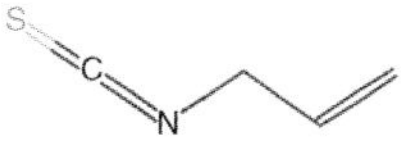	Brassicaceae	Generalists

Benzoxazinoide glycoside (see Figure 2.29)	Benzoxazine (see Figures 2.29 and 3.12)	Maize	Generalists
Phlorizidin 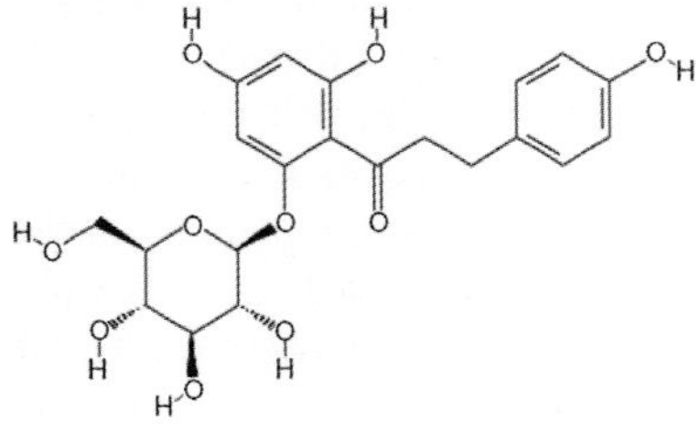	Phloretin 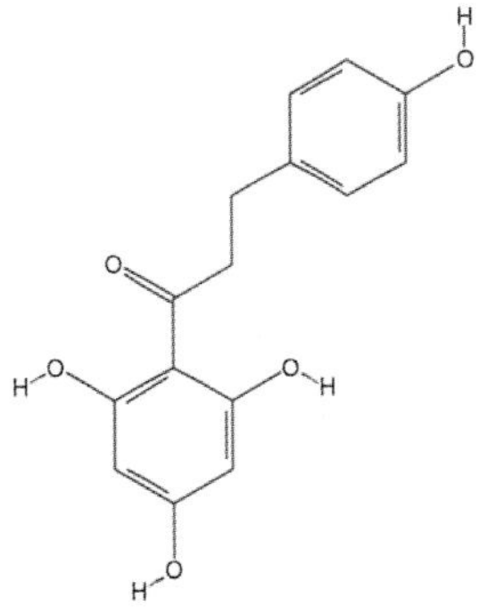 o-Quinone 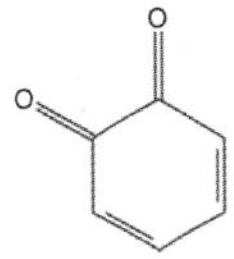	Apple	*Venturia inaequalis*
Arbutin 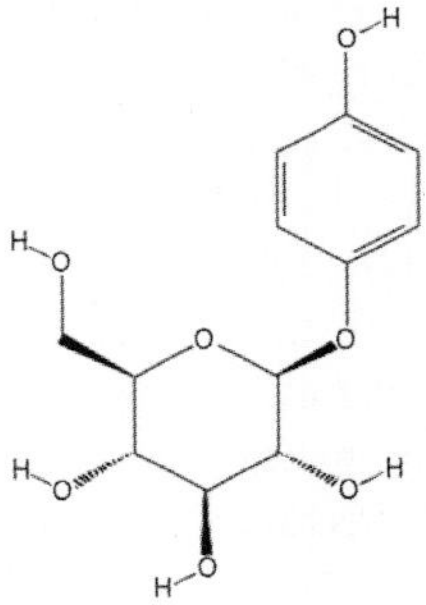	p-Hydroquinone 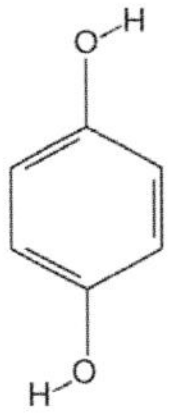 p-Quinone 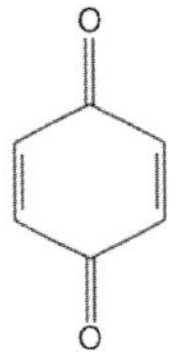	Pear	*Venturia pirina*
Alliin (see Figure 2.39)	Allicin 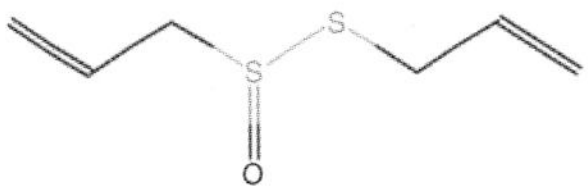 Diallylsulfide 	Onion species	Fungi, bacteria *Phytophthora cactorum*

Consequently, release of the toxic compound must be controlled only after pathogen or herbivore attack, and an effective detoxification mechanism must also be present in the plant (see Figure 1.12). The sulfur-containing compounds of garlic and onion (e.g., alliin) must also be converted enzymatically to the active antimicrobial compounds, for example allicin and diallylsulfide. Phytoanticipins often occur in specialized cells or organelles within the plant, and many are directed towards the surface of a plant tissue. The biologically active compounds can be found in the trichomes, laticifers, and also simply in the outer cell layers. Within the cells they are typically present in vacuoles which may be decomposing after pathogen attack and thus liberating the compounds. Proanthocyanins also belong to the preformed defense compounds, and are mostly located in the apoplast of a cell so that their antimicrobial function is directed towards defense from the outside. Their role has been demonstrated for poplar plants overexpressing a gene for leucoanthocyanidine reductase, which plays an important role in the synthesis of proanthocyanins (Figure 1.8). The transgenic plants were more resistant against the fungus *Marssonina brunnea*. Proanthocyanidins, which are located in seed heads, have antimicrobial activity and can be used to protect the seeds.

Often, the inactive forms of phytoanticipins are the glycosylated versions (Table 2.10), where the sugar moiety is cleaved by β-glucosidases to yield the active compound. However, for the majority of compounds, such as the benzoxazines, deglycosylation is the step that leads to the active compound whereas for glucosinolates an additional step is necessary (see Figure 1.12). This reaction is either protein-dependent, or is a non-enzymatically catalyzed chemical reaction of the aglycone to the toxic compound(s). Detoxification of the active benzoxazines is possible by the addition of a glucosyl moiety via UDP-glucose glycosyltransferases (Figure 2.29; see Section 3.1).

Saponins are glycosylated, preformed defense compounds that are widely distributed in the plant kingdom and can be divided into three major structural groups: triterpenoid saponins, which also act in insect defense (Section 2.5.3); steroids; and steroidal glycoalkaloids. Triterpenoid saponins are found primarily in dicotyledonous plants, whereas steroid saponins occur mainly in monocots. Oats (*Avena* sp.) are unusual because they contain both triterpenoid and steroid saponins. Steroidal glycoalkaloids are found primarily in members of the family Solanaceae. The preformed defense compound against the fungus *Gaeumannomyces graminis*, the causal agent of the "take all disease" of wheat, is the saponin avenacin A, which is synthesized in oat, but not wheat. Other saponins occur in different plant species and vary slightly in their structure. Some saponins from Solanaceae have a nitrogen-containing heterocycle and could therefore also be classified as alkaloids. Examples include the antifungal compounds α-tomatin or α-chaconin in tomato and potato, respectively. The levels of tomatin are particularly high in the leaves, flowers and green fruits of tomato, and these compounds can be degraded by resistant pathogens such as *Botrytis*, which cause gray mold of tomato. The pathogens possess a glucosidase specific for the side chain of saponin of the host plant. *In-vitro* experiments have indicated that the fungal pathogens of tomato are considerably more tolerant to the compound than are the nonpathogens. If the compound cannot be degraded by the pathogen, then the plant will show resistance.

The triterpenoid saponin of oat, avenacin A, can be found in the outer layers of the root cortex and also the lateral roots initials, so that the fungus comes into contact soon after entering the root system. This leads to resistance of the plant, because the phytoanticipin can integrate into the membranes of the fungus (see below). As wheat plants cannot synthesize this saponin, they are therefore subject to fungal colonization by *G. graminis*. However, there are some fungal strains which have developed a resistance against the saponin; these possess a glucosidase (avenacinase) that is specific for cleaving avenacin A, so that these strains can also colonize oat plants (Figure 2.28).

The function of saponins relates to their structural similarity to membrane sterols. In fungi, the major sterol is ergosterol, which is integrated into membranes to enhance the stability of the phospholipid bilayer. However, if a saponin such as avenacin A is integrated into the membrane, the membranes lose their functionality, which is harmful to the invading organism (Figure 2.28). The nonglycosylated form of the saponin resembles the fungal steroids and is not toxic. The detoxification mechanism functions generally by means of a specific glucosidase enzyme secreted by the fungus.

Other examples of phytoanticipins are the glucosinolates of Brassicaceae, which occur in the vacuoles of specialized cells and are thus compartmented against the enzyme that catalyzes the conversion to the active component. After decomposition of the tissue, which is often directly caused by a penetrating pathogen or chewing insect, the two components are brought together and the toxic (iso-)thiocyanates are formed (Figure 1.11). A detailed description of the glucosinolate–myrosinase system and its localization is provided in Section 1.1.2. The "mustard oil bomb" is mainly directed against generalist pathogens or herbivores, but examples also exist of an efficient protection against specialist pathogens. Recently, indole glucosinolates and a specific myrosinase called PEN2 were identified as components of a defense reaction against fungi (Figure 2.29). Metabolism of the compound 4-methoxy-indole-3-methylglucosinolate led to the formation of callose, which in turn prevented entrance of the fungal hyphae, but it did not result in the normally formed degradation products of indole-3-methylglucosinolate. Interestingly, the 4-methoxy glucosinolate was induced after pathogen attack. Indole-3-methylglucosinolate itself was converted to the isothiocyanate by the myrosinase PEN2. In the presence of glutathione or ascorbic acid, the reaction leads to the formation of indole-3-methylamine and raphanusamic acid, or to ascorbigen and thiocyanate. The thiocyanate can be detoxified by methylation. The 4-methoxy-indole-3-methylglucosinolate is not converted to the amine, so that detoxification by the pathway via thiocyanate is not possible for this indole glucosinolate. The reactions leading to callose formation, however, are not yet clear. It was also shown that not only glucosinolates but also their desulfo-derivatives can be metabolized to toxic compounds after attempts of colonization by phytopathogenic fungi. The desulfo-glucosinolates are actually intermediates in the biosynthesis pathway. Nitriles, which are nontoxic intermediates after degradation

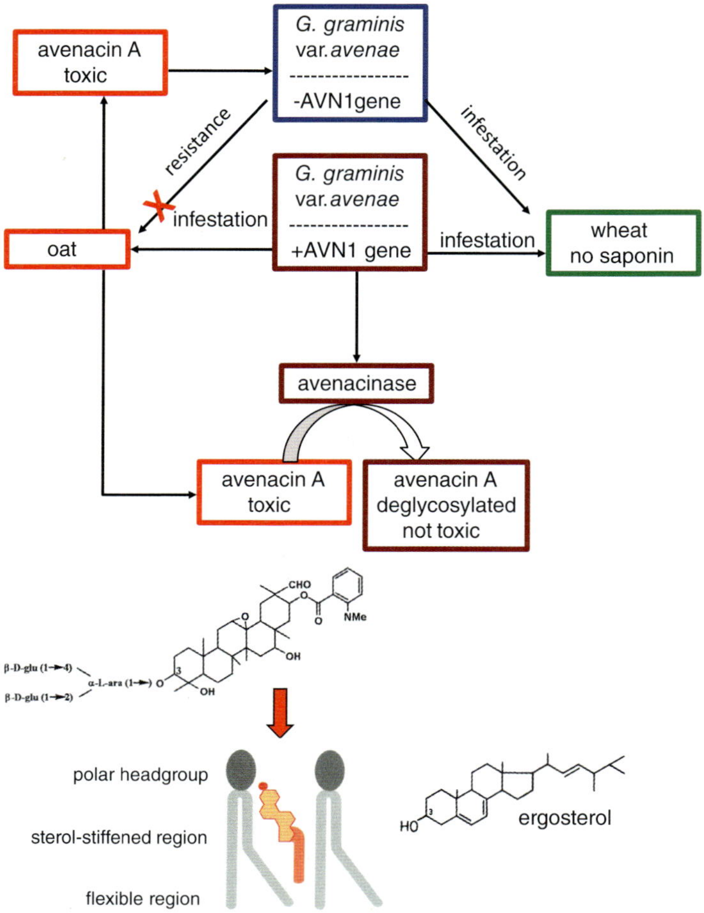

Figure 2.28 The saponin avenacin A is a preformed defense compound. The connection between the detoxification of avenacin A and pathogenicity of the fungus *Gaeumannomyces graminis* var. *avenae* against oat and wheat are shown. Plants, which do not synthesize saponins are susceptible to the fungus, whereas plants making avenacin A are resistant. However, fungal strains have developed an enzyme, avenacinase, which can cleave the side chain of the saponin, resulting in detoxification of the compound. These fungal strains can also colonize oat plants. The lower part of the figure shows the possible mechanism of integration of saponins in fungal membranes, as well as the structures of avenacin A and the main fungal sterol, ergosterol. Chemical structures are from The PubChem Project.

of glucosinolates, can also be converted to toxic chemicals. The mustard oils can bind to the SH-groups of proteins and inhibit their activities (see Section 3.2), and can also be integrated into membranes and increase their permeability; this results in a growth-inhibitory effect on the microbes.

A sequential action of the phytoanticipins indole glucosinolates (Table 2.10) and the phytoalexin camalexin (Table 2.11) was active against infection by the oomycete *Phytophthora brassicae* on *Arabidopsis*. While a deficiency in one of the

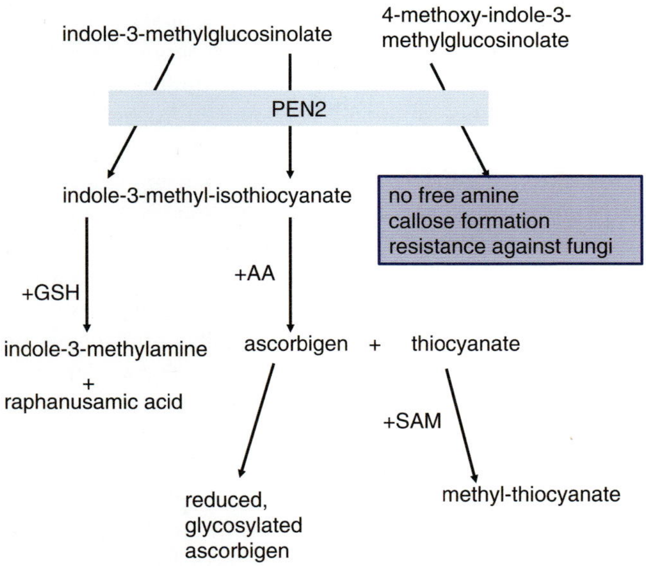

Figure 2.29 Metabolism of different indole glucosinolates to toxic defense compounds. Indole-3-methylglucosinolate (glucobrassicin) can be converted to isothiocyanates by the myrosinase PEN2. Further reaction products are formed in the presence of gluthathione or ascorbic acid, where indole-3-methylamine and raphanusamic acid or ascorbigen and thiocyanate are formed. Thiocyanate can be further methylated; SAM = *S*-adenosylmethionine. 4-Methoxy-indole-3-methylglucosinolate is not converted to an amine. Callose is produced via as-yet unknown mechanisms.

pathways did not lead to changes in resistance reactions, blocking both pathways resulted in more plants being susceptible to the oomycete infection. These results indicated that camalexin and indole glucosinolates are involved in resistance against *Phytophthora brassicae*. Observations on the tissue structures indicated that the hydrolysis products of indole glucosinolates are involved in inhibiting the early steps of colonization because, in mutants, more epidermal penetrations were observed. Camalexin controls the later steps of pathogenicity. Similarly, two pathways connected to indole derivatives are involved in the resistance of *Arabidopsis* against the soilborne fungus *Verticillium longisporum*. This fungus colonizes through the roots, but moves through the vasculature to the leaves, where the disease symptoms become visible. After colonization of the roots, no indole glucosinolate breakdown products were accumulated, but levels of indole-3-carboxylic acid and camalexin were increased. Only the phytoalexin inhibited the growth of *Verticillium longisporum in vitro*. While the genetic disruption of one pathway, as described also for *Phythophthora brassicae*, did not result in any changes in disease symptoms, the blockade of both pathways rendered the plant more susceptible to the fungus. These compounds are thought to be involved in the early resistance response of *Arabidopsis* to *Verticillium longisporum* during root infection.

The indole glucosinolates might also act as pathogenicity factors. Clubroot disease, caused by the obligate biotrophic protist *Plasmodiophora brassicae*, results in root colonizations of Brassicaceae that transform into large tumorous swellings (the so-called "clubs") in the final stage of disease development. As plant tumors are associated with high levels of growth-regulating hormones (e.g., auxins and cytokinins), it was hypothesized that indole glucosinolates, by being degraded to nitriles and further to the auxin IAA, could serve as an auxin source. According to this hypothesis, compartmentation between the glucosinolates and myrosinase would be eliminated upon infection and invasion by the pathogen, and this would then lead to higher levels of auxin. In consequence, correlative evidence was collected for high indole glucosinolate levels and the susceptibility of plants to the clubroot pathogen. The hypothesis was corroborated by showing that garden cress (*Lepidium sativum*) seedlings would develop tumor-like structures when treated with indole-3-acetonitrile and IAA. When several *Brassica* cultivars that differed in their glucosinolate composition were compared, in many cases low indole glucosinolate levels correlated with fewer disease symptoms. In *Arabidopsis*, mutants with lower indole glucosinolate levels also showed a reduction in disease symptoms. However, the analysis of mutants in which the indole glucosinolate pathway was completely blocked showed no differences in disease symptoms, which indicated that the plant is able to synthesize the auxin needed for club development by using different pathways. It should be noted that the phytoalexin camalexin was not involved in the defense reaction against *Plasmodiophora brassicae*, even though the compound was seen to accumulate in the clubs.

Other indole derivatives, the benzoxazinoids DIBOA and DIMBOA, occur mainly in grasses and protect the plants against a wide range of pathogens and herbivores. Like the glucosinolates and cyanogenic glycosides, these compounds constitute a group of compounds active against generalist pests, and can react with the SH-groups of glutathione or proteins (Figure 2.30). A reduction in glutathione levels has dramatic consequences on the antioxidative potential of a given organism. Another toxic effect can be caused by the alkylation of pest-derived proteins, as these compounds may inhibit the plasma membrane H^+-ATPases of competing plants and thus also possess allelopathic potential (Section 2.6). The formation of conjugates with thiol group-containing enzymes has been demonstrated for an intermediate of DIMBOA synthesis and glutathione transferases of *Arabidopsis*. The intermediate DIM-IQ is highly reactive and can bind to the SH-groups of proteins; the formation of a di-conjugate is mainly preferred (Figure 2.30). N-substituted analogs of DIMBOA form the same DIM-IQ intermediate and react with nucleophiles in the same manner as DIMBOA itself. In order to prevent these toxic reactions, DIMBOA can be reduced in activity by deglycosylation; however, the N-methoxy derivative HDMBOA, which also accumulates after insect attack, cannot be deglycosylated by insects and is therefore highly toxic. Thus, those insects which are adapted to plants containing DIMBOA are deterred from plants with high HDMBOA levels. Benzoxazinoid derivatives have additional roles in the attraction of zoophagous insects in plant defense against herbivorous pests (Section 3.1).

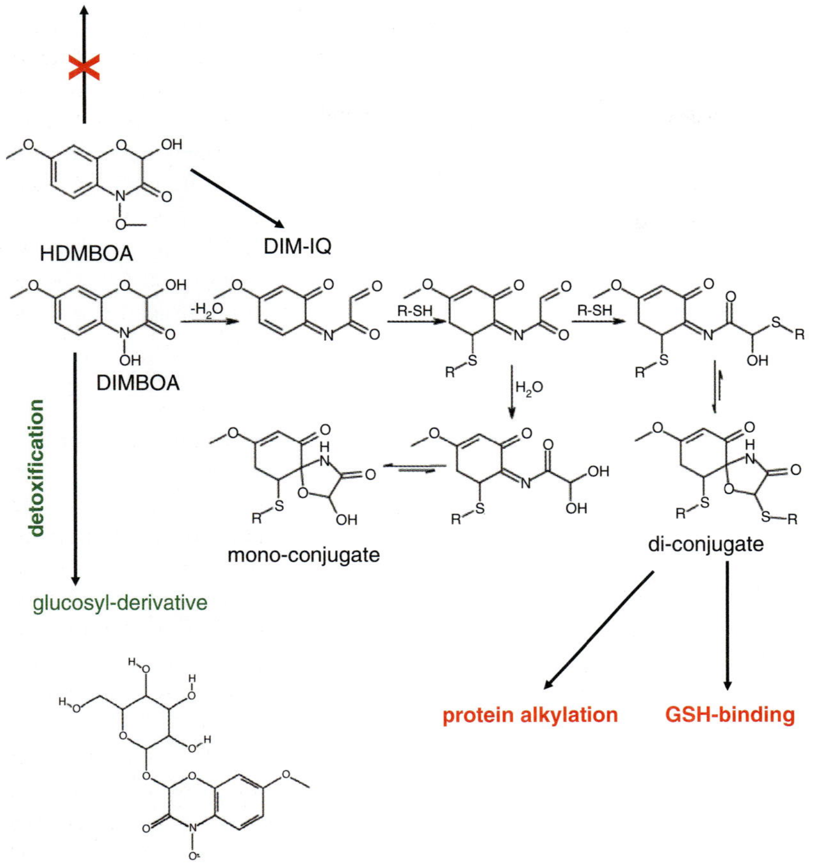

Figure 2.30 Formation of conjugates with thiol groups by benzoxazinoids. The conjugates can influence the redox potential of a cell in a negative way and also interact with SH-groups of proteins. Detoxification (green) by some adapted insects is only possible for DIMBOA, but not HDMBOA (red). DIMBOA = 2,4-dihydroxy-7-methoxy-1,4-benzoxazin-3-one; DIM-IQ = reactive intermediate; GSH = glutathione; HDMBOA = *N*-methoxy derivative of DIMBOA. Modified from Dixon, D.P. *et al.* (2012) *Phytochemistry*, **77**, 171–178.

2.5.1.2 **Induced-Defense Response**

Induced-defense compounds can establish a structural as well as a chemical defense in plants after pathogen attack. Lignin and suberin can be induced to form barriers against pathogens, while the polymer callose may be synthesized to prevent the penetration of fungal hyphae in resistant plants. Typical low-molecular-weight compounds with antimicrobial potential are the phytoalexins, which differ widely in their chemical structure. Some examples of phytoalexins in different plant species are summarized in Table 2.11. Interestingly, some of these (e.g. resveratrol) are suspected of affecting human health (see Chapter 6).

Table 2.11 Examples of typical phytoalexins from different plant species. In each case an example of a plant in which the phytoalexin occurs is given. The respective chemical structures are also shown. Chemical structures are from The PubChem Project.

Plant species	Substance	Chemical class
Arachis hypogaea	Arachidine	Stilbenes
	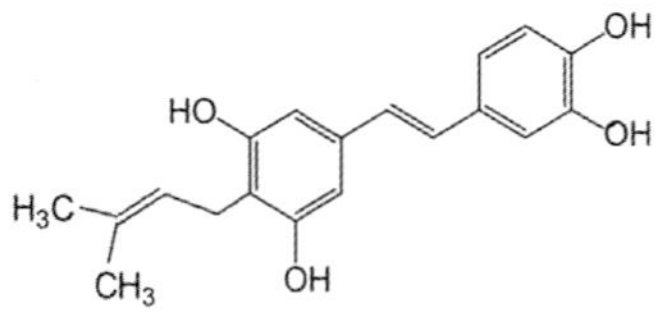 	
	Arahypine	
	Aracarpine	
	Resveratrol (structure in Figure 1.7)	
Arabidopsis thaliana	Camalexin	Indole derivative
	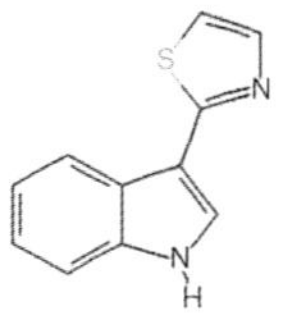 	
Brassica juncea	Cyclobrassinin sulfoxide	Indole derivative
Brassica napus	Brassilexin	Indole derivative
	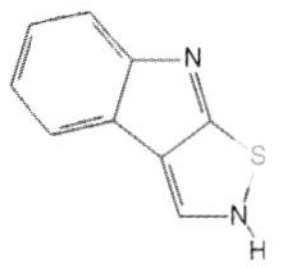 	
	Brassicanal A	
	Brassinin	
	Cyclobrassinin	
	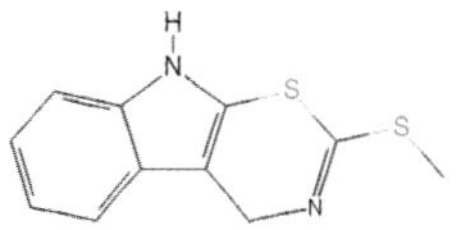 	
	4-Methoxybrassinin	
Brassica oleracea	4-Methoxybrassinin	Indole derivative
Capsicum annuum	Capsidiol	Sesquiterpene
	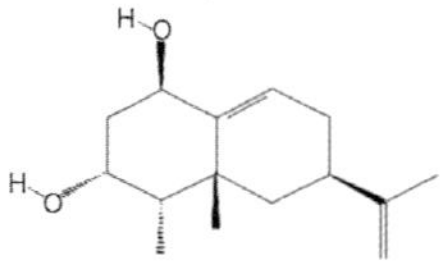 	

Eschscholzia californica	Sanguinarine 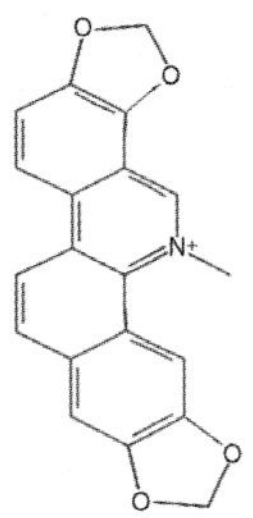	Benzophenanthridine alkaloid
Glycine max	Glyceollin 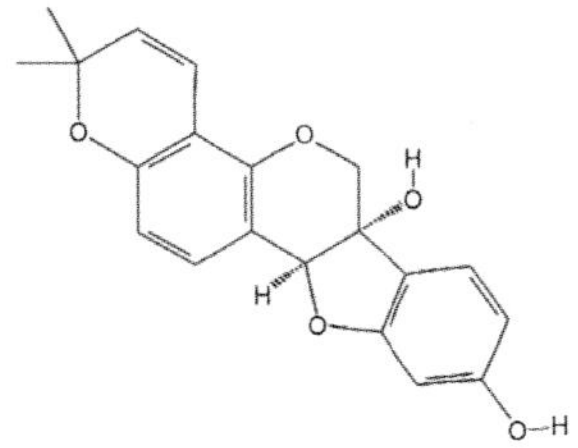 Coumestrol 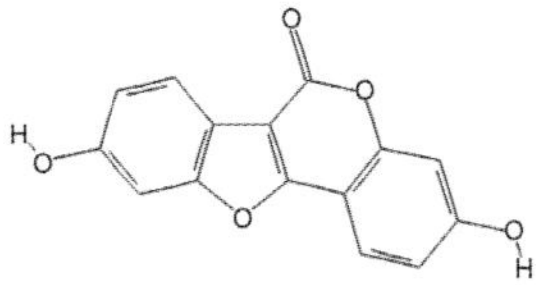	Pterocarpane Coumestane
Gossypium hirsutum	Gossypol 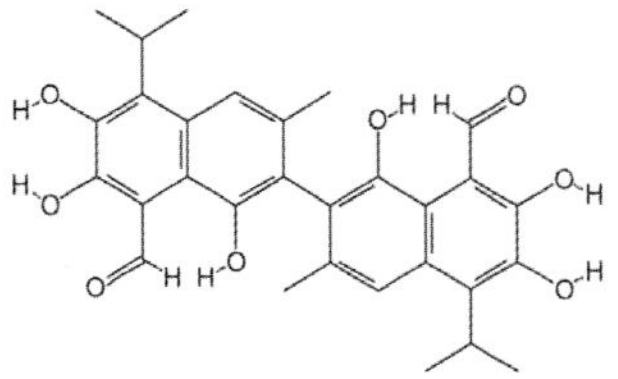	Sesquiterpene
Ipomoea sp.	Ipomoeamaron 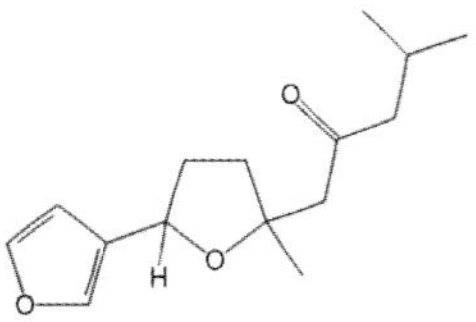	Furanosesquiterpene
Lablab niger	Kieviton 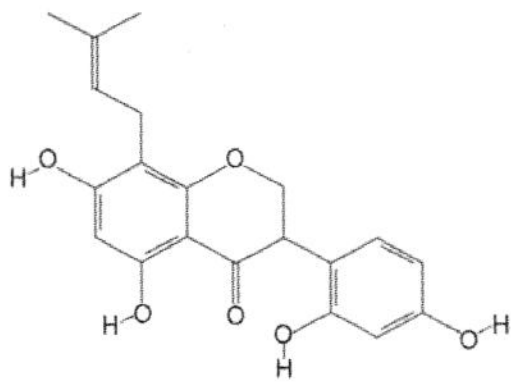	Isoflavanone

(continued)

Table 2.11 (*Continued*)

Plant species	Substance	Chemical class
Macroptilium atropurpureum	Kieviton	Isoflavanone
Nicotiana tabacum	Chlorogenic acid 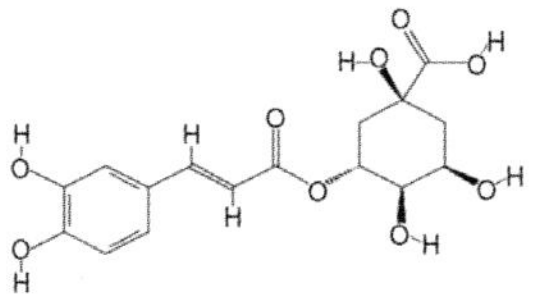	Phenylpropanoid
	Lubimin 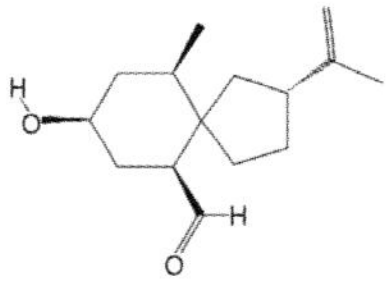	Sesquiterpene
	Rishitin 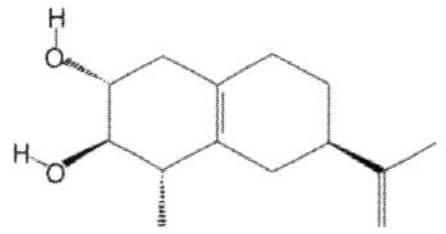	Norsesquiterpene
	Scopoletin 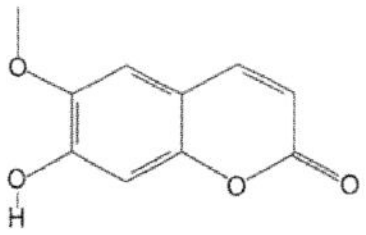	Coumarin
Orchis militaris	Orchinol 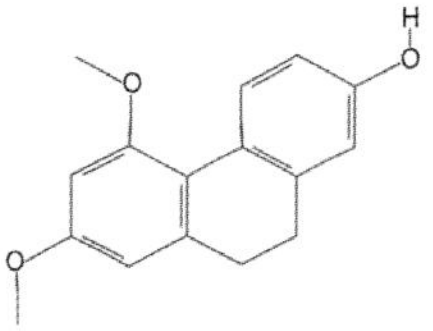	Dihydrophenanthrene
Oryza sativa	Momilactone B 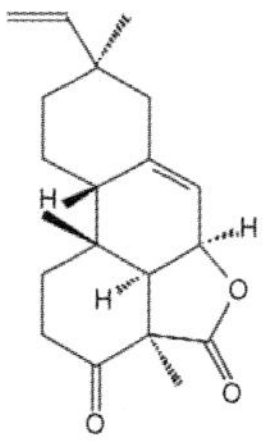	Diterpene
	Phytocassane	Diterpene
		Flavonoid

	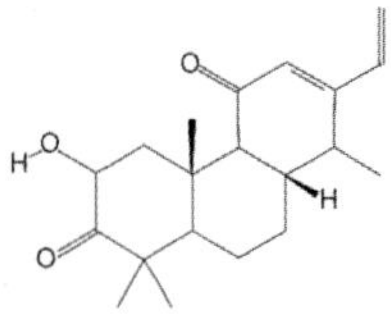	
	Sakuranetin	
	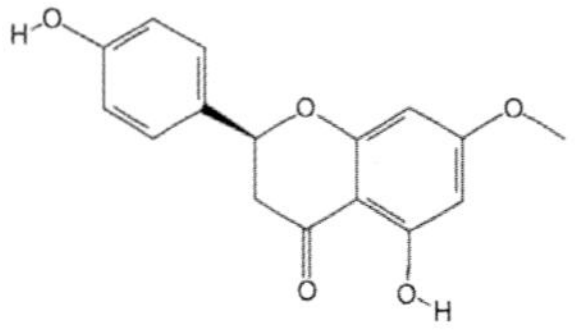	
Petroselinum crispum	Bergapten	Furanocumarin
	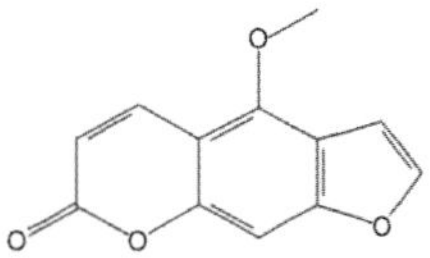	
Phaseolus aureus	Kieviton	Isoflavanone
Phaseolus vulgaris	Coumestrol	Coumestan
	Kieviton	Isoflavanone
	Phaseolin	Pterocarpane
	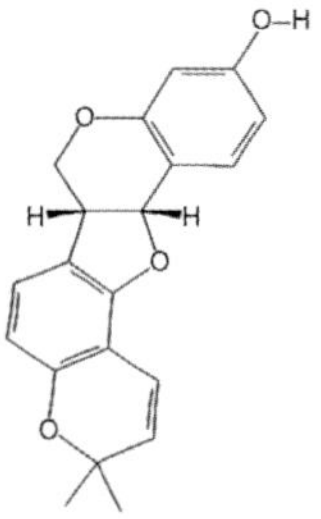	
Pisum sativum	Pisatin	Pterocarpane
	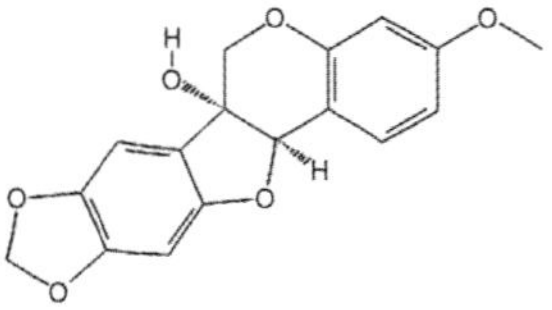	
Psoralea corylifolia	Psoralen	Furanocumarin
	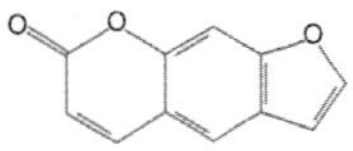	
Solanum lycopersicum	Lubimin	Sesquiterpene
	Rishitin	Norsesquiterpene
	Falcarinol (structure in Table 1.1)	Polyacetylene

(continued)

Table 2.11 (*Continued*)

Plant species	Substance	Chemical class
Solanum tuberosum	Chlorogenic acid	Phenylpropanoid
	Lubimin	Sesquiterpene
	Rishitin	Norsesquiterpene
Sorghum bicolor	Apigenidin (structure in Figure 2.33)	Deoxyanthocyanidine
Vitis vinifera	Resveratrol	Stilbene
Zea mays	Kauralexin	Diterpene
	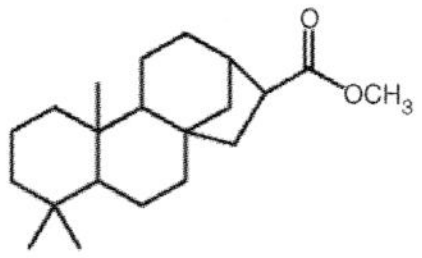	Sesquiterpene
	Zealexin	
	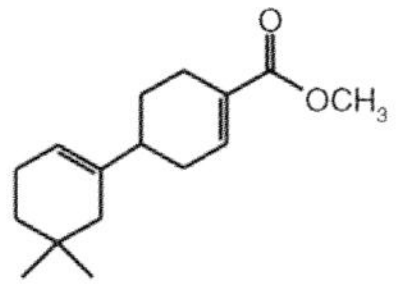	

The mechanism of inhibiting microbial growth is quite different for the individual compounds, most likely due to their wide diversity in chemical structure. A connection between the hypersensitive reaction induced by an oxidative burst of ROS (including H_2O_2) and the synthesis of a phytoalexin (glyceollin of soybean) has been demonstrated in a series of experiments (Figure 2.31). The determination of glyceollin in a susceptible and resistant interaction of soybean with two different races of *Phytophthora megasperma* f.sp. *glycinea* showed that such accumulation only occurred in the interaction with the incompatible race, leading to resistance of the plant. The H_2O_2 accumulation correlated with the increase of glyceollin in the incompatible interaction, whereas neither glyceollin nor H_2O_2 were synthesized in large amounts in the compatible interaction. Finally, it was shown that treatment with H_2O_2 led to an accumulation of glyceollin in uninfected plants, whereas treatment with water did not elicit phytoalexin synthesis. Cell death specifically prevents the colonization by biotrophic pathogens, which require living plant tissues for their life-style.

Phytoalexins of the indole type are derived from the same precursors as the indole glucosinolates (Section 1.1.2, Figure 2.4). Whereas, in *Arabidopsis* one major compound, camalexin, has been detected to date, other *Brassica* species possess a wide range of these compounds, including brassicanal A, brassilexin, brassinin, 4-methoxybrassinin, or cyclobrassinin and its sulfoxide-derivative. Not all known *Brassica* phytoalexins have been found in all species; rather, the major ones were detected in *Brassica napus* and *B. rapa*. The antimicrobial activity of

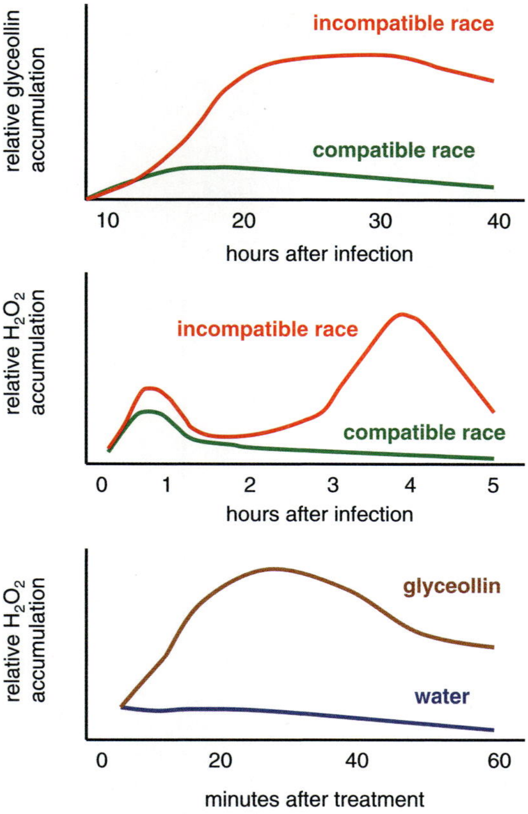

Figure 2.31 Relative glyceollin accumulation in a susceptible (compatible) and resistant (incompatible) interaction with two races of *Phytophthora megasperma* f.sp. *glycinea* in soybean (*Glycine max*). The upper panel shows glyceollin accumulation, and the middle panel H_2O_2 accumulation. The red line symbolizes the incompatible interaction, and the green line the compatible interaction with the two bacterial races. The lower panel shows glyceollin synthesis after treatment of healthy plants with H_2O_2. The brown line shows H_2O_2 accumulation after glyceollin treatment, and the blue line after water treatment as control.

these compounds is most likely due to their potential for destabilizing the pathogen membranes.

Alkaloids can also act a phytoalexins, as shown for the benzophenanthridine alkaloids of the Californian poppy plant (*Eschscholzia californica*), which can be induced by elicitors and other signaling molecules of plant defense, such as jasmonic acid. Insects are repelled by these compounds because of their feeding-deterrent properties (Section 2.5.3). The substances have multiple effects on microbial pathogens, including intercalation into nucleic acids, increase of membrane permeability, and the inhibition of SH-dependent enzymes.

Flavonoids also function as antimicrobial compounds. In particular, millet (*Sorghum bicolor*) contains the red-colored phytoalexins luteolinidin and

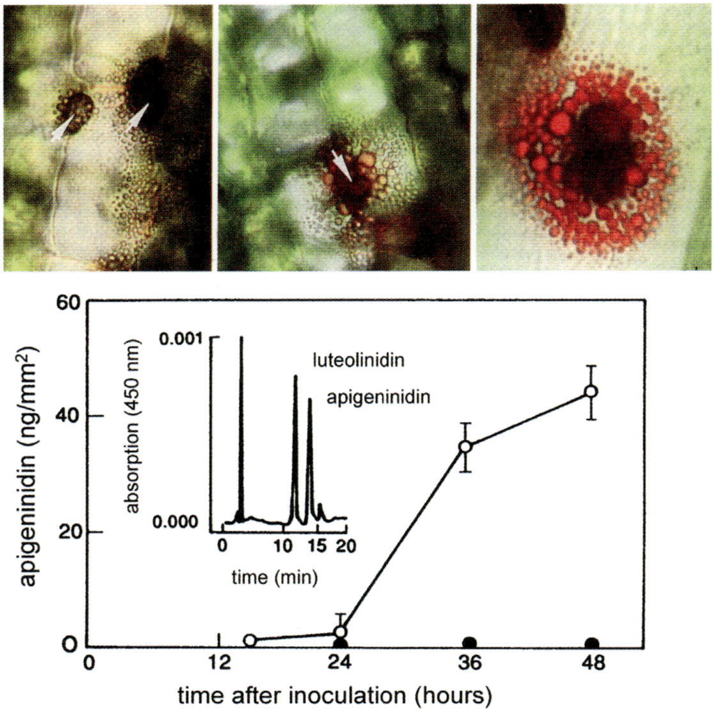

Figure 2.32 The red-colored phytoalexins of millet (*Sorghum bicolor*). The structures of the two compounds can be found in Figure 2.33. The photo shows the accumulation of the phytoalexins in vesicles close to the fungal entrance site. In addition, the induction of two major phytoalexins is shown after inoculation with the fungus *Colletotrichum graminicola* over time. Figure modified from Hipskind, J.D. *et al.* (1990) *Physiol. Mol. Plant Pathol.*, **36**, 381–396. Photograph from Snyder, B.A. and Nicholson, R.L. (1990) *Science*, **29**, 1637–1639 (see Further Reading). Reprinted with permission from The American Association for the Advancement of Science.

apigeninidin, which show fungicidal effects at very low concentrations. Following infection with fungal pathogens, these compounds are highly induced, accumulate in special vesicles, and are then transported to the entrance site of the fungus (Figure 2.32), such that high concentrations are achieved at the infection site. Indeed, concentrations of apigeninidin accumulated in the vesicles have reached 160 mM, while 9 μM apigeninidin was sufficient to inhibit hyphal growth of the fungus *Colletotrichum graminicola*. Thus, the concentration of phytoalexin in the plant was about 18 000-fold higher than was needed for its inhibitory effect.

Sakuranetin from rice is also a flavonoid derivative (Table 2.11), and its antimicrobial activity can be attributed to its binding to DNA and proteins. The hydrophobicity of flavonoids, and hence their ability to interact with biological membranes, is one of the main determinants of their biological activity. The number of hydroxyl groups is also responsible for the compound's polarity, and binding to different membrane areas is influenced by these properties. Flavonoid molecules can be localized in the hydrophobic region or in the polar membrane interface, depending on the hydroxyl groups. A specific interaction between flavonoids and some integral membrane proteins is also possible, but this would lead to an altered or loss-of-function of these proteins. An example of the integration of secondary plant metabolites into membranes is also provided in Section 3.2. These various interactions with different target molecules are further described in Chapters 5 and 6.

The levels of phenolic compounds in potato (*Solanum tuberosum*) were found to differ upon infection with two pathogens, the oomycete *Phytophthora infestans* and a phytopathogenic bacterium. This points to a role for various flavonoids in the differential resistance responses. The major flavonoids rutin (quercetin-3-*O*-rutinoside) and nicotiflorin (kaempferol-3-*O*-rutinoside) are synthesized by the plant after inoculation with *P. infestans*, whereas both were induced to a lesser extent after bacterial infection. The synthesis of antimicrobial rutin was accompanied by high levels of chlorogenic acid, another phenol with a high antimicrobial activity. As the differential resistance response for five potato cultivars correlated with the amount of phenols synthesized, the formation of chlorogenic acid can be used as physiological marker to identify further resistant cultivars.

Flavonoids might show antimicrobial potential also by influencing biofilm formation by bacteria. The infection of a host with bacteria is facilitated if they have formed a so-called "biofilm", for which they require signaling molecules that include also so-called "quorum sensing" compounds. Within the biofilm a large population is brought together, and this is advantageous for bacteria living in consortia compared to individuals. Flavonoids were shown to have an inhibitory effect on the formation of bacterial biofilms, and thereby to play a role in inducing plant defense. This effect is also of interest for humans, examples being reductions in the numbers of bacteria on the teeth, or adhering to the bladder wall. Thus, cranberry (*Vaccinium* sp.) juice with its high flavonoid content can be recommended to support treatments during inflammation of the bladder.

The concentration of phytoalexins in agricultural and crop plants can be raised artificially in three ways: (i) by breeding; (ii) by genetic engineering (see Section 1.3.3); and (iii) by eliciting specific inhibitors of enzymatic activities.

As an example of the latter approach, elicitation has been used successfully to increase the resistance of plants against different pathogens and insects. Such a targeted approach to alter the biosynthetic route of different flavonoids was used in Rosaceae, by treating the plant with the growth inhibitor prohexadione-calcium. This compound is an inhibitor of ascorbic acid-dependent dioxygenases, and inhibits biosynthesis of the plant hormones gibberellin and ethylene. In addition, it inhibits flavonole-3-hydroxylase (Figure 1.8), so that the flavonole naringenin will accumulate. However, if the plant possesses an unspecific flavonol reductase, then additional compounds that are not present without treatment will be synthesized (Figure 2.33). Although apple and related species possess such an enzyme, it is not present in *Arabidopsis*. It was shown that in several plant species a flavonoid, luteoforol, with a high structural similarity to the *Sorghum* phytoalexins, would accumulate after treatment with prohexadione-calcium, rendering the plants more

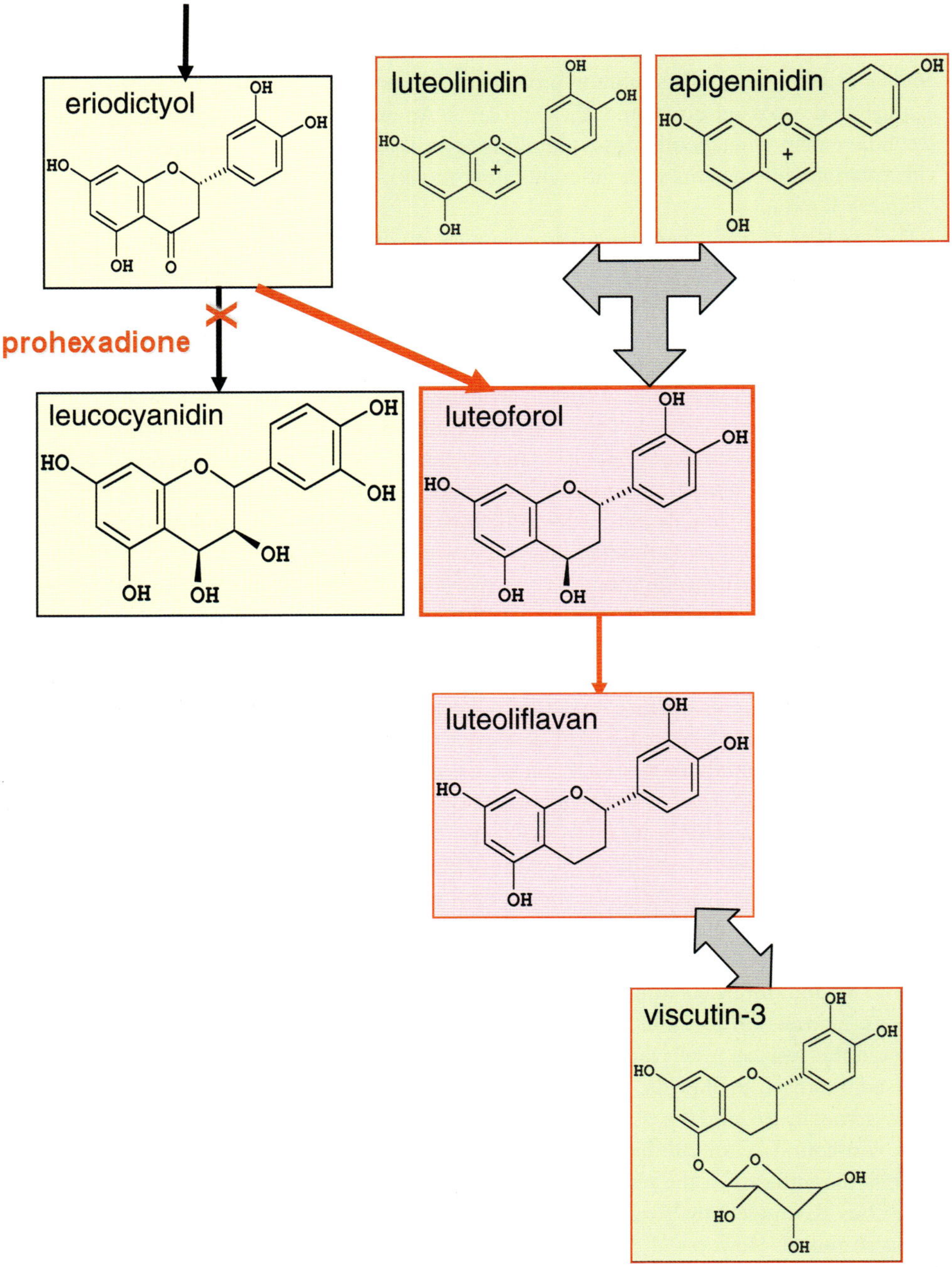

Figure 2.33 Effect of treatment with prohexadione-calcium on the accumulation of different flavonoids. By inhibition of leucocyanidine synthesis it is possible that 3-deoxyflavonoids are formed, if an unspecific reductase is present in the plant. These compounds have structural similarity to the *Sorghum* phytoalexins and an insecticidal compound from mistletoe. Structures from the normal pathway are shown in yellow, and compounds after inhibition of the pathway

Table 2.12 Inhibition of symptom development in many plant–parasite interactions after treatment with prohexadione-calcium. The number of tested interactions with higher tolerance and the number of interactions without effects are given.

Parasite	No. of host–parasite interactions investigated	No. with positive effect	No. without effect
Bacteria	7	6	1
Oomycetes	2	1	1
Fungi	8	6	2
Insects	7	7	0

resistant to various pathogens (Table 2.12). In addition, a derivative with similarity to the insecticidal glycoside viscutin of mistletoe (*Viscum album*) was found (luteoliflavane). In the majority of the plant–pest interactions investigated after prohexadione-calcium treatment, an antimicrobial or insecticidal effect was identified which resulted in a higher tolerance/resistance of the plant (Table 2.12).

The binding of flavonoids to the DNA of various organisms has been described *in vitro*. As this observation is in accordance with the localization of flavonoids in the nuclei of plants and of animals (see Section 3.2), flavonoids might be directly involved in transcriptional control (though this has not yet been demonstrated). The antiviral effect observed for flavonoids could be due to a direct binding to double-stranded nucleic acids. The compound methylquercetagetin, from the Croatian endemic flower *Centaurea rupestris*, showed a high antiviral potential. Several flavonoids (including quercetin) were able to bind *in vitro* with high affinity to artificial double-stranded DNA or RNA, single-stranded RNA, and also to viral RNA. This binding was increased by complex formation with rare earth metals, such as lanthanum. However, it is not clear whether such complex formation could occur also in the plant.

2.5.1.3 **Systemic Acquired Resistance**

Systemic acquired resistance (SAR) is a phenomenon in plants caused by the inoculation of a tissue with a (weak) pathogen and the induction of resistance responses in plant parts above this inoculation site. SAR can be promoted by secondary metabolites. Whereas the primary signals are considered to be plant hormones (e.g., the methyl ester of salicylic acid) that are synthesized in the infected tissues and transported to the systemic organ (Figure 2.27), secondary metabolites can assume this role in some cases. In conifers, an abietane diterpenoid (dehydroabietinal) was found to be a potent inducer of systemic resistance against

by prohexadione-calcium in red. Structures with fungicidal and insecticidal properties from other plant species are shown in green, and show homology of the novel flavonoids formed. In most investigated plant–pest interactions, an inhibitory effect of the treatment on colonization of the host plants was found (Table 2.12). Based on Rademacher, W. *et al.* (2006) *Acta Hort.*, **727**, 97–106.

insects, with this substance accumulating to a large extent in the resin. In *Arabidopsis*, treatment with dehydroabietinal resulted in an induction of gene expression of critical components for systemic resistance. Another secondary metabolite most likely involved in *Arabidopsis* systemic resistance response was identified as azelaic acid. Previous mutant experiments had indicated that azelaic acid might be involved in the production and/or transport of the mobile SAR signal which, in many plant species, is the methylester of salicylic acid. Pipecolic acid, a degradation product of the amino acid lysine was also placed into the group of compounds involved in SAR signaling in *Arabidopsis*. Here, translocated pipecolic acid might induce its own synthesis, because transcripts encoding a protein involved in the formation of pipecolic acid can be induced by the compound. It is to be expected that the number of secondary metabolites identified as being involved as signals in SAR will increase in the future.

2.5.2 **Antimicrobial Potential in Plant Protection (Biofumigation)**

The potential to use secondary plant metabolites in plant protection can be explained by using glucosinolates as an example. The concept of biofumigation does not use single compounds or mixtures, but rather relies on the total plant as a continuous source of the chemicals, in combination with certain soil properties (Figure 2.34). Glucosinolate-containing plants are grown and then plowed under the soil surface, so that they constitute a depot of glucosinolates in the soil; the glucosinolates can then be either directly converted to the toxic metabolites by plant myrosinase (see also Figure 1.12), or allowed to leach into the soil where they will be degraded by soil microbes containing glycosidases that cleave the glucosinolates unspecifically. The spreading and leaching of the compounds within the soil leads to a large area being treated very effectively. An alternative approach would be to transfer seed meals of glucosinolate-containing plants directly into the soil; these meals would in turn influence the nematode populations, demonstrating that biofumigation can also affect the numbers of plant parasitic herbivores in the soil. The efficiency of nematode reduction was seen to depend on high levels of glucosinolates and the presence of myrosinase in the seed meals. Whilst the feasibility of these practical applications has yet to be tested, an additional benefit would be that the plant material would constitute a fertilizer in the soil for the next crop.

2.5.3 **Invertebrate Herbivores**

Plants have developed many different strategies to cope with herbivores. Hence, the roles of secondary metabolites in defense against invertebrate herbivores will be described in the following sections (the effects on vertebrates are briefly discussed in Section 3.2). Secondary metabolites involved in the defense responses against insects can act as repellents and as feeding deterrents or insecticides.

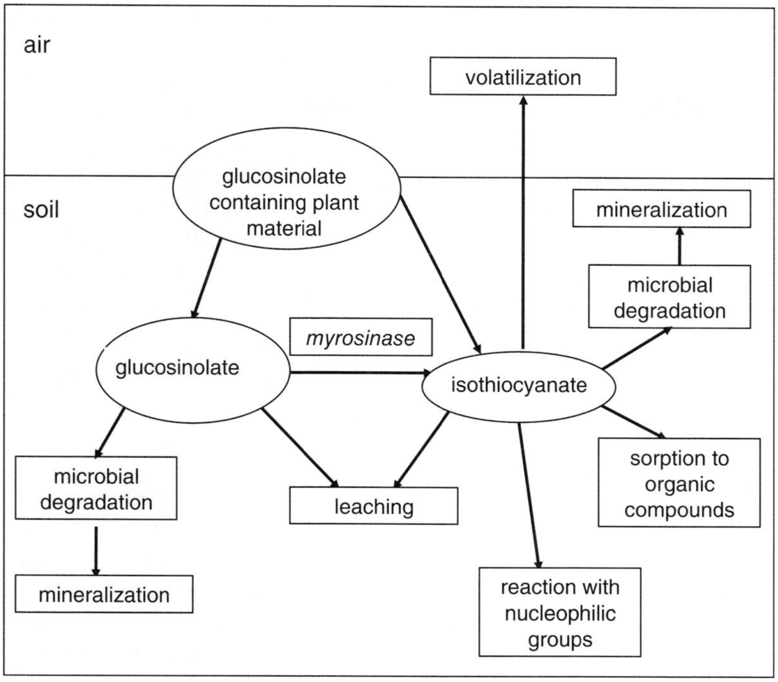

Figure 2.34 Biofumigation as a method to control plant pathogens in the soil. Glucosinolate-containing plants are grown on fields and then plowed into the soil. Here, the glucosinolates are constantly converted to their toxic metabolites by either plant myrosinase or glycosidases of microbes in the soil. The glucosinolates can be regarded as a depot for antimicrobial compounds, which leach continuously into the soil. Modified from Gimsing, A.L. and Kirkegaard, J.A. (2009) *Phytochem. Rev.*, **8**, 299–310.

2.5.3.1 **Insects**

During its lifetime, a plant will generally be confronted by insects classified as either "generalists" or "specialists," depending on the range of plant species to which they are adapted:

- *Generalists* have a broad host spectrum but are not well adapted, so that they are usually restricted by general defense mechanisms such as preformed defense compounds or morphological means.
- *Specialists* are more adapted to their host in terms of nutrition and offspring production, so that the defense mechanisms which are effective against such a pest need to be more specific.

Clearly, different defense strategies need to be established against the two different groups. For the recognition of specific pests, the plants need to be able to detect specific insect patterns (Figure 3.2). A generalist and a specialist are shown, together with the response of the same plant species to the respective pest, in Table 2.13. In some cases the response against the two different insects is very

Table 2.13 Generalists versus specialists. For selected plant species the different defense responses against a specialist and a generalist in terms of secondary metabolite induction are given. GSL = glucosinolate.

Plant	Herbivore	Generalist	Specialist	Response
Arabidopsis thaliana	Aphid	*Myzus persicae*	*Brevicoryne brassicae*	*GSL:* specialist elicits small induction
Brassica oleracea	Aphid	*Myzus persicae*	*Brevicoryne brassicae*	*GSL:* pattern dependent on water status of plant
Arabidopsis thaliana	Butterfly caterpillar	*Spodoptera exigua* *Spodoptera littoralis* *Trichoplusia ni*	*Pieris rapae*	*GSL:* generalist induces total GSL; specialist induces specific indole glucosinolates *Volatiles:* parasitoid insects are attracted similarly; only differences observed in the pest belongs to a different group
Brassica nigra	Butterfly caterpillar	*Mamestra brassicae*	*Pieris rapae* *Plutella xylostella*	*GSL:* indole glucosinolates induced after *Mamestra* and *Pieris* inoculation, but not after *Plutella*; no common response after specialist or generalist attack
Brassica nigra	Butterfly caterpillar	*Trichoplusia ni*	*Pieris rapae*	*GSL:* sinigrin *Trichome:* difference in induction by specialists and generalists is the number of trichomes; also dependent on leaf position
Sinapis alba	Butterfly caterpillar	*Spodoptera frugiperda*	*Athalia rosae*	*GSL and myrosinase:* both induced, but more strongly by specialist, which can sequester GSL
Plantago lanceolata	Butterfly caterpillar	*Junonia ceonia*	*Spilosoma congra*	*Iridoid-Glycosides:* higher levels induced by specialist, which can sequester the compounds; adaptive defense against generalists
Zea mays	Leaf beetle	*Diabrotica balteata*	*Diabrotica virgifera*	*Volatiles:* specificity for the attraction of parasitoids; specialist induces much more (E)-β-caryophyllene than generalist
Nicotiana tabacum	Butterfly caterpillar	*Helicoverpa amigera*	*Helicoverpa assulta*	*Alkaloids:* nicotine more strongly induced by specialist than generalist *Polyphenols:* polyphenoloxidase lower in specialist compared to generalist interaction *Lignin:* peroxidase stronger induced by specialist

different, but in other cases there is only a gradual difference or no difference at all. In some cases an additional stress is needed to elicit the defense response. For example, the induction of glucosinolates in *Brassica oleracea* is dependent not only on a pest but also on an environmental factor, namely the water status of the plant. Thus, an abiotic stress situation is responsible for the differential response of the plant to the two pests. Another example shows an induction of trichomes on leaves after colonization by a specialist but not a generalist, and this was dependent on the position of the leaf on the plant. Specialists are better adapted to the chemical weapons of a plant, and can often circumvent these toxic compounds (see also Section 3.1). However, there is no uniform answer of a plant to any pest, so that the induction of a certain class of secondary metabolites cannot be attributed to a specific adaptation against herbivore attack.

The accumulation of secondary defense metabolites depends on the way in which insects feed on the host plant. Phloem-sucking insects elicit a different response compared to chewing and biting insects (Table 2.14), which cause large wounds on the host plant and also interact with the tissue by secreting their spit into the host (Figure 2.36). Sucking insects can avoid toxin-containing cells or tissues (Figures 3.5 and 3.6). Therefore, the defense responses of the plants to the feeding behaviors must be adapted. For differences elicited by generalists or specialists, no general hypothesis on the outcome of the plant defense reaction can be proposed (Table 2.13). Sucking insects cause different responses from chewing insects (Table 2.14); the latter cause a stronger and complex response while the response to the former is more subtle. The attraction of parasitoid insects by volatile emission is induced by chewing pests. Thus, sucking insects can be assigned to the group of specialists, because they have acquired the sieve tubes as a nutritional source.

The specific mixture of glucosinolates within a plant or plant organ is a crucial factor for the development of feeding strategies by insects, and also has consequences on the defense responses of the plant. Therefore, it is not only the concentration of the plant defense compound in an organ but also its spatial distribution; in a leaf this may be critical when explaining the feeding strategy of herbivores, and whether they feed from the edge or exclusively in the middle of a leaf. For the plant it is advantageous to have high levels of glucosinolates in the leaf margin, because insects crawl onto the leaf surface from the fringe of a leaf. When this mechanism is employed, the insects will be in direct contact with the defense compounds, when they begin to feed on the leaf margin. Some generalists are able to avoid those regions with high glucosinolate levels in the tissue. For example, the generalist lepidopteran larva *Helicoverpa armigera* (see Table 2.13) avoids the midvein and periphery of *A. thaliana* rosette leaves, and feeds almost exclusively on the inner lamina (Figure 2.35). This feeding pattern was attributed to glucosinolates because the feeding behavior was not evident in a myrosinase mutant that lacked the ability to activate glucosinolate defenses by hydrolysis. Therefore, the larva must either be able to sense the glucosinolates (Figure 3.10), or the first contact with glucosinolate breakdown products leads to an avoidance reaction. The spatial distribution of glucosinolates in *Arabidopsis* leaves at a fine

Table 2.14 Sucking versus chewing insects.

Plant	Sucking insects	Chewing insects	Response of plant
Arabidopsis thaliana	*Myzus persicae* *Brevicoryne brassicae*	*Spodoptera exigua* *Pieris rapae* *Plutella xylostella*	*GSL:* aliphatic GSL induced by chewing insects, downregulated by sucking pests[a)]
Arabidopsis thaliana	*Myzus persicae* *Brevicoryne brassicae*	*Spodoptera exigua* *Pieris rapae*	*GSL:* aliphatic GSL induced by sucking insects, chewing insects upregulate aliphatic and indole GSL; *Pieris rapae* dos not cause any induction[a)]
Arabidopsis thaliana	*Myzus persicae*	*Spodoptera exigua* *Pieris rapae* *Plutella xylostella*	*Volatiles:* Parasitoid insects preferred plants which were damaged by chewing insects
Gossypium hirsutum	*Bemisia tabaci*	*Spodoptera exigua*	*Volatiles:* strongly induced by chewing insects, not by sucking pests; the same was observed for pigment-containing glandular cells
Plantago lanceolata	*Dysaphis* cf. *Plantaginea*	*Grammia incorrupta* *Heliothis virescens*	*Secondary metabolites:* general upregulation of many metabolites by chewing insects
Zea mays	*Rhopalosiphum maidis*	*Spodoptera littoralis*	*Volatiles:* induction of many compounds by chewing insects, while after attack from sucking insects no detectable emission was found
Solanum lycopersicum	*Macrosiphum euphobiae*	*Spodoptera exigua*	*Herbivore resistance:* prior sucking by aphids did not cause resistance to *S. exigua*, while prior feeding by *S. exigua* could induce resistance against the same species
Solanum tuberosum	*Myzus persicae*	*Leptinotarsa decemlineata*	*Volatiles and precursors:* stronger induction by sucking insects[b)]

For selected plant species the defense reactions with respect to secondary metabolites are given in response to sucking and chewing insect species. GSL = glucosinolates.
a) Datasets are from different studies using a similar experimental approach, but with contradicting results.
b) Exception from the hypothesis that sucking insects cause less damage than chewing pests and thus elicit a weaker defense response.

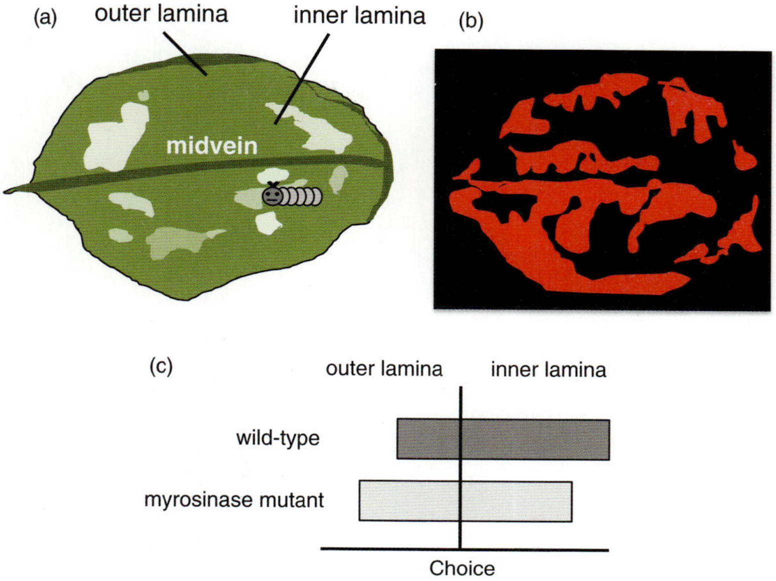

Figure 2.35 Distribution of glucosinolates in different leaf parts, as determined by mass spectrometry in the tissue, and effects on insect behavior. (a) Feeding pattern of *Helicoverpa armigera* larvae on mature rosette leaf of *Arabidopsis thaliana* after 5 h; (b) Mass spectrometry imaging of *Arabidopsis* leaves. The ion intensity map of 4-methylsulfinylbutylglucosinolate, created from about 100 000 MALDITOF/MS spectra, is shown. The red color indicates high glucosinolate levels; (c) Feeding preference of second-instar *H. armigera* larvae for the outer and inner lamina of *Arabidopsis* wild-type and myrosinase mutant leaves. As the two major myrosinase genes are not functional in this mutant, the glucosinolates cannot be converted to the active compounds after insect feeding. Based on Shroff, R. *et al.* (2008) *Proc. Natl Acad. Sci. USA*, **105**, 6196–6201.

scale was measured for the major individual glucosinolates (including 4-methylsulfinylbutylglucosinolate) using matrix-assisted laser desorption/ionization-time-of-flight mass spectrometry, from which ion intensity maps were constructed (Figure 2.35). The major glucosinolates were found to be more abundant in the tissues of the midvein and the periphery of the leaf than at the inner lamina. For the plant, protection of the leaf margin has another advantage, as the integrity of the margin is important for the mechanical stability of the leaf. Moreover, the midvein is important for the transport of nutrients within the plant, and must also be protected.

The plant can differentiate between mechanical wounding or wounding caused by chewing insects (Figures 2.36; see also Figure 3.2). This can be achieved by the identification of molecules present in the saliva of the insect. Experiments have shown that wounding elicited a specific reaction and a defined set of volatile compounds was emitted, whereas the pattern was different after the simultaneous application of insect saliva into the wound. Some of these volatile signals can be used to attract zoophagous insects or parasitoids to the plant, which can then help

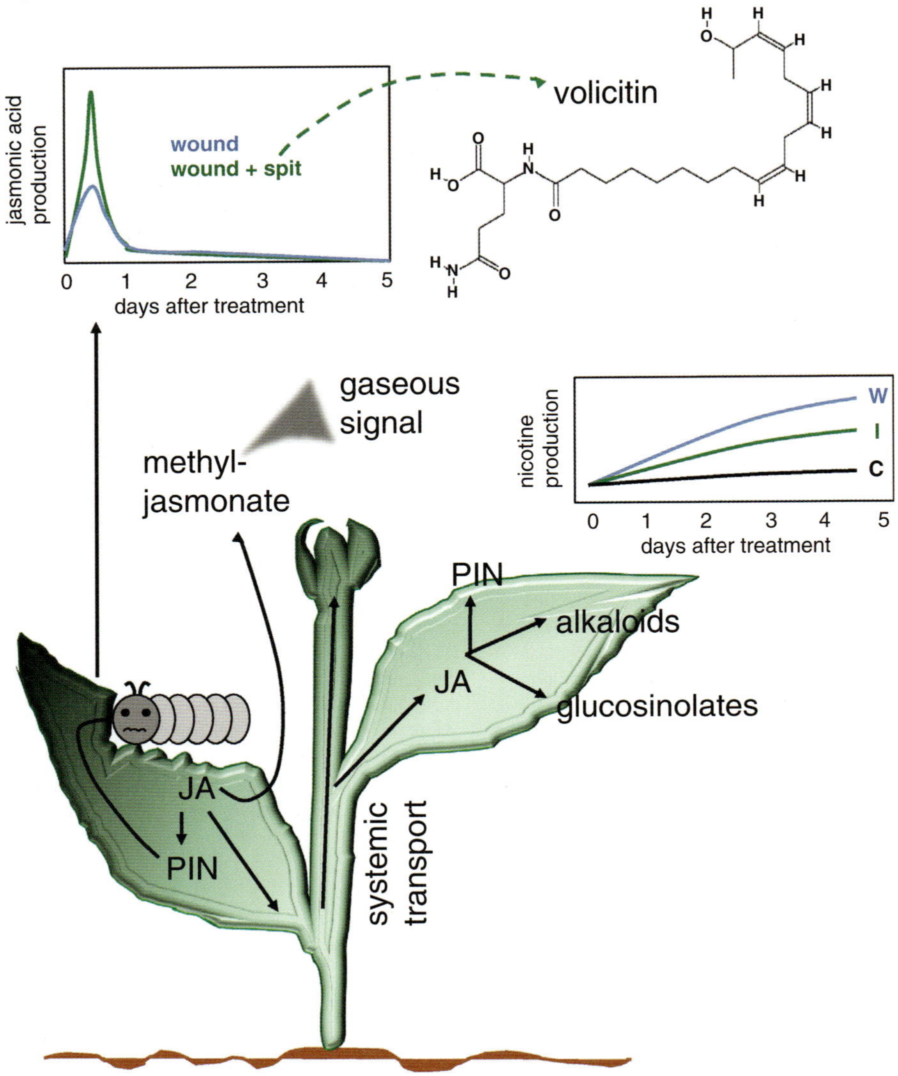

Figure 2.36 The plant defense is differentially regulated after wounding or damage by insects, using jasmonic acid as signal. The production of local jasmonic acid leads to the increase of protease inhibitors (PIN), which reduced protein digestion in the insect gut. Jasmonic acid can also be transported to systemic leaves, where it induces the same defense reaction. The synthesis of secondary metabolites such as glucosinolates or alkaloids (e.g., nicotine) is also induced, which further prevents damage by insects. Jasmonic acid becomes more volatile by methylation, and this leads to defense activation also in neighboring plants. The plant can distinguish between wounding and insect chewing by recognizing components in the saliva of the insects, such as volicitin (*N*-(17-hydroxylinolenoyl)-L-glutamine). W = wounding, I = insect (in this case a caterpillar), C = Control. Based on Riddiford, L.M. (1994) *Adv. Insect Physiol.*, **24**, 213–227. Chemical structure is from The PubChem Project.

to defend against chewing pests (see also Section 3.1). Jasmonic acid as a plant signal is also more strongly increased after insect saliva is applied to the plant (Figure 2.36). The plant is able to discriminate between wounding and insect damage by recognizing specific compounds of insect spit, such as volicitin (*N*-(17-hydroxylinolenoyl)-L-glutamine). The wounding may also induce the synthesis of secondary metabolites (e.g. in tobacco, alkaloid synthesis may be induced), and these metabolites act as feeding deterrents for many insects (and also vertebrates), thereby protecting the plant. The alkaloid nicotine also has insecticidal properties. Similar induction patterns have been observed for glucosinolates, the synthesis of which is inducible by jasmonates. Protease inhibitors are also synthesized to prevent the digestion of proteinogenic nutrients taken up by the insects. Finally, jasmonic acid can be methylated and, as a volatile compound, induce the defense responses described above in neighboring plants; this leads to a better protection of the whole plant population.

Although the alkaloid demissin from *Solanum demissum* also protects against insect attack, the mechanism employed differs from that of alkaloids which act as feeding deterrents. In contrast to the structurally related solanin from *Solanum tuberosum*, which has inhibitory activity on fungi as saponin (see Section 2.5.1), demissin interacts with the ecdysone synthesis of the insects and so can block metamorphosis (Figure 2.37).

Flavonoids can influence resistance against aphids by an as-yet unknown mechanism. Correlative evidence derives from the accumulation of higher levels of quercetin and isorhamnetin in an aphid-resistant line of *Vigna ungiculata*, compared to a susceptible line (Figure 2.38). As no concentration differences of kaempferol were shown in the two lines, it was concluded that this flavonol is not important for aphid resistance. The application of quercetin and isorhamnetin, but not of kaempferol, resulted in reduced offspring numbers of the aphids.

Flavonoids may affect the number and type of blood cells in different insect species, with reduced numbers of hematocytes being observed after treatment with genistein. *Schistocerca gregaria*, *Periplaneta americana* and *Dysdercus cingulatus* each showed no consistent changes in cell numbers after isoflavonoid application. With regards to the differentiation of cells into prohematocytes, plasmatocytes and granulocytes, genistein showed the largest effect on the increase in plasmatocytes of *Periplaneta americana*, whereas the hematocytes of *Schistocerca gregaria* were reduced in number. In the case of *Dysdercus cingulatus*, only small effects on total cell numbers were found.

Saponins, which are triterpene glycosides, possess insecticidal activity apart from the fungicidal activity described in Section 2.5.1 (Figure 2.28); thus, plants with high levels of saponins are protected against insects. *Barbarea vulgaris* was used to isolate the saponins oleanolic acid-cellobiose and hederagenin-cellobiose, which resulted in an *in-vitro* resistance against the larvae of *Phyllotreta nemorum* and *Plutella xylostella*. As *Barbarea vulgaris* is the only known species among the Brassicaceae in which saponins have been detected, this points to a late evolution of saponin synthesis in this species, at least after the glucosinolates had evolved.

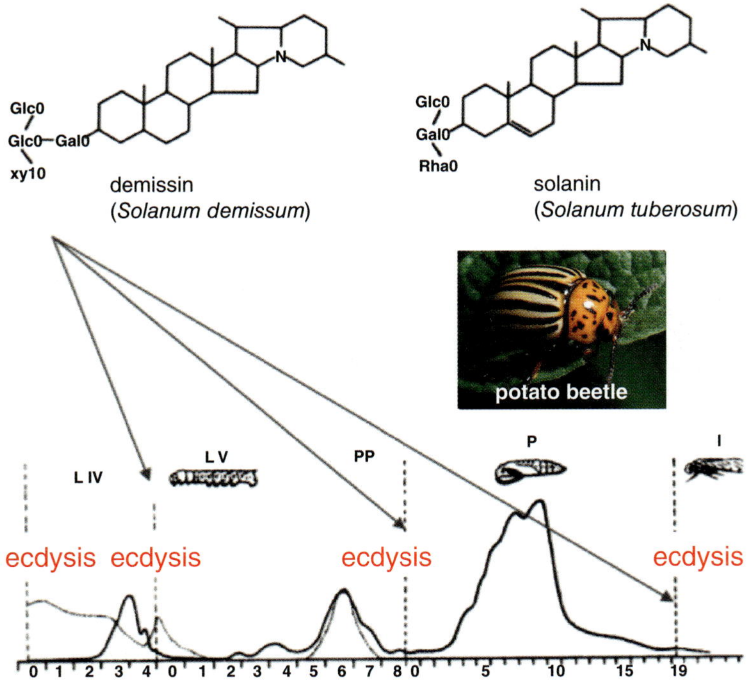

Figure 2.37 The alkaloid demissin (right) protects *Solanum demissum* from insect damage by blocking ecdysone synthesis, which leads to a block in metamorphosis. The structurally closely related solanin (left) from *Solanum tuberosum* does not show this effect. Shown are the titers of molting hormone (---) and juvenile hormone (...) during the fourth and fifth larval stages (L IV, L V) and the metamorphosis of *Manduca sexta*. I = imago, P = pupa, PP = pharate pupa. The time for development is given in days for the respective stage. Similar protection exists also against the potato beetle shown in the photograph. Fotolia.com, © mhp.

Volatiles also play important roles as defense compounds in plant–insect interactions. Indeed, the increase of a single compound in the herbivore-induced volatile bouquet of *Nicotiana attenuata* plants was shown to result in an increased predation rate of *Manduca sexta* eggs by a parasitoid insect (Section 3.1); a decreased oviposition rate of the adult moths was also observed. Both factors together reduced the number of herbivores on the plants. It is well known that many insect pests select their hosts based on visual, and especially chemical, stimuli. As the nutritional status of the plant also has an effect on secondary metabolite synthesis (Section 2.3.5), it was concluded that the use of fertilizers would lead to altered responses of the plant to pathogens and herbivores. The patterns of isothiocyanates and other volatiles derived from glucosinolates, which are emitted from *Brassica napus* leaves, were different when fertilized with nitrogen. In consequence, the number of pests was also altered because the well-fertilized host plants were less attractive to the herbivores tested. This also has practical

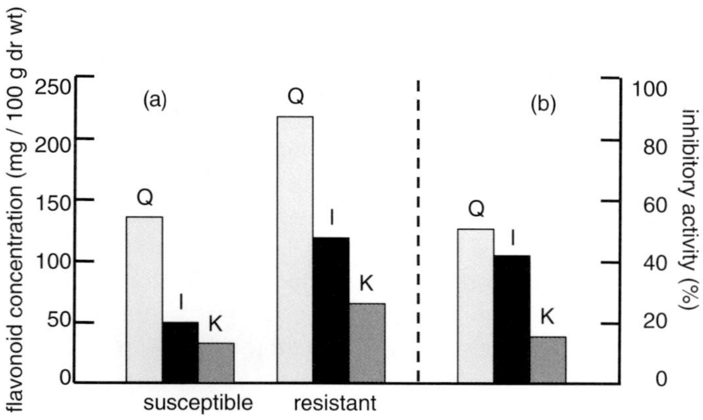

Figure 2.38 Effect of three different flavonoids on aphid resistance of bean plants. (a) Flavonoids in two near isogenic lines of *Vigna unguiculata*, one susceptible and one resistant to aphids. Q = quercetin, I = isorhamnetin, K = kaempferol; (b) Inhibitory effect of flavonoids (0.1 mM) on the oviposition of aphids (*Aphis fabae*).

applications because the use of fertilizers could substitute for the use of pesticides, although the experimental data have to be acquired for each plant–herbivore system individually. It is not only the leaves but also the roots that can emit volatiles in response to insects; for example, *Brassica nigra* roots induce the emission of volatile compounds after infestation with the *Delia radicum* root fly larvae. It was proposed that the emission of volatiles could be used to detect actively feeding larvae, because such emission terminated when the larvae had pupated.

2.5.3.2 **Phytopathogenic Fungi Alter the Chemical Defense of Plants Against Insects**

As freely moving organisms, insects can choose attractive host plants and avoid unattractive and toxic plants. This plays an important role in the interaction not only for pollination but also for the recognition of host plants for feeding or oviposition. The chemical characteristics of a plant are altered after pathogen invasion, because novel compounds are synthesized that are involved in defense responses, and such changes can also be perceived by insects. These changes in parameters can influence whether the pests accept a given plant for nutrition, or not. The details of some mechanisms for the recognition and detoxification of secondary metabolites by insects are provided in Section 3.1. On the basis of changes as described, even within a single plant population, the attraction of individual plants can be changed, with consequences of herbivore infestation.

The oviposition by plant parasitic insects is reduced on leaves of plants previously infected with pathogenic fungi. The leaf beetle *Gastrophysa viridula* showed a different feeding and oviposition pattern on leaves of the host plants *Rumex obtusifolius* and *Rumex crispus* if the plants had been infected previously with the rust fungus *Uromyces rumicis*. Similarly, Chinese cabbage (*Brassica rapa*) leaves, which were inoculated on 10–15% of their leaf surface area with the fungus

Alternaria brassicae, were avoided by females of the beetle *Phaedon cochleariae*. In this case, the glucosinolates induced after *Alternaria*-inoculation played a major role in the behavior of the beetles. Clearly, secondary metabolites are not only directly involved in attraction or defense against herbivores but also play an indirect role as toxins in defense against the parasite progeny.

A systemic reaction of the plant is also observed; for example, in *Brassica rapa* an increased resistance against the larvae of *Plutella xylostella* was observed on the leaves if the roots had been infected by the fungus *Leptosphaeria maculans*. At only one day after infection the larvae showed a reduced feeding on the leaves compared to controls, and the high levels of glucosinolates in the leaves of infected plants were shown to correlate with the feeding behavior of the moth larvae. However, it cannot be excluded that the leaves were simply deemed as "not fit" by the pests, which would be disadvantageous for feeding of the progeny within the population. In line with this hypothesis, the larvae of the moth *Spodoptera exigua* showed a clear preference for the leaves of their host plants (peanut) without disease symptoms after infection with the fungus *Sclerotium rolfsii*.

2.5.3.3 **Biological Plant Protection Against Insects with Secondary Metabolites**

Previously, thousands of plants containing compounds with insecticidal potential have been identified. Although nicotine is a powerful insecticide (see Section 3.1), its practical use has been banned since the 1970s due to its high toxicity. However, other plants have been discovered, such as the neem tree (*Azadirachta indica*), in which the main insecticidal component is azadirachtin, a triterpene (Figure 2.39). The insecticidal properties of neem extracts have long been known, as the major component interferes with larval development (see Figure 2.37), with crippled wings being observed on the insects. The progeny is also affected in neem-treated insects; it was observed that female potato beetles feeding on leaves treated with neem extract laid only 10% of their eggs on the leaves, compared to 100% egg-laying by beetles feeding on untreated leaves. In India, the branches of the neem tree have been used traditionally in dental care, due to their high antimicrobial potential.

The protective functions of certain plant compounds are used to develop insecticides for commercial use. For example, the pyrethrins (Figure 2.39) from *Chrysanthemum* species provided the chemical scaffold for the development of the contact insecticide pyrethrum. This compound is, unfortunately, nonselective and therefore also a toxin for beneficial insects.

2.5.3.4 **Nematodes**

Plant parasitic nematodes can also be controlled by secondary metabolites. One prominent example is that of sulfur-containing thiophenes from marigold (*Tagetes*) species (Figure 2.39), although the different marigold species vary considerably in the bioactive compounds they produce. These substances are secreted from glands on the leaf margins and also from the roots, so that herbivores avoid these plants. One highly active compound is alpha-terthienyl, which is secreted into the

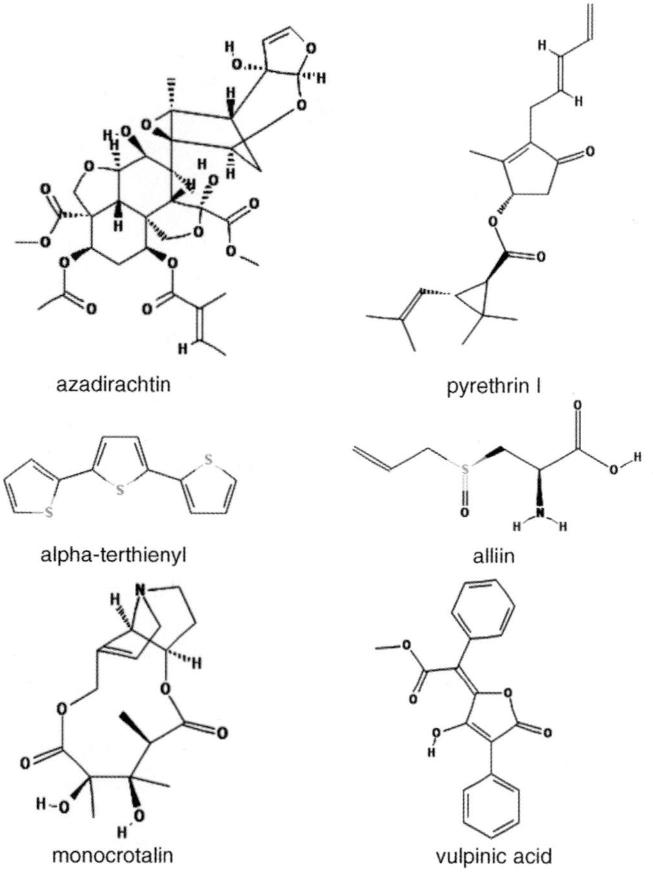

Figure 2.39 Examples of some secondary metabolites with strong inhibitory effect on some herbivore classes. Some metabolites are used in biological pest control. Pyrethrin is the active principle for the insecticide pyrethrum. Terthienyl is a nematicidal compound in *Tagetes* species, Monocrotalin and alliin are also nematicidal substances, while vulpinic acid from lichens is active against slugs. Chemical structures are from The PubChem Project.

rhizosphere and attracts nematodes to the plant. Subsequently, when infected, the plant changes the metabolite profile to more toxic compounds that kill the nematodes when they come into contact with these cells. Effects on hatching as well as retardation of the life cycle have also been observed for these substances. Unfortunately, these compounds act in species-specific fashion, and have only limited effects on some nematode species, while others are not affected at all. *Tagetes* plants may be planted between susceptible hosts to reduce the nematode population, or they may even be grown on fields with high nematode infestations to reduce insect numbers for one vegetation period.

Other compounds with nematicidal potential are the sulfur-containing compounds from onion (*Allium* species), which also have antimicrobial activities (Table 2.10). An effect on nematodes was also reported for some cyanogenic glycosides, such as dhurrin, isothiocyanates, alkaloids (monocrotalin), polyacetylenes, phenols, and terpenes (Figure 2.39). It is interesting to note that many compounds with nematicidal properties are sulfur-containing. Polyphenols can be used to build a barrier against nematodes.

2.5.3.5 **Snails/Slugs**

The damage caused by slug feeding on plants is immense. The leaves which are consumed preferentially have a low level of deterring secondary metabolites, because these are crop plants bred to meet the taste of human consumers. This implies that slugs can perceive these compounds and circumvent bitter-tasting plants. Slugs are sensitive to phenols present in potato tubers. *Deroceras reticulatum*, one of the major potato pests, avoids potato tubers with high levels of phenols and phenol oxidase activity, but will feed preferentially on tubers with low phenol contents. Similarly, coniferous trees with high α-pinene concentrations are avoided by slugs. Cyanogenic glycosides are also active against slugs, because toxic cyanide is generated during the feeding process and can directly inhibit ATP generation in the herbivore mitochondria.

Complex plant extracts, which contain many different bioactive metabolites, are also effective against slug feeding. However, of nine different extracts from various plant species known to possess high secondary metabolite concentrations, only extracts of the species *Saponaria officinalis* and *Valerianella locusta* were effective against the slug, *Arion lusitanicus*. The tests were carried out on oilseed rape leaves treated with the extracts. As lichens are highly resistant against various herbivores (including slugs; Section 2.4.4), extracts of these organisms were tested against *D. reticulatum*. The highest potential to prevent feeding by slugs was found for an extract of the lichen *Letharia vulpina*, and the active compound was identified as vulpinic acid (Figure 2.39), which also showed activity as a single compound. This point was important, as single compounds are often less active than a complex plant extract, and identification of the active principle is often hampered by the complexity of the extract.

2.5.4 **Parasitic Plants**

Parasitic plants depend on the physical interaction with their host plants, from which they receive assimilates and other nutrients, as their own ability to perform photosynthesis is either reduced or is completely absent. About 3000 species of parasitic plants are known, ranging from trees to small weeds. Some interactions guarantee a protected habitat, an example being the epiphytic parasite mistletoe (*Viscum album*). The plant parasites can be grouped into 40% shoot and 60% root parasites (Table 2.15). Hemiparasites are able to conduct some photosynthesis, because they contain a small amount of chlorophyll (e.g., mistletoes), whereas

Table 2.15 Plant families with parasitic representatives. These families include only parasitic genera, with a few exceptions such as the Convolvulaceae and the Lauraceae. The numbers of parasitic genera in this family are shown; it is also indicated whether the genus contains root or shoot parasites, and also whether photosynthesis is still present to some extent in the hemiparasites, or whether the parasite is completely dependent on the host in the case of holoparasites.

Family	No. of parasitic genera	Remarks	Shoot parasite	Root parasite	Hemiparasite	Holoparasite
Apodanthaceae	3		x			x
Balanophoraceae	17			x		x
Convolvulaceae	Cuscuta	Only one genus parasitic, root (holo-) as well as shoot (hemi-) parasites	x	x	x	x
Cynomoriaceae	1			x		x
Cytinaceae	2			x		x
Eremolepidaceae	3		x		x	
Hydnoraceae	2			x		x
Krameriaceae	1			x	x	
Lauraceae	Cassytha	Only one genus parasitic	x		x	
Lennoaceaae	2			x		x
Loranthaceae	73	Root as well as shoot parasites	x	x	x	
Misodendraceae	1		x		x	
Mitrastemonaceae	1			x		x
Olacaceae	14			x	x	
Opiliaceae	10			x	x	
Orobanchaceae	90	Hemi- as well as holoparasites		x	x	x
Rafflesiaceae	3		x			x
Santalaceace	35	Root as well as shoot parasites	x	x	x	
Schoepfiaceae	1			x	x	
Viscaceae	7		x		x	

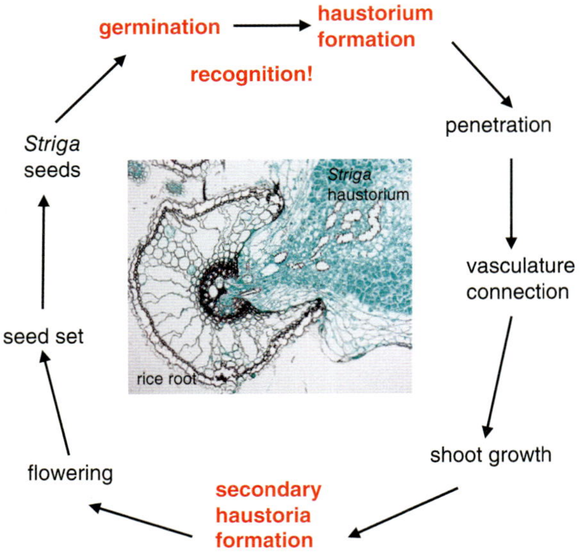

Figure 2.40 The life cycle of *Striga*. The stages, which are dependent on plant signals, for example secondary metabolites, are shown in red. The figure was modified according to, and the picture taken from Yoshida, S. and Shirasu, K. (2012) *Curr. Opin. Plant Biol.*, **15**, 708–713 (see Further Reading). Reprinted with permission from Elsevier.

holoparasites have no chlorophyll and are therefore completely dependent on their host for carbon nutrition. Plant families in which parasitic representatives are present are listed in Table 2.15. Most shoot parasites are still hemiparasitic, and the majority of root parasites is holoparasitic, though there are some exceptions. The genus *Cuscuta* shows the most diverse adaptations to the host, due to the presence of both root and shoot parasites, as well as hemi- and holoparasitic behaviors. *Cuscuta reflexa* represents a hemiparasitic shoot parasite, while *Cuscuta europea* is a holoparasitic root parasite. The largest family of parasitic plants, with 90 genera, are the Orobanchaceae, all of which are root parasites; this genus contains 70 hemiparasitic but only 20 holoparasitic organisms.

Among the plant parasites that cause the greatest economic damage is the genus *Striga* (witchweed). This is a typical holoparasitic root parasite that colonizes crops such as maize (*Zea mays*), sorghum (*Sorghum bicolor*), and legumes such as cowpea (*Vigna unguiculata*). The parasite thus impacts food resources mainly in Third World countries that are already under considerable strain. The life cycle of *Striga* is shown in Figure 2.40. For germination, the plant requires signals from the host that are similar to the germination and branching signals involved in arbuscular mycorrhizal formation (Section 2.4). In the case of *Striga* and *Orobanche* (broomrape), these signals are known to be terpene plant hormones called strigolactones; however, some *Striga* and *Orobanche* species have specifically evolved the use of other compounds as germination signals. *Orobanche cumana*, a parasite that specifically infects sunflowers, recognizes a sesquiterpene lactone that is structurally distinct from strigolactones, for germination. *Striga* can also recognize phenolic compounds. *Phelipanche ramosa*, a recently identified parasite to oilseed rape (*Brassica napus*), germinates in response to isothiocyanates released from glucosinolates of the host. Originally, such signals could be produced by the plant to defend itself against competing plants or herbivore attack, but the parasitic plants, in turn, have evolved receptors for these secondary metabolites for host recognition.

The next stage, haustorium formation, also requires signals from plants. Haustoria are employed as the tissue connection at the host–parasite interface to facilitate nutrient uptake from the host plant (Figure 2.40). From the parasite, the haustoria connect to the vasculature of the host plant (the same situation applies to shoot parasitic plants). The haustoria from parasitic plants have highly modified xylem cells but lack phloem; they should not be confused with the haustoria of phytopathogenic fungi and arbuscular mycorrhiza (where they are termed arbuscules), which extend into the host cell without penetrating the plasma membrane, although their function in nutritional exchange is similar. A stem parasite, *Cuscuta pentagona*, uses a blend of volatile compounds, such as β-phellandrene, β-myrcene and α-pinene, for locating its host.

Another example of a parasitic plant is the facultative parasite *Triphysaria versicolor*, which is able to recognize a range of compounds from the host plants, including anthocyanins. Some anthocyanins are able to increase the number of haustoria in *Triphysaria versicolor* (Figure 2.41). A comparison between the structurally similar molecules peonidine and pelargonidine showed that recognition by the parasite is quite specific, and haustoria are formed only in the presence of peonidine. Haustoria formation is also concentration-dependent. A methanolic extract from maize was able to induce haustoria formation in similar fashion as a root extract from *Arabidopsis*, which indicated a rather broad host range of the parasite. However, haustoria can also be induced by the simple quinone 2,6-dimethyl-benzoquinone, which acts as a redox signal.

Other parasitic plants themselves produce secondary metabolites that are not involved in host–parasite recognition, but are intended to attract pollinators (Section 2.2.3). For example, the roots of the genus *Hydnora* produce directly white fleshy flowers, which have a foul odor to attract beetles for pollination. Similarly, the flowers of *Rafflesia*, a genus which comprises species with the largest flowers in the world, produce volatiles that smell distinctly like carrion in order to attract pollinating flies. The plant itself lacks stems and leaves and consists, besides the flowers, only of thin, filamentous haustoria which, due to their small size, can easily be confused with fungal hyphae. To further mimic carrion, the flower is also darkly pigmented.

To protect themselves from pathogens and herbivores, parasitic plants are able to synthesize defense compounds such as viscotoxin, an antimicrobial peptide from mistletoe. Other genera, such as dodder or Devil's thread (*Cuscuta*), Indian paintbrush (*Castilleia*) or beard-tongue (*Penstemon*), can take up secondary defense compounds from their hosts, most likely directly via the vascular connections. As

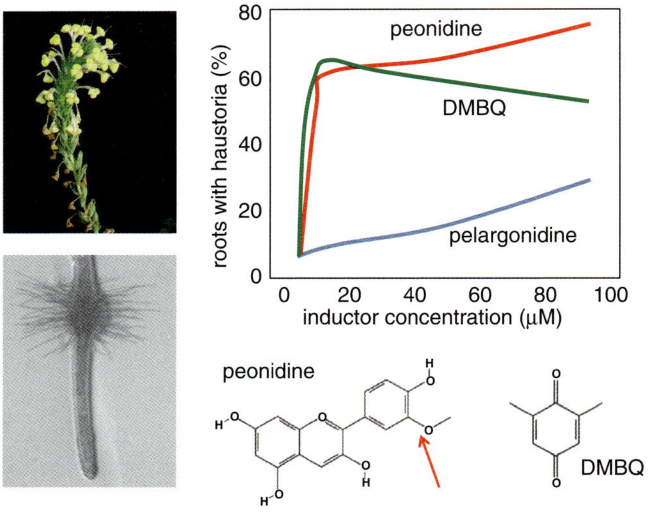

Figure 2.41 Induction of haustoria on roots of the parasitic plant *Triphysaria versicolor* by anthocyanins. The upper image shows the flowers, and the lower image the roots with haustoria. A comparison between the structurally similar molecules peonidine and pelargonidine shows selectivity of the recognition mechanisms by the parasite. The functional group in peonidine marked by a red arrow is missing in pelargonidine. Formation of haustoria is also concentration-dependent. A good inductor of haustoria is the quinone DMBQ (2,6-dimethyl-benzoquinone), which acts as redox signal. Pictures courtesy of J. Yoder, University of California, USA. Chemical structures are from The PubChem Project.

many alkaloids are found among these compounds, the pattern of alkaloids in the parasitic plant would appear to be very similar to that of the host plant. Typical examples are the quinolizidine and pyrrolizidine alkaloids, which are not synthesized in the parasitic plant, but can be found in high concentrations there (Figure 2.42). In the case of alkaloids, the defense reaction is mainly directed against insects.

2.6 Allelopathy

The term allelopathy derives from the Greek *allelon*, meaning "from each other," and *pathos*, which means "suffer." The definition of allelopathy includes the interaction between organisms producing and releasing chemical compounds directly. The compounds can also be produced, released, and then converted to toxic compounds after they have been secreted into the environment. The inhibitory effect of allelochemicals may be on the same species or on a different species, and also on microorganisms and fungi. In a broad sense, all secreted defense compounds are allelochemicals, but in a more narrow sense the term is used especially to

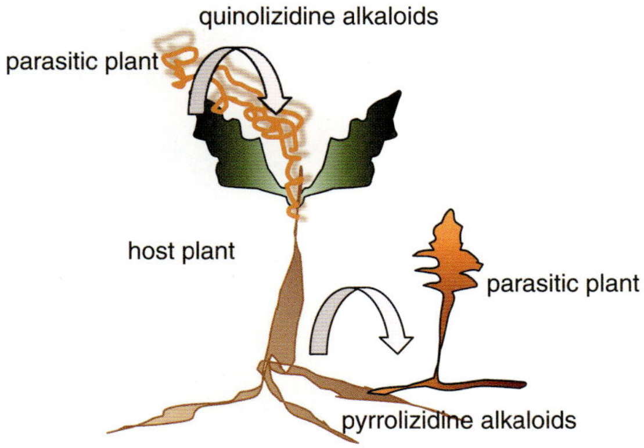

Figure 2.42 Some parasitic plants can accumulate alkaloids, which are derived originally from their respective host plants. Species of the genus *Cuscuta* (dodder) take up quinolizidine alkaloids from their host plant *Spartium iunceum* (Fabaceae); the patterns of pyrrolizidine alkaloids in some *Castilleia* (Indian paintbrush) and *Penstemom* (beard-tongue) species also resemble those of their hosts.

describe the inhibition of organisms that compete for the same habitat. The allelopathic effects on the same organism, as occur during the cultivation of one plant species over a long time period, is termed "autotoxicity." This phenomenon should prevent competition by the same species for the same growth space. Allelochemicals are products of plant secondary metabolism; their concentration and properties in soil are determined by the intensity of their secretion and their degradation by soil microbes. Soil properties such as pH value also play an important role in the function of allelochemicals. The identification of allelochemicals has led to the discovery of many novel compounds for biological plant protection. These compounds are not always secreted throughout the entire life cycle of a plant, and they may be synthesized in only some plant organs. The levels of allelopathic compounds are usually higher in younger plants, although there are exceptions where the mature plant in particular synthesizes large amounts of certain compounds. Such synthesis is usually downregulated in senescing plants or organs, however.

It should be noted that allelochemicals are often only protective against plants which are not natural to the habitat. For example, juglone (Figure 2.43) has germination- and growth-inhibiting effects only on plants not native to the original geographic range of the producing walnut tree (*Juglans regia*). Similarly, plant growth under Eucalyptus trees is inhibited by the allelochemicals secreted by these trees only outside of Australia, because the plants in non-native habitats could not adapt to these compounds.

One well-studied example of allelochemicals is the germination inhibitor juglone, from walnut. The glycoside of juglone is synthesized in the green tree crown and then washed out by rain into the soil, where it is metabolized by

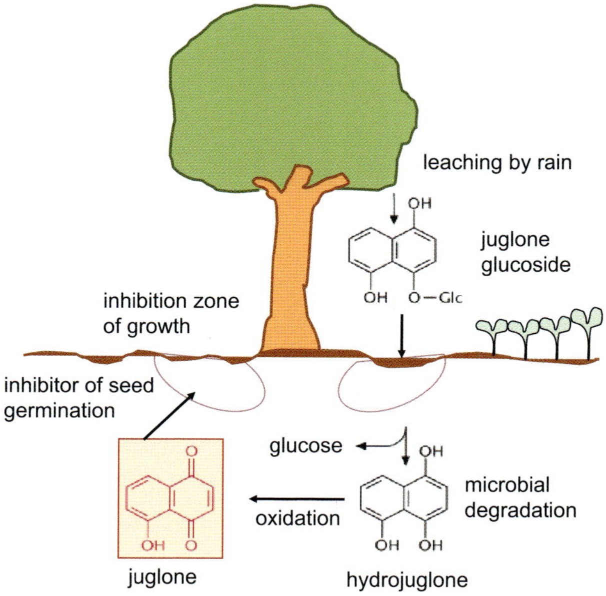

Figure 2.43 Synthesis and metabolism of the germination-inhibiting compound juglone of walnut trees (*Juglans regia*). The glycoside is synthesized in the leaves, washed out by rain, and converted by soil microbes to the active principle. Modified from *Strasburger Lehrbuch der Botanik*, 36. Auflage, Spektrum-Verlag.

glucosidases from microorganisms to form the toxic juglone (Figure 2.43). As juglone has a germination-inhibiting effect at only low concentrations, an inhibitory zone is formed around the tree. The seed shells of the walnuts also contain, besides tannins, a growth-inhibiting compound. Juglone also has both antibacterial and fungicidal potential.

Volatiles can also serve as allelochemicals, not only through direct effects in the air but also in the soil. The growth-inhibitory effect of purple sage (*Salvia leucophylla*) on its surrounding vegetation is brought about by monoterpenes, which are emitted at high concentrations and may precipitate with the nightly dew (Figure 2.44). A concentration effect leads to a zone with no growth, followed by a zone with meager growth, and finally a zone with normal growth of other plants. The major components are camphor and 1,8-cineol. Indirectly, the airborne monoterpenes are responsible for the growth-inhibitory effects of the sage vegetation. However, if the sage plants outnumber the other plants, the concentration of monoterpenes in the air may reach a critical level, and in hot weather the terpene mixture may ignite spontaneously, such that the fire will extinguish the dominant sage vegetation. The seeds of other species, of which germination had been inhibited by the terpenes, may then germinate and the vegetation will be renewed. This phenomenon has been observed in the shrublands of the Sierra Nevada.

As shown in this second example, it is not only single components that can elicit an allelopathic effect but also a combination of several compounds, with

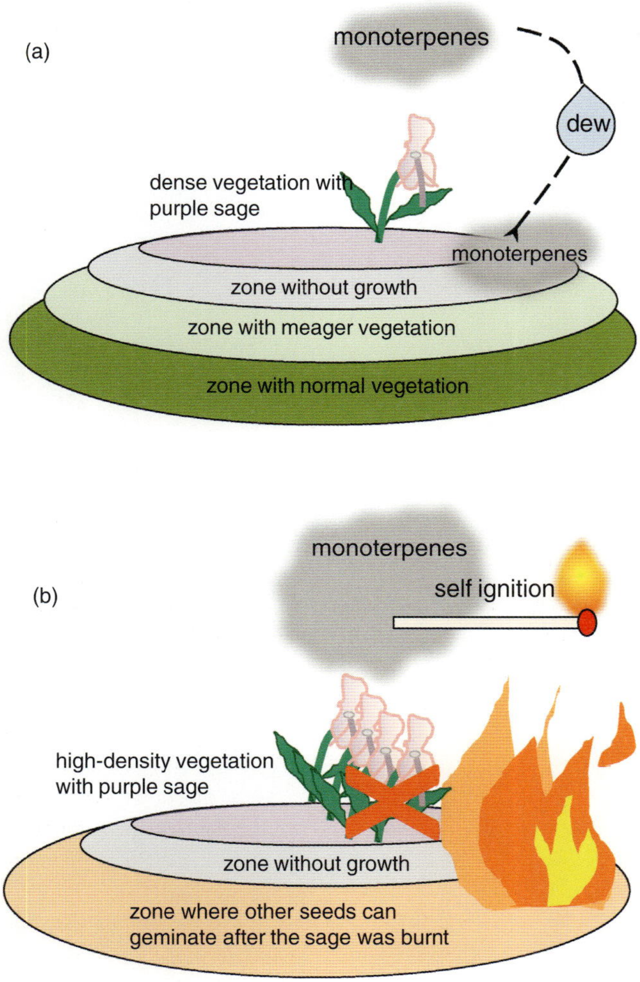

Figure 2.44 Terpenes as allelochemicals, and their role in an ecosystem. Purple sage (*Salvia leucophylla*) plants emit high concentration of monoterpenes. (a) The monoterpenes reach the soil, together with the nightly dew. They are growth-inhibitory, so that there is a zone with no growth around the sage plants, followed by zones with meager growth, and then normal growth; (b) If the population of sage plants becomes too dense and the emission of monoterpenes rises above a critical level, the terpene mixture can ignite spontaneously, which results in destruction of the sage plants by the fire. In consequence, other seeds can then germinate in the soil without the inhibitory monoterpenes.

mixtures being more potent than individual compounds. Other allelopathic substances include tannins, cyanogenic glycosides, glucosinolates, different flavonoids, and phenolic acids such as ferulic, *p*-coumaric, syringic, vanillic and *p*-hydrobenzoic acids.

3
Biological Effects on Microorganisms and Animals

3.1
Coevolution

Plants and insects have for a long time shared their habitats during the course of evolution. At the start of this coevolution, simple plant destroyers in the Devonian period were able to use lower land plants for feeding (Figure 3.1), but evolution to a greater degree of specialization was observed during the Carbon and Permian periods, with the development of simple sucking insects on pteridophytes, conifers, cycads, and gingkos. The development of leaf-chewing and mining insects occurred subsequently, between the Permian and Jurassic periods. Finally, pollen- and nectar-consuming insects emerged during the Jurassic and Cretaceous periods, and this in turn resulted in the evolution of modern pollinators of angiosperms in the Tertiary period. A higher degree of specialization of pests then emerged, together with an increased plant diversity. Most of the types of herbivore that evolved during these periods remain abundant today.

A hypothesis on the biochemical coevolution of species takes these arguments further. Based on the observation that angiosperms and insects are characterized by a high level of diversity, it was postulated that these organisms are situated in a type of coevolutionary race that can result in the formation of novel species. According to this hypothesis, the secondary metabolites of plants play major roles as both attractants and also as deterrents. The prediction is that new compounds must be developed by the plants in order to overcome the resistances of insects, and that the latter will need to adapt to the novel metabolites if they are to maintain their ecological niche. One argument in favor of this hypothesis is that there is a remarkable coincidence in the development of plant species and those insects which are specialists on these plants.

From these observations it can be concluded that plants, along with their pests, adapt during long-term periods. There is, however, a "tug-of-war" between the species for a selective advantage. Secondary metabolites serve not only as mediators in the chemical warfare in this coevolution, for example in recognition and/or detoxification by parasites, but also in the development of defense reactions against these pests on the plant side. Plants must be able to identify their invader(s)

Plant Natural Products: Synthesis, Biological Functions and Practical Applications, First Edition.
Herwig O. Gutzeit and Jutta Ludwig-Müller.

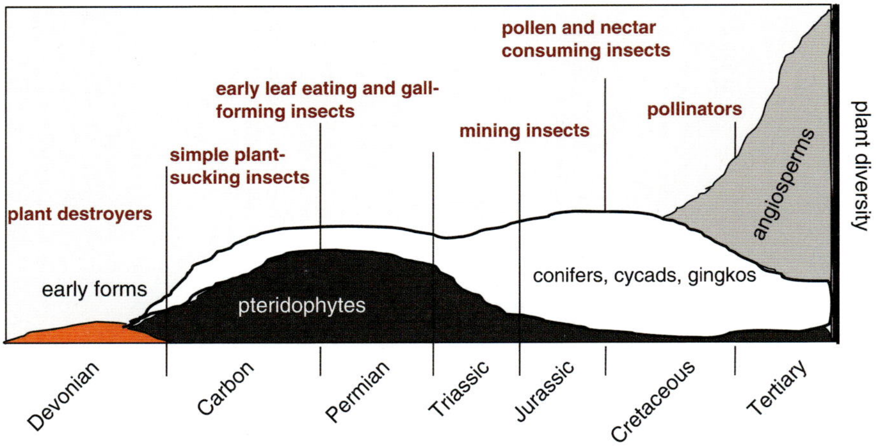

Figure 3.1 Plants and insects share a long history of over 400 million years of coevolution. Modified from L. Shyamal, Wikimedia (File: InsectPlantEvol.svg; May 2007).

by using receptor proteins that have been evolved to recognize a diverse range of signaling molecules – so-called "elicitors" – from pathogens and herbivores. Wounding produces signals that can also be detected by plants (Figure 3.2). The two recognition factors of herbivores or wounding are important to answer the question of whether an attack was made by a living organism, or whether an abiotic factor has created a wound. In addition, the variety of herbivores attacking a plant always causes different wounding patterns by the way in which they feed on the plant (see Section 2.5.3). In the interplay of plants and pests, the outcome is determined by the "power of weapon" and/or chemical innovation, and this situation has been coined the "race of arms." As the plant is not without protection against the vast array of pathogens present, defense reactions are mounted against the parasites. The plant possesses many receptors that are capable of recognizing quite diverse signaling molecules, such as pathogen-, microbe-, herbivore- and damage-associated molecular patterns that in turn lead to the so-called PAMP-triggered immunity (PTI). In this scenario, the plant will be resistant against the parasite. Parasites can now develop tolerance or resistance mechanisms in response to a defense reaction of the plant by using suppressor molecules (also called effectors), which are able to suppress the defense responses of the plant; in this case a successful colonization by the pathogen is again possible. However, these effectors can be recognized by a different set of plant receptors and elicit an effector-triggered immunity (ETI) where, again, the plant is resistant and the pathogen cannot colonize the host. Further, the pathogens can again synthesize effector molecules so they are able to suppress this second defense response and prevent colonization of the plant. This has been coined the "zig-zag model," and is symbolized by the circular arrow in Figure 3.2 to indicate that both partners are able to interfere with the resistance response of the other partner in the interaction between host and pathogen during evolution.

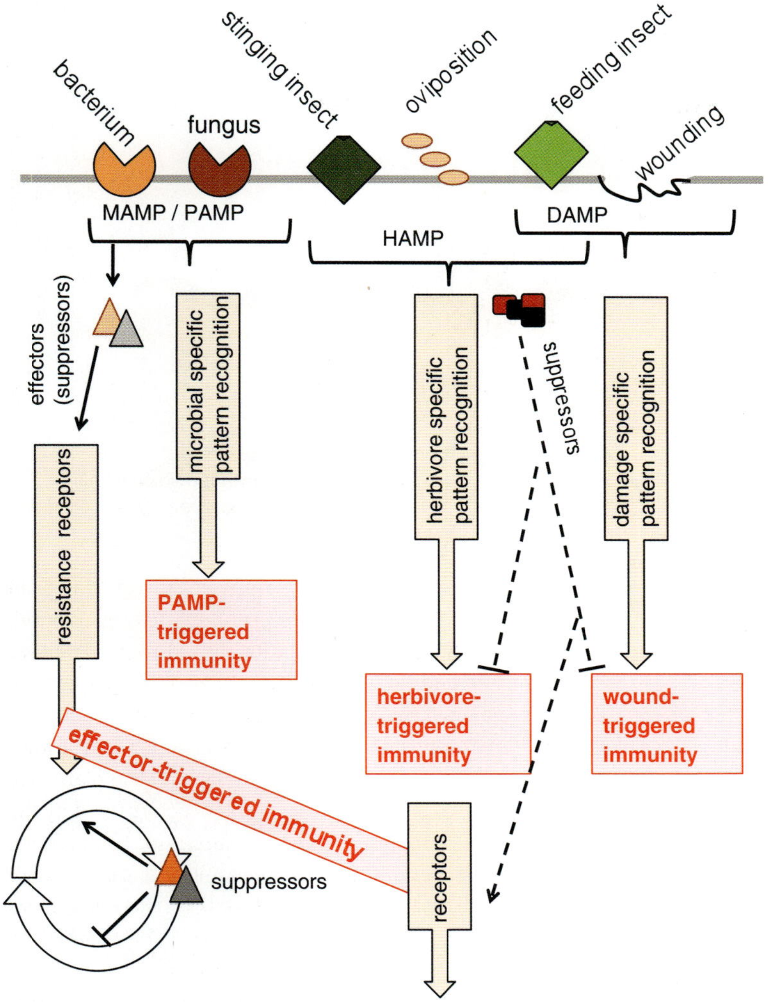

Figure 3.2 Perception of diverse signals by plants generated by microorganisms, herbivores and wounding, and the development of adaptation mechanisms by the host plant and the parasite during evolution. PAMP = pathogen-, MAMP = microbe-, HAMP = herbivore-, DAMP = damage-associated molecular patterns are recognized by plant receptors, which elicit an immune response. This defense reaction can be suppressed by effector molecules of the pathogens. These effectors are again recognized by a different set of receptors, leading to another round of defense reactions. Novel effectors can evolve, which can again inhibit the plant defense (symbolized by the circular arrow).

Plants are able to secrete up to 40% of their photosynthetic products into the rhizosphere, which then becomes an attractive place for microbes; consequently, the microbial density around a plant is much higher than in the surrounding soil. Since in this microbiome both beneficial and pathogenic microbes are present around a plant, the plant requires active mechanisms to control these microbes.

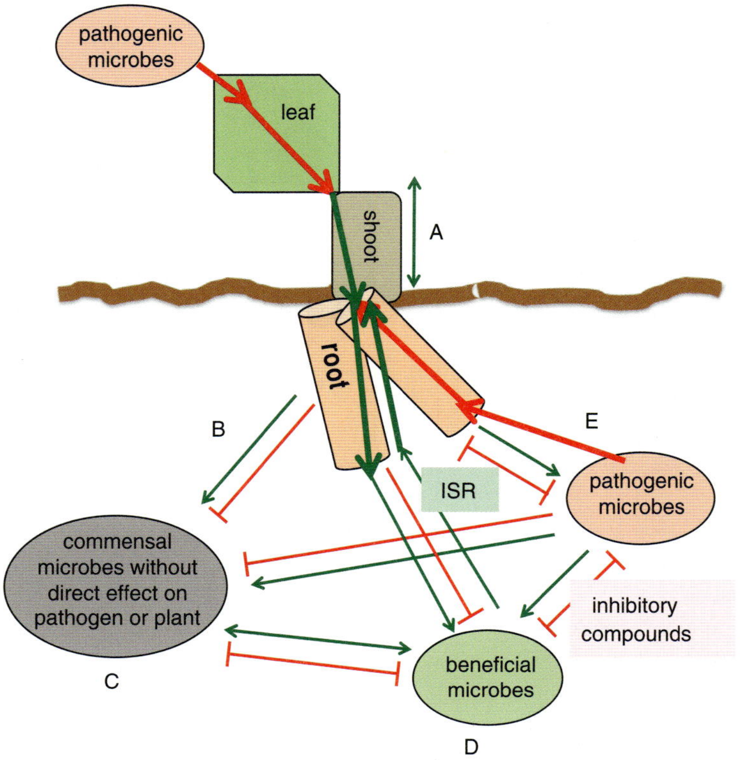

Figure 3.3 Plants actively shape their root microbiome. The microbiome is the community of microbes present in a certain area. Interactions in the rhizosphere through secretion of inhibitory (red) or stimulatory (green) compounds; plants can also recruit or activate beneficial microbes. (A) Induction of resistance in below-ground plant parts spreads to above-ground parts and vice versa; (B) Root exudates and other rhizodeposits stimulate and/or inhibit various microbes; (C) Most microbes will neither affect the plant nor the pathogen directly, but they will affect other microbes; (D) Either promote the plant directly or protect the plant by inhibiting deleterious microbes or through induced systemic resistance (ISR); (E) damage the plant through infection or production of phytotoxic compounds. Modified from Berendsen, R.L. *et al.* (2012) *Trends Plant Sci.*, **17**, 478–486.

The microbiome is the community of microbes present in a certain area, in this case in the vicinity of the plant root. However, by employing secondary metabolites plants can actively influence their root microbiotic flora (Figure 3.3). These allelochemicals (see Section 2.6) are not only generally synthesized by the plant and then secreted; rather, the composition of compounds in a given plant can be altered by pathogens or beneficial microbes. For example, the roots can secrete either inhibitory or stimulatory compounds which act on microbes by either increasing or decreasing their levels; alternatively, these compounds may attract certain microbes whereas others will be deterred. In consequence, the beneficial microbes might indirectly elicit a plant defense response – termed “induced systemic resistance” – or the pathogenic organisms might elicit directly a defense response. Again, the metabolites may be secreted into the soil and alter the rhizospheric community. It is interesting to note that differences between plant genotypes in a single gene can have a significant impact on the rhizospheric microbiome. One single metabolite – in this case a single specific glucosinolate – from a transgenic plant had changed the microbiome completely. Similarly, the invasion of above-ground plant organs can lead to the induction of secondary metabolites as defense compounds, and these could also be transported to below-ground plant parts, thereby contributing to changes in the composition of soil microbes.

Especially, the recognition of molecules from insect saliva is a great achievement of plants in their adaptation to the environment. Of course, insects synthesize these molecules not for recognition by the plant but rather because they are involved in the insect's primary metabolism. One well-known example of this is the conjugated fatty acid volicitin (Figure 2.36), which can be recognized by specific plant receptors and determines that the plant can distinguish between wounding and insect attack. For the insects, however, the compound is essential, and it was shown that in the moth *Spodoptera litura* it plays a role in the regulation of glutamine levels. These compounds are not synthesized in all lepidopteran species; occasionally, these substances act as suppressors of plant defense responses (Figure 3.2). Such conjugated fatty acids of the specialist herbivore *Manduca sexta* are able to suppress the production of nicotine, but not the emission of volatiles to attract parasitoid or zoophagous insects.

An impressive example of the interaction between insects has been observed for a wild tobacco species (*Nicotiana attenuata*). When a plant is in its natural habitat, one of the major threats is to be infected by a parasite (Section 2.5.3). Transgenic tobacco plants, where the synthesis of toxic secondary metabolites (e.g., nicotine) is turned off, are quite attractive for caterpillars (Figure 3.13). A similar effect has been observed in different varieties of tobacco with natural variations in nicotine levels. However, one toxic compound is normally insufficient to protect the plant, and therefore these toxic secondary metabolites are supported by other mechanisms in the defense against herbivores. This is necessary, because the parasites have also adapted to the plant's strategies. For example, a caterpillar feeding on plants with high concentrations of protease inhibitors (see Figure 2.36) might need more of these leaves for sufficient nutrition, because the protease inhibitor prevents digestion. Among the defense compounds are not only those directed against the herbivore, but also volatiles, which can attract predatory insects and thus reduce the plant parasite population (Figure 3.12). The jasmonate-dependent plant defense cascade must also be intact (Figure 2.36). Certain cicadas of the species *Empoasca* can actually test tobacco leaves for the presence of this defense mechanism, by using their mandibles to feed only on leaves without defense reactions. The insect's behavior is independent of the presence or absence of nicotine, which is induced by the jasmonate-dependent signaling pathway.

3.1.1 Adaptation Mechanisms of Microbes and Herbivores

Organisms are exposed to many influences in their given environment, amongst which are many secondary metabolites (Figure 3.4). In this scenario, both above-

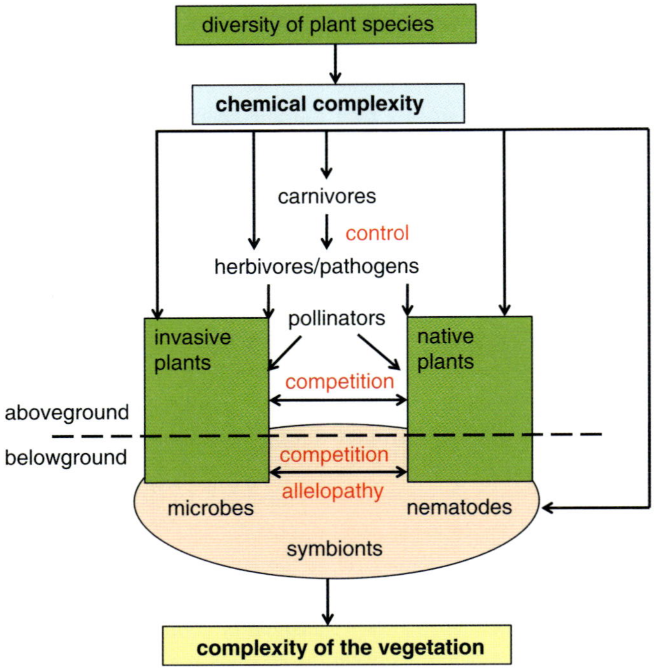

Figure 3.4 Secondary metabolites in the environment. The composition of metabolites in plants is influenced by pathogens and herbivores. On the other hand, the plant influences the environment by secretion and emission of chemicals. In the soil, microbes can alter the compounds by their own metabolism. These metabolites can attract beneficial organisms or act in the defense against pathogens; they could also act on a different set of species than the originals. In the atmosphere, the emitted compounds attract pollinators and also zoophagous insects. Leaf metabolites can play a direct role in defense responses as toxic compounds. This chemical complexity is the result of the diversity of the plant species, which are responsible for the complexity of the total vegetation.

ground and below-ground interactions must be taken into account. The specific flora is determined by the composition of other vegetation, with native plants being in competition with invasive species, both above and below ground. In order to compete successfully for the habitat, allelochemicals are synthesized and either secreted or emitted. Secondary metabolites attract pollinators as well as parasites; invasion by the latter will result in changes in the plant's metabolite pattern that can in turn influence other organisms such as pathogenic or symbiotic microbes (Figure 3.3). The rhizosphere is also the habitat for nematodes, which encompass many plant parasitic species. These may be controlled by applying nematicidal compounds such as the alpha-terthienyl from *Tagetes* species (Section 2.5.3). It is possible for all organisms that live in this specific environment to adapt to the plant-specific metabolites, and to achieve this they have developed a plethora of different mechanisms (Table 3.1).

Table 3.1 Mechanisms for the adaptation of specialized insects and microorganisms to plant secondary metabolites. Herbivores have developed a larger range of different mechanisms to cope with the plant's defense responses than microbes.

Organism	Adaptation to host plant	Adaptation strategy
Insects/Pollinators	Recognition of nutritional sources	
Insects/Herbivores	Recognition of the host plant for nutrition and/or oviposition	
	Avoidance of toxic plants or plant organs	Feeding on parts without secondary metabolites
	Cutting out tissues or cells filled with secondary metabolites, such as laticifers or resin ducts	
	No uptake or fast passage of the intestinal tract	
	Uptake is followed by detoxification mechanisms	Hydroxylation
		Conjugation
		Elimination
	Uptake and accumulation	Specific compartments for distribution
		Evolution of insensitivity
	Utilization of secondary metabolites	Defense against zoophagous insects (Digitalis glycosides, cyanogenic glycosides, glucosinolates, pyrrolizidine alkaloids etc.)
		Signaling molecules, pheromones (pyrrolizidine alkaloids)
		Morphogen (pyrrolizidine alkaloids)
Microorganisms	Inactivation of allelochemicals	
	Secretion of chemicals	
	Evolution of insensitivity	

3.1.1.1 Fungi

The effect of some secondary metabolites in plant protection, for example biofumigation (see Section 2.5.2), is not always as strong as might be predicted, even though the compounds have a high antimicrobial potential. This is due to the fact that pests have adapted to these compounds (Table 3.1). One strategy used by many microorganisms is the metabolic detoxification of plant secondary compounds. For example, an isolate of the fungus *Aspergillus flavus*, identified from soil, showed a strong resistance to growth inhibition in the presence of glucosinolates. This resistance was based on the presence of two detoxifying enzymes for the glucosinolates, an arylsulfatase and a β-thioglucosidase with unusual metabolic properties. The thioglucosidase from the fungus was able to convert the glucosinolates directly to nitriles and not into the more toxic (iso)thiocyanates, as could the plant myrosinase. For other substances such as the benzoxazines, detoxification via glycosylation is possible (see Figure 2.29).

The fungus *Leptosphaeria maculans* is able to metabolize the most important phytoalexins of the host genus *Brassica*. The biotransformation of brassinin and brassicanal A allows the specialized fungus to colonize the host roots, which is not possible for generalists lacking this detoxification mechanism. This principle can be found in many plant–pathogen interactions, so that the property to detoxify phytoanticipins or phytoalexins is an important pathogenicity mechanism (see Figure 2.28).

A second mechanism for detoxification is the secretion via transport. For example, the fungus *Botrytis cinerea* was shown to have an ABC transporter that was capable of transporting the phytoalexin camalexin out of the cells. The presence of this transporter was determined to be an essential pathogenicity factor for colonization of the host plant *Arabidopsis*, whereas mutant strains for the respective gene were not able to colonize the host plant. In a plant mutant without camalexin the fungal mutant strain was able to invade again.

Finally, the development of insensitivity against fungicidal compounds is possible. An example of this is resistance against the fungicidal strobilurins, which originally was not obtained from a plant but was isolated and identified from the fungus *Strobilurus tenacellus*. For practical application, strobilurins were later chemically modified to increase their stability in the soil. Their fungicidal effect is due to an inhibition of the cytochrome *bc* complex of the mitochondrial electron transport chain at the inner mitochondrial membrane. Simple point mutations in the gene for cytochrome *b*, which led to the exchange of a glycine to alanine in the protein, resulted in an insensitivity to the fungicide. The development of insensitivities to plant secondary metabolites is also possible.

3.1.1.2 Insects

Insects have to deal with several different physico-chemical structures of plants, which have been developed by plants to defend themselves against herbivores (Figure 3.5). Among these structures are thorns (causing avoidance responses), trichomes (containing chemicals), wax and walls, as well as latex and resins. These different barriers demand a variety of adaptation responses by the insects.

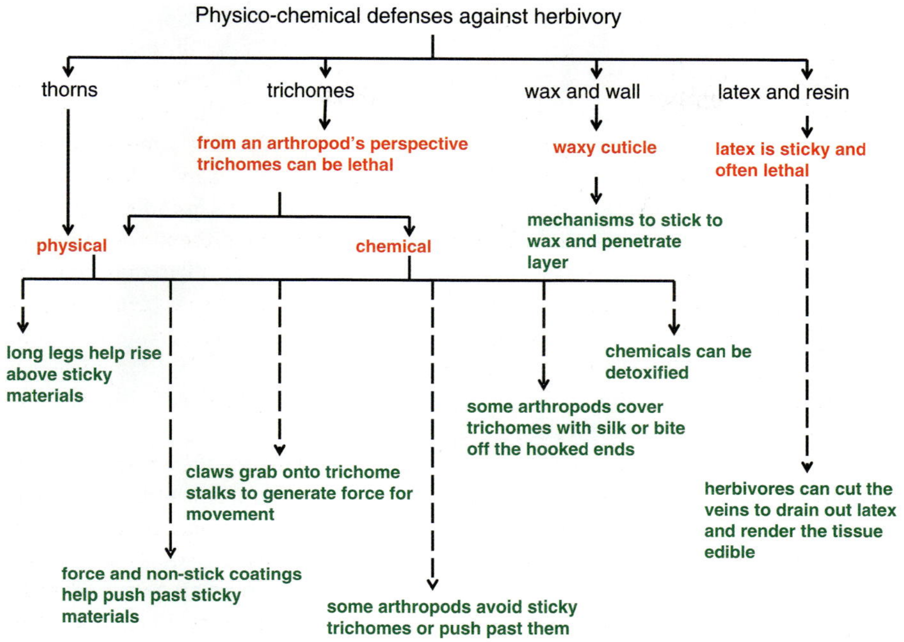

Figure 3.5 Adaptation mechanisms of insects to physico-chemical plant surface structures to avoid toxic metabolites. The consequences for a non-adapted insect are shown in red; various possible adaptations are shown in green.

Probably the easiest way to cope with defense compounds is avoidance, which can be achieved by the mechanisms outlined in Table 3.1 and Figure 3.5. Some examples of the adaptation mechanisms summarized in Figure 3.5 are shown in Figure 3.6. Trichomes (which can be a lethal mechanical and chemical trap) and thorns can be avoided by the insect having long legs, which help it to rise above any sticky materials. Such an adaptation also helps to prevent opening of the trichomes by wounding, which would result in the contact with secondary metabolites. Force and nonstick coatings of tarsi also help the insect to push past sticky materials, and some arthropods can avoid sticky trichomes completely. The most elaborate way to deal mechanically with trichomes is to develop claws, which allow the insect to grab onto the trichome stalks and to generate force for movement. By applying force the trichomes are bent away so that the insect can pass. Some arthropods cover trichomes with silk, or bite off the hooked ends. However, if the cuticle is designed to make it difficult for an insect to walk on the leaf surface, then the adaptation strategy must be opposite – that is, mechanisms to stick to wax must be developed and eventually allow penetration of the wax layer. The preference of insects to feed on leaf parts with low concentrations of secondary metabolites has been described for glucosinolates (Figure 2.35). The chemicals liberated from trichomes can be detoxified, while herbivores can also cut the veins to drain out the latex and render the tissue edible (Figure 3.6).

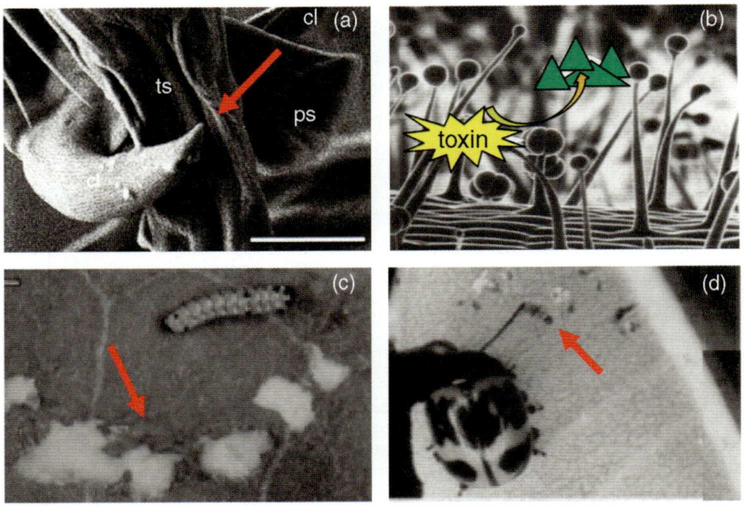

Figure 3.6 Some examples to illustrate different strategies to avoid the plant's defense mechanisms (shown in Figure 3.5). The red arrows point to events described in the text. (a) Some arthropods avoid sticky trichomes or push past them; here, a claw is shown grabbing a trichome base. Photograph from Voigt, D. and Gorb, S. (2010) *Arthropod-Plant Interact.*, **4**, 69–79 (see Further Reading). Reprinted with permission from Springer; (b) Trichomes from the wild potato *Solatium berthaultii* can release chemical deterrents to arthropods, which in turn can be detoxified by the insect specialists. Photograph from Gibson, R.W. and Pickett, J.A. (1983) *Nature*, **302**, 608–609 (see Further Reading). Reprinted with permission from Nature Publishing Group. Latex can be avoided through trenching (c) or vein biting (d). Photographs from Dussourd, D. and Eisner, T. (1987) *Science*, **237**, 898–901 (see Further reading). Reprinted with permission from The American Association for the Advancement of Science.

The data listed in Table 3.1 indicate that insects can not only detoxify the metabolites but also use them for their own metabolism. The ability to sequester plant metabolites and turn them into defense compounds represents a high degree of specialization and adaptation by the insects. Several selected interactions will be described in the following sections, demonstrating different adaptive mechanisms. A number of taxonomically unrelated specialized insect herbivores sequester pyrrolizidine alkaloids from their food plants. Many butterflies and moths (Lepidoptera) and certain chrysomelid leaf beetles (Coleoptera) are known to store and utilize plant-acquired pyrrolizidine alkaloids for defensive purposes. Larvae of the moth *Tyria jacobea* store pyrrolizidine alkaloids without changing them, and thus become unpalatable for zoophagous enemies. In order to demonstrate this unattractiveness to predators, all developmental stages are brightly colored.

There may be a flow of plant-produced defensive compounds through three trophic levels – that is, from the plant producer via the herbivore to its predator. Ladybirds (*Coccinella septempunctata*) grazing on the aphids can ingest the pyrrolizidine alkaloids from these specialized pests, which take up alkaloids from the plant via phloem sucking. Three subfamilies of the butterfly family Nymphalidae

(Danainae, Ithomiinae, and Arctiidae) synthesize pheromones during the larval stage, using the alkaloids that they have taken up from the plants as precursors; however, these larvae also store the plant metabolites as protection against predators. Other *Danaus* species metabolize the pyrrolizidine alkaloids of the plant to the compounds danaidone, danaidale and hydroxy-danaidale, all of which are components of liquid secretions via small hairs. These compounds function as pheromones and are thus important for reproduction. A further adaptation by the insects is the use of these alkaloids as precursors in the biosynthesis of substances that the insects need for pattern formation, the so-called "morphogens." These compounds form gradients within the body, which then leads to morphogenesis and to the differentiation of organs.

Some insects employ glucosinolates from their host to protect themselves against predators. Alternatively, these compounds can also be metabolized or excreted. Both, chewing and sucking insects feed on glucosinolate-containing plants (see Table 2.14), with the glucosinolates being taken up from different plant tissues, typically from the leaves by larvae and caterpillars or from the phloem by aphids (Figure 3.7). In the aphid *Myzus persicae*, the glucosinolates taken up from the phloem are secreted unaltered via honeydew. The glucosinolates are not degraded through the stabbing movement of phloem-sucking aphids in the plant, but are ingested without chemical modification. For the general inactivation of the intact glucosinolates in insects, β-thioglucosidases similar to plant myrosinases are involved. The presence of a myrosinase has been demonstrated in the aphid *Brevicoryne brassicae* (Figure 3.7). Sulfatases and nitrile specifier proteins have also been isolated which can contribute to the degradation of glucosinolates. The evolution of a specific mechanism is dependent on the feeding behavior of the pest and, in consequence, on the metabolic fate of the glucosinolate after ingestion. If the glucosinolates are already degraded while the herbivore is feeding on the plant, the toxic isothiocyanates are further metabolized and detoxified to nitriles by the insect. Typical herbivores which wound leaves by chewing are *Plutella xylostella* and *Pieris rapae*; the former species can degrade glucosinolates in the gut via the action of sulfatase, while the latter species produces mainly nitriles due to the presence of a nitrile-specifier protein. Both metabolites are then excreted. While most herbivores have these degradation enzymes within the gut, the myrosinase of *Brevicoryne brassicae* is located in the muscles, so that the intact glucosinolates must be transported in the hemolymph before being brought together with the enzyme for degradation.

In the case of *Pieris rapae* an additional adaptation mechanism has been elucidated. If the insect takes up benzylglucosinolates from the host plant, the so-called "mustard oil bomb" can be converted to a "cyanide bomb," which is even more toxic for possible predators. *Pieris rapae* possesses enzymes for the synthesis of cyanide directly from glucosinolate degradation products. The nitrile, which is generated from glucosinolates via a nitrile-specifier protein (see Section 1.1.2), is not further converted to carboxylic acids; rather, analogous to the degradation of cyanogenic glycosides, it is converted enzymatically to cyanide. However, if the cyanide levels increase too much in the insect the cyanide may be detoxified

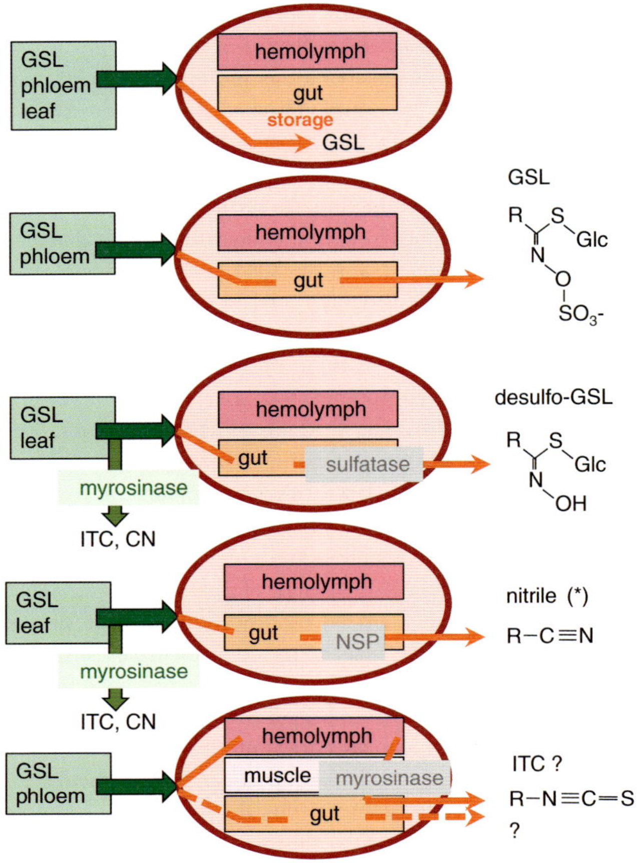

Figure 3.7 Glucosinolates (GSL) are taken up, metabolized and/or excreted from various plant parasitic insects. The first scheme shows the accumulation of glucosinolates without metabolism or excretion, which deters predation of the insects. In the second scheme the situation in aphids such as *Myzus persicae* is shown, where the glucosinolates are excreted in the honeydew as intact molecules. In the third scheme, which shows the typical situation for butterfly larvae such as *Plutella xylostella*, the glucosinolates are metabolized by a sulfatase in the gut of the larvae. In *Pieris rapae* (fourth scheme) the glucosinolates are metabolized in the presence of a nitrile-specifier protein to nitriles, which are less toxic. The aphid *Brevicoryne brassicae* (fifth scheme) has a myrosinase, which can degrade the glucosinolates.

enzymatically to β-cyanoalanine (see Figure 1.12). The larvae are able to grow on transgenic *Arabidopsis* plants containing the cyanogenic glycoside dhurrin, which is normally highly toxic to species feeding on Brassicas. This shows that the larvae have adapted to novel ecological challenges by exploiting new potential hosts, and use the plant secondary metabolites to their own advantage.

The alkaloid nicotine is used by larvae of the hawkmoth *Manduca sexta,* a herbivore on wild tobacco (*Nicotiana attenuata*), for their own defense against predators (Figure 3.13). The specialist parasites can degrade nicotine by specific cytochrome

P450-dependent monooxygenases, and their metabolism of nicotine is actually much more efficient than that of humans. However, the larvae also serve as a source of nutrition for specialized lizards that feed more on larvae without or with low nicotine levels, but avoid larvae containing large amounts of nicotine. Hence, the accumulation of nicotine provides a good protection against predators. As nicotine is also toxic to the larvae, there needs to be a delicate balance between the degradation of the toxic compound in the insect and the need to maintain a certain concentration as feeding deterrent.

Larvae of *Chrysomela* species (Coleoptera) have developed a defense mechanism which depends on the secretion of toxic compounds from glands. These insects are specialized on a few plant families such as Salicaceae, Brassicaceae or Ranunculaceae, and the defense reaction is based on the *de-novo* synthesis of monoterpene iridoids by the larvae. These compounds are built from host metabolites such as *O*-glycoside-salicin as precursors, which are taken up and transported in the larvae. After deglycosylation, the aglycone is further metabolized enzymatically to the respective aldehyde, salicylaldehyde. Only a small modification of the basic structure will result in different transport velocities in the insects, so that further modifications of the compound will result in either a fast secretion or an accumulation.

3.1.1.3 **Vertebrates**

Some plants and fungi produce substances which are toxic (Section 4.2.2) and deter herbivores from feeding on the respective plant, although only a few examples exist of defined chemical deterrents for mammalian herbivores. The brown-eyed rockrose (*Cistus ladanifer*) produces exudates with flavonoids such as apigenin and 3,7-di-*O*-methyl kaempferol, which strongly inhibit the Ca^{2+} ATPase activity of the sarcoplasmic reticulum. The activity of the enzyme is essential for relaxation of the skeletal mouth musculature, and it has been suggested that this inhibitory activity deters herbivores such as rabbits from feeding on this plant.

One peculiar way to handle the "mustard oil bomb" (see Figure 1.11) has been developed by a small desert mouse (spiny mouse) from the Israeli Negev Desert, in conjunction with the bushy flowering plant known as "sweet mignonette" (*Ochradenus baccatus*), which belongs to the glucosinolate-producing Resedaceae family. The compartmentation of the "mustard oil bomb" is neither cell- nor organelle-specific, but is in fact tissue-specific (Figure 3.8). Within the berries of *O. baccatus* the myrosinase is located in the fruity pulp, while the glucosinolates are concentrated in the seeds. If the fruit is eaten as a whole, then the two components come together in the mouth of the feeding animal, such that the mustard oils (isothiocyanates) are formed. This not only confers a bitter taste but the compounds are also toxic. Normally, the plant is protected against animals that would eat the berries, but the specialized desert mouse feeds only on the pulp and spits out the seeds. This means that the two components of the mustard oil-forming system are kept separate and, as a consequence, the fruit is seen as tasty food by the little rodent and can also be properly digested. This behavior has been identified in three rodent species (*Acomys cahirinus*, *Acomys russatus* and *Sekeetamys*

Figure 3.8 A desert mouse can avoid the "mustard oil bomb". By compartmentation of glucosinolates (GSL) and myrosinase in different tissues inside the fruit, the fruit is not attractive to herbivorous animals. Myrosinase is located in the pulp, and GSL in the seeds. After eating the whole fruit the isothiocyanates are formed, which are bitter and toxic. This specialized rodent spits out the fruit and thus avoids the metabolism of the GSL. The plant also has the advantage that the seeds are dispersed in the surroundings, usually in shady locations.

calurus). The aim of the plant is that the berries are eaten only by a few animals, which do not retain the seeds; rather, as the seeds are spat out they will be distributed in the environment such that the "pest" has become a beneficial animal that distributes the seeds. The advantage for the plant is that the seeds are carried away to shady places, which the rodents prefer. Shady locations are better for seed germination because there is more moisture than in sunny areas, and other adjacent plants will protect the seedlings against high irradiation. Another benefit was that the germination rate of the seeds spat out by the mice was higher than that of seeds still in the pulp, which might explain why so many seeds have a bitter taste. Some vertebrates may have adapted to these tastes and distribute the seeds of other plants in a similar fashion. For the plant, this is advantageous in two ways: (i) it ensures protection against most herbivores; and (ii) it provides an optimized seed dispersal system. For the animal, the advantage is the availability of fruit that are normally avoided by other herbivorous competitors.

There is, however, a pitfall of the animal behavior for the plant. It was observed that, occasionally, the rodents would return some hours later and then eat the seeds, when the myrosinase-containing pulp was no longer present in their mouth. In this case, the glucosinolates in the seeds could not be degraded and so would not be metabolized to the bitter and toxic mustard oils. In laboratory experiments, mice would feed on both the fruit and the seeds together if the "mustard oil bomb" had been deactivated in the plant.

3.1.2
Attraction of Insects

Some examples of the high degree of specialization of insects are provided in the following sections. It should be noted that the examples in Sections 2.2.3 and 2.2.4 also describe coevolutionary aspects in the biology of pollination and seed dispersal more from the plant's point of view. However, in both cases the plant and the respective beneficial organism had to undergo a high degree of adaptation during evolution.

Next to volatiles emitted to attract animals for pollination and reproduction, these volatile metabolites may have additional functions. For example, they may act as cues for insects to find plants as nutritional sources via flower or leaf volatiles. The synthesis of volatiles is advantageous to the plant because they can serve as defense signals in herbivory. During evolution, specialists have employed these volatiles for their own benefit to identify host plants for both oviposition and/or feeding; moreover, nonvolatile compounds can also be perceived. The insects have evolved mechanisms to sense the blend of secondary metabolites from a specific plant in a given situation, and thus have the ability to recognize the plant species in question (see Figure 3.11).

It is not only the flowers of angiosperms but also the sporophytes of mosses that can be fertilized by insects attracted to these plants (Figure 3.9). In several sets of experiments, two springtail species (*Folsomia candida* and *Sinella curviseta*) and two moss species (*Ceratodon purpureus* and *Bryum argenteum*) were used to determine if the springtails could recognize specific blends emitted by the moss plants. First, the male and female plants of *C. purpureus* were shown to emit different volatile blends. Second, the increase in springtail numbers resulted in more fertilized moss plants using two species in "microcosm" experiments. Third, olfactometer measurements on springtails showed that the two species preferred the scent emitted by female moss plants over that emitted by male plants. Despite their lack of flowering structures, mosses – in analogy to flowering plants – utilize volatile scents as cues to manipulate arthropod behavior, which in turn results in an increased moss fertilization. As a result of this, a plant–pollinator-like relationship has evolved between these two groups of organisms.

The specific patterns of glucosinolates in a leaf can be recognized by the tarsi of some insects, which either feed or reproduce on glucosinolate-containing plants. Figure 3.10 shows a scanning electron microscopy image of the ventral side of a prothorax tarsus of a female *Delia radicum*, displaying several such sensors. The D-sensors are able to react to the presence of glucosinolates; the extracellular response of a D3-sensillum to various stimuli is shown at the right-hand side of Figure 3.10. Here, the first stimulus is a solvent control, the second is pure glucobrassicin, and the other two lanes relate to extracts

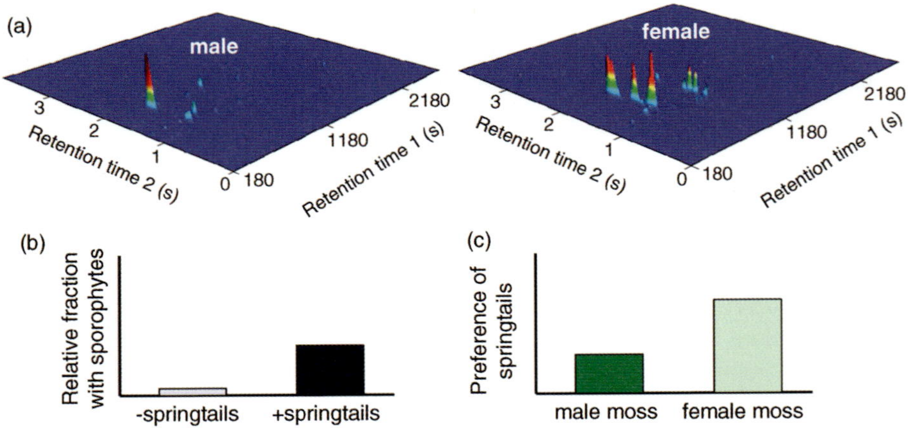

Figure 3.9 Gender-specific volatile profiles in a moss (*Ceratodon purpureus*) and preference of pollinating springtails for females. (a) Representative two-dimensional GC × GC–TOFMS chromatograms of volatile compounds from intact shoots of a reproductive male and a reproductive female of the moss *Ceratodon purpureus*. Color-coded concentration differences are shown, with red representing the highest; (b) Springtails enhance fertilization in moss microcosms: fertilization success in *C. purpureus* and *Bryum argenteum* microcosms, measured as the fraction of microcosms that developed sporophytes; (c) Springtails prefer female moss: the fraction of *C. purpureus* samples chosen by springtails in preference assays of male versus female samples in an olfactometer. Part (a) from Rosenstiel, T.N. *et al.* (2012) *Nature*, **489**, 431–433 (see Further Reading). Reprinted with permission from Nature Publishing Group. Parts (b) and (c) redrawn from the same publication.

from leaves: one from a light-grown plant and another from a dark-grown plant. The glucosinolate solution and the extract from a leaf of a light-grown plant are well recognized by the sensillum.

The perception of metabolite patterns can be taken even further. For the fruit fly *Drosophila melanogaster* (albeit not a plant pathogenic organism), it was shown that different brain areas can be activated based on the attractiveness of a compound or mixture (Figure 3.11). The terpene linalool triggers a negative response because the fly is not attracted by this compound, whereas the application of 3-methylthio-1-propanol serves as a positive signal. This signal is perceived in different areas of the brain. Next to the perception of individual compounds, the combination of different substances in a mixture is important and can be recognized. In volatile mixtures, which can be classified as either attractive or unattractive to insects, the individual compounds can be compared with one another. This results in a certain pattern where the presence or absence of a compound is important for the decision of the insect to be attracted to a plant, or not.

Although it is generally assumed that many flowering species may have converged on a similar scent profile to attract a specific pollinator or pollinator class, the identification of compounds that mediate the innate responses, and how learning alters the behavior of the pollinating insects, remain unclear. The olfactory

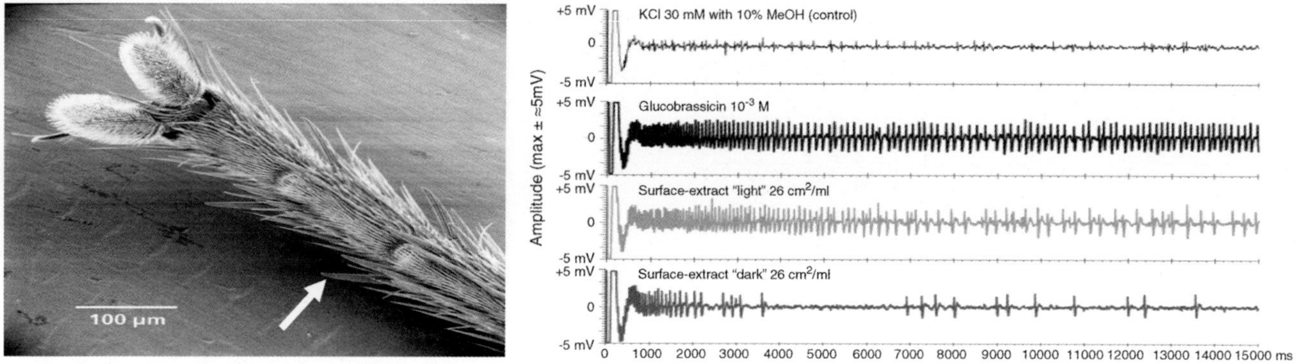

Figure 3.10 Insects can sense glucosinolates. Left: Scanning electron microscopy image of the ventral side of a prothorax tarsius of a female *Delia radicum*. The D-sensors can react to glucosinolates, as shown at the right. The extracellular response is shown for a solvent control, glucosinolate standard solution, and leaf extracts from light- and dark-grown plants, respectively. Courtesy of E. Städler, Germany.

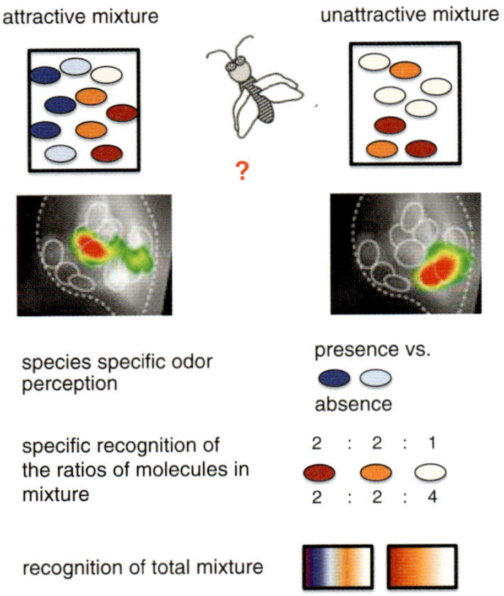

Figure 3.11 Perception of attractive and unattractive compounds or volatile mixtures by insects. The herbivores can recognize the mixtures using different combinations, but they can also react to single compounds. Center: Two images of a fruit fly brain (*Drosophila melanogaster*), which was activated by a positive and negative compound. The left image shows active glomeruli stimulated with a deterring compound (linalool); the right image shows the same brain after stimulation with an attractive compound (3-methylthio-1-propanol), both of which activate different brain areas. Picture of *Drosophila* brain courtesy of Max-Planck-Institut für Chemische Ökologie, Jena, Germany.

bases of these responses in a naturalistic context also remain poorly understood. The hawkmoth *Manduca sexta* has a wide geographic distribution and is an important pollinator for many night-blooming flowers, but can learn through olfactory conditioning to visit other nectar resources. An analysis of the flowers that are generally attractive to moths showed that the scents have all converged on a similar chemical profile that is uniquely recognized in the moth's antennal olfactory lobe. Although many of the hawkmoth-visited flowers differed qualitatively and quantitatively in their scent profiles, their scents were often dominated by oxygenated aromatic compounds, especially methyl benzoate, benzyl alcohol, and benzaldehyde. By contrast, flowers visited by other pollinator taxa were dissimilar in their scents in comparison to the hawkmoth-visited flowers. Experiments with untrained moths showed that they attempted to feed from flowers which emitted aromatic-rich scents. The moths also showed an ability to learn through olfactory conditioning to utilize other floral resources; however, such flexibility does not extinguish the innate olfactory preferences. The choice for certain night-blooming plants that exhibit characteristics typical of moth-pollinated flowers is mediated by an octopamine-associated modulation of antennal-lobe neurons. The ability to recognize two different stimuli through separate olfactory channels – one involving an innate response and the other a learned association – allows the moth to adapt to a dynamic floral environment, without losing its grade of specialization.

It is not only insects that are directly attracted by volatiles to plants, as plants also emit these signals to "cry for help." A single volatile compound (or also a mixture) can be quite attractive for parasitoid or zoophagous insects, which in turn helps to control the herbivore population. These predatory insects either feed on the herbivore or its offspring, and are thus involved in a direct reduction of herbivore numbers, or they are parasites to the herbivorous insects or mostly to their offspring, notably the larvae where they deposit eggs. In the latter case, the herbivore population is diminished in the second generation (Figure 3.12). In this example it is not only a single plant that is protected but also the neighboring plants. Compounds falling into this category include the indole derivative DIMBOA from maize, and also terpenes (Figure 3.12). DIMBOA must be activated from the glycosidic nonactive precursor (see Table 2.10) and thus can be classified as a phytoanticipin. The volatile compound DIMBOA is active in the aboveground communication between plants and zoophagous insects, and can also be emitted from the roots of maize plants into the rhizosphere, where the compound attracts beneficial microorganisms. The bacterium *Pseudomonas putida*, which has a growth-promoting effect on surrounding plants, is chemotactically attracted by DIMBOA; however, the bacterium can avoid the toxic effect of the compound by metabolic conversion. The beneficial bacteria are not attracted to maize plants, in which DIMBOA synthesis has been silenced.

A functional approach with elm trees (*Ulmus* sp.) has employed inhibitors of terpenoid synthesis, namely cerivastatin and fosmidomycin. Following treatment with these compounds, the parasitoids were shown to be less strongly attracted to trees with lower terpene synthesis and subsequent emission. The oviposition of the parasitoid *Oomyzus gallerucae* on the beetle larva was strongly reduced if insufficient terpenes were emitted, and this resulted in a higher number of elm leaf beetles *Xanthogaleruca luteola* on the leaves of the tree. Among the terpenes necessary to attract these parasitoid insects, (*E*)-4,8-dimethyl-1,3,7-nonatriene and (*E*)-β-caryophyllene were effective, in addition to "green-leaf volatiles" such as (*E*)-2-hexenal (Figure 3.12). Further investigations on individual compounds led to the conclusion that, in this case, a single compound – dimethyl-nonatriene – was the attracting component in the mixture, and not the complete blend of compounds. A background metabolite mixture used as control failed to attract the parasitic wasp, whereas the single component on top of the background emission of a healthy plant attracted the parasitoid.

A highly adapted behavior of caterpillars was observed on different oak (*Quercus robur*) trees. One group of trees was susceptible to the oak leaf roller (*Torthrix viridana*), whereas the other group showed some degree of tolerance. A comparison showed that the moth could use the original defense mechanism of the oak and turn it into an advantage for itself. Although the sensitive leaves contained less toxic compounds than the tolerant leaves, the sensitive leaves were shown to emit herbivore-induced plant volatiles to attract the zoophagous insects involved in plant defense. Female moths were attracted by the blend of these volatiles,

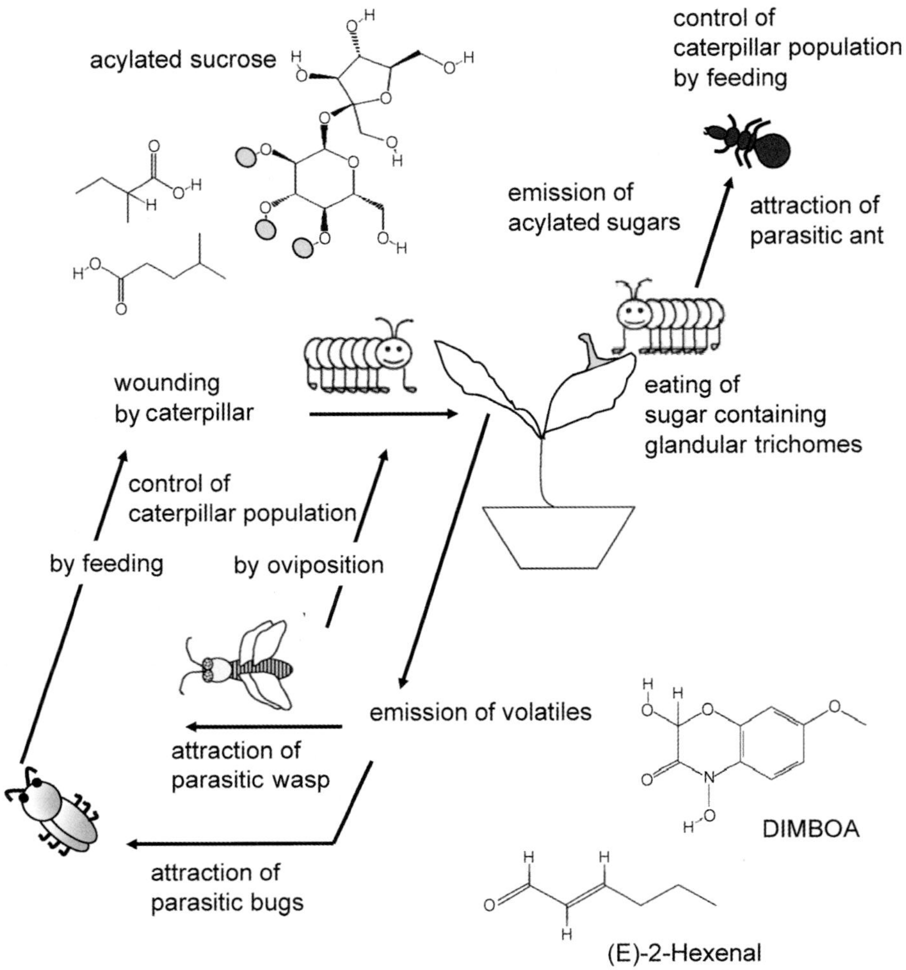

Figure 3.12 Plants attract zoophagous or parasitoid insects in the defense against herbivorous insects. The attraction is mediated by emission of a variety of volatile compounds. The chemical structures of the compounds DIMBOA (2,4-dihydroxy-7-methoxy-1,4-benzoxazin-3-one) and (*E*)-2-hexenal, which are used by the plants to attract the predatory insects, are shown as examples. The zoophagous insects can either feed directly on herbivores or their offspring, or they are parasites to the herbivore, that is, for oviposition into the insects or their larvae. The attraction of caterpillars by acylated sugars located in trichomes is also a strategy used by the wild tobacco (*Nicotiana attenuata*). The caterpillar eats the leaves containing the sugars, and is thus attractive to a parasitic ant, which then feeds on the caterpillar. Two of the four possible modifications of the sugar moiety by (secondary metabolite) side chains (2-methylbutanoic acid; 4-methylpentanoic acid) are shown here; the trichome is not to scale compared to the leaf. The circle on the sugar residue symbolizes one acyl group position. These strategies represent many different plant species. Chemical structures are from The PubChem Project.

including high levels of (*E*)-4,8-dimethyl-1,3,7-nonatriene and (*E*)-β-ocimene, to find their host plant for feeding, but were deterred by volatiles emitted from the tolerant leaves, including the sesquiterpenes α-farnesene and germacrene. The moth adapted to volatiles from a sensitive plant and used them for recognition of the host tree. As the leaves are also used for oviposition, the signal is also important to show the females those trees with low levels of toxic compounds in the leaves, so that their offspring can feed from nontoxic plants. Hence, the strategy of tolerant oaks exhibiting directly herbivore-repellent volatile emissions instead of the high emissions of predator-attracting compounds by susceptible oaks appeared to be a better mechanism for avoiding defoliation.

In many plant–insect interactions, "green volatiles" are emitted by leaves and recognized by zoophagous insects. Wild tobacco (*Nicotiana attenuata*) is able to recognize compounds from insect spit (see Section 2.5.3) in order to mount a specific defense response. In addition, these plants can emit volatiles also to attract parasitoids to eliminate the herbivores (Figure 3.12). One major compound is (*E*)-2-hexenal, which attracts those predatory bugs feeding on the larvae of the moth *Manduca sexta*. The plant will survive if it can send out these signals, whereas the herbivore is reduced in numbers by the zoophagous bugs. In contrast to the situation where a parasitoid is using the larvae for oviposition, the effect on the larval population can be seen much more rapidly in the case of direct feeding. These parasitoid insects are highly important for ecological systems, because they can reduce or eliminate many plant parasites in a "top-down" control mechanism. If such a predator is not available for a herbivorous insect, the control is much more difficult for the plant, because the population can increase dramatically in size over a vegetation period. One such example is the parasite of chestnut trees (*Aesculus hippocastanum*), the horse-chestnut leaf miner (*Cameraria ohridella*), which is economically very important because it has very few natural enemies. Zoophagous or parasitoid insects are also used for biological plant protection.

The glandular trichomes of tobacco serve not only as depots for secondary metabolites (such as the insecticidal nicotine); rather, they also accumulate acylated sugars which are synthesized from sucrose by a modification of side groups with branched aliphatic acids (Figure 3.12). The latter must be created by using secondary metabolite pathways. Both, nicotine and acylated sugars together form a highly effective defense strategy of wild tobacco (Figures 3.12 and 3.13). The larvae of *Manduca sexta* are attracted by the sugar-containing trichomes as a nutritional source, and feed on these leaves, ingesting the acylated sugars; in fact, the larvae were observed specifically to lick the surface of the trichomes, whereby they took up the compounds and also secreted them. Surprisingly, a predatory ant (*Pogonomyrmex rugosus*) can use these secreted acylated sugars, which constitute the typical body odor of the larvae, as recognition signals. These chemicals alter the body odor of the moths involuntarily in a way that the larvae become attractive for their predators; the specialized ants can then feed on the larvae (Figure 3.12). This represent one possibility of reducing herbivores that are not deterred by the otherwise toxic nicotine, such as the hawkmoth larvae.

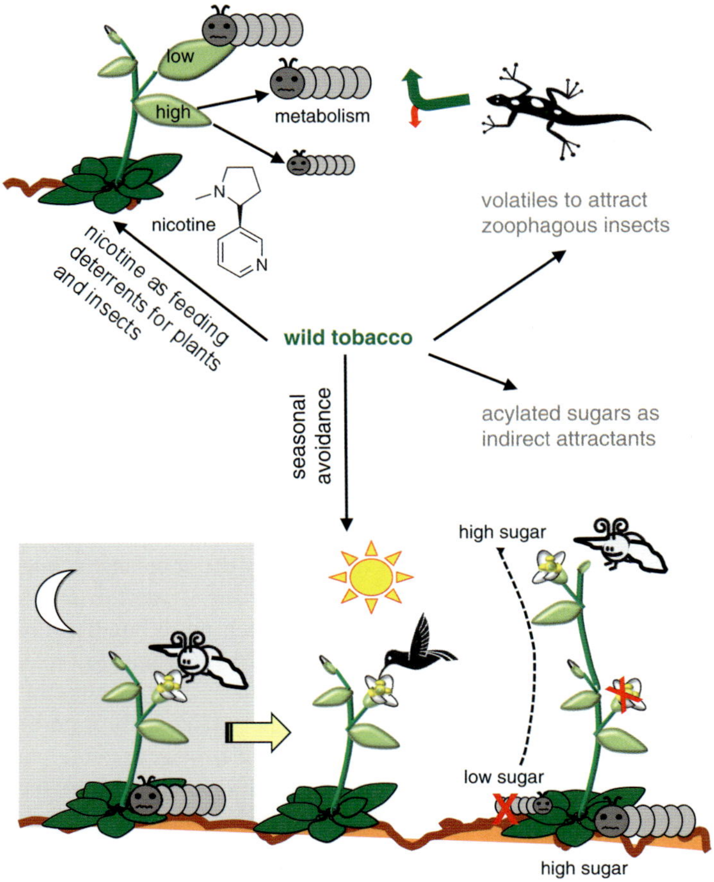

Figure 3.13 The wild tobacco (*Nicotiana attenuata*) and its various complex defense strategies against insect herbivores. Top: Nicotine as a feeding deterrent for the plant and larvae. While the plant is protected by high nicotine levels in leaves, there are adapted herbivores such as the larvae of the hawkmoth *Manduca sexta*, which can efficiently metabolize nicotine. Larvae with low nicotine levels, however, are attractive to predators, such as lizards, which preferentially eat larvae with low levels of nicotine (green arrow). Thus, a higher nicotine level would be beneficial for the larvae (red arrow), but they remain small because of the toxic alkaloid. Bottom: Seasonal and diurnal changes are used to avoid herbivores. The hawkmoth is a night-active insect, which can perceive the volatiles emitted from the flowers of the wild tobacco by its antennae and then pollinate the flower during the night. However, since the moth lays eggs on the leaves and the larvae subsequently feed on the tobacco plant, the beneficial effect can turn into damage. In that case, the tobacco can change the flowering time from night to day, when hummingbirds are attracted for pollination. The sugar content can also vary during the period of pollination, so that lesser larvae feed on the plant and more assimilates can be used for flower and nectar production. The strategies, as described in Figure 3.12, are shown in gray. Chemical structure is from The PubChem Project.

While the ability to metabolize nicotine allows the larvae to feed on tobacco plants, an accumulation of the alkaloid in the larval body also results in protection against predators. A further adaptation of the plant to the moth larvae could occur via differential gene induction after contact. An attack by *Manduca sexta* larvae, or simply the addition of larval oral secretions to mechanical wounds, will suppress the wound-induced transcription of the nicotine biosynthetic genes and in turn the accumulation of nicotine, but sustain the release of volatiles. Moreover, because the parasitoids of *Manduca sexta* are known to be negatively affected by the nicotine in their larval hosts, the downregulation of nicotine induction while maintaining volatile emissions to attract parasitoids may also allow wild tobacco plants to optimize the function of their indirect defense. It should be noted that the moth is also the nightly pollinator of wild tobacco flowers (Figure 3.13).

Plants can punish pollinators which are too "greedy." The homeostasis between fertilization (a beneficial trait for the plant) and feeding by larvae of the pollinator (a damaging trait) is quite delicate, but can be avoided by wild tobacco plants by exchanging the pollinators. Normally, the tobacco plants are pollinated by the nightly active hawkmoth; however, the insect also lays eggs on the leaves, which results in damage caused by the feeding larvae after hatching. If too many larvae are present, the plant switches the flowering time to daytime, which means that different pollinators are needed. During the day, hummingbirds are attracted by the nectar of the flowers, and they too pollinate the tobacco. However, the nectar of wild tobacco also contains nicotine, the normal levels of which serve as a deterrent for the hummingbirds, so the pollination strategy would be poor. However, the hummingbirds have adapted to low nicotine levels, such that they can approach the flowers. In fact, the levels of nicotine in individual flowers have been shown to differ strongly and the birds try to identify those flowers with a low nicotine concentration in the nectar. As a consequence they will test many flowers, which proves advantageous for the plant as more flowers will be pollinated. In addition, the genetic variability of a population with varying nicotine levels in the nectar was much higher, and this resulted – at least potentially – in a better performance for the offspring of those plants in a changing environment. This high variability in nicotine levels is most likely controlled by microRNAs (see Figure 1.16), as a reduction in the enzyme dicer (which is involved in the genesis of microRNAs) resulted in a low variability of nicotine levels in the nectar.

Plants react to high numbers of larvae by changes in their sugar accumulation patterns in a seasonal fashion. The sugars are stored in the roots until the larval development is completed; only then are the sugars transported to the flowers for nutrition, when the larvae are no longer present. These observations must be appreciated in the context of the secondary metabolites, which are synthesized in flowers to attract insects, and also substances in the nectar. If the nectar is more attractive to insects (sweet), then the plants will be pollinated much more efficiently.

The options that plants possess to adjust the pattern of their secondary metabolite patterns is highly complex, and allows adaptation to changes in the environment, including pathogens and pests. The parasites have also evolved many mechanisms to cope with the defense mechanisms of plants. The coevolution between plants and invertebrates has taken place over a much longer time period than with vertebrates, and consequently the better adaptation strategies are found in invertebrates whilst mammals must be regarded as generalists in terms of plant metabolites. Plant secondary metabolites also have many effects on vertebrates, including humans, as described in the following sections.

3.2
Effects on Vertebrates: An Overview

3.2.1
Structural and Functional Diversity of Natural Substances: Challenges and Promises

Animals have learned to cope with the huge number of secondary plant metabolites available, with numerous examples existing to illustrate the coevolution of plants and animals in a particular ecological niche. Evolutionary pressure has presumably been particularly intense between animals that depend on plants as a food source and have a long evolutionary history, such as phytophagous nematodes or insects (see Figure 3.1). The species-specific interactions between animals and plants who have to defend themselves against herbivores, and the inventive strategies that animals employ to cope with the mixture of secondary metabolites of the host plants, highlight the dynamics of the evolutionary interaction. Although plants have played a major role in the evolutionary history of vertebrates, there is little preference for a particular plant species as a food, and consequently vertebrates are mostly generalists (for definitions, see Section 2.5.3).

How do humans fit into this picture? Mammals are well equipped to cope with the different plant secondary metabolites, and during the course of evolution have acquired a wide range of catabolic enzymes that accept secondary plant metabolites as substrates, so as to minimize toxicity (the mechanisms and dynamics of mammalian metabolism with respect to xenobiotics of plant origin are detailed in Section 4.1.). Some of the plant polyphenols contained in the human diet are thought to be health-promoting, and evidence for this claim is discussed in Chapters 5 and 6. There appears to be no selective advantage for plants in being "healthy" for humans. In fact, if there are any beneficial effects of polyphenols, this rather reflects the successful biological adaptation of humans to most secondary metabolites in vegetarian food products. At an early stage in human history, the cultivation of suitable plants as major food sources and their selection for taste and productivity gave rise to a unique situation in evolutionary history. Most recently, with the advance of genetic technologies and the option to produce transgenic plants with desirable properties, humankind has acquired the option to interfere with evolutionary processes. While the new genetic tools hold great promise, the future will demonstrate if there are also unexpected risks associated with the application of these new technologies.

Today, there are estimated to be 131 000 natural substances in terrestrial plants, but for most of these compounds there has been no contact with humans in their evolutionary history. The enormous biodiversity of tropical forests and other ecosystems worldwide provides almost unlimited resources for biological and medical investigations. Only recently have marine ecosystems been discovered as an additional source for bioactive substances, and today about 22 000 marine products have been identified, though doubtless many more remain to be discovered. The bioactivity of many substances and their structural diversity represents a valuable asset for the development of drugs in the treatment of diabetes, cancer, and diseases of the cardiovascular system.

Although plant polyphenols comprise only a very small fraction of the human diet, their complexity is enormous. The total polyphenol uptake is estimated to be about 1 gram per day, and fruits and vegetables account for about 30% of this. Clearly, depending on the dietary habits, major differences are to be expected between individuals and between different ethnic groups. The consumption of soy and its consequences (see Section 5.2) represents an excellent example of the importance of dietary polyphenol content in human health.

Natural substances of the different chemical classes (see Sections 1.1 and 6.2) often differ by, for example, only a methyl or a hydroxyl group. Among the flavonoids such small structural modifications are common, yet the substances differ remarkably with respect to their biological activities. Each plant species is characterized by a specific set of secondary metabolites. Knowledge of the biological activities associated with the consumption of specific plants, or the use of plant extracts for the treatment of diseases, forms part of human culture. Traditional medicine (ethnomedicine) has exploited the health benefits of medicinal plants for thousands of years, and this old knowledge is a valuable source of information that may help in the development of new pharmaceuticals. Today, the molecular mechanisms of the beneficial effects of plant compounds are under investigation for different medical indications in a variety of current research projects (see Chapter 6).

The synthesis of secondary metabolites is energy-requiring for the plant and subject to selective pressure (see Chapter 1). Among the large number of possible low-molecular-weight organic substances in a virtual chemical space, the biologically relevant substances comprise only a small fraction defining a biological space. As a result, the random synthesis of small organic molecules is unlikely to generate biologically meaningful compounds. In contrast, the natural compounds of plants are designed by evolutionary processes, and hence the chance that such a substance will interact with a chosen target protein is comparatively high. Flavonoids, with more than 5000 identified substances, represent a quantitatively and qualitatively important group of natural compounds, and their interactions with cellular components and the resultant biological effects have been studied intensively. A single substance such as quercetin, for example, will interact with dozens (if not hundreds) of different proteins in a human cell. The promiscuity of phytochemicals with different target proteins is a pharmacological problem, but at the same time it offers chances for the development of new drugs and therapies (see discussion in Section 6.3). Essentially all major cellular functions can be affected by natural substances which interact not only with proteins but also with lipids, carbohydrates, and nucleotides (RNA and DNA), and may also chelate essential ions (Figure 3.14).

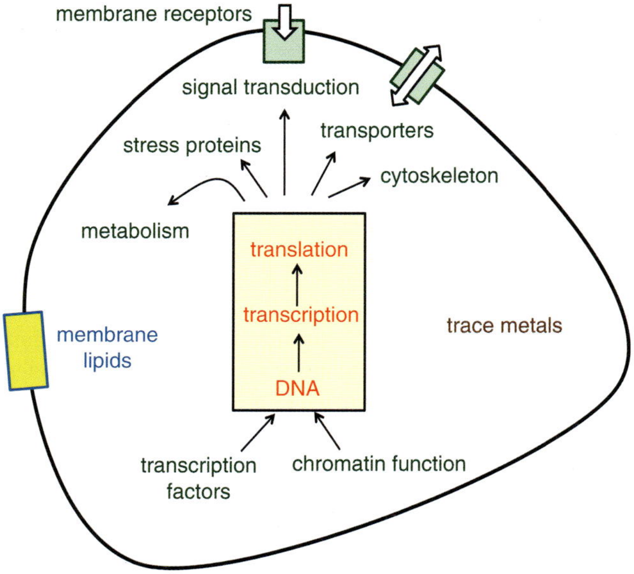

Figure 3.14 Cellular targets of natural substances. The genetic information of the genome is transcribed into mRNA and finally translated into proteins (red script). These essential processes are targets of some natural substances. The regulation of the genetic activity (chromatin structure and transcriptional regulation) may also be affected by phytochemicals. Some major cellular functions (green script) depend on the activity of various proteins whose activity may be modulated by natural substances. Membrane proteins involved in intercellular communication and transport proteins controlling cellular homeostasis are molecular targets of some natural substances. In addition, some substances may chelate essential trace metals (brown script) or integrate into the lipid phase of cell membranes (blue script). The latter reactions will modulate cellular reactions in complex ways.

Many proteins have been conserved during the course of evolution, and this is particularly true for proteins with essential functions. For example, histones are highly conserved structural proteins of the chromatin, while actin and tubulin – the principal components of the cytoskeleton (microfilaments and microtubules) – are structurally related between invertebrates and mammals. The consequence of this is that inhibitors of highly conserved proteins are likely to be active in many animals and also in humans. Peptide and steroid hormones have a long evolutionary history, and this also holds true for the synaptic transmitters of the nervous system (see Section 5.3). While the chemical structure of the signals has been largely conserved, the biological processes controlled by the respective signals may not have been conserved, and the same signal is used in several biological processes. As a result, the biological effects of phytochemicals may produce distinct biological reactions in different animal species, although the target proteins may be related.

One curious property of some natural compounds is that the dose–response curve is biphasic (see Section 5.1.5). The biological effect of a low dose may be distinctly different from the effect caused at a high dose of the respective compound. One reason for the complexity of such a response may be due to effects on several relevant target proteins to which the substance binds, with specific and different binding affinities. While the low-dose effect may be beneficial (for example by eliciting a cellular stress response), a higher dose of the same compound may be cytotoxic.

3.2.2 Molecular Interactions with Cellular Components

3.2.2.1 Protein Targets

The interaction of natural substances with proteins such as enzymes or receptors has been studied in great detail, and efficient assays to quantify the effect on the protein's activity are available. In many cases, these studies are driven by suspected health benefits. As common components of food, flavonoids have been particularly well studied, and all major enzyme systems are targets of specific flavonoids. Of particular relevance are the effects on catabolic enzymes, as their activity determines the toxicokinetics of natural substances and pharmaceuticals (see Chapter 4). Furthermore, natural substances interfere with endocrine and paracrine signaling (see Chapter 5) and affect regulatory processes, either on the level of organs or in the entire organism.

On the cellular level, inhibitors of the cytoskeleton are particularly instructive examples of bioactive natural compounds (Figure 3.15). Microfilaments

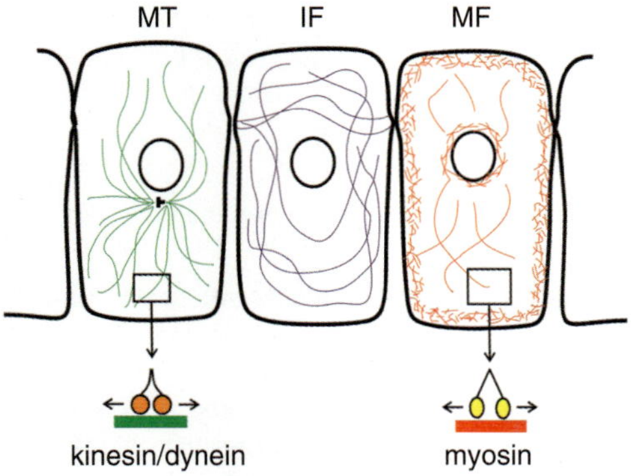

Figure 3.15 Cytoskeletal elements are target proteins of many natural substances. The principal cytoskeletal elements include microtubules (MT), intermediate filaments (IF), and microfilaments (MF) which are drawn schematically in three epithelial cells in their typical cellular orientation. Motor proteins can associate with microtubules and microfilaments and move along these filaments in both directions due to flexible domains that can bind reversibly to the cytoskeletal elements and step forward as a result of energy-requiring conformational changes. Myosin motor proteins associate with actin of microfilaments, while dynein and kinesin associate with the tubulin units of microtubules. Natural substances target cytoskeletal proteins as well as the associated motor proteins (see Table 3.1) and inhibit essential cellular functions such as the transport of different cargos in the cell and cellular motility.

(polymerized from actin monomers) and microtubules (polymerized from tubulin monomers) are each essential for vital processes such as cell migration, cell division and intracellular transport. The inhibitors of these processes are toxic. Numerous plants produce efficient inhibitors of these cytoskeletal elements, and interfere with the dynamics of polymerization and depolymerization. Some inhibitors favor the stability of polymers while others prevent polymerization or promote instability (Table 3.2). The functions of the microtubule and microfilaments depend largely on force-generating molecular motor proteins, which also are targets of natural substances. The motor proteins myosin (which binds to actin) and dynein and kinesin (which bind to tubulin) have numerous essential functions, including the intracellular transport of vesicles and of macromolecules. All motor

Table 3.2 Examples of natural compounds affecting cytoskeletal elements. See Figure 3.2 for structural organization of microfilaments (MF), microtubules (MT) and intermediate filaments (IF). Many of these substances are important tools in the research on cytoskeleton-dependent processes and some of the inhibitors are also important pharmacological drugs.

Inhibitor	Source of natural compound and biological activity
Tubulin	
Epothilones	Soil bacterium *Sorangium cellulosum*. Stabilizes MT and induces MT polymerization (Epothilone B)
Paclitaxel	Bark of the Pacific yew tree *Taxus brevifolia*. Stabilizes MT
Colchicine	Autumn crocus *Colchicum autumnale*. Inhibits MT polymerization
Vincristine and vinblastine	Vinca alkaloids of the periwinkle plant *Catharanthus roseus*. Inhibit MT assembly and favor MT disruption
Curvulin	*Pericallis steetzii* (syn. *Senecio steetzii*) endemic on Canary Islands. Inhibits MT assembly
Tubulin-dependent motor protein	
Adociasulfate-2	Marine sponge *Haliclona* sp. Kinesin inhibitor
Intermediate filaments	
Withaferin A	Medicinal plant *Withania somnifera*. Affects vimentin function
Actin	
Cytochalasin B	Metabolite of fungi including *Helminthosporidium dematioideum*. Inhibits assembly and disassembly of MF
Jasplakinolide	Marine sponge *Jaspis splendens*. Stabilizes MF
Phalloidin	Death cap *Amanita phalloides*. Stabilizes MF (see Figure 4.18)
Latrunculin	Marine sponge of the genus *Latruncula*. Prevents actin polymerization
Swinholide	Marine sponge *Theonella swinhoei*, Destabilizes MF
Actin-dependent motor protein	
Pentrabrompseudilin (PBP) and pentachlorpseudilin (PCP)	Marine bacteria *Alteromonas luteoviolaceus* (PBP) and *Actinoplanes* sp. (PCP). Inhibitors of some myosin isoforms.

proteins require ATP as an energy source for their activity. Many important energy-requiring processes like cell motility and changes in cellular shape, as well as muscle contractions, depend on the activity of myosin. The important role of myosin in mammalian cells is reflected by a large number of related myosin proteins (isoforms) with different molecular and functional properties.

The third type of filamentous structures in cells is represented by a heterogeneous group of intermediate filaments that have stabilizing functions; the term "cytoskeletal element" best fits these polymers. The spatial organization of the intermediate filament vimentin is disturbed in human fibroblasts by Withaferin A. Substances inhibiting intermediate filament function are rare. Presumably, the toxicity would be low in many cases, as intermediate filaments constitute a large and diverse group of proteins; indeed, in many cell types the functions of a particular intermediate filament protein may be substituted by other members of the respective protein family.

Natural compounds that target the function of the cytoskeleton nicely demonstrate the ambiguous biological nature of their activity; they are toxic in animals (including humans) and presumably contribute to the chemical defense of the plant against enemies. On the other hand, such compounds have important medical applications, and Paclitaxel and other microtubuli-targeting drugs are now widely used in cancer therapy on the basis of their cytostatic properties. Inhibitors of myosin, the molecular motors of actin, may be useful drugs to inhibit the migration of tumor cells and the formation of metastasis, or to prevent the migration of parasites (e.g., malaria) into the host cell.

The chemistry of natural substances and the properties of the protein-binding site largely determine the type of interaction. A noncovalent and reversible interaction can involve hydrogen bonds, hydrophobic interactions, or ionic bonds (Figure 3.16). In some special cases covalent bonds may form, but this is often a potential hazard as it may lead to nonfunctional proteins or mutations in the genome. How specific are the interactions of natural compounds with their target proteins? Are there well-defined binding sites, or is there an unspecific "stickiness" of, for example, flavonoids with their binding partners? If this were the case, it could explain the numerous target proteins which have been identified for quercetin and many other well-studied flavonoids. Clearly, some interactions have a very low affinity, and in many experiments the reported effects were observed only at very high concentrations which are never reached *in vivo*. However, in many well-studied cases there is ample evidence for specific interactions with enzymes, hormone receptors and other target proteins. The molecular analysis using biophysical techniques allows the analysis of the nature of the bond between the respective substance and the target protein. X-ray crystallography is the technique of choice to obtain structural information with high resolution about the binding site (Figure 3.17). The isolated protein (in this example, myosin) is crystallized in the presence of the inhibitor (pentachlorpseudilin in this case) and the formed crystal is subjected to high-energy X-rays. The three-dimensional structure of the substance, and its interaction with neighboring amino acids of the target protein, can then be computed from the diffraction pattern. By using

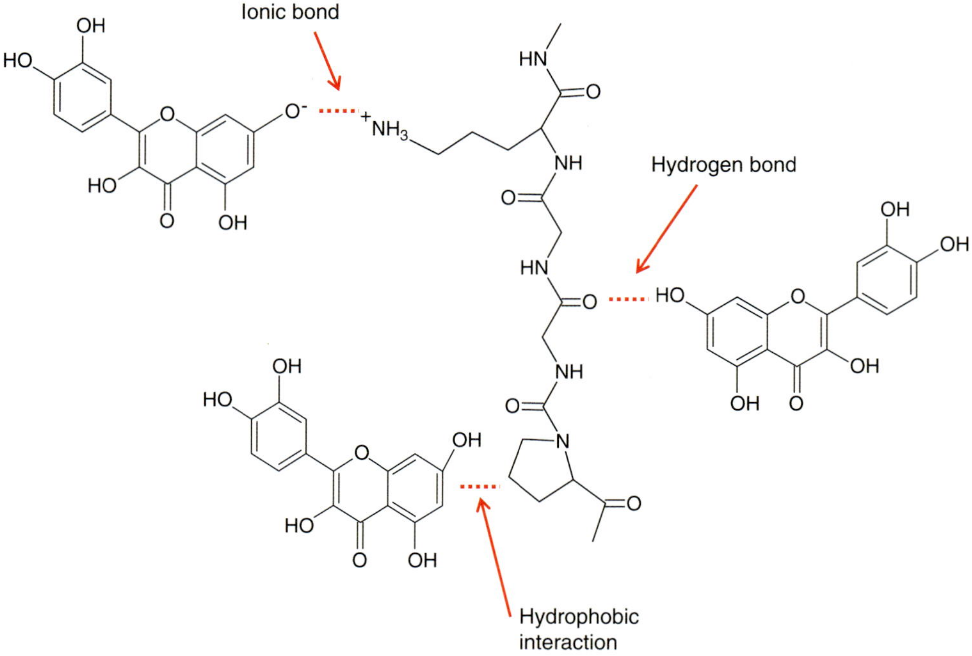

Figure 3.16 Binding modes of polyphenols to proteins. Polyphenols interact with proteins in different ways: hydrogen bonds, ionic bonds, or hydrophobic interactions may contribute to the interaction of natural substances with different amino acid residues in the proteins. The principle of the interactions is shown with lysine and proline as examples for amino acids which form bonds with a flavonoid (quercetin). Redrawn and modified from Le Bourvellec, C. and Renard, C.M. (2012) *Crit. Rev. Food Sci. Nutr.*, **52**, 213–248.

this technology the nature of the bonds can be analyzed, and the dimensions of the binding pocket in the protein defined. Such structural information is very valuable for the development of pharmaceuticals (see Chapter 6).

Some substances react chemically with functional groups of the target protein. For example, electrophilic compounds (which are common among natural substances and their metabolites) possess an oxidative potential and may react with cellular proteins. The cysteine residues of proteins may be subject to redox-reactions, which alter the protein conformation and act like switches that control the activity of the respective protein in an on/off fashion. Such reactions control the activity of proteins acting as redox-sensors (see Section 5.1 for molecular mechanisms) and of many other regulatory proteins. A "tasty" example of such a reaction is provided by the pungent sensation of some spices. Allyl isothiocyanate (in mustard oil) or cinnamaldehyde (in cinnamon oil) react with cysteine residues of an ion channel protein, and this mediates the taste sensation (Figure 3.18). The covalent bond with cysteines of the protein known as transient receptor potential ankyrin 1 (TRPA1) alters the conformation and allows an influx of cations into the neuron, which in turn elicits an action potential. It should be noted that the

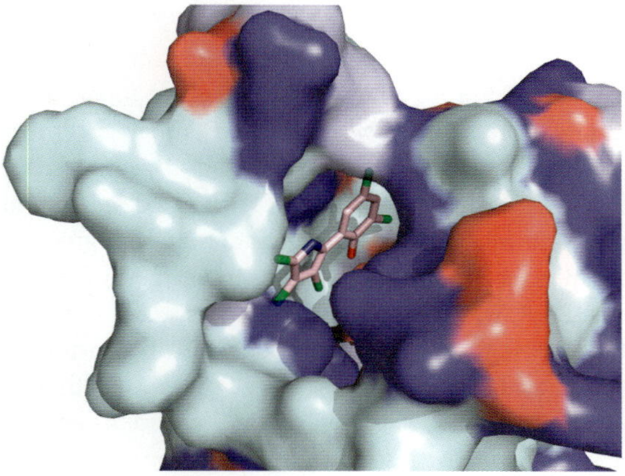

Figure 3.17 The interaction of a natural substance with a protein can be analyzed using X-ray crystallography. The image shows a close-up view of the allosteric pocket of myosin-2 occupied by the marine natural product pentachloropseudilin. The atom charges on the surface are shown in different colors (blue, positive; red, negative; cyan, neutral). These differences are the result of atomic interactions which lead to asymmetric charge distributions (van der Waals forces). The affinity of halogenated pseudilins for this allosteric pocket depends on the nature of the halogen substituents and the binding pocket's polarity, which can vary greatly between different myosin isoforms. Polar hot spots in the binding pocket are involved in maintaining hydrogen bonds to the inhibitors. Clusters of apolar residues interact via multiple hydrophobic interactions with the chlorine substituents. Courtesy of D. Manstein, Institut für Biophysikalische Chemie, MH Hannover, Germany.

cysteine residues are located in the intracellular domain of the membrane protein, and hence passage of the substance through the cell membrane is a precondition for the reaction. Many natural substances (particularly when they are not conjugated with sugars) are sufficiently hydrophobic to cross cell membranes.

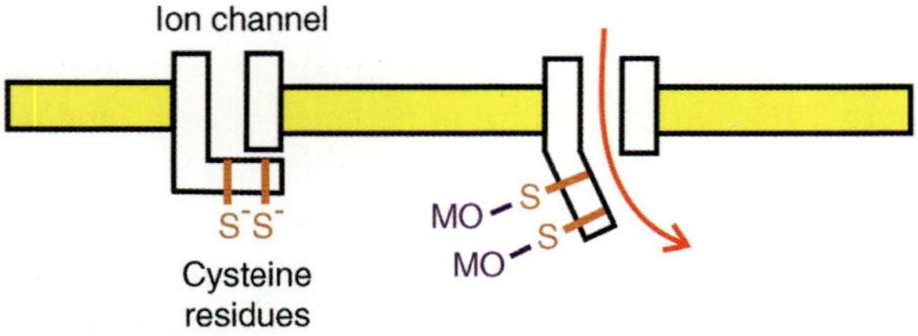

Figure 3.18 The pungent sensation of spices is mediated by the reaction with an ion channel. The TRPA1 membrane receptor belongs to the family of transient receptor potential (TRP) ion channels and is activated by various environmental irritants and by natural compounds present in spices. Electrophilic natural substances such as cinnamaldehyde or mustard oil (MO) were shown to form covalent bonds with cysteine residues of the intracellular domain of the transmembrane ion channel. As a result, the protein is activated by an induced conformational change and cations enter the cell and trigger a neural activity which elicits the pungent sensation. Based on Macpherson, L.J. *et al.* (2007) *Nature*, **445**, 541–545.

3.2.2.2 Interaction with Nucleic Acids

Natural substances such as flavonoids may affect the conformation and activity of chromatin by interaction with DNA-binding proteins. Telomerase and topoisomerase, for example, are flavonoid target proteins; however, flavonoids may also interact directly with DNA and a mutagenic activity could be detected *in vitro* in some cases (see Chapter 4). Mutations in critical regulatory genes favor the development of uncontrolled cell proliferation, and may promote the initiation and progression of cancer. Flavonoids may also interact directly with single-stranded (ss) RNA as well as with double-stranded (ds) RNA, which is involved in gene-specific silencing and has an important regulatory function. The interaction may also play a role in the defense against foreign nucleic acids and viruses (see Section 2.5.1). Although the interaction of flavonoids with nucleic acids has been studied *in vitro*, it remains to be seen if the interaction plays a role *in vivo* also in plants and animals.

3.2.2.3 Interaction with Lipids

Many polyphenols, such as flavonoids, can pass readily through cell membranes and enter all cellular compartments. Highly lipophilic substances will integrate into the lipid bilayer and lower membrane fluidity; evidence for this has been obtained with several flavonoids. The spice curcumin (Figure 3.19) is the active component of turmeric powder, and has been shown to integrate first into the outer layer of artificial biomembranes and, at higher concentrations, also into the inner lipid layer. The change in the biophysical properties of the cell membrane is likely to have profound effects on the activity of membrane-bound proteins. Curcumin has been claimed to have anticancer, antibiotic, antiviral, antidiabetic and anti-inflammatory properties; some of these claimed health benefits may be due to the interaction with the lipid bilayer of the cell membranes.

3.2.2.4 Interaction with Carbohydrates

Polyphenols also interact in complex ways with carbohydrates. This interaction is of practical importance as it influences the turbidity, taste and color of beverages produced from natural products. Cyclodextrins are cyclic oligosaccharides and are known to interact strongly with polyphenols such that complexes may be formed with, for example, curcumin or resveratrol. The type of bonds are the same as described for proteins (Figure 3.16), but the relative importance of each binding mode is not known. Polyphenols also bind to soluble polysaccharides such as starch or pectins. Immobilized methylcellulose on plates or beads can be used as a stationary phase in chromatographic techniques to separate different polyphenols on the basis of their different interactions with the immobilized polysaccharides. The intensity of the carbohydrate–polyphenol interaction (and the binding to other macromolecules) depends on many factors, such as solvent composition, pH, ionic strength, and solvent temperature. Apart from their role as structural components and as an important source of energy, carbohydrates also have important functions in cell recognition, signaling, and cell adhesion. However, whether flavonoids or other natural substances interfere with such important biological processes is currently largely unknown.

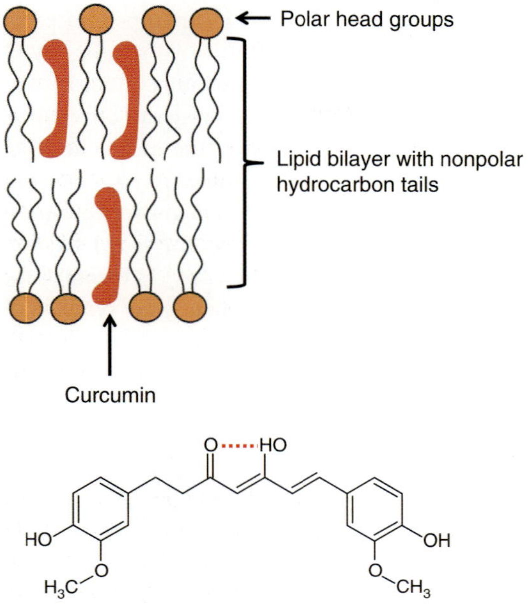

Figure 3.19 Hydrophobic substances may integrate into the lipid bilayer of cell membranes. The lipid bilayer of a cell membrane is shown schematically with the typical structure of phospholipids consisting of a polar head group and apolar hydrocarbon tails. The hydrophobic curcumin, a major component of turmeric, may associate with the hydrocarbons first in the outer lipid layer and, at higher concentrations, also in the inner layer. The presence of curcumin leads to altered physical properties of the membrane. Based on Barry, J. *et al.* (2009) *J. Am. Chem. Soc.*, **131**, 4490–4498.

3.2.2.5 **Metal Chelators**

Trace elements have essential functions in all organisms. Most of these elements are present as cations and are required as structural components for macromolecules with various catalytic and regulative functions. The biologically important trace elements include iron (Fe), copper (Cu), zinc (Zn), cobalt (Co), selenium (Se), molybdenum (Mo), manganese (Mn), chromium (Cr), iodine (I), and fluorine (F). Some elements, such as iron, copper and zinc are important factors for controlling the cellular redox state (see Section 5.1). The essential element Zn may serve as a representative example of the important biological role of trace metals. Zn^{2+} is present in an estimated 300 mammalian enzymes, and is also bound to special motifs ("zinc finger") in important transcription factors. Because of these essential functions a tight control of the intracellular Zn^{2+} concentrations is essential, and most cellular Zn^{2+} (the same holds true for other metals) is bound tightly to proteins, with a much smaller fraction being only loosely bound or present in free ionic form. The free Zn^{2+} concentration in the cytoplasm is in the range of picomolar to nanomolar, while the total concentration of cellular Zn^{2+} (including the bound metals) is in the range of 100 to 300 μM. In the case of Zn^{2+}, 14 specialized transmembrane proteins are involved in a controlled increase in the cytoplasmic Zn^{2+} concentration, while another 10 ion pumps located in mammalian cell membranes control Zn^{2+} efflux.

Flavonoids, resveratrol, catechins and their polymers (proanthocyanidins) and other polyphenols are able to bind trace metals, and in this way can affect their availability in the cell. The question remains as to whether the metal-chelating property of, for example, quercetin is beneficial or undesirable. To date, no convincing evidence has been acquired from *in vivo* experiments to answer this question. It has been claimed that the removal of free Fe^{2+} and Cu^{2+} by chelation reduces oxidative stress (see Chapter 5). On the other hand, Zn^{2+} is essential for some enzymes in antioxidative defense mechanisms, and hence the removal of free Zn^{2+} may have a prooxidative effect. If polyphenols bind trace metals with high affinity, it is conceivable that some enzymes are unable to recruit the essential trace metal which they require for their activity.

Changing metal concentrations may have profound effects on the pattern of transcription. For example, grape-seed proanthocyanidin extract and epigallocatechin-3-gallate bind Zn^{2+} and lower its concentration so that the transcription pattern resembles the condition of experimental Zn^{2+} depletion. The extract apparently binds to Zn^{2+} with a higher affinity than does the fluorescent chelator zinquin, which is often used to monitor the concentration of free Zn^{2+} in the cytoplasm. In contrast to these data, the Zn^{2+} concentration was found to be increased in the cytoplasm after the treatment of cultured human cells with procyanidins or with resveratrol (which also chelates Zn^{2+}). Recent evidence has suggested that polyphenols stimulate Zn^{2+} uptake, but the molecular mechanism involved is not understood. Zn^{2+} is also known to have an insulin-mimetic effect due to an induced phosphorylation of the insulin receptor and the activation of downstream signaling cascades. Interestingly, grape-seed proanthocyanidine extract has an insulin-mimetic effect by promoting the phosphorylation of the insulin receptor and triggering the ligand transduction cascade.

A few examples may illustrate the role of Zn^{2+} for important developmental process, though similar arguments could be made for other trace elements. Particularly in germline cells, the Zn^{2+} concentration is apparently critical; indeed, the element was found to be required for the progression of meiosis in mouse oocytes. Zn^{2+} accumulates in the mammalian testis, and the metal has been shown to be essential for spermatogenesis. In a testis organ culture system of the Japanese eel (*Anguilla japonica*), the metal was found to be particularly concentrated in spermatozoa as well as in spermatogonial stem cells (oogonia type A). Interestingly, spermatogonia are particularly resistant to experimentally induced oxidative stress, due most likely to elevated Zn^{2+} levels and the resultant high activity of the Cu- or Zn-containing isoforms of the enzyme superoxide dismutase (SOD), which specifically scavenges superoxide anion radicals and produces oxygen and hydrogen peroxide (see Chapter 4). Zn^{2+} also plays a role in human stem cells, and the loss of pluripotency of embryonic stem cells was found to correlate with an increase in nuclear Zn. Other trace metals have similarly important functions, and their chelation is also likely to be physiologically relevant.

3.2.3 Visualization of Natural Substances in Living Cells

In some exceptional cases the fate of natural substances in the body can be observed directly in the microscope, and it is possible to acquire important information on the uptake, distribution, metabolism and biological activity of the particular compound. Modern microscopic imaging techniques allow the visualization of weak signals, and with the advance of confocal laser scanning microscopy three-dimensional images can be obtained and specific parameters of interest quantified. These techniques will become increasingly important, and a few examples here will illustrate the potential of *in-vivo* imaging. However, *in-vitro* techniques are indispensable when the molecular reaction mechanisms of a natural compound are in focus, and many examples of this are provided in Chapters 4 and 5.

The direct visualization of radiolabelled flavonoids in the mammalian brain or other tissues has been achieved using positron emission tomography (PET) with ^{11}C-labeled compounds. The rapid uptake and distribution of nobiletin, a polymethoxylated citrus flavone, in the rat brain has been studied in this way. Radioactively labeled natural compounds are also useful for analyzing the distribution and metabolism of the respective substances *in vivo*.

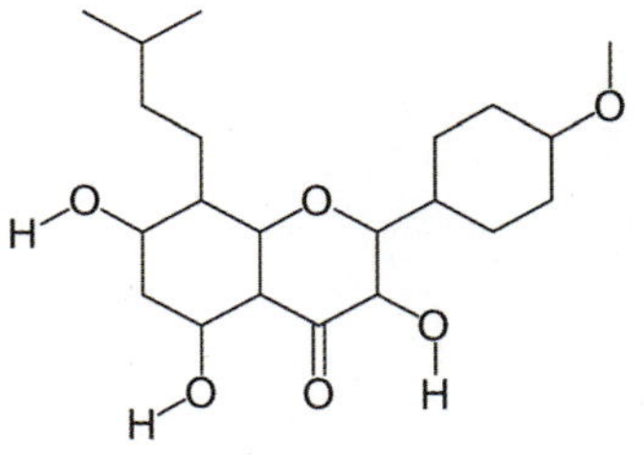

Icaritin

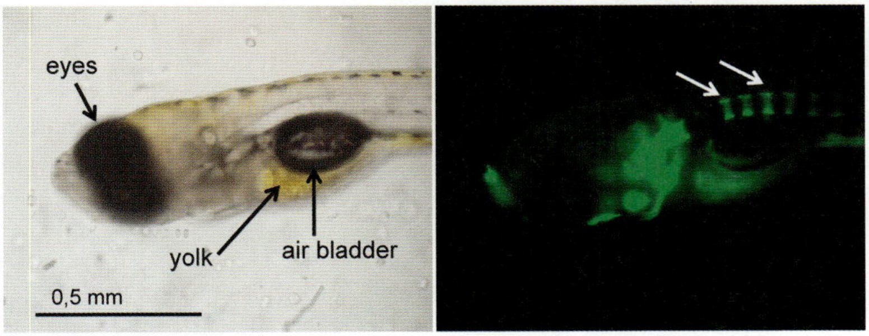

Figure 3.20 Staining of bone structures of the zebrafish larva with a fluorescent flavonoid. The bony skeleton of zebrafish larvae can be visualized with the fluorescent flavonoid icaritin. The photographs (bright-field and fluorescence microscopy) show the anterior part of a translucent 3-day-old zebrafish larva which was treated with 1 μM icaritin for 5 days. The prenylated flavonoid icaritin from the Chinese medicinal herb *Herba epimedii* shows strong fluorescence at the mineralized bone structures of the head. The arrows point at the differentiating vertebrae. Photographs courtesy of S. M.-Y. Lee, Institute of Chinese Medical Sciences, University of Macau, Macao, China.

In some cases, the natural compound of interest is sufficiently fluorescent to follow the distribution *in vivo*. Fish larvae (zebrafish or the medaka) allow direct observations due to their transparency, and such experiments provide information on the toxicity and pharmacokinetic properties of the compound. For example, flavonoids from *Herba epimedii*, a medicinal herb used in China against osteoporosis and other indications, can be followed *in vivo* in zebrafish larvae. The prenylated flavonoid icaritin elicits a bone-specific fluorescence and allows the development of the bony skeleton of the species to be monitored (Figure 3.20). In most cases, as the metabolites are not fluorescent, the fluorescence intensity provides information on the dynamics of metabolic activity (Chapter 4).

Some flavonols (e.g., quercetin) are weakly fluorescent, and this fluorescence is enhanced when the substance binds to a protein and the conditions for fluorescence resonance energy transfer (FRET) are met. The biophysical basis of this effect has been studied using human (or bovine) serum albumin, which binds hydrophobic substances. Many other polyphenol-binding cellular proteins possess the same property and will enhance the flavonol fluorescence; this in turn permits the distribution of target proteins in vital cells to be studied, using fluorescence microscopy. The large polyploid cells of *Drosophila* follicles are well-suited to carry out such experiments (Figure 3.21). The nuclei fluoresce brightly,

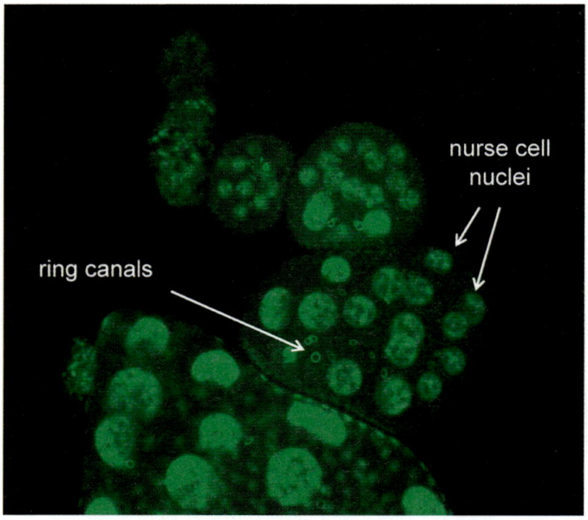

Figure 3.21 Quercetin elicits fluorescence preferentially in nurse cell nuclei of *Drosophila* follicles. *Drosophila* follicles normally do not fluoresce, but after incubation with quercetin for a few minutes the large polyploid nuclei fluoresce brightly. When quercetin binds to cellular proteins, the endogenous week fluorescence of the flavonoid is enhanced so that the major target proteins can be visualized. Each follicle consists of 15 large nurse cells and the oocyte which is nursed by the synthetically active nurse cells. All 16 cells are interconnected by cytoplasmic bridges and hence form a syncytium. The cytoplasmic bridges are lined by rings of actin (ring canals) and these structures also fluoresce (arrow). The fluorescence in the cell cytoplasm is comparatively weak. From Gutzeit, H.O. *et al.* (2004) *Biochem. Biophys. Res. Commun.*, **318**, 490–495 (see Further Reading). Reprinted with permission from Elsevier.

while the cytoplasm only shows a weak and diffuse fluorescence, with the exception of actin-containing rings that line the cytoplasmic bridges connecting the giant cells. The preferred nuclear localization of flavanols in plant and animal cells was also shown by using advanced microscopic techniques such as two-photon excitation and fluorescence lifetime imaging. The reason for this cellular distribution of these flavonoids is not clear. Histones and nuclear actin have been suggested as possible target proteins, but a direct interaction with DNA or RNA (see above) is another possibility.

Quercetin and other flavonoids profoundly affect the transcription pattern. This well-studied effect is due either to a direct interaction with the transcription machinery, or to a modulation of cytoplasmic signals which indirectly control the transcription pattern. One elegant approach to visualizing the effects on transcription is to use transgenic reporter strains. The induced expression of a luciferase gene or of intrinsically fluorescent proteins (e.g., green fluorescent protein; GFP) under the control of a specific promoter allows the possible effects of natural substances on the expression of genes to be monitored. A large number of reporter strains in fish are available for such studies. Translucent fish (larvae or mutant adults) provide promising disease models for the screening of natural substances or pharmaceutical drugs (see Chapter 6). Genetic techniques and advanced *in-vivo* imaging techniques are powerful tools that contribute to the present understanding of the biological activities of natural substances *in vivo*, and also help to accelerate drug development (Chapter 6).

4
Metabolism and Toxicity of Natural Substances in Mammals

4.1
Metabolism of Natural Substances in Mammals

4.1.1
Principles of Pharmacokinetics

Secondary plant metabolites taken up with food or with medicinal plant preparations are "foreign" substances (xenobiotics) as they are not normal molecular components of the human body (endobiotics). In this chapter, the fate of natural compounds in the body will be followed, and some of the experimental approaches used to assess and quantify the toxicity of some particularly interesting xenobiotics will be reviewed.

Toxicokinetics is concerned with the analysis of the uptake, distribution, metabolism, and excretion of xenobiotics (Figure 4.1). In pharmacological studies, the term pharmacokinetics is commonly used, and in view of the health benefits of natural substances and their importance for drug development this term will be used in the following sections. The biological activity of a substance is largely determined by its concentration in the blood (bioavailability), and hence this is an important parameter for the pharmacokinetic assessment. The biological reactions in the different tissues of an organism, the molecular reactions, and possible induced adverse effects are subject of toxicodynamics or, from a pharmacological perspective, of pharmacodynamics (Figure 4.1). Whilst, in the case of pharmacokinetics, the fate of the respective substance is the focal point, pharmacodynamics deals with the consequences of being exposed to xenobiotics (see Chapters 5 and 6).

Polyphenols and other natural substances that are ingested with food are subject to intensive metabolism. Organic small-molecular-weight compounds become rapidly modified and degraded by numerous enzymes. These processes affect endobiotics such as neurotransmitters and hormones, as well as xenobiotics which are taken up in food either voluntarily or involuntarily. Apart from secondary plant metabolites, the cocktail of xenobiotics to which humankind is exposed to includes pharmaceuticals, pesticides, and environmental pollutants. The metabolic modification is often referred to as "biotransformation," of which two phases

Plant Natural Products: Synthesis, Biological Functions and Practical Applications, First Edition.
Herwig O. Gutzeit and Jutta Ludwig-Müller.

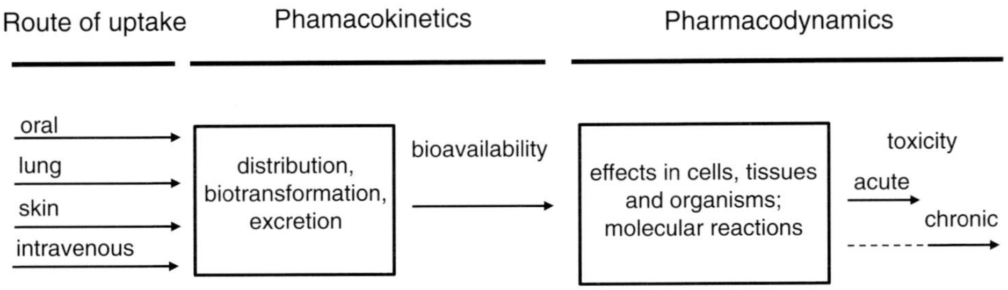

Figure 4.1 Physiological processes from the uptake of a natural substance to the biological effect *in vivo*. Pharmacokinetics is concerned with the uptake, distribution, and metabolism of a xenobiotic, while pharmacodynamics deals with the biological effects of the compound in the organism and also includes the analysis of possible acute or chronic toxicity.

may be distinguished. In Phase I reactions, functional groups (e.g., —OH, —SH, $—NH_2$, or —COOH) are added to a substance and, if it is sufficiently polar, it may be excreted at this point. However, in most cases the functional groups are used to attach hydrophilic molecules to the metabolite such as glucuronic acid, sulfate, amino acids, glutathione, or acyl groups; the latter reactions are referred to as Phase II metabolism. The thus-generated water-soluble metabolites are finally excreted through the kidney or bile. However, this sequence of reactions is not always strictly followed, and a great spectrum of possible chemical reactions is available. Typically, numerous products will be generated from a single pure compound, due to the low substrate specificity of Phase I and Phase II biotransformation enzymes. The metabolic reactions must be viewed as an efficient evolutionary mechanism to detoxify xenobiotics, but in some cases substances of low toxicity may be converted to highly toxic materials.

Numerous enzymes are involved in the biotransformation process. Of particular importance in Phase I metabolism are the families of cytochrome P450 enzymes (abbreviated CYP), which are involved in the oxidative metabolism of xenobiotics and endobiotics. The important enzyme CYP3A4 may serve as an example of the nomenclature. Here, the Arabic numeral '3' designates the family to which the gene belongs, the subfamily is specified with a capital letter 'A,' and the specific gene with another numeral '4.' To date, 57 CYP genes have been identified in the human genome and classified into 18 families. Although CYP enzymes are abundant in the liver, these enzymes are also expressed in the intestine and in other organs. They are characterized by a heme group with a centrally positioned Fe that is essential for the enzyme's function. In a typical reaction, a CYP enzymes acts as monooxygenase and uses O_2 to oxidize an organic substrate (RH) in the presence of NADPH:

$$RH + O_2 + NADPH + H^+ \rightarrow ROH + H_2O + NADP^+$$

Not all reactions catalyzed by CYP enzymes are oxidations. Under hypoxic conditions some CYP enzymes may also act as reductases. The removal of

Table 4.1 Typical biotransformation reactions. The listed enzymes are representatives of a particular type of reaction and do not represent a single specific enzyme. In particular, cytochrome P450 (CYP) monooxygenases and aminooxidases encompass numerous different reactions by enzymes with different substrate specificities. Among the different Phase II metabolic reactions, glucuronidation is the dominant pathway for many xenobiotics.

Phase I metabolic reactions	
Chemical reaction	**Enzymes**
Oxidation	CYP monooxygenase
	Aminooxidase
	Alcohol dehydrogenase (ADH)
	Xanthine oxidase
Reduction	CYP reductase
Hydrolysis	Esterase
	Amidase
	Epoxide hydrolase
↓	
Phase II metabolic reactions	
Methylation	Methyltransferase
Sulfation	Sulfotransferase
Acetylation	*N*-Acetyltransferase
Glucuronidation	UDP-Glucuronosyltransferase (UGT)
Glutathione conjugation	Glutathione *S*-transferase (GST)
Glycine or glutamine conjugation	*N*-Acetyltransferase (NAT)

halogens from polyhalogenated compounds is a physiologically important example of this type of reaction. Other enzymes are also involved in Phase I reactions, such as alcohol dehydrogenase and xanthine oxidase. Aminooxidases comprise a heterogeneous group of enzymes and include flavone-containing monooxygenases which are able to oxidize thiols and disulfides. Finally, esterases, amidases and epoxide hydrolase may also be classified as Phase I metabolic enzymes (Table 4.1).

The introduction of functional groups into substrate molecules during Phase I metabolism lays the ground for the following type II reactions. Among the possible different enzymatic modifications, the formation of a covalent bond with glucuronic acid (glucuronidation) by the enzyme uridine 5′-diphospho-glucuronosyltransferase (UGT) represents a major metabolic pathway for flavonoids and other dietary polyphenols such as catechins. There is also an endocrinological aspect to UTG activity: steroids are eliminated by glucuronidation, and daidzein (but not genistein) was reported to stimulate the glucuronidation of estradiol by UTG1A1. Several other enzymatic reactions (Table 4.1) lead to the typical spectrum of metabolites that can be identified in the urine or feces.

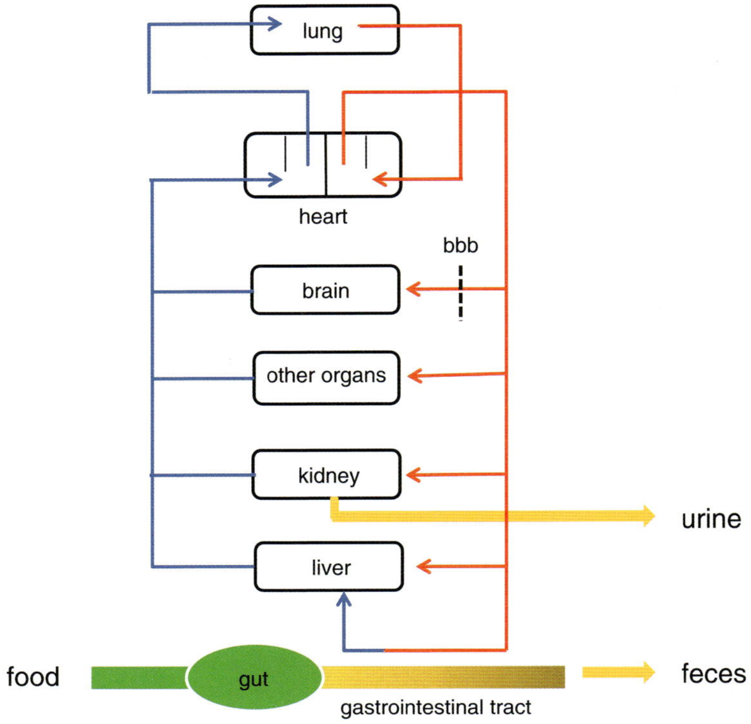

Figure 4.2 The human cardiovascular system and the blood supply of the intestine. In the schematic drawing of the human blood circulation, oxygenated blood (red) and oxygen-poor blood (blue) is distinguished. Compounds taken up by the gut epithelium are transported via the hepatic portal vein directly to the liver, where intensive metabolism of the substances takes place (first-pass metabolism). Having passed the liver the compounds and their respective metabolites enter the systemic circulation and become distributed in the entire body. Access to the brain is restricted by the blood–brain barrier (bbb), depending on the chemical properties of the respective compounds. The bulk of the natural substances and their metabolites are excreted in the urine and in the feces (yellow/brown).

4.1.1.1 Absorption, Distribution, Metabolism, and Excretion

Flavonoids can serve as typical examples of the metabolism of natural compounds ingested with food. In the intestine and in the colon, substances are taken up by the gut epithelium and passed on to the blood circulation (Figure 4.2). The draining vessel, the hepatic portal vein, transports the compounds first to the liver, where the concentration of biotransformation enzymes is high. As a result, the composition of natural substances and metabolites present in the hepatic portal vein differs significantly from that in the systemic circulation after passage through the liver (this is the "first-pass metabolism"; see Figure 4.2). The biotransformation results in an increased solubility of the xenobiotics, and the metabolites are finally excreted either with the feces or in the urine (via the kidneys). These processes of absorption, distribution, metabolism and excretion (collectively

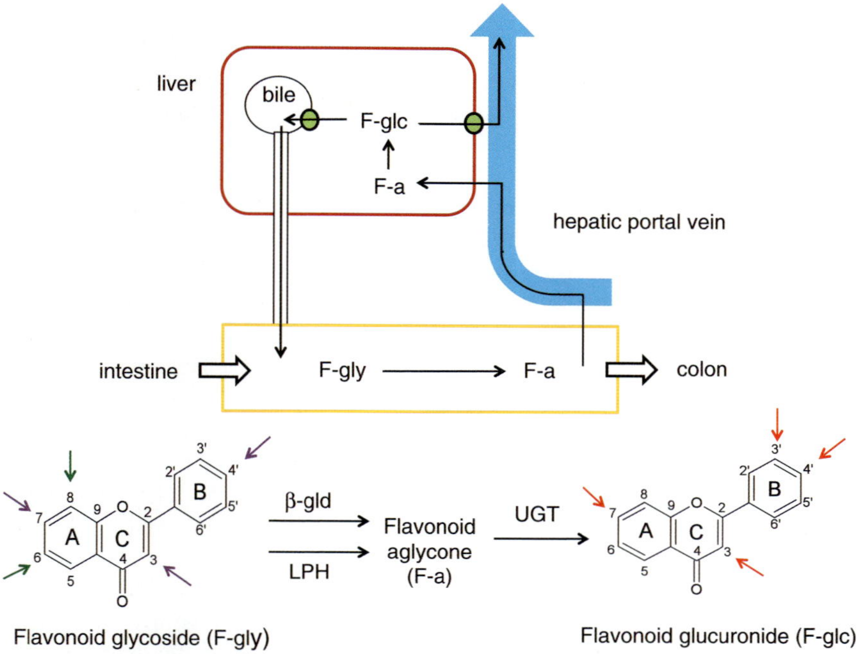

Figure 4.3 Sequence of typical first-pass metabolic reactions. Flavonoids are common natural substances in the diet and serve as an example of typical biotransformation reactions. Flavonoids are mostly glycosylated (F-gly), and the sugar moiety is removed enzymatically in the small intestine. The flavonoid aglycones (F-a) readily pass the cell membranes of the epithelial cells, enter the hepatic portal vein, and reach the hepatocytes of the liver. As many flavonoids possess functional hydroxyl-groups, Phase II enzymes can glucuronidate the molecule directly (F-glc). This hydrophilic molecule is unable to pass cell membranes, and hence the excretion into the bile ducts or into the systemic circulation requires the activity of transporter proteins (green circle). The flavonoid glucuronides can be either excreted in the urine or transported into the bile ducts and back to the intestine (enterohepatic circulation) and finally excreted with the feces. The enzymatic reactions are shown schematically in the lower part of the figure. The basic flavonoid structure (without hydroxyl groups) is shown, with arrows indicating the most common positions of glycosylation. Data from Cuyckens, F. and Claeys, M. (2005) *J. Mass Spectrom.*, **40**, 364–372. Dark green arrows show positions where glycosides can be attached by C-bonds, while purple arrows show sites for O-linked glycosides. The sugars are removed enzymatically either by β-glucosidases (β-gld) or by lactase phlorizin hydrolase (LPH). The aglycone is then typically glucuronidated in the liver by the enzyme UDP-glucuronosyltransferase (UGT); possible linkage sites are indicated by red arrows. Data from Day, A.J. *et al.* (2000) *Free Radical Biol. Med.*, **29**, 1234.

abbreviated ADME) largely determine the biological effects of a particular compound in the body. At this point, some basic principles that apply to natural substances as well as to pharmaceutical drugs will be outlined before specific metabolic pathways of particularly interesting natural compounds are discussed.

A major metabolic pathway of ingested flavonoids is illustrated in Figure 4.3. The flavonoids are typically present in glycosylated form; for example, in soybeans the β-glycosides of genistein and daizein prevail and only a few percent (on the order of 4–6%) are present without the sugar moiety (aglycone). Mono-, di-, tri-, or tetrasaccharides consisting of pentoses or hexoses are typically present in food. The sugar moieties may consist of glucose, arabinose, galactose, rutinose, the disaccharide neohesperidose or others sugars which may be attached via a carbon (C)-bond or oxygen (O)-bond to the flavonoid structure (Figure 4.3). The glycosides are not efficiently taken up by the intestine, but the bulk of the flavonoid glycosides is hydrolyzed in the intestine by β-glucosidases and by the activity of the enzyme lactase phlorizin hydrolase (LPH). The generated hydrophobic aglycones diffuse readily into the gut epithelium, without having to be transported actively across cell membranes. The aglycones reach the liver, where Phase II biotransformation reactions make the substances more hydrophilic. Glucuronidation of the flavonoid molecule at specific attachment sites (Figure 4.3) is a typical Phase II reaction, and these metabolites dominate in the vascular system while the aglycones and glycosides are present only in small concentrations. For example, after a soy meal 7-*O*-glucuronides account for 90% of the metabolites. Of the total daidzein and genistein content, 73% and 71%, respectively, are excreted as glucuronides in the urine. Sulfates and methylated forms are also typical metabolites.

The bioavailability of a substance is an important pharmacological parameter, and refers to the concentration in the serum and the change over time after uptake of the compound. As a consequence of this definition, the bioavailability of a substance is 100% after injection into a blood vessel; however, when the substance is ingested the bioavailability is typically much lower due to the first-pass metabolism. Inhaled substances taken up through the lungs are not subjected to first-pass metabolism, and from the pulmonary artery they are rapidly distributed through the body via the systemic circulation. For this reason, bioactive substances such as nicotine in cigarette smoke or Δ-THC in joints (see below and Section 5.3) reach the brain rapidly.

The bioavailability of a substance depends on a number of physiological parameters. The activity of transporter proteins for substances that are unable to diffuse through cell membranes is one factor that affects the bioavailability. Hydrophilic compounds must be transported actively into the cell by uptake transporters and, similarly, transport in the reverse direction out of the cell requires the activity of efflux transporter proteins. For example, flavonoid glucuronides are unable to diffuse through membranes, and hence these important Phase II metabolites must be actively transported out of the liver cell (Figure 4.3). Such transporters for flavonoid glucuronides or glycosylated compounds are present in the gut epithelium, the liver, and other organs. The uptake transporters in hepatocytes or renal tubule cells mostly belong to the solute carrier (SLC) family of transporter proteins. The excretion of hydrophilic metabolites of flavonoids or other natural substances in the kidneys from the renal tubule cells into the tubular lumen requires the activity of pumps which mostly belong to the ATP-binding cassette (ABC) transporter family. The energy provided by ATP allows the substances to be pumped against a concentration gradient, and because of the activity of these uptake and efflux pumps the concentrations of metabolites in bile and urine are typically much higher than in the serum. If the activity of the transporter proteins in the liver or kidney is reduced by chemical inhibitors or genetic polymorphisms (see below), the bioavailability of a natural compound may be increased considerably.

The excretion of natural compounds and their metabolites from the liver into the bile duct is an important protective mechanism. This enterohepatic circulation of substances (from gut to liver, then back via bile to the gut) favors the excretion and limits the uptake and distribution of xenobiotics in the systemic circulation of the body. Elimination with the feces is a major route of flavonoid excretion (Figures 4.2 and 4.3). Another protective mechanism limits the bioavailability of substances which have been taken up by the gut epithelium. Transporter proteins in the apical membrane of the gut epithelial cells (enterocytes) act as gatekeepers, and transport potentially harmful substances back into the gut lumen. In this way the efflux pumps can limit the access of xenobiotics to the blood circulation.

Because of the defense mechanisms against the uninhibited uptake of xenobiotics into the systemic circulation and the intensive biotransformation reactions, the concentration of flavonoids and other natural substances in the serum is low (typically nanomolar concentrations). The hydrophobic natural substances are mostly bound to proteins, and only a fraction is present as free aglycones in the blood. For example, 99% of quercetin is bound to plasma proteins, with albumin being the most important binding partner. This association affects the uptake of flavonoids in the different tissues, as well as the route of clearance of their metabolites. Binding to human serum albumin (HSA) is reversible, and the binding affinity of the flavonoid to HSA depends on the pH, temperature, and concentration of the binding partners. Furthermore, each flavonoid and its metabolites will have specific affinities to HSA; for example, glycosylation tends to decrease the affinity while hydroxylation has the opposite effect (Figure 4.4). The binding affinity to

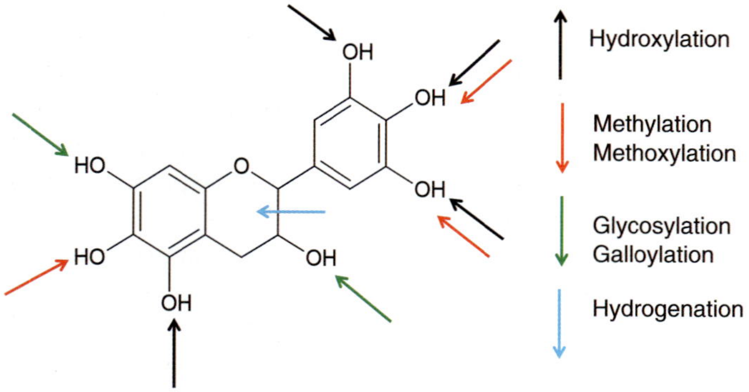

Figure 4.4 Chemical modifications of polyphenols affecting the binding affinity to plasma proteins. Binding of the ingested hydrophobic substances to albumin and other plasma proteins affects the pharmacokinetics and pharmacodynamics of the compounds. Using the flavonoid scaffold as example, the type of modification is shown and color-coded arrows show the site of the respective modification. Increased binding affinity to plasma proteins is indicated by arrows pointing upwards (right margin) or downwards in the case of lowered affinity. Redrawn from: Xiao, J. and Kai, G. (2012) *Crit. Rev. Food Sci. Nutr.*, **52**, 85–101.

serum proteins can greatly affect the biological activities of xenobiotics, since usually only the free aglycone is the pharmacologically relevant substance. The importance of albumin can easily be demonstrated in cell culture. Human cells are routinely cultured in medium containing fetal calf serum (FCS), which contains a large amount of albumin. Without FCS, the cells can survive for some time but do not proliferate. If quercetin ($\geq$20 μM) is added to such cultures the flavonoid is cytotoxic and many cells die due to an activated cellular suicide program (termed apoptosis). This effect is attenuated in the presence of added albumin as most of the quercetin is bound by albumin and the concentration of free flavonoid is greatly reduced. Similar results can be obtained with other natural substances that interact with serum proteins.

The bioavailability of quercetin or other natural substances taken up with food can be quantified by applying a high-performance liquid chromatography (HPLC) analysis of the serum. Typically, the serum concentration rises sharply within a few hours but, after having reached a maximum, the bioavailability decreases first rapidly and later more slowly; the dynamics of this process depends on the degree of glycosylation and a number of other factors. It must be borne in mind that the determined bioavailability is only relevant for blood cells and the endothelial cells that are in direct contact with the serum. Other cells are bathed in interstitial fluid, which accounts for 15% of the body weight in adults, and the concentration of a natural substance in this tissue fluid may differ from that in the serum. Tissue-specific differences must also be taken into account; this is particularly relevant for organs that are protected by special barriers, such as the blood–brain barrier (Figure 4.2). In the brain, the endothelium of the vessels is sealed by tight junctions, and the close association of the endothelial cells prevents the free passage of many xenobiotics from the blood into the central nervous system.

Metabolic reactions may occur in different organs in a tissue-specific manner. Some tissues that possess β-glucuronidase activity may even reverse biotransformation reactions, and the biologically active aglycone may be generated by the deconjugation of a flavonoid glucuronide. Recent evidence has suggested that the flavonoid glucuronides may act as a reservoir and that tissue β-glucuronidases may produce the bioactive aglycone locally. For example, quercetin-3-glucuronide was suggested to lower blood pressure and to restore endothelial dysfunction in hypertensive animals; however, the positive effect of the glucuronide was abolished when β-glucuronidase activity was inhibited.

The metabolism of endosymbiotic bacteria of the gut is of particular relevance. It is estimated that the human gut is colonized by more than 400 bacterial species, though few of these have been isolated and studied in detail. The importance of these gut-colonizing microorganisms in terms of human health has become clear during recent years. Although the gut microbiota is composed of populations of different microorganisms, for historic reasons they are referred to collectively (but incorrectly) as the gut "microflora." The bacterial species present differ widely between individuals, depending on the genetic constitution of the host and the diet. Developmental aspects are also apparent: during pregnancy, the child's gut is essentially sterile, but after birth and the first solid food a succession of different bacterial communities colonizes the gut until the age of about 2 years, when the final gut microbiota is established. As a result of this a specific set of bacterial metabolites will be produced in infants and adults, and these individual differences may have consequences for health.

The flavonoid daidzein provides an example of metabolic differences due to variations in the gut microflora. The isoflavonoid is a major component of soy, along with genistein. Two metabolites of daidzein – equol and *O*-desmethyl-angolensin (*O*-DMA) – show greater estrogenic activities than the parent compounds in some assays, and the ability to generate these metabolites is an important issue when possible health benefits of these compounds are considered (see Section 5.2). As germ-free animals and young infants do not produce these metabolites, the gut microbiota is thought to be solely responsible for the metabolic conversion. The ability to convert daidzein into equol differs among human populations. The large majority of individuals (80–90%) can metabolize daidzein to *O*-DMA, but for equol the figure is considerably lower (30–50%); moreover, there are considerable differences in these values between Western and Asian populations. The metabolism of lignans also seems to depend on the metabolic activity of microorganisms. The enterolignans formed are thought to have beneficial effects in some types of cancer, and the metabolites generated in the liver, gut and other organs are characterized by their own specific biological activities. Hence, the biotransformation reactions not only facilitate the excretion of these substances but also add to the complexity of their biological responses. For this reason, it is often difficult to distinguish if it is the administered substance or a metabolite that produces a biological reaction in the organism.

4.1.2
Case Studies of Selected Natural Compounds

A complete survey about the metabolisms of natural substances is beyond the scope of this chapter. However, the following examples will summarize some of the metabolic reactions that are characteristic of the respective chemical class of compounds. Those substances have been given preference which are of particular interest in later discussions with regards to their biological activity (see Chapters 5 and 6). Structurally similar molecules not mentioned in this section are likely to be subject to similar catabolic reactions.

4.1.2.1 Quercetin

The flavonol quercetin is chosen as an example of typical flavonoid metabolism because it occurs commonly in fruits and vegetables and is a component of many medical concoctions and dietary supplements. Quercetin has been extremely well studied, with almost 900 publications per year (880 listed in the PubMed library for 2011); clearly, the limit of 10 000 published articles with some relevance to this natural compound will soon be reached. The main reason for this considerable scientific interest lies in the medical promise of the compound, since it has been claimed that quercetin may have positive effects for cardiovascular diseases and cancer.

The daily intake of quercetin in the diet has been estimated at between 5 and 40 mg per day, although with a particularly flavonol-rich diet this amount may be increased by an order of magnitude. The substance is present in the diet mostly in glycosylated form, and only a small fraction is consumed as the aglycone. Different sugar moieties may be attached to quercetin at different positions, and glycoside oligomers contribute to the heterogeneity of the quercetin derivatives in different food sources. Despite the relatively high amounts of quercetin ingested daily, the plasma levels remain low (<100 nM) due to an intensive first-pass metabolism. When the diet was enriched with quercetin (1 g per day for 28 days), a plasma concentration of 1.5 μM was obtained. Similar data were published for other flavonoids. From a pharmaceutical perspective, these concentrations are rather low as most biological effects using *in-vitro* test systems were obtained at much higher concentrations. For this reason, efforts have been made to increase the bioavailability experimentally, either by chemically conjugating quercetin to other molecules, or by encapsulating quercetin into nanoparticles or liposomes. Intravenous injection of the aglycone represents a simple, but inconvenient, way to increase the bioavailability of the bioactive aglycone.

Quercetin is subject to typical Phase I and Phase II biotransformation reactions (as noted above). Phase I reactions for quercetin include oxidation, methylation, hydroxylation and sulfation (Figure 4.5), while reactive metabolites may also be formed (e.g., semiquinone radicals and quinone methides) that are suspected of being involved in relevant biological reactions (Section 4.2.3). Other flavonoids may

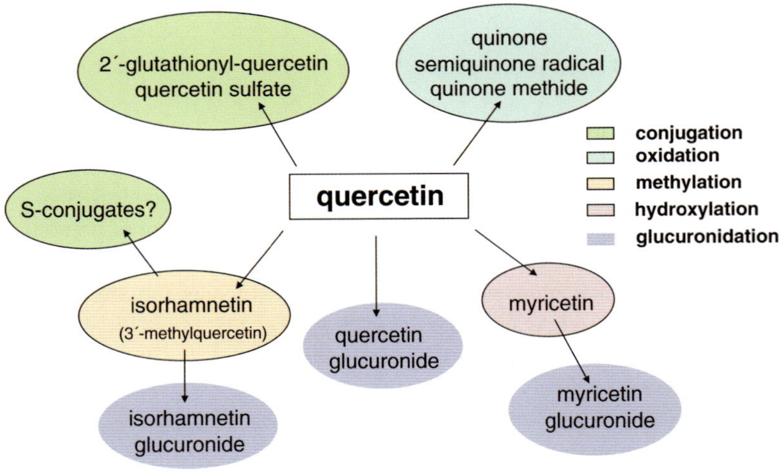

Figure 4.5 Major Phase I and Phase II metabolic reactions of the flavonoid quercetin. The various metabolites produced by different biotransformation reactions are represented by different colors, according to the type of chemical reaction which gives rise to the respective compound. Some reactions produce other flavonoids as metabolites (isorhamnetin, myricetin) which possess different biological and pharmacological properties.

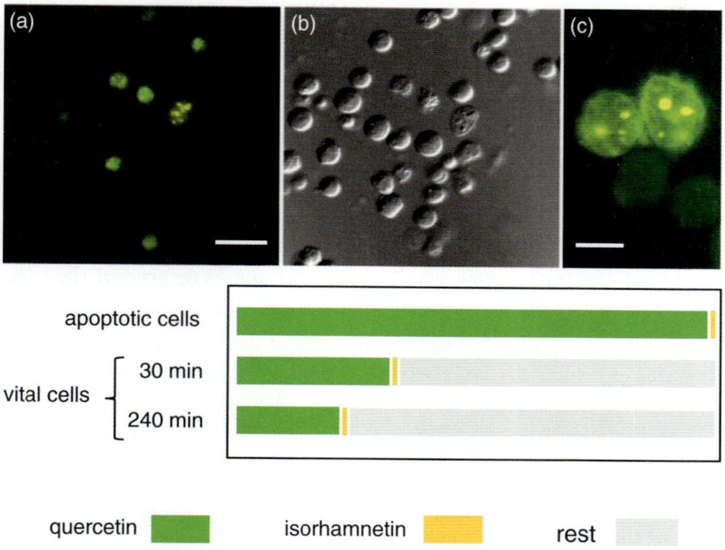

Figure 4.6 Visualization of quercetin metabolism in human cancer cells. Human acute myeloid leukemia cells (HL-60) were treated with 100 μM quercetin for 30 min and viewed in a fluorescence microscope (a, c) and using differential interference contrast microscopy (b). As shown previously in *Drosophila* follicles (see Figure 3.17), the intrinsic quercetin fluorescence is also enhanced in human HL-60 cells upon binding to a target protein. However, only apoptotic cells maintain the fluorescence for some time and show bright foci (c), while vital cells lose fluorescence rapidly (compare a with b). The content of quercetin and its metabolites in apoptotic and vital HL-60 cells was quantified using HPLC. Relative concentrations of quercetin and metabolites were calculated by their peak areas with reference to the known concentration of a standard substance. Based on these values, the percentages of quercetin, isorhamnetin and nonfluorescent metabolites were calculated. In apoptotic cells, the quercetin concentration remained high and only a small amounts of isorhamnetin and metabolites were detected. In vital cells, however, the concentration of quercetin decreased during 30 and 240 min incubation. Scale bars: (a) 30 μm; (c) 10 μm. Data from Ludwig-Müller, J. *et al.* (2005) *Biol. Chem.*, **386**, 279–283; Gutzeit, H. *et al.* (2005) *Croatica Chemica Acta*, **78**, 337–342.

be the products of such reactions (isorhamnetin and myricetin), and these substances have distinct biological properties. Finally, glycosides (mostly glucuronides) are formed as typical Phase II metabolites ready to be excreted via the urine. The half-life of quercetin and its metabolites in the body was determined as between 11 and 28 h, although large interindividual variations were typically observed.

In some exceptional cases the conversion of quercetin to its metabolites can be visualized using fluorescence microscopy. Quercetin is weakly fluorescent, and upon binding to a target protein a strong fluorescence is elicited which can be visualized in the microscope (see Figure 3.21). The distribution of fluorescence in the cells indicates the distribution of target proteins in a cell. When human leukemia cells are incubated with quercetin the cells – and, in particular, the nuclei – fluoresce strongly for a short time but the fluorescence then decreases rapidly (Figure 4.6). Only in cells which were no longer metabolically active and destined to die by apoptosis did the fluorescence remains strong, which indicated that an intact metabolism was responsible for the decreasing fluorescence in vital cells. Quantification of the changing concentrations of quercetin and its metabolites showed that the fluorescence and quercetin concentration decreased in parallel; the metabolite levels were increased at the same time but did not elicit any fluorescence on binding to a target protein. After 30 min, only about one-third of the initial quercetin concentration in the cells was detectable in vital leukemia cells. Visualization of the metabolic processes is only possible if certain structural features of the ligand compound are met, which is the case for some strongly hydroxylated flavonols.

4.1.2.2 **Catechins and Proanthocyanidins**

Flavan-3-ols (commonly referred to as flavanols) are important phenolic dietary components in, for example, fruits, vegetables, cocoa, tea, and red wine. The (epi)catechins belong to this class of flavonoids, and because of suspected health benefits they are of particular interest (see Chapter 6). Flavanols are typically present in polymerized form, and in many foods large polymers with more than 10 elements prevail. The oligomeric and polymeric flavan-3-ols are known as proanthocyanidins (or condensed tannins). After ingestion, the bulk of the monomeric forms is absorbed in the small intestine, and then subjected to biotransformation reactions such as glucuronidation, sulfation, and methylation. The oligomeric and polymeric forms cannot be absorbed in the small intestine. At least 90% of the dietary flavan-3-ols pass to the colon without any modification, where they are subject to intensive microbial metabolism (Figure 4.7). The microorganisms present in the colon are not well-characterized, and in view of the large differences in the intestinal microbiota among individuals it seems likely that the efficiency to metabolize proanthocyanidins also varies greatly. Typical microbial end-products include valerolactone derivatives, which can be further metabolized to various phenolic acids (Figure 4.7). These metabolites can be absorbed by the epithelium of the colon and finally excreted in the urine. Some of the beneficial effects ascribed to the consumption of green tea or cocoa may be due the metabolites of proanthocyanidins. For example, phenolic acids were reported to have an anti-inflammatory effect and to inhibit the growth of pathogenic bacteria.

4.1.2.3 **Resveratrol**

Scientific interest in resveratrol and its bioavailability has increased enormously during recent years when the beneficial effects of resveratrol for preventing cardiovascular diseases became apparent (see Section 6.1.2). Resveratrol is present in grapes in both *trans* and *cis* forms (Figure 4.8); *cis*-resveratrol is rather instable and hence is not commercially available, but the isomer is detected in the urine and may possess biological activity. In addition, glucose bound to resveratrol (also known as piceid) is present in grapes and might be hydrolyzed in the gut to yield *trans*-resveratrol.

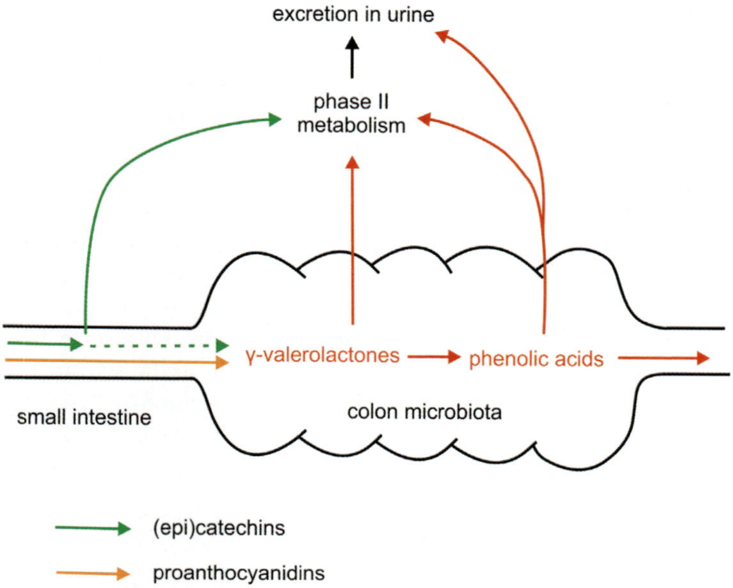

Figure 4.7 Role of the gut microbiota for the uptake and metabolism of flavanols. Catechins belong to the class of flavane-3-ols (flavanols), which are important dietary components. Monomeric catechins are mostly absorbed in the small intestine and subjected to biotransformation reactions. In contrast, the polymeric forms (proanthocyanidins) are degraded by microorganisms in the colon. The typical reaction products (red script) are mostly taken up and modified further by biotransformation enzymes before the metabolites are excreted in the urine. Some metabolites are also excreted with the feces.

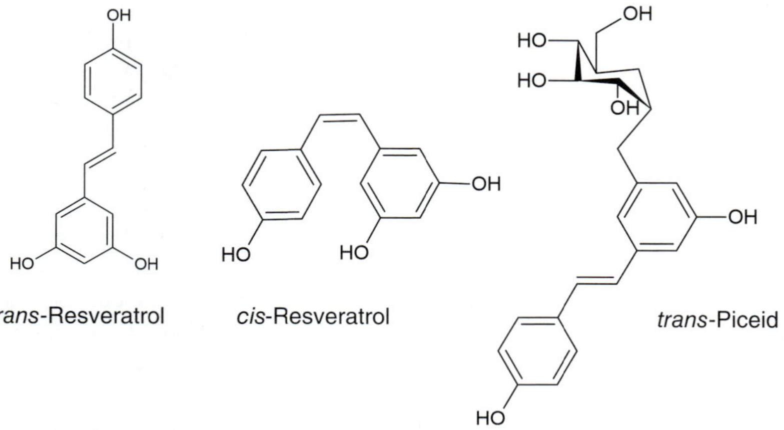

Figure 4.8 Resveratrol and derivatives. The three substances are present in grapes and are thought to be responsible for the health benefits of grape products.

The maximum plasma concentration of resveratrol is reached at about 30–60 min after ingestion. Apparently, the substance is rapidly absorbed and metabolized to *trans*- and *cis*-glucuronides and sulfates which are mostly excreted via the kidneys. The bioavailability of resveratrol is poor, and only very low serum levels (a few nanomolar at best) can be detected after administration. This has prompted various attempts to increase the bioavailability by chemical modifications, or by using different carrier molecules. The reported low concentration of free resveratrol in the serum may be misleading, however, as more than 90% of the substance is bound to macromolecules in the plasma, thus providing a reservoir for the compound. Resveratrol is rather insoluble in water (due to its hydrophobicity), and binds not only to albumin but also to fatty acids, lipoproteins (low-density lipoprotein) and hemoglobin in the serum. The toxicity of resveratrol in animals and humans is low, and nephrotoxicity in rats was only observed at extremely high doses (1 and $3\,g\,kg^{-1}$ per day). At a dose of $0.3\,g\,kg^{-1}$ per day, which would correspond to 21 g per day for a person of 70 kg body weight, there was no apparent toxicity. A similar amount of resveratrol would be present in 2100 l of red wine (ca. $10\,mg\,l^{-1}$), while for grape juice the required volume would be 10-fold greater.

Several questions remain to be answered regarding resveratrol. First, it is unknown whether resveratrol itself and/or its metabolites produce the observed effects *in vivo*. For example, a hydroxylated form of resveratrol (piceatannol) is suspected to contribute to the biological effects, while little is known of the biological activity of piceid as a component of grape products. Finally, the hydrophobicity of resveratrol permits interaction with membrane lipids and membrane proteins; this effect is thought to modulate the function of membrane receptors and may well have significant biological effects.

4.1.2.4 **Morphine**

The opioids morphine and codeine are important medical drugs that are used widely because of their powerful analgesic properties (see Section 5.3.3). Morphine is relatively water-soluble due to the presence of two OH-groups (Figure 5.33), and is readily absorbed in the alkaline milieu of the small intestine. When morphine is administered orally, the maximal serum concentration is reached within 1 h, but if administered intravenously the peak is reached within 5 min. In the systemic circulation the concentration of morphine remains relatively low after ingestion, due to an intensive first-pass metabolism; ultimately, only 8–10% of the morphine is excreted unchanged in the urine. The most important metabolic pathway for morphine is glucuronidation in the liver and also in other tissues, which facilitates excretion through the kidney (Figure 4.9). Several different UGT enzymes are involved in this reaction. Typically, close to 90% of a single dose of morphine is excreted in the urine within 72 h. Morphine sulfate has also been detected as a metabolite, as well as the biologically inactive normorphine, which is generated from morphine via N-demethylation. The metabolites are all considered biologically rather inactive, except for morphine-6-glucuronide (M6G), which has similar analgesic properties to the parent compound.

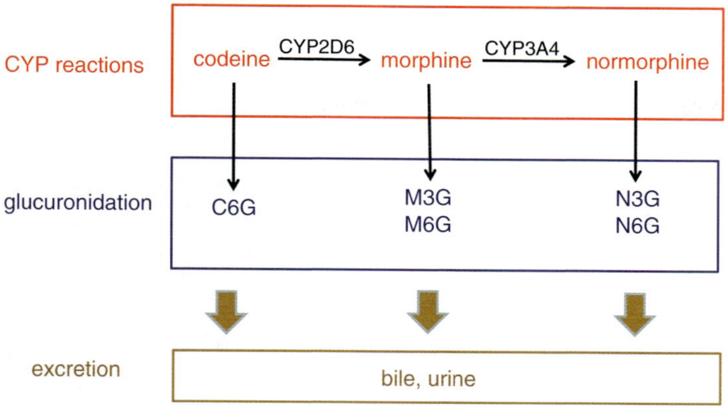

Figure 4.9 Metabolism of morphine. Different CYP enzymes can convert codeine (C) into morphine (M) and normorphine (N). The main metabolic pathway for the three substances is glucuronidation. C6G: Codeine-6-glucuronide; the other metabolites are abbreviated correspondingly.

Codeine is widely used as a remedy against cough. The demethylation of codeine by CYP2D6 leads to the formation of morphine (Figure 4.9); however, large interindividual variations in CYP2D6 activity have been reported and the correct dosing of patients is, therefore, a medical problem (see Section 4.1.5). Other variables that contribute to the individual variation of morphine metabolism include the patient's age, the ethnic background, and physiological differences in hepatic and renal functions.

In order to exert an analgesic effect, morphine must reach the brain by crossing the blood–brain barrier. Apparently, morphine and M6G are both able to pass the barrier as both have been detected in the central nervous system. Some specific transporter proteins are held responsible for the passage of these compounds into the brain, which has its own metabolic capacity to convert morphine into the glucuronide metabolite.

4.1.2.5 **Nicotine**

Since smoking is known to strongly increase the risk for lung cancer, the metabolism of nicotine – the main alkaloid component of tobacco – has been studied intensively. The pharmacokinetics of nicotine has some special features which distinguishes this compound from the previously discussed natural substances. The nicotine present in cigarette smoke is readily absorbed in the alveoli of the lung, and reaches the brain in only 10–20 s, which is faster than could be achieved with an intravenous injection. Thus, the dopaminergic reward system (see Section 5.3) is activated almost instantaneously. If nicotine is administered orally (e.g., in solution), absorption in the alkaline milieu of the small intestine is much slower and takes about 1 h. Nicotine is also absorbed efficiently through the skin, a fact that must be considered when nonsmokers are exposed to cigarette smoke.

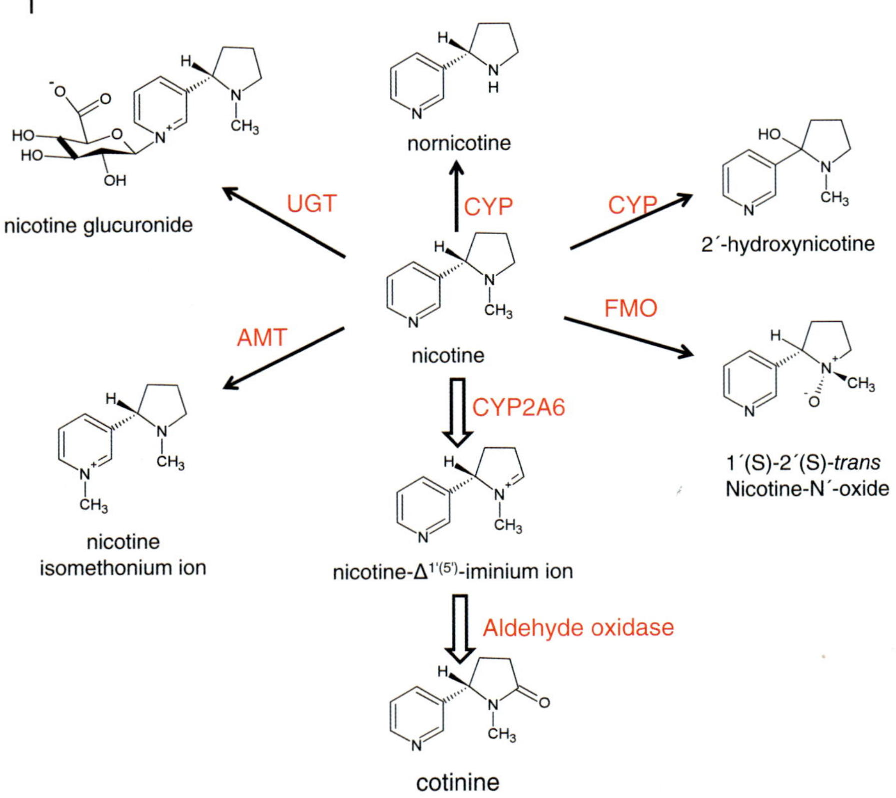

Figure 4.10 The metabolism of nicotine. The formation of cotinine is catalyzed efficiently by CYP2A6 and aldehyde oxidase, and this represents the most important pathway (thick arrows). Other CYP enzymes produce further metabolites such as 2′-hydroxynicotine and nornicotine. Glucuronidation (UGT) plays a role as well as oxidation via a flavin-containing monooxygenase (FMO). Methylation by the enzyme amine N-methyltransferase (AMT) leads to the formation of the nicotine isomethonium ion. Data based on Hukkanen, J. *et al.* (2005) *Pharmacol. Rev.*, **57**, 79–115.

The concentration of nicotine in the serum of regular smokers is in the order of 0.1 μM, but heavy smokers may exceed this concentration by far. Cotinine, the main metabolite of nicotine, is present in much higher serum concentrations (an average of 1 μM for regular smokers). Following one puff of a cigarette the serum concentration of nicotine declines over the following 20 min as the substance is distributed in the body to different organs. The subsequent release of nicotine from these tissues is comparatively slow, so that even at night the plasma levels remain relatively high (about one-third of the peak concentrations during the day). The change in cotinine bioavailability is even slower, as the compound has a longer half-life than nicotine. In effect, smokers are "dosed" with nicotine during the entire day, and even at night.

The main metabolic routes and numerous metabolites of nicotine have been identified (Figure 4.10). Cotinine is by far the most important metabolite (ca. 75%

of all excreted metabolites) and is generated in two steps: (i) conversion of nicotine to the nicotine iminium ion by a cytochrome P450 enzyme (CYP2A6); and (ii) conversion of the nicotine iminium ion catalyzed by aldehyde oxidase to yield cotinine. Individuals who carry a homozygous deletion of the *CYP2A6* gene have very low cotinine levels in the serum, and inhibitor studies have confirmed the importance of the enzyme. Interestingly, the *CYP2A6* gene can be induced by estrogen. Studies using estrogen-responsive cell lines have shown that transcription of the *CYP2A6* gene is stimulated by the estradiol receptor α (ERα). In keeping with these data, women were found to metabolize nicotine and cotinine faster than men, while women taking oral contraceptives metabolize nicotine very rapidly. However, flavonoids with estrogenic properties, such as genistein, daidzein or glycitein (see Section 5.2) inhibit CYP2A6 activity and hence slow down the C-oxidation of nicotine.

The intermediate compound, the nicotine iminium ion (Figure 4.10), is an alkylating agent and might affect the activity of proteins relevant for the initiation of diseases. In another oxidative reaction, the metabolite nicotine *N'*-oxide is generated by a flavin-containing monooxygenase (FMO3). In nonoxidative pathways, both nicotine and cotinine and their metabolites may become glucuronidated and excreted in the urine. Nicotine can also be metabolized to nornicotine by the activity of a demethylase which belongs to the CYP family of enzymes. Some other metabolites which are produced only in small amounts are suspected to contribute to the health risks. For example, the substance 4-(methylnitrosamino)1(3-pyridyl)-1-butanone (NKK) is one of many other carcinogens in cigarette smoke.

4.1.3 Metabolic Bioactivation

The use of natural plant products as chemical weapons against other plants, bacteria, insects and other animals, and the adaptation of the target organisms in an evolutionary process, have already been discussed (see Sections 2.5, 2.6, 3.1, and 3.2). During the course of evolution, vertebrates have acquired an arsenal of metabolic enzymes that modify and inactivate potentially harmful compounds. However, in some cases the metabolic reactions produce the opposite effect, and this results in the generation of toxic compounds. The metabolic conversion of an inactive precursor to an active compound is known as "bioactivation." Natural substances that are able to react with DNA or proteins after bioactivation are present in some food and herbal concoctions. The bioactivation of some xenobiotics by CYP enzyme activity (in particular, CYP1A1) has been reported to increase the incidence of some types of cancer.

The following example illustrates the hazard associated with the bioactivation of a natural compound. Aristolochic acid is synthesized in *Aristolochia clematitis* and some other *Aristolochia* species. These plants, which grow in wheat fields in the Danube basin, acquired some inglorious fame when they were found to cause "Balkan endemic nephropathy." The aristolochic acid present in food was found to be the cause of this disease, and also of the development of colorectal cancer.

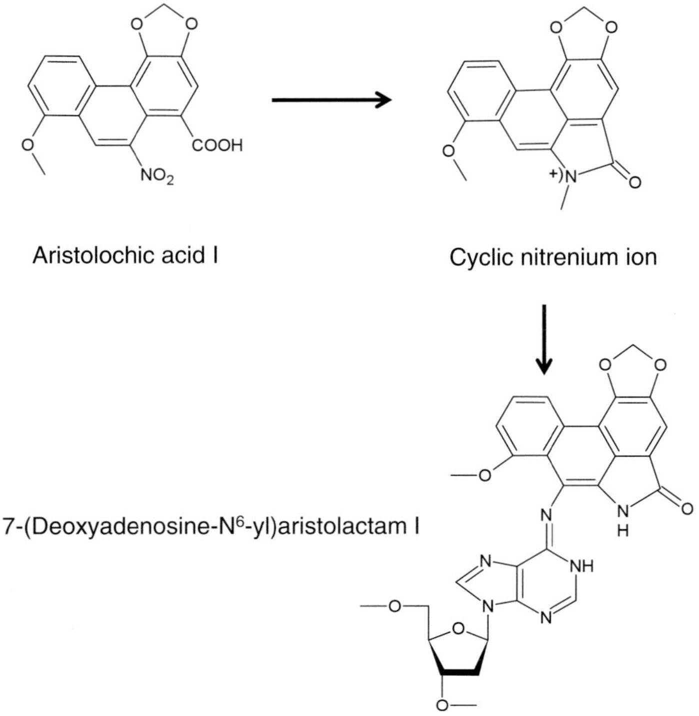

Figure 4.11 Bioactivation of aristolochic acid. Aristolochic acid is a toxic compound in various *Aristolochia* plant species, and is a component of some medicinal herbal concoctions. After ingestion, the substance is bioactivated by cyclooxygenase I (COX-1) or CYP1A1/2 which catalyze the nitro reduction giving rise to the cyclic nitrenium ion. The reactive product can bind covalently to DNA, and the DNA adducts formed promote carcinogenesis. Based on Chen, C.H. *et al.* (2011) *Chem.-Biol. Interact.*, **192**, 161–176.

Aristolochic acid is a component of several Chinese herbal medicines, and has also been used in other countries to treat arthritis, gout, and rheumatism. When the toxicity of aristolochic acid became apparent the preparations were withdrawn from the market in several countries. The metabolism of aristolochic acid involves its reduction to the reactive cyclic nitrenium ion (Figure 4.11); under anaerobic conditions this reaction is catalyzed by hepatic CYP1A1/2 or by xanthine oxidase, cyclooxygenase I and other peroxidases. The cyclic nitrenium ion is then able to form covalent bonds with guanine or adenosine and form stable DNA adducts. This may lead to point mutations and the activation of oncogenes which promote the development of cancer.

Aflatoxins are metabolic products of the fungus *Aspergillus flavus*, and are bioactivated to form highly toxic compounds *in vivo*. The substance aflatoxin B1 is a potent carcinogen in many species of fish, birds and primates; only mice are less susceptible. The fungus is widely distributed in the environment and may contaminate nuts, wheat and other food, particularly under humid storage conditions.

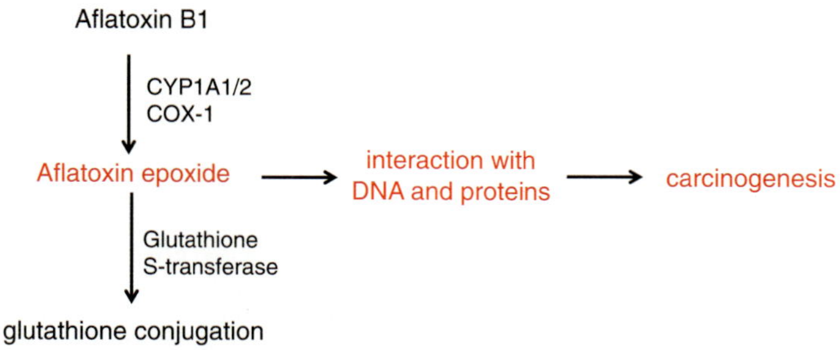

Figure 4.12 Metabolism of carcinogenic aflatoxin in food. Aflatoxins are produced in fungi (*Aspergillus flavus*) that may contaminate stored food. The enzymatic bioactivation leads to the formation of a toxic intermediate (epoxide) which is able to react with DNA. Glutathionylation of the epoxide has a protective effect and converts the reactive intermediate to a product that is unable to react with DNA.

The cancerogenicity of aflatoxin B1 is due to its metabolic conversion to an epoxide which may alkylate the bases of the DNA and also react with proteins. The enzyme CYP1A1 is mainly responsible for the epoxidation reaction in humans. Conjugation with glutathione appears to be a major detoxification pathway (Figure 4.12). Mice were found to have a particularly efficient glutathione *S*-transferase enzyme, and this may account for the relative resistance of this species against aflatoxin-induced cancer. Birds are particularly sensitive to aflatoxin poisoning due to an efficient epoxidation by several CYP enzymes. The species-specific bioactivation of the toxin illustrates the possible pitfalls that may be encountered in toxicity testing with experimental animals (see Section 4.2).

Another example of Phase I metabolic conversion into a reactive substance is provided by teucrin A and teuchamaedrin A from germander (*Teucrium chamaedrys*), which is used as a herbal remedy against gout, asthma, cough, depression and other diseases. The substances are converted by CYP3A4 to reactive epoxides, which can then interact with this CYP enzyme or with epoxide hydrolase. Severe hepatotoxicity is a typical reaction to these two diterpenoids.

Alkenylbenzenes are components in basil, fennel, tarragon and cinnamon, and are common components in food and beverages. CYP biotransformation enzymes (in particular CYP1A2) are held responsible for the conversion into reactive metabolites, which can crosslink with DNA. The formation of reactive metabolites has also been reported for common flavonoids such as quercetin. The formation of quinones and quinone methides is thought to be responsible for the mutagenic property of quercetin which has been detected using *in-vitro* test systems. The bioactivation of natural substances to reactive intermediates is not necessarily a harmful reaction, since such reactive compounds may elicit an (anti)oxidative adaptive reaction with beneficial effects (see Section 5.1). Bioactivation reactions can also be exploited to the benefit of a patient. Some pharmaceuticals are administered as inactive prodrugs that need to be converted to the active drug by metabolic enzymes.

4.1.4 Effects on Enzymes Controlling Biotransformation and Bioavailability

Natural substances are not only substrates of metabolic reactions but may also affect the process itself by interacting with key enzymes of Phase I and II metabolism. In some cases the interaction of natural substances in the diet has far-reaching consequences, as these interactions may strongly affect the bioavailability of a natural substance, as well as the metabolism and bioavailability of pharmaceuticals.

Some compounds affect the microbiota in the colon, which in turn may alter the bioavailability of ingested natural substances. Bacterial metabolic enzymes may be inhibited or, if the ingested substances possess antibiotic activity, this will change the species composition of the microbiota in the gut. The use of commercial antibiotics has a long-term effect on the composition of bacterial populations. The microbiota even reacts to nutritional changes from a high-fat/low-fiber diet to low-fat/high-fiber diet, with the differences becoming apparent within a day.

Proteins that mediate the transport of hydrophilic molecules across the otherwise impermeable cell membrane are important targets, and their inhibition by natural substances can significantly alter the distribution of these compounds in the body. Numerous substances in herbal tea, vegetables or herbs have been reported to inhibit transporter proteins belonging to the ABC family of efflux transporters. The P-glycoprotein belongs to this class of transporters, and is expressed in the liver and the intestine. Many flavonoids have been reported to affect ABC-mediated transport (see Section 1.2.5). Some substances inhibit the ATPase activity of the transporters, while others act as competitive inhibitors with other substrates. Interestingly, some flavonoids (e.g., quercetin, kaempferol) show a biphasic response and cause either an inhibition or an activation of the P-glycoprotein transporter, depending on their concentration. If the activity of the P-glycoprotein transporter is inhibited, then the bioavailability of the substance transported by this protein will be increased. However, some flavonoids modulate the activity of the transporter protein, as well as the activity of CYP enzymes (such as CYP3A4), and it is often difficult to distinguish the contributions of both proteins to the determined bioavailability of the compound. Another class of transporters, the multidrug resistance (MDR) transporters, affect the bioavailability of substances by limiting the access of xenobiotics into the systemic circulation.

The CYP enzymes of Phase I metabolism are common target proteins of flavonoids. For example, *in-vitro* assays with the important enzymes CYP3A4 and CYP2C9 have revealed that inhibitors of these enzymes are frequently found in flavones, coumarins, flavonols, isoflavones, and flavanols. For most assayed substances the inhibition was weak, though some substances with an IC_{50} $<5\,\mu M$ were also identified (Table 4.2). Whether this inhibitory activity, when quantified by *in-vitro* tests, would affect the bioavailability of xenobiotics *in vivo* remains to be demonstrated in clinical studies.

CYP enzymes are not only the main actors in Phase I metabolism; rather, they also play a role in important biochemical pathways such as steroid biosynthesis.

Table 4.2 Selected natural substances which inhibit the important biotransformation enzymes CYP3A4 and/or CYP2C9. Data from Kimura, Y. *et al.* (2010) *Food Chem. Toxicol.*, **48**, 429–435.

Classification	Compound	IC_{50} (μM)		Food or plant source(s)
		CYP 3A4	CYP2C9	
Coumarin	Bergamottin	0,3	0,4	Lemon, grapefruit, bergamot, lime
	Imperatorin	0,5	7,0	Lemon, lime, angelica
	Dicoumarol	23,2	0,5	Moldy sweet clover
Chalcone	Phloretin	5,0	3,7	Apple
Flavone	Amentoflavone	0,07	0003	*Ginkgo biloba*
	Chrysin	0,9	6,5	Propolis
	Apigenin	0,4	6,4	Bell pepper, garlic, celery, orange,
	Acacetin	1,2	4,2	Fennel
Flavonol	Galangin	3,0	7,5	Propolis

Recent evidence has suggested that the maternal exposure of rats to some natural substances from raspberry leaves can have longlasting effects on CYP activity in the offspring. The mechanism of this effect is unclear, but the long-term effect suggests an epigenetic basis and resembles the well-studied transgeneration effect observed with estrogenic natural compounds (see Section 5.2.4).

4.1.4.1 **Medical Importance**

Medication is standardized to a "normal" metabolizer, so that the desirable bioavailability is achieved. However, two factors may lead to strong deviations: (i) the genetic variability of individuals (Section 4.1.5); and (ii) the modulation of metabolic activity by natural substances present in the diet. In such cases the bioavailability of medical drugs may be strongly affected, and the useful therapeutic concentration may be exceeded or may not be reached. In particular, substances that inhibit the activity of CYP3A4 reduce the metabolic clearance of more than 50% of clinically prescribed pharmaceuticals. Such undesirable interactions are scored as "adverse drug reactions" (ADRs).

Grapefruit juice is a well-studied example of drug metabolism being significantly modulated by natural compounds. Certain components of grapefruit juice have been shown to inhibit transporter proteins (MPR2 and OATP), while the flavonoid components of the juice such as naringenin, quercetin and kaempherol are known inhibitors of various CYP enzymes. However, most relevant to this effect is bergamottin, a furanocumarin that interacts with the important enzyme CYP3A4. Due to the monooxidase activity of the enzyme, bergamottin is converted into a reactive intermediate that is able to react with the heme group, or with amino acids that are essential for the catalytic activity of the enzyme. This results in an irreversible inactivation of the enzyme that is referred to as

Table 4.3 Common fruits interfere with biotransformation enzymes that metabolize medical drugs. Natural substances in common fruits and vegetables inhibit (or stimulate) the activity of numerous Phase I biotransformation enzymes, and this has consequences for the metabolism and bioavailability of many pharmaceuticals. Each of the listed enzymes was shown experimentally either *in vitro* or *in vivo* to interfere with the metabolism of at least one commonly administered medicine. OATP: organic anion transporter polypeptide; MRP: multidrug resistance-related proteins. For other abbreviations, see Table 4.1. Examples from: Rodríguez-Fragoso, L. *et al.* (2011) *J. Food Sci.*, **76**, R112–R124.

Food source	Affected target proteins
Grapefruit	Inhibits: CYP1A2, CYP3A4, MRP-2,OATP-B, P-glycoprotein
Sevilla orange	Inhibits: CYP3A4, OATP-A, OATP-B, P-glycoprotein
Tangerine	Stimulates: CYP3A4; inhibits P-glycoprotein
Grapes	Inhibits: CYP2E1, CYP3A4
Mango	Inhibits: CYP1A1, CYP1A2, CYP2C6, CYP2E1, CYP3A1, P-glycoprotein
Apple	Inhibits: CYP1A1, OATP family
Broccoli	Inhibits: CYP1A1, CYP2B1/2, CYP2E1, CYP3A4, MRP-1 and 2
Cauliflower	Inhibits: CYP1A1, CYP2B1/2, CYP2E1, CYP3A4, MPR-1 and 2
Tomato	Inhibits: CYP1A1, CYP1B1; stimulates: CYP2E1

"mechanism-based inactivation." The pharmacokinetics of numerous commonly administered medicines have been shown to be affected by grapefruit juice; among these are psychotropic drugs, immunosuppressants, antihistamines, antibiotics, and antiarrhythmic drugs. Considering the importance of a correct dosage when treating life-threatening diseases, the danger of adverse drug reactions is very apparent. Bitter oranges (Sevilla oranges) also contain bergamottin and inhibit the important CYP3A4 enzyme, but other types of orange do not. Is bergamottin in grapefruit juice a special case? The effect is particularly strong and has been studied in detail, but inhibitors of Phase I enzymes and transporters are common in various fruits and vegetables. Indeed, evidence is steadily accumulating that such compounds in many fruits and vegetables affect numerous biotransformation enzymes (Table 4.3).

Is it advisable to avoid foods which contain such inhibitors? Clearly, the health-promoting effects of fruits and vegetables are unquestionable, and for healthy persons there is no reason for concern; for example, the consumption of grapes and grape juice is apparently beneficial (see Section 6.1.2). Resveratrol is the bioactive component which is thought to be mainly responsible for the positive effects, and yet this compound is an active CYP3A4 inhibitor similar to bergamottin. The CYP3A-mediated oxidation of resveratrol leads to a reactive *p*-benzoquinone methide metabolite that binds covalently to CYP3A4 and irreversibly inactivates the enzyme. Whether this reaction poses a problem for some medication remains to be shown, but the documented beneficial effects of grape juice have shown that, at least for most people, the possible inactivation of CYP3A4 *in vivo* is not relevant. Knowledge regarding possible interactions between food components and critical medications is, however, an issue that both doctors and patients must consider.

Table 4.4 Incidence of polymorphisms for some important biotransformation enzymes. For abbreviations, see Table 4.1. Data from Aktories K., Förstermann, U., Hoffmann, F., and Starke K. (2009) *Allgemeine und spezielle Pharmakologie und Toxikologie*. Elsevier, Urban and Fischer, Munich.

Enzyme	Incidence of deficient metabolizers in Europeans (%)
CYP2A6	1–2
CYP2D6	5–10
CYP2C9	≈2
CYP2C19	2–5
ADH 2	5–20
UGT1A1	5–7
GST	≈38 (GST T1), 30–60 (GST M1)
NAT II	≈50

4.1.5 Genetic Polymorphisms

4.1.5.1 Genetic Polymorphisms of Key Metabolic Enzymes

In all studies concerning the bioavailability and metabolism of natural substances and pharmaceuticals, large interindividual differences were consistently observed. Dietary habits and variations in the bacterial populations of the gut contribute to the observed metabolic differences. An additional factor which has come into focus in recent years are polymorphisms of genes that can directly or indirectly affect the metabolism of xenobiotics; this important pharmacological field of research has been termed "pharmacogenetics." Genetic polymorphisms of receptors, ion channels or other target proteins which affect general physiological processes or signal transduction pathways of medical importance are beyond the scope of this discussion, and in this section only those polymorphisms that directly affect the biotransformation of some selected natural compounds will be discussed. Although mutations in such genes are common, it is only when an allele in at least 1% of a population is affected that this is considered a polymorphism of the respective gene (Table 4.4). The genetic variants do not necessarily have significant physiological effects, but numerous cases have been described in which polymorphisms lead to an insufficient metabolism of a particular drug of medical importance. According to the ability to metabolize a particular substrate, it is possible to distinguish among poor metabolizers (PMs), extensive metabolizers (EMs), and ultra-rapid metabolizers (UMs).

The activities of transporter proteins affect the pharmacokinetic properties of specific drugs, and hence polymorphisms of the respective genes may affect processes such as:

- uptake of the compound by the gut epithelium;
- excretion via the liver (enterohepatic circulation) or kidney; and

- passage of the substance through physiological barriers such as the blood–brain barrier. Transporter proteins can either mediate transport of the drug across the boundary (the endothelium) into the brain, or its active exclusion from the brain by transporting the compound back into the vessel.

Polymorphisms of transport proteins are a common occurrence with CYP and UGT enzymes having long been the focal point of research, and consequently genetic variations have undergone intensive investigation. The human genome codes for 17 UGT isoforms which are grouped into two gene families. Some mutations affecting important isoforms may lead to severe diseases; for example, the degradation of hemoglobin in the red blood cell involves formation of the metabolite bilirubin. In the liver the yellowish bilirubin is glucuronidated by UGT and excreted. However, if this process is inhibited or is inefficient, then the serum concentration of bilirubin will be increased (hyperbilirubinemia) and jaundice will occur as a result. Other cases have been reported in which polymorphisms of UGT genes are associated with diseases. For example, variants of the UGT2B17 gene increase the risk to develop prostate cancer.

Polymorphisms can have very different genetic bases. For example, a mutation in the promoter or in the regulative sequences upstream of a CYP gene can affect its expression either negatively or positively (Figure 4.13, case B). The altered enzyme concentration in the cells will affect the bioavailability of the compound that is metabolized by the respective CYP enzyme. If, in an individual, the CYP gene is duplicated and the enzyme activity is extremely high, then that person would classify as UM and treatment with a drug would be ineffective due to its very low bioavailability (Figure 4.13, case C). At the other extreme, little or no enzyme is produced due to a mutation in the coding region (Figure 4.13, case D), defective splicing, or a deletion of the CYP gene. In such a case the patient would be classified as PM and the medication adjusted accordingly. If the status is not known and the patient is overdosed, the consequences may be severe; such cases are scored as "adverse drug reactions" (ADRs).

The enzyme CYP3A4 is implicated in the metabolism of more than 50% of all prescribed drugs, and CYP2D6 and CYP2C9 follow suit with 25% and 20%, respectively. Deletions of the gene coding for the important enzyme CYP3A4 have not been reported, but several polymorphisms leading to large variations in enzyme activity have been detected. Some of this variability seems to be due to mutations in transcriptional regulators (e.g., PXR or NR112) that are members of nuclear receptor superfamilies. The enzymes CYP3A5 and CYP3A7 are closely related to CYP3A4, and polymorphisms in these genes may affect the overall activity of the CYP3A subfamily. The CYP3A5 gene is expressed in only 10% of Europeans. Little is known regarding the consequences of polymorphisms for the metabolism of dietary polyphenols. Some substances which are of medical interest, such as caffeine, nicotine and morphine, are exceptions and data relevant to human health will be briefly reviewed in the following subsections.

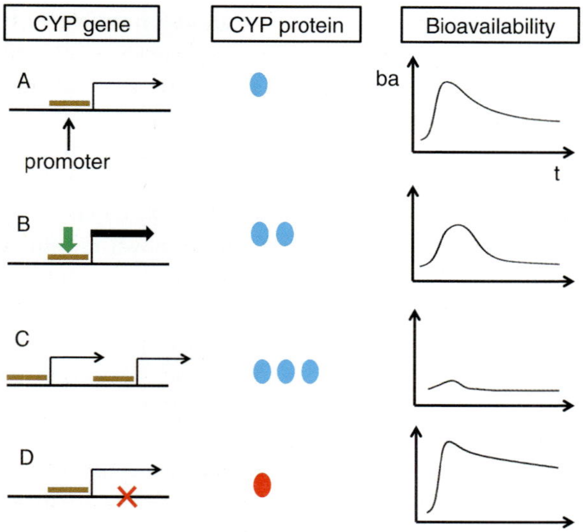

Figure 4.13 Polymorphisms of a CYP gene affect the bioavailability of a substance. The enzyme activity of a CYP biotransformation enzyme may be strongly affected by mutations. In the three examples (B–D) the changes in enzyme activities compared to the normal level (A) and the resulting consequences for the bioavailability (ba) of a substance are illustrated. (A) In the normal situation (no mutation) the promoter controls the faithful expression of the respective CYP protein. The concentration of the substance in the serum rises after its ingestion, reaches a maximum, and finally declines with time (t) due to the CYP activity; (B) A mutation affecting control of the promoter may lead to a decreased or increased expression of the enzyme (as in the example shown), resulting in a higher CYP activity and a lower bioavailability of the compound. This effect is even stronger in the case of gene duplication (C). Finally, a mutation may affect the coding region of the gene (D) and abolish the enzyme's activity. The result is an increased and prolonged bioavailability of the compound, which is substrate of the respective CYP enzyme.

4.1.5.2 **Caffeine**

Coffee consumption and caffeine metabolism is studied as a model for addiction, and since it is the most widely used beverage worldwide the risks and benefits are of particular scientific and medical interest. The enzymes CYP1A1 and CYP1A2 are key to the metabolism of the various polyphenols in coffee. CYP1A1 is known to metabolize polycyclic aromatic hydrocarbons (PAH) that are produced when the coffee beans are roasted. CYP1A2 is the most important metabolic enzyme for caffeine (1,3,7-trimethylxanthine), and demethylates this natural substance in two consecutive steps to yield 1-methylxanthine. The enzyme also metabolizes about 15% of clinical drugs and endobiotics such as steroids, melatonin, and retinoids. Polymorphisms of CYP1A2 are of medical interest as coffee consumption has been claimed to reduce the risk for Alzheimer's disease, Parkinson's disease, and type II diabetes. On the other hand, coffee consumption has been associated with increasing blood pressure and myocardial infarction. Experimental data related to the effect of caffeine on blood pressure have identified a short-term effect that lasts for a few hours, although when such caffeine-mediated effects were followed for some weeks the blood pressure of nonsmoker volunteers was neither increased nor decreased. It should be noted that smoking induces CYP1A2, and this effect must be taken into account in the analysis. Genistein, a major flavonoid component in soy (see Section 5.2) has the opposite effect and causes a decrease in CYP1A2 activity, which in turn affects caffeine metabolism *in vivo*. Other biological effects complicate the analysis; for example, caffeine is known to inhibit important neural signaling pathways by blocking adenosine A_{2A} receptors (ADORA2A2). In fact, the analysis of ADORA2A2 polymorphisms has indicated that the acute effects on blood pressure are mediated (at least in part) by these pathways.

It has been claimed that coffee drinking reduces the risk for Parkinson's disease (PD). When ADORA2A and CYP1A2 polymorphisms were analyzed, the former were shown to be protective for PD but the association with coffee consumption was weak. However, the incidence of PD in individuals with CYP1A2 variant alleles was significantly lower compared to controls. This gene is not the only cytochrome P450 gene which influences the risk for this disease; for example, polymorphisms for the gene CYP2D6 have a greater risk for PD, a finding that was attributed to the insufficient metabolism of pesticides.

4.1.5.3 **Nicotine**

About 75% of nicotine is metabolized to cotinine by the consecutive activities of CYP2A6 and aldehyde oxidase (see Figure 4.9). Poor metabolizers with CYP2A6 polymorphisms smoke fewer cigarettes than EMs, have a decreased risk of dependence, and show fewer withdrawal symptoms. There is also evidence for polymorphisms that affect the glucuronidation of nicotine and cotinine, and this contributes to the wide variations in the metabolic clearance of these substances. Ethnic differences in CYP2A6 activity indicate underlying genetic polymorphisms.

4.1.5.4 **Codeine and Morphine**

Morphine is an effective analgesic and is widely used to treat pain (see Section 5.3.3). The metabolism of opioids has been studied intensively because these compounds are characterized by a narrow therapeutic index (the therapeutic index defines the useful concentration window of a drug between the lowest effective dose and the concentration which results in toxicity; for details, see Section 4.2.1.). A narrow therapeutic index of a drug requires accurate dosing, and this can be critical if the polymorphisms of metabolic enzymes influence the drug's bioavailability. Unexpectedly high serum levels of the opioid would be toxic and result in respiratory depression that may be fatal. In contrast, an insufficient dose of morphine would provide no relief of pain. In view of the millions of cancer patients treated with opioids in advanced stages of the disease, metabolic anomalies are of concern. This problem is illustrated by the finding that morphine dosage requirements for patients receiving hip replacement therapies varied almost 40-fold.

Individual differences in the experience of pain may be due to polymorphisms of the μ opioid receptor or other genes concerned with neural signaling and signal transduction, though polymorphisms in metabolic enzymes are most relevant to the observed variable response. The gene CYP2D6 is particularly important and converts codeine by O-demethylation to morphine (Figure 4.9). More than 100 allelic variants have been described, ranging from nonfunctional alleles to duplications with increased CYP2D6 activity. In Caucasian populations the percentage of PMs for CYP2D6 is particularly high (6–10%), while the incidence in other ethnic groups is lower. Some UMs were identified who possessed up to 13 copies of the gene. Pharmacokinetic studies using human liver microsomes as a test system, as well as clinical studies, have been carried out. The subsequent *in-vitro* data acquired showed that the CYP2D6 enzyme kinetics between EMs and PMs may differ between 10- and 200-fold, depending on the type of opioid substrate used. Some reported cases of CYP2D6 polymorphisms may illustrate the importance of the enzyme. For example, the child of a breastfeeding woman receiving codeine treatment suffered from morphine intoxication. A life-threatening morphine intoxication was also observed in a patient carrying several CYP2D6 alleles. In these cases, the patients were UMs. Other studies showed that PMs for CYP2D6 are underrepresented among opioid-dependent individuals, and codeine treatment in such individuals is ineffective. These and many other observations show clearly the clinical relevance of polymorphisms of this CYP enzyme, and the need to genotype patients in critical cases before treatment. There is some evidence that polymorphisms of an important transporter protein (ABCB1) may modulate the analgesic effects of morphine. Efflux transporters limit the passage of substances across the blood–brain barrier by pumping the respective substances out of the endothelium back into the blood circulation. However, polymorphisms concerning the genes of transporter proteins have little predictive value and the published data do not provide a coherent picture.

4.1.5.5 Pharmacogenomics and Personalized Medicine

Until now, discussions have centered on polymorphisms in single genes which are involved in the biotransformation of natural compounds or drugs, and their impact on human health. However, xenobiotics typically affect several genes or gene products, and the natural substances are metabolized by several enzymes. To meet this degree of complexity the analysis of a pathway, rather than the activity of a single gene, has come into focus in recent years. Technological achievements in modern genetics have allowed the mapping of human genome variations. Whole-genome sequencing and genome-wide association studies are often referred to as pharmaco*genomics* (as opposed to pharmaco*genetics*). The whole-genome sequencing of 179 individuals from four populations revealed 15 million single nucleotide polymorphisms, one million short insertions and deletions, and 20 000 structural variants (data provided by the 1000 Genomes Project Consortium). Although most of these differences have no practical relevance the study results showed that, on average, each person carries about 250–300 loss-of-function variants in annotated genes, and 50–100 of those are implicated in inherited diseases.

Pharmacogenomic approaches can provide the technical tools for a personalized medicine that takes different metabolic efficiencies into account, and individual medications based on identified polymorphisms that affect the pharmacokinetics is a goal within reach. These modern techniques would also facilitate an analysis of the metabolism and biological effects of natural substances present in food.

4.2 Toxicity

4.2.1 Basic Toxicology

Toxicology is the science of poisonous substances, their effects on organisms, and the treatment of poisoning. This interdisciplinary subject includes aspects of pharmacology, pathology, physiology, biochemistry, cell biology, and genetics. The specific demands for a toxicological evaluation have led to a specialization in pharmacology, food science, forensic medicine, and in environmental science. The established test systems and risk evaluation have been adapted to the requirements of the respective area of interest. At this point, toxicity will be viewed from a biological perspective and the current approaches to evaluate the toxicity of natural substances will be discussed.

The concentration of a toxic substance (c) that is taken up by an organism and the length of exposure (t) to the respective substance are important factors that determine the degree of toxicity. In the early twentieth century, the German chemist Fritz Haber was studying the toxicity of poisonous gases and suggested that the product of $c \times t$ is a constant (this is Haber's rule). For many poisons the situation is more complex, however, and the route of uptake, the bioavailability and the pharmacodynamics have strong influences on the toxicity of a substance.

It is important to distinguish between *acute* and *chronic* toxicity:

- If the exposure is a single event, then *acute toxicity* will become apparent shortly after exposure to the respective compound. Acute intoxication in humans accounts for an estimated 5–10% of all cases requiring medical emergency treatment. The reason for this may be poisoning with medical drugs, occupational exposure to chemicals, or poisoning with household chemicals.
- In contrast, *chronic toxicity* is characterized by the repetitive uptake of a substance over a longer period of time (see Figure 4.1). Typically, chronic toxicity is brought about by cumulative effects, and a single dose is ineffective.

The consumption of poisonous mushrooms is an excellent example of acute toxicity when, typically, the toxicity is readily apparent and requires immediate treatment. In an exceptional case, however, mushroom poisoning may also have the features of chronic toxicity. Repeated consumption of the Common roll-rim

(*Paxillus involutus*) can elicit an autoimmune reaction after months or even years (Paxillus syndrome), and in severe cases antigen–antibody complexes on red blood cells may lead to hemolysis and kidney failure. Unfortunately, because of the long latency period between ingestion and the appearance of symptoms, this mushroom was formerly considered edible. It is often difficult to draw a clear distinction between acute and chronic toxicity, and for this reason further categories of subacute and subchronic toxicity have been introduced. These can be distinguished according to the dynamics, the symptoms that develop, and the time schedule of the applied toxicity tests.

Chronic toxicity is often difficult to identify as the harmful effects of toxic substances need to be monitored over a long period of time. Environmental chemicals such as pesticides and industrial chemicals are of major concern, though components in food and beverages may also produce chronic toxicity. The long latency period before toxicity manifests in the development of a disease is one reason for the widespread acceptance of the known risks taken with the regular consumption of excessive amounts of alcohol or smoking cigarettes. In such cases, the development of cancer and abnormal development or reproductive failure may be the result of chronic toxicity.

The response of an organism to increasing concentrations of a biologically active substance is described in the dose–response relationship. It has long been known that toxicity depends critically on the dose of a substance – a point initially raised by the medieval physician Paracelsius in the sixteenth century. Since it is critically important to define the useful concentration range of a drug for therapeutic applications, toxicity tests are carried out at an early stage of drug development. The dose that is lethal to 50% of experimental animals (LD_{50}; Figure 4.14) reflects the acute toxicity of the substance in that species, but this has little prognostic value for toxicity in humans. The LD_{50} for humans is only known for some accessible drugs and poisons, and from cases of deliberate or accidental intoxication. In pharmacological research, the ED_{50} (when 50% of treated individuals show an effect) and the TD_{50} (toxic dose in 50% of cases) are more relevant data that provide information on the safety of a compound for pharmacological applications. Of course, the criterion to score an effect as toxic when determining the TD_{50} must be defined for each study. The concentration difference between the TD_{50} and the ED_{50} is referred to as the therapeutic index (TI) (Figure 4.14). In order for a drug to be considered as "safe," a large TI is clearly advantageous. The dose–response curve may not necessarily be sigmoidal, as shown in Figure 4.14; rather, the response will depend to some extent on the nature of the target molecules in the organism. For example, the target may be a membrane receptor or an enzyme, the reaction may be reversible or irreversible, and synergistic or antagonistic effects must also be taken into account. A further complication is introduced when a substance interacts with many target molecules in the organism. Furthermore, if the substance of interest reacts chemically with other compounds taken up at the same time, this may be reflected in the TD_{50} or ED_{50}.

The Lowest-Observed-Effect-Level (LOEC) and the NOEC (No-Observed-Effect-Level; Figure 4.14) define biological reactions in the low concentration range

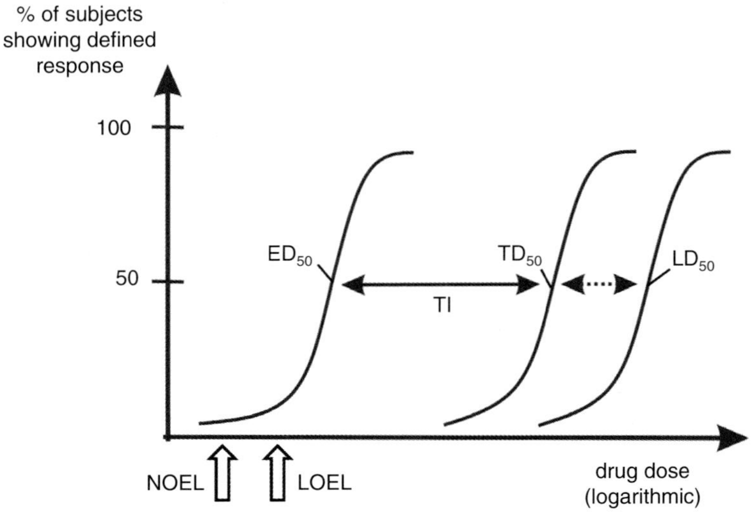

Figure 4.14 Quantitative toxicological evaluation *in vivo*. The experimentally determined lethal dose (LD_{50}) for mice or other experimental animals gives a first estimate of the toxicity of a compound. In pharmacological studies a useful concentration range can be defined by the therapeutic index (TI). The concentrations of the ED_{50} (when 50% of treated individuals show an effect) and the TD_{50} (the toxic dose in 50% of cases) delimit the TI. At very low concentrations a Lowest-Observed-Effect-Level (LOEL) and a No-Effect-Level (NOEL) can be determined. Such data are particularly important in the assessment of environmental hazards.

which are of interest when the respective substance is present in the environment, or in regularly consumed food and there is a suspected risk for chronic toxicity. The toxicological evaluation of food components is particularly difficult. Natural substances typically interact with several target molecules and elicit complex reactions. A possible consequence is that low-dose and high-dose effects may be observed (see Chapters 5 and 6) that are distinct and may even produce opposite effects. Today, in commercial food production, an estimated 3000 additives (synthetic or natural compounds) require toxicological evaluation. The NOEC, which is usually determined in animal experiments, is taken as a reference point; however, to account for possible individually different sensitivities to the respective compound a safety factor is introduced (1/100 of the NOEC), and on that basis an "acceptable daily intake" (ADI) is calculated. The danger of chronic food intoxication by food additives is minimized by this precaution.

4.2.1.1 **Cytotoxicity**

When the toxicity of a natural substance is unknown, a cytotoxicity test with an established cell culture is usually carried out first. A large collection of human cell lines is available for this purpose, and numerous different cell types are

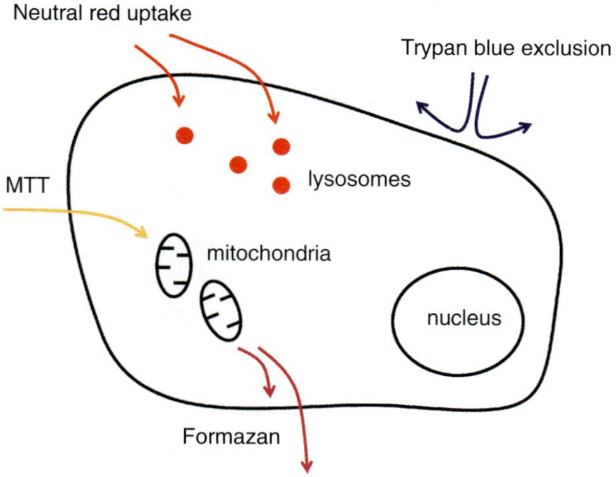

Figure 4.15 Popular tests to determine the cytotoxicity of a compound. Vital cells are able to transport Neutral red into the cell in an active ATP-requiring process. The yellow MTT, when added to the culture medium, is converted to the purple formazan by mitochondrial dehydrogenases. In these tests the presence of a formed dye in the cell is quantitated. The dye Trypan blue is unable to penetrate intact cell membranes, so that an absence of the dye is indicative of vitality. Damaged cells are readily identified by their deep blue color.

represented among these lines. Several simple assays are commonly used to test the vitality of cells in the presence of a test compound (Figure 4.15):

- The *Trypan blue exclusion test* differentiates between cells with an intact cell membrane and nonvital cells with a damaged cell membrane. In the former case, the dye is excluded from the cell, whereas in cells with a compromised membrane the dye will stain the cytoplasm an intense blue color, and this can be easily detected in the microscope.
- In the *Neutral red test*, the dye is only taken up by vital cells in an energy (ATP)-requiring process. The dye accumulates in the lysosomes of vital cells due to their low pH, becomes charged, and remains trapped in these vesicles. The amount of dye present in the cells allows a quantification of the fraction of vital cells in a cell population.
- The *MTT test* offers an alternative method to quantify the vitality of cells. The test makes use of the enzymatic activity of dehydrogenases in the mitochondria, where the soluble yellow dye (3-(4,5-dimethylthiazol-2-yl)-2,5-diphenyltetrazolium bromide) is converted to the insoluble and violet formazan. This reaction product is then solubilized and quantified spectrophotometrically.

Flow cytometry offers another elegant technique for quantifying cytotoxicity, by allowing different parameters to be monitored such as the size, granularity and fluorescence of single cells. In this technique, cells in a physiological buffer are forced to pass in a jet stream through a narrow capillary, and the parameters of interest are measured using laser beams and suitable detectors. A standard application is to determine the cell cycle phase of individual cells in a cell culture. The DNA content in each cell (haploid amount = C) is determined by measuring the quantity of a fluorescent DNA-binding dye in a cell. Propidium iodide is a dye with such properties, and is often used to determine the fraction of cells in G_1 phase (normal diploid 2C DNA content), in S phase (between 2 and 4C due to DNA synthesis), and in the G_2/M phase (4C) in preparation for cell division. A substance that arrests the cell cycle at a specific phase has a cytostatic effect, and the phase in which cells are arrested can be determined using flow cytometry. Very often, the toxicity not only affects the cell cycle but also induces cell death. During normal development, and also under adverse conditions, cells can be eliminated in a controlled process of self-destruction, termed apoptosis. Here, the cellular reaction involves an enzyme-controlled fragmentation of the DNA, which allows the process to be quantified using flow cytometry as the amount of DNA in these cells will be lower (<2C) than the normal diploid content.

Cell-cycle analysis can be further refined and the dynamics of cell proliferation studied over several days by means of flow cytometry. In the following experimental example the effects of quercetin (Q) and 8-prenylnaringenin (8-PN) on cultured human cells were quantified (Figure 4.16). The cells were stained with carboxyfluorescein (CFSE), a fluorescent dye that binds to protein components and has no apparent toxicity. All of the cells in a culture were first uniformly labeled with CFSE, at which time most were in G_1 phase of the cell cycle (Figure 4.16). The cells were then cultured under standard conditions for four days; at each cell division the dye was distributed equally to the daughter cells, so that each cell obtained 50% of the fluorescence. By the second and third divisions the fluorescence of the daughter cells had decreased to 25% and 12.5%, respectively. After four days the culture had become heterogeneous with respect to the CFSE content. In flavonoid-treated samples the differences were markedly increased; for example, at day 2 the quercetin-treated cells lagged behind and had hardly proliferated at all. In addition, some cells were characterized by an abnormally low DNA content (hypodiploid) which was indicative of cell death by apoptosis. By day 4, the 8-PN-treated cells showed a considerable heterogeneity in terms of their proliferation dynamics. It became apparent that cells would react differently to natural substances, depending not only on the cell-cycle phase but possibly also on other physiological differences.

The complexity of cellular reactions to a natural substance is increased when the activities of several target proteins are affected, and this is typically the case (see Section 6.3). As a result, toxicological evaluation is often difficult due to complex reaction patterns. For example, a biphasic dose–response curve of cytotoxicity may be observed that may have resulted from the interaction of two toxic substances which interfere with different cellular pathways. Typically, H_2O_2 added to the

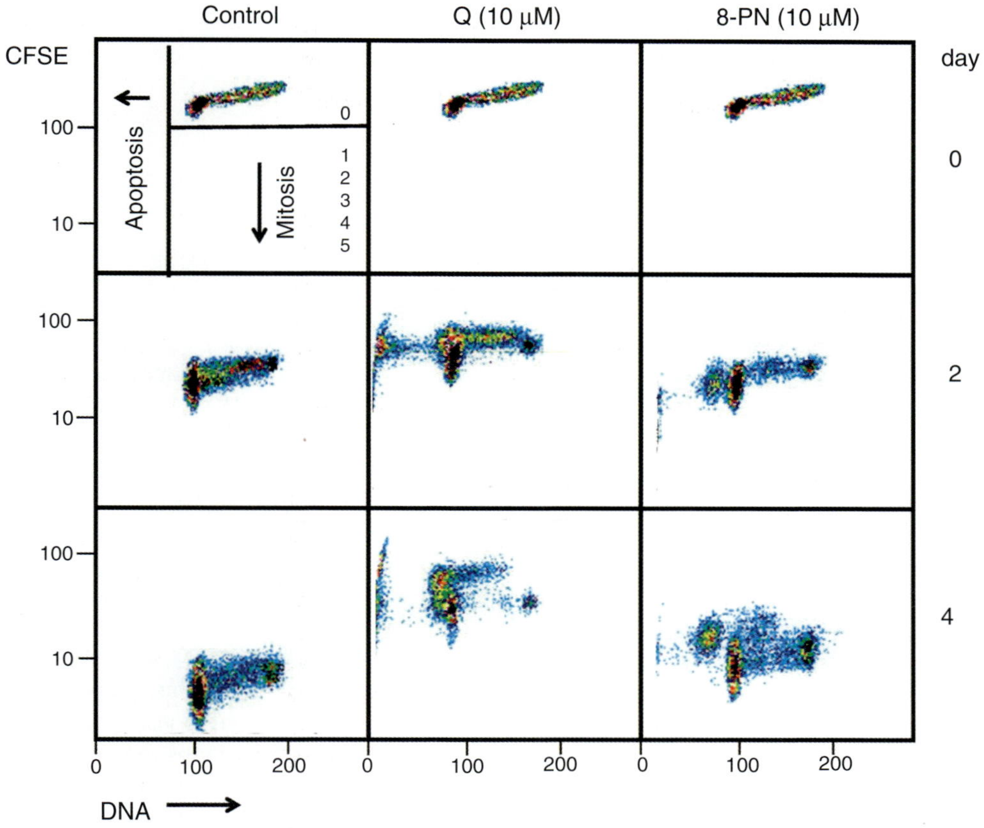

Figure 4.16 Effect of quercetin (Q) and 8-prenylnaringenin (8-PN) on cell proliferation, assayed by flow cytometry. Human leukemia cells (HL-60 cell line) were labeled with the fluorescent dye carboxyfluorescein (CFSE) and Q or 8-PN (10 μM) added to the culture medium. Aliquots were removed from this culture (day 0), and again after 2 and 4 days of culture, fixed, and the DNA stained with propidium iodide. The DNA content (red fluorescence) and CFSE content (green fluorescence) were determined simultaneously. With each cell division the amount of CSFE per cell is halved, so that the dynamics of cell proliferation in the entire culture can be analyzed. In the control culture (day 0, no flavonoid added) most cells are in the G_1 phase and the CFSE labeling is homogeneous. When the cells replicate their DNA (S phase) and in the following G_2/M phase the cells increase slightly in size, which is reflected by an increasing CFSE signal. In control cultures all of the cell divide but do not proliferate in synchrony so that, with time, the culture becomes more heterogeneous. In the flavonoid-treated samples, however, some cells divide while others do not, so that different subpopulations can be distinguished. Some cells undergo apoptosis, which results in the fragmentation of the DNA and a reduced DNA content (see marked area at day 0). S. Tokalov and H.O. Gutzeit, unpublished results.

culture medium of human cells is cytotoxic at a concentration of 500 μM and, as a result, the vitality of human cultured cells is low (Figure 4.17). High concentrations of quercetin are also cytotoxic ($EC_{50} = 48.5$ μM). However, when both substances were added to the culture and the H_2O_2 concentration was kept constant,

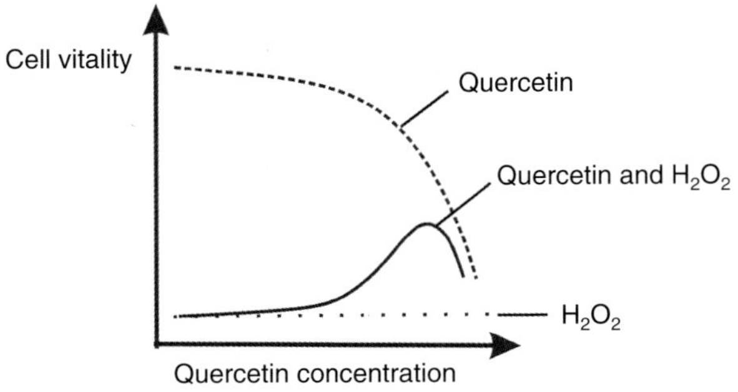

Figure 4.17 Cytoprotective effect of quercetin. At high concentrations (500 μM) H_2O_2 is cytotoxic, and increasing quercetin concentrations are also toxic. However, the effect of H_2O_2 is mitigated at a certain quercetin concentration, so that the fraction of cells that survive H_2O_2 treatment is increased. Apparently, quercetin has a cytoprotective effect within a certain concentration range. Data redrawn from Seibert, H. *et al.* (2011) *Food Chem. Toxicol.*, **49**, 2398–2407.

the quercetin concentration gradually increased and a surprising effect was observed, in that at a quercetin concentration of about 5–20 μM some cells survived despite the presence of a toxic concentration of H_2O_2. This cytoprotective effect may have been due to an induced antioxidative stress–response by the flavonoid (see Section 5.1.5), although other physiological reactions may also have accounted for the observed effect.

4.2.2 Toxicity of Selected Natural Substances

4.2.2.1 Animal Poisons

Many animals produce poisons in order to defend themselves against predators, or the predators themselves may have developed poisons to immobilize their prey. As the focus of this book is plant products, animal poisons will be discussed only briefly. The few examples of particularly toxic substances produced by animals show that there are no principal differences between plants and animals in the development of biologically active secondary metabolites. During evolution, numerous poisonous substances have been created by different groups of organisms, ranging from bacteria to vertebrates. In most cases there is no evolutionary history of a toxin; rather, the substances have been developed independently in the different taxa or species.

Among the examples of poisonous animals, snakes come first to mind because the poisoning of humans by snake bites is frequent and the number of deaths worldwide is estimated to exceed 40 000 each year. More than 400 poisonous species belong to two large families: the Elapida (including cobras); and the Viperida (including the common adder and rattlesnakes). The poison is not a single active substance but typically contains a cocktail of different proteins that act synergistically. For example, hyaluronidases are generally present which depolymerize mucopolysaccharides, which in turn facilitates the spreading of toxic enzymes such as phospholipase A_2 and proteases. The phospholipases damage cell membranes and promote inflammatory reactions. Proteases are particularly important in the poison of Viperida, where the enzymes may lead to a fall in blood pressure and the local destruction of tissues that results in bleeding and pain. A special case here is the peptide α-bungarotoxin of some Elapida, which is a potent inhibitor of the acetylcholine receptor (Section 5.3.6) and inhibits the propagation of nerve signals at the neuromuscular junction, such that the prey is immobilized. Among vertebrates there are poisonous species in amphibian and reptiles (other than snakes), while even some mammal species have developed poisons. The Australian Platypus (duckbill) and the Northern short-tailed shrew (*Blarina brevicauda*) are examples of poisonous mammals. Among the invertebrates, poisons have been developed in scorpions, spiders, and insects. The poison of bees contains a mixture of various polypeptides and enzymes including hyaluronidases and phospholipase A (resembling the poison of Viperida) and histamine as the mediator of an inflammatory response.

In some marine species highly effective poisons have been developed which are not proteins and may be regarded as equivalent to secondary metabolites in plants. For example, tetrodotoxin from the pufferfish (*Fugu*) blocks Na^+ channels and is a highly efficient nerve poison. In mice, the LD_{50} of tetrodotoxin (oral uptake) is only $334\,\mu g\,kg^{-1}$. Other poisonous marine animals include numerous fish species, some jellyfish, echinoderms, mollusks, and annelids.

4.2.2.2 Poisons in Plants, Fungi, and Bacteria

Among the more than 500 000 known plant species, about 10% are estimated to contain toxic components, and only a small fraction of those (ca. 1000) have been characterized. A few biologically interesting examples have been chosen to illustrate the enormous chemical diversity of plant toxins in the following sections. Some examples of particularly toxic substances are listed in Table 4.5; most of these poisons, such as the toxins ricin, taxin, physostigmine and atropine, produce acute toxicity, although in some cases the symptoms develop after a delay of a few days, as in the hemolytic-uremic syndrome (HUS) caused by shiga toxin (Table 4.5) or after poisoning with α-amanitin. The symptoms of other substances become apparent after weeks or even years, as with the regular consumption of nicotine or opioids (see Section 5.3); the carcinogens aristolochic acid and aflatoxin also fall into this category. Currently, the most toxic substances known – botulinum toxin and tetanus toxin – are produced by bacteria. The toxicity of botulinum toxin (as determined by the LD_{50} in μg/kg; intravenous) and of the well-known cobra neurotoxin differ by a factor of about 250 000. In many cases toxins may serve as useful drugs for the medical treatment of specific diseases, when the biological activity of the toxin can be used to counteract a dysregulated physiological process.

Table 4.5 Examples of poisons produced in plants, fungi, and bacteria. LD_{min}: minimal lethal dose, i.v.: intravenous; p.o.: per os (uptake by mouth), parenteral: uptake not through the digestive tract (thus avoiding first-pass metabolism). The concentrations given for LD_{50} or LD_{min} may vary depending on the experimental design of the study and the animal strains used. As a result, the data given in the literature often vary considerably. Toxic concentrations for humans are also approximations.

Toxic substance	Plant	Symptoms of intoxication	Toxicity
Ricin	*Rhicinus communis, Croton* sp.	Diarrhea, thrombosis, kidney inflammation, circulatory collapse	LD_{min} (human): 20–30 mg kg^{-1} (p.o.) and 22 µg kg^{-1} (i.v. or inhalation); LD_{50} (mouse) 12 µg kg^{-1} (i.v.)
Aristolochic acid	*Aristolochia clematis*	Nephropathy. Urothelial, colorectal and other types of cancer	LD_{50} (rodents): 56–203 mg kg^{-1} (p.o.); chronic toxicity (cancer) >0.1 mg kg^{-1} (rats)
Taxine	*Taxus baccata*	Diarrhea, dizziness, bradycardia and finally circulatory collapse and respiratory paralysis	LD_{min} (human): approximately 3.0–3.6 mg kg^{-1} (p.o.); LD_{50} (rat) 4.5 mg kg^{-1} (i.v.)
Atropine	*Atropa belladonna*	Anticholinergic activity. At high concentration tachycardia, restlessness, spasms, hallucinations, death by respiratory paralysis	LD_{min} (human, adult): about 100 mg
Physostigmine	*Physostigma venenosum*	Stimulates parasympathetic activity; vomiting, diarrhea, spasms, respiratory paralysis.	LD_{min} (human): about 6–10 mg (p.o.)
Colchicine	*Colchicum autumnale*	Vomiting, diarrhea, anemia, kidney failure, death (after about 3 days) by cardiac failure, respiratory failure or sepsis.	LD_{min} (human, adult): 20 mg
Morphine	*Papaver somniferum*	Overdose: hypotension, bradycardia, asphyxia and respiratory depression. Chronic toxicity: addiction	LD_{min} (human, adult): >0.1 g (parenteral)
Tetrahydrocannabinol (Δ^9-THC)	*Cannabis sativa*	Overdose: fright, paranoia, hallucinations.	LD_{50} (mouse): 42 mg kg^{-1} (i.v.), 482 mg kg^{-1} (p.o.); LD_{min} (Rhesus monkey): 128 mg kg^{-1} (i.v.)

Nicotine	*Nicotiana tabacum*	Acute toxicity: stimulation of cholinergic neurons, tachycardia, hypertension, vomiting. Chronic toxicity: cardiovascular diseases, Cancer of the respiratory tract due to carcinogens in cigarette smoke.	LD_{min} (human, adult): 50 mg (parenteral)
	Fungi		
α-Amanitin	*Amanita phalloides*	Vomiting, diarrhea, (latency period 8–24 h). Later liver and kidney necrosis.	LD_{min} (human, adult): 0,1 mg kg^{-1}
Aflatoxin	*Aspergillus flavus*	Carcinogen (particularly liver tumors)	TD_{50} (rodents): 1 μg kg^{-1} induces tumors in 50% of animals
	Bacteria		
Botulinum toxin	*Clostridium botulinum*	Muscle weakness, visual and speech disturbances, respiratory paralysis, heart failure.	LD_{50} (human) 1 ng kg^{-1} (parenteral), 1 μg kg^{-1} (p.o.).
Tetanus toxin	*Clostridium tetani*	Increased muscle tone, tetanus, death by suffocation	LD_{min} (human): 2.5 ng kg^{-1}
Shiga toxin	*Shigella dysenteriae*, Related toxins in some pathogenic strains of *Escherichia coli* (EHEC)	Nausea, colic, fever, vomiting. A complication is the Hemolytic uremic syndrome (HUS) which involves lysis of red blood cells and kidney failure	In severe HUS cases the infection may be fatal

4.2.2.3 Taxus Alkaloids

Yew trees of the genus *Taxus* produce interesting toxic alkaloids in their bark, seeds, and leaves. One particularly important substance among these taxanes is paclitaxel, which is better known under the tradename Taxol®. This substance stabilizes microtubules and may hence interfere with the essential dynamics of the polymerization/depolymerization of the functional mitotic spindle. As a result, cells exposed to paclitaxel are arrested during mitosis and are unable to divide, and this cytostatic effect forms the basis for paclitaxel being used to treat various types of cancer. The intensive use of paclitaxel in cancer therapy has made it difficult to meet the demand for this compound, as the bark of *Taxus baccata* contains at most 0.06% paclitaxel. Hence, the modern production of paclitaxel involves the synthesis of a precursor molecule in cell cultures (using bioreactors; see Section 1.3) and a chemical modification to yield the active compound. Apart from paclitaxel, other major components of yew trees are Taxin A and Taxin B; the latter is cardiotoxic due to its inhibitory effects on Ca^{2+} and Na^{+} channels. When ingested, yew causes dizziness, vomiting and convulsions and, at higher concentrations, bradycardia and, finally, heart failure. The amount of taxins present in 50 to 100 leaves of *T. baccata* may be fatal. The toxicity of yew extracts has been known since Roman times.

4.2.2.4 Phallotoxins and Amatoxins

Some basidiomycete fungi of the genus *Amanita* produce extremely poisonous substances which account for most cases of mushroom poisoning. The "death cap" (*Amanita phalloides*) synthesizes a number of cyclic octapeptides (amatoxins), with α-amanitin being the most prominent member. This compound inhibits RNA polymerase II, and hence mRNA transcription. The efficient inhibition of protein synthesis is highly cytotoxic, and organs with strong metabolic activities such as the liver or kidneys are affected first (Table 4.5). The toxicity of α-amanitin has been exploited recently to kill tumor cells. In this case, the poison was coupled to an antibody directed against a human epithelial cell adhesion molecule which was overexpressed in some cancers. This allowed α-amanitin to be targeted to the tumor cells, where the poison is taken up into the cells by endocytosis. The advantage here is that normal cells in the tissue, which do not express the cell adhesion protein, are not affected.

Cyclic heptapeptides with other biological properties are also synthesized in another *Amanita* species (*Amanita virosa*). One prominent member of this group of phallotoxins is phalloidin, which binds to polymerized actin and inhibits the activity of this important cytoskeletal protein. Phalloidin is most valuable in basic research as an actin inhibitor and when it is conjugated with a fluorochrome, the actin cytoskeleton (microfilaments) can be visualized in cells (Figure 4.18).

4.2.2.5 Ricin

The oil prepared from the castor oil plant *Ricinus communis* is used as mild laxative due to the presence of ricinoleic acid, an unsaturated omega-9 fatty acid that promotes intestinal functions by activating the prostaglandin E_2 receptor. When

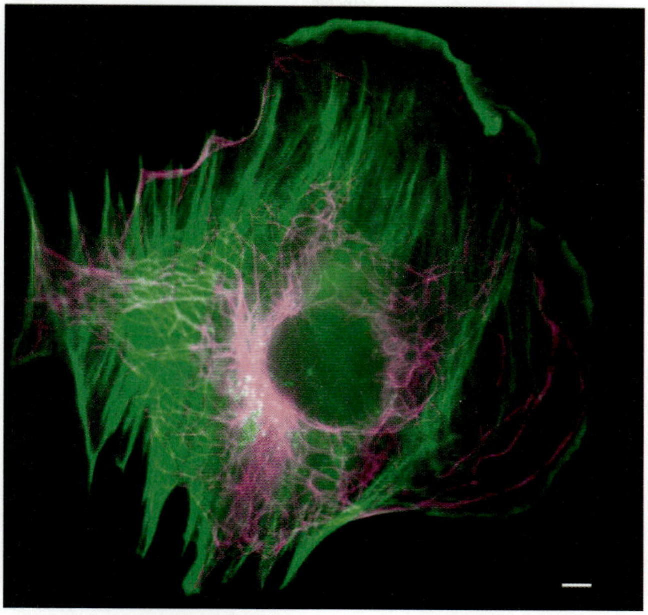

Figure 4.18 Visualization of the actin cytoskeleton with phalloidin. A fibroblast of the mouse cell line Swiss 3T3 was treated with fluorescence-labeled phalloidin, which binds to polymerized actin. Fluorescent microfilament bundles (green) transverse through the cell and are also present in specialized cortical extensions (lamellipodia) which provide the driving force for the migration of the cell on the substrate. The tubulin cytoskeleton is visualized by fluorescence-labeled antibodies (magenta). The microtubular network is particularly dense around the nucleus. Scale bar: 5 μM. Photograph courtesy of A. Verkhovsky, Laboratory of Cell Biophysics, EPFL Lausanne, Switzerland.

preparing ricin oil it is essential that the extract is heated in order to inactivate the highly toxic protein ricin in the castor beans. The protein consists of two polypeptide chains (A and B) connected by a sulfur bridge. Ricin targets protein synthesis, and its inhibitory effect on this vital process is the reason for its extreme toxicity. The two chains cooperate in the procedure; chain B binds to the cell surface and promotes the internalization of the protein, while chain A acts as a highly specific N-glycosidase and inactivates the ribosomes by hydrolyzing the N-glycosidic bond of an essential adenine residue. The resultant depurination inactivates the ribosome and, in turn, inhibits protein synthesis. In terms of potency, a single ricin molecule is thought to be sufficient to kill one cell!

If castor beans are ingested, the intestinal tract is first to be affected. Ricin cytotoxicity is mediated by a necrotic degeneration of the gut epithelial cells, which leads to heavy bleeding. In severe cases of ingestion other organs are also affected, and the victim will die after a few days. However, if ricin is inhaled or injected into a blood vessel, then the toxicity is dramatically increased (Table 4.5). In the past, ricin has been used in assassinations and efforts have been made in several countries to develop a biological weapon based on its extreme toxicity. Similar poisons have been developed by other organisms; for example, the protein Abrin from the Jequirity beans (*Abrus precatorius*) exploits the same reaction pathway.

4.2.2.6 **Bacterial Toxins**

The Shiga toxin, a protein named after the Japanese bacteriologist Kiyoshi Shiga, is produced by some bacteria and elicits ricin-type reactions. When ingested with contaminated food, the bacterium *Shigella dysenteria* causes dysentery, as suggested by its name. Some strains of *Escherichia coli*, a common intestinal bacterium, have become highly pathogenic due to the production of a shiga toxin-like protein. One well-known example of this is the enterohemorrhagic *E. coli* (EHEC) strain which cause severe dysentery and the destruction of colon tissue. As EHEC is invasive, severe complications can lead to the destruction of red blood cells and kidney failure (hemolytic-uremic syndrome) which can be fatal. Several serious epidemic outbreaks of EHEC infections due to contaminated food have been reported, most recently in Northern Germany in 2011.

Some bacterial toxins do not penetrate the cell membrane of humans or animals cells, but act extracellularly (as an exotoxin). For example, the protein streptolysin O of some *Streptococcus* strains binds to cholesterol and induces the formation of pores in the cell membrane of erythrocytes, so that the cell cannot maintain an osmotic balance and it lyses. In addition to the hemolytic activity, the protein is also strongly immunogenic and mediates inflammatory reactions. Another exotoxin, the alpha-toxin of *Staphylococcus aureus*, is also able to form pores in the cell membrane. Most cases of food poisoning are due to bacterial contaminations with either *Salmonella* or *Staphylococcus* strains.

Bacteria have been highly inventive in producing toxins that affect vital cellular functions, with protein synthesis and the integrity of the cell membrane being very effective targets for toxins. The essential signal transduction pathways in eukaryotic cells are also common targets. Of particular importance are the signaling pathways mediated by GTP-binding proteins (known as G proteins), which are associated with membrane receptors (G protein-coupled receptors; GPCRs). Pertussis toxin and cholera toxin both target these signaling pathways.

Finally, botulinum toxin should be mentioned as it is one of the most potent poisons known. The protein, which is produced by the ubiquitous anaerobic bacterium *Clostridium botulinum*, inhibits release of the transmitter acetylcholine from the presynaptic nerve at the neural or neuromuscular synapse (Section 5.3.6; Table 4.5). A suitable antidote in this situation is treatment with an antibody that binds to, and neutralizes, the toxin. Botulinum toxin is also used for various cosmetic applications (under the trade name Botox), and also to treat some diseases that require a reduced acetylcholine signaling (see Section 5.3.6). Acetylcholine release is also blocked by tetanus toxin that is produced by another *Clostridium* strain (*C. tetani*) and, as the name suggests, causes tetanus. Compared to botulinum toxin, other sets of neurons are affected by tetanus toxin and hence the symptoms will differ in these two cases.

4.2.3
Strategies and Difficulties in Toxicological Evaluations

The toxic substances discussed so far have been well known for centuries, and the molecular mechanisms of their activities have recently been elucidated. The main challenge of modern toxicology is to identify weakly toxic substances in the environment or in food which are a threat to human health. The need to evaluate such chemicals in the environment and, above all, the requirements of rigorous tests of new pharmaceuticals in the preclinical phase, have spurred the development of suitable test. Some experimental strategies and common experimental protocols that are used to define the toxicity of medical drugs, and can also be used to assess natural substances, will be briefly discussed in the following sections.

A reasonable strategy for the toxicological evaluation of a compound involves first of all some simple and inexpensive experiments; only at the final stages of investigation are *in-vivo* tests indispensable (Figure 4.19). In the past, numerous structural and toxicological data have been gathered about many organic molecules (including natural compounds), and this knowledge can be used to predict possible mutagenic and carcinogenic properties. The *in-silico* analysis is based on structure–activity relationships (SAR) and, by using specific models, the available software attempts to connect a known chemical structure with a specific biological activity. Such *in-silico* studies are not substitutes for the *in-vitro* and *in-vivo* experiments, but help in the experimental design and interpretation of the results. The first experimental tests are usually carried out with established human cell cultures. If at least one of the tests suggests genotoxicity, then *in-vivo* tests will follow and, for example, induced chromosome aberrations will be quantified. If further evidence substantiates the genotoxic potential of a substance and suggests cancerogenic potential, a two-year *in-vivo* test using rodents can be applied to clarify

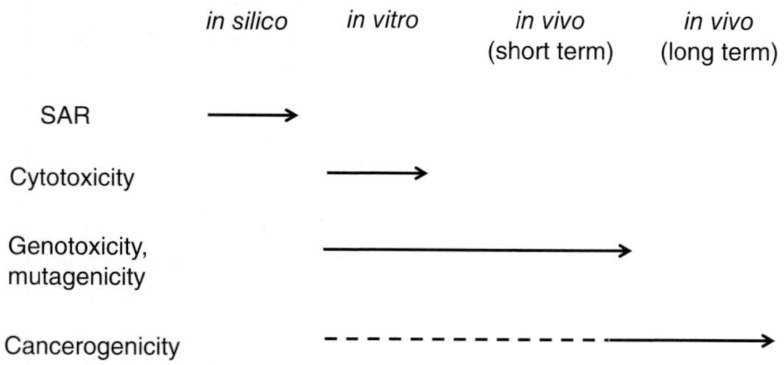

Figure 4.19 Course of toxicity testing from *in-silico* analysis to *in-vivo* tests with animal models. Toxicological studies are usually performed in a sequence of steps. First, toxicity is predicted by comparison with data from other substances with similar chemical scaffold (SAR = structure–activity relationship). Fast and inexpensive *in-vitro* tests can then be carried out, followed by more expensive and labor-intensive assays using animal models to determine possible genotoxic and carcinogenic activity.

the possible hazard for mammals. This scheme (Figure 4.19) can be modified in many ways, depending on the purpose and the chemical substance of the toxicological evaluation. The aim of some current research programs is to develop new test protocols in order to reduce the required number of animals, and to limit the pain inflicted on experimental animals.

4.2.3.1 **Mutagenicity and Cancerogenicity**

The demonstration of the mutagenic and cancerogenic activities of a natural substance is a special challenge, and highly sensitive and reliable tests are essential to detect carcinogens. Substances which significantly enhance the normal mutation rate will promote the development of cancer, as the functional dysregulation of critical genes forms the basis of the process of cancerogenesis. A number of *in-vitro* and *in-vivo* test systems have been developed to evaluate the cancerogenic potential of substances by identifying mutagens and carcinogens. Although most carcinogens have also been found to be genotoxic, some are not and only affect critical signal transduction pathways. This classification depends to a large extent on the test system applied, and although describing the outcome of the tests the underlying molecular mechanisms are, in most cases, not resolved by such experiments. However, the biochemical reactions which lead to the initiation or promotion of cancerogenicity are of great importance for a realistic risk assessment.

In order to determine the cancerogenicity of a substance, the two-year rodent bioassay is the standard test. The long-term exposure is required because of the typical lag phase before a tumor becomes detectable. A large number of animals is also required to obtain statistically significant results and, typically, high doses of the test substance are used to induce measurable effects. Whether such tests are relevant to humans is often questionable, however, as some tumors are species-specific and develop only in rodents. As such a labor-intensive bioassay would cost up to US$ 1 million for each chemical, these disadvantages continue to spur the search for alternatives. Unfortunately, the *in-vivo* tests are difficult to replace because the pharmacokinetics and bioavailability of the test substances is an important issue that cannot be simulated *in vitro*.

During recent years, transgenic animal models have been introduced which offer greater sensitivity for carcinogens, such that fewer animals are required. Frequently used transgenic strains include the rasH2 model (with an introduced Ras proto-oncogene) and the hemizygous p53 mouse model and other transgenic strains with defective DNA repair genes. The p53 tumor suppressor gene and the Ras oncogene are frequently mutated in human cancers. In these models, the tumor incidence after chemical induction by carcinogens is substantially increased. Transgenic animal models are also extremely useful for mechanistic studies of carcinogens. Strains in which reporter genes are expressed when cancer-related genes become activated are also promising disease models. Finally, transgenic strains that lack major biotransformation enzymes (P450 knockouts) are helpful in studies addressing the relevance of biotransformation reactions for cancerogenicity.

The metabolic differences between humans and rodents are of concern, and represent a major reason for the poor predictive value of rodent models. Because of this limitation the important human enzyme CYP3A4, which is involved in more than 50% of clinically used drugs, has been introduced into the mouse genome (CYP3A4 "humanized mice"), and this strain can be used in mutagenicity studies of compounds in which the activity of CYP3A4 is indicated. Transgenic mouse strains for many other relevant genes are available. Lower vertebrate model organisms such as the zebrafish (*Danio rerio*) or the medaka (*Oryzias latipes*) offer additional options for fast and efficient *in-vivo* toxicological assays over one or more generations. A number of useful transgenic strains have been established in these species.

Short-term tests that save time and money are carried out routinely, but each of the available tests has certain strengths and drawbacks. Genotoxicity can be evaluated by scoring the induced chromosomal aberrations such as aneuploidy, translocations and similar cytologically detectable anomalies. The mutagenicity at a single susceptible locus can also be quantified when a mutation leads to inactivity of the gene product so that, for example, the lack of an enzyme activity can be measured. In all of these tests false-positive results pose a problem and must be taken into account. An efficient bacterial test with *Salmonella typhimurium* is widely applied using strains deficient in histidine synthesis. The dose-dependent induction of revertants forms the basis of this popular mutagenesis test (Ames test); however, the main shortcoming of this test is that the human biotransformation enzymes are not present in bacteria and cytoplasmic extracts from mammalian cells must be added to circumvent this problem. Approximately 80% of mutagens identified in the Ames test were shown also to be mutagens in vertebrates.

Suitable cell cultures which show contact inhibition can also be used to study induced cancerous transformations. Whereas, normal cells stop proliferating when contact is made with neighboring cells and form a confluent single-cell layer, cancer cells lose this property and the cells pile up to form lumps. The formation of these foci can be quantified after about 6–8 weeks of culture as a measure of tumorous transformation. Unfortunately, the emergence of culture-dependent false-positive scores may be a serious problem in these tests.

The newly developed genetic techniques allow the analysis of complex DNA sequences with great efficiency (genomics), while the effects of natural substances on mRNA expression patterns can be examined using analytical chip technology. The parallel analysis of many proteins (proteomics) and of key metabolic substances (metabolomics) also facilitates toxicological analyses on the molecular level. The use of these new approaches has led to the term "toxicogenomics" being coined. Despite these advances, *in-vivo* testing cannot be replaced. Experiments with human cell cultures face the problem that the available cell lines are either immortalized (by genetic manipulation) or are mutant in oncogenes and/or tumor suppressor genes, and hence do not necessarily respond to a carcinogen as would normal tissues. An additional problem is that the pharmacokinetic properties of a particular substance of interest cannot be simulated *in vitro*.

The toxicological evaluation of quercetin illustrates the typical toxicological approach, and the possible pitfalls. Quercetin and other flavonoids have been suggested to be mutagenic and this prompted intensive toxicological research. In this case, the risk assessment is of particular importance as quercetin is a common component of foods such as fruits and vegetables and is also present in many beverages such as black tea, red wine, and various fruit juices. In some countries quercetin is also an approved food additive (in Japan) and a component of dietary supplements (in the US), and consequently the total daily uptake for some consumers may be 200–500 mg. The question is, are such high concentrations of quercetin toxic?

Many flavonoids have an antioxidative effect *in vitro*, and this property may be marketed as beneficial for health. The antioxidant and prooxidant activities of quercetin have been the subjects of several studies, and it appears that quercetin may have both properties, depending on the experimental conditions and the redox-state of the cell. A mild prooxidative activity may have beneficial effects through the induction of a hormesis (see Section 5.1.5 and Figure 4.17). However, the oxidative degradation of quercetin may result in the formation of a free semiquinone radical (Figure 4.5), after which prooxidative quinone methides may ultimately form that can bind to the DNA and may cause mutations. Indeed, the mutagenicity of quercetin has been demonstrated in bacterial systems (Ames test), and also in several *in-vitro* test systems with hamster, mouse, and human cells. In these tests, chromosomal aberrations, micronucleus formation and single strand breaks were among the recorded genetic defects.

Numerous studies have been conducted to demonstrate the mutagenicity (and also possible cancerogenicity) of quercetin *in vivo*, both in animal experiments and in clinical studies. In almost all studies no evidence was found of any adverse effects when quercetin was administered orally. Apparently the *in-vitro* tests do not reflect the *in-vivo* situation. The first and most important factor is presumably the rapid metabolism of quercetin after ingestion (see Section 4.2.3), with an efficient first-pass metabolism and degradation of the flavonoid by the intestinal bacteria each contributing to a low bioavailability. The intracellular redox-state is tightly controlled (see Section 5.1.2), and prooxidative metabolites may be detoxified rapidly via endogenous mechanisms. Some evidence also suggests that quercetin–DNA adducts may be formed only transiently and do not necessarily result in irreversible genetic damage.

4.2.3.2 **Reprotoxicity and Developmental Toxicity**

Early embryonic development and the subsequent phases of tissue differentiation are particularly sensitive to substances that interfere with the complex signaling network during these processes. In humans, the first eight weeks after fertilization are particularly critical. The developing organs may differ in their sensitivity to a toxic substance, while the periods of sensitivity are typically organ-specific and reflect the timing of critical differentiation processes. Interference with the developmental processes may result in malformations, and such teratogenic effects are of major concern in toxicity testing. The pharmacological scandal during the early

1960s with the drug thalidomide (Contergan^{TM}) showed the importance of adequate testing for developmental toxicity. Pregnant woman who took thalidomide between the 35th and 48th days of pregnancy to treat morning sickness and sleeplessness gave birth to children with malformed extremities. Many thousands of children were affected, and about 40% of the victims died before their first birthday.

The question is: "Are natural compounds also able to interfere with embryonic development?" The toxicity of cyclopamine and the epigenetic transgeneration effects of phytoestrogens (see Section 5.2) show that vertebrate embryogenesis may well be affected by natural substances which interfere with critical signaling during development. However, these effects have been detected fortuitously and were not the result of systematic screening. In view of the numerous food additives used today and changing dietary habits, tests for the chronic toxicity of natural substances affecting developmental processes will become increasingly important in future.

The germline cells which become specified during embryogenesis in vertebrates give rise to gametes in the reproductive organs of both sexes, and transmit the genetic and epigenetic information to the next generation. The integrity of this information is extremely important but difficult to assay, and requires long-term *in-vivo* toxicity tests to be conducted. Whilst a two-generation life-cycle test can be used in rats to cover all critical phases of development and reproduction, experiments using nonhuman primates must, on scientific grounds, be considered for critical compounds of medical interest.

In order to partially replace long-term *in-vivo* testing, a number of *in-vitro* assays have been developed which allow specific critical phases of the life cycle to be assessed (Table 4.6; Figure 4.20). The *in-vitro* culture of rodent embryos covers an important developmental phase. Other tests use primary cells from the limb mesenchyme or embryonic stem cells to test for possible effects of toxic substances on cellular differentiation processes. Possible effects on gamete differentiation and fertility have also been developed by which it is possible to assess the differentiation of gametes and the functional integrity of gametes in an *in-vitro* fertilization assay (Table 4.6). Whilst a combination of these tests can provide valuable information, they cannot fully substitute for the *in-vivo* life cycle tests.

4.2.3.3 **Experimental Variables and Limitations of Test Systems**

The choice of animal species in which toxicological tests are carried out is important, due to the different activities of Phase I and Phase II biotransformation enzymes. Tests with rodent models are time-consuming, costly, and have limited prognostic value for humans. Tests with other animal model systems have been considered, and for routine toxicological screening the life cycle test with zebrafish (*Danio rerio*) as a model organism has been established. For example, the cardiotoxicity of drugs can be screened in the zebrafish successfully, as the transparent nature of the fish allows a direct observation of the heartbeat. Moreover, some of the electrophysiological properties of the zebrafish heart resemble the human heart more closely than the mouse heart. Small and well-characterized fish species

Table 4.6 Test systems to evaluate developmental and reproductive toxicity. The numbers in brackets refer to the phase of the life cycle indicated in Figure 4.20.

Test	Scored endpoint	Phase of reproductive cycle
***In vivo* tests**		
Rat (OECD 414): fetal development	Litter size, gender ratio, fetal survival, weight of brain, spleen, thymus	Prenatal development (1,2)
Rat (OECD 416): two-generation test	F1 and F2 litter size, gender ratio, normal organ development. F1: normal sexual differentiation and maturation	Development and reproduction (1)
Zebrafish (*Danio rerio*): whole life cycle test	Mortality, hatching rate, body length, development of offspring	Development and reproduction (1)
Fertility (*in vitro*)		
Follicle bioassay (FBA); mouse	Differentiation of pre-antral follicles up to ovulatory stage	Folliculogenesis and oogenesis (2)
In-vitro fertilization assay (bIVF); bovine	Penetration of bull spermatozoa into matured oocytes; formation of male and female pronuclei	Fertilization (3)
Embryonic peri-implantation assay (MEPA); mouse	Survival on day 8. Hatched blastocyst stage	Development from zygote to blastocyst (4)
Embryonic development (*in vitro* and *ex vivo*)		
Whole-embryo culture (WEC); mouse, rat	Growth and morphology of embryos (*ex vivo*)	Early embryonic development (4,5)
Embryonic stem cell test (EST); mouse	Inhibition (IC_{50}) of the differentiation of embryonic stem cells into contracting cardiomyocytes	Embryonic tissue differentiation (5)
Micro mass test (MM); mouse, rat	Limb mesenchyme: formation of foci	Embryonic tissue differentiation (5)

(zebrafish and medaka) also allow the large-scale screening of substance libraries or of plant extracts, and these options are important for the approach of reverse pharmacology (see Section 6.3). Further test systems with other vertebrates such as frogs and chicken have been developed to screen for developmental toxicity and to minimize species-specific effects. Tests with experimental animals have failed to predict human toxicity in a surprisingly high number of cases. In fact, based on tests with rodent species alone, human toxicity was predicted faithfully in only 43% of cases, though when all animal species were taken together the predictability level rose to 71%. The consequences of this unsatisfactory situation is that ADRs frequently occur and 23% of all registered drugs are ultimately retracted.

It is not only the differences between species but also interindividual variability that must be borne in mind within a population of experimental animals of the

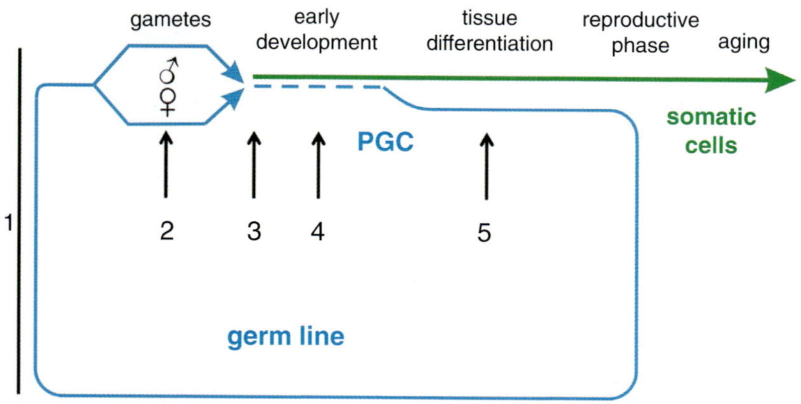

Figure 4.20 Tests to assay developmental and reproductive toxicity. Several *in-vitro* and *in-vivo* tests are used to assess the toxicity of substances for developmental processes and reproductive functions. Reproduction entails the faithful transmission of the individual's genome to the next generation. The primordial germ cells (PGC) are the founder cells of the germline (blue) during embryogenesis. The cells migrate into the developing gonads, where the gametes differentiate in both sexes. The somatic cells (green) differentiate during embryogenesis into different tissues and organs. After the reproductive phase, aging accelerates. In toxicity testing either the entire life cycle is assayed (indicated by the vertical line labeled 1) or selected stages of development for which assays have been established. The numbers refer to the tests listed in Table 4.6 which are used to assay processes at different stages of development (arrows).

same strain. Typically, age, gender, environmental factors and genetic differences may each contribute to individually different responses to a test compound. Other variables may involve the endocrine system, the regulation of blood pressure, and the physiological response of the heart. The existence of such variations leads to an increase in the number of animals required for *in-vivo* tests in order to obtain statistically significant results.

Innovative *in-vitro* test systems with organ-specific cell types will improve toxicity testing in the future. Both, primary and immortalized human cells of various tissues are available, while the use of embryonic stem cell cultures induced to from organ-specific cell types and microengineered organ models represent promising options for the future. Such test systems may supplement, and also partly substitute, the current routine *in-vivo* tests in terms of speed, cost-efficiency, and predictive value.

5
Examples of Physiological Effects of Natural Substances Present in Food and in Medicinal Plants

5.1
Are Antioxidants Good for Health?

5.1.1
Generation and Homeostatic Control of Reactive Oxygen and Nitrogen Species

The oxygen homodimer (O_2) in the air is the fuel for energy production in aerobic organisms, and in this form oxygen is only moderately reactive. However, some oxygen-containing molecules are highly reactive and are able to damage cellular macromolecules such as proteins, lipids, or DNA. These oxygen-containing molecules are referred to as "reactive oxygen species" (ROS; see Section 2.3.1). Their concentration in the body may vary depending on the metabolic activity, and the health and lifestyle of a person, as pollution, smoking, radiation and some drugs are known to increase levels of ROS in the body. ROS-induced molecular damage is thought to be involved in the development of major diseases such as diabetes, cancer, and also cardiovascular and neurodegenerative diseases. Under these pathological conditions elevated levels of ROS are typically observed in the tissues. It appears plausible that lowering ROS concentrations by treatment with antioxidants would have a positive effect on health, and today many natural antioxidants are commercially available in drugstores and have been appraised as health-promoting substances. However, is this claim scientifically sound? The sources of ROS and their normal biological roles, as well as the consequences of abnormally high ROS levels (oxidative stress), will be discussed in this chapter.

Viable cells produce different ROS in varying quantities, depending on the tissue and the metabolic activity of the organism. However, apart from ROS, reactive nitrogen species (RNS) are also produced under normal physiological conditions. Nitrogen monoxide ($NO^\bullet$) is a well-known member of this group of compounds (Table 5.1). Typically, these compounds are radicals that are characterized by the presence of one or more unpaired electrons, and this property makes them highly reactive. Superoxide ($O_2^{\bullet-}$), an important member of the ROS family, is produced in the mitochondria during the process of aerobic respiration when oxygen is used in the electron transport chain to generate energy in the form of ATP. While most electrons are used to reduce oxygen and to form water, a small fraction of electrons

Plant Natural Products: Synthesis, Biological Functions and Practical Applications, First Edition.
Herwig O. Gutzeit and Jutta Ludwig-Müller.

Table 5.1 Important reactive oxygen species (ROS) and reactive nitrogen species (RNS).

Singlet oxygen	$^1O_2$	No radical, relatively stable but highly reactive
Superoxide radical	$O_2^{\bullet -}$	Mitochondria prominent source, signaling molecule, generates other ROS
Hydroxyl radical	$^{\bullet}OH$	Highly reactive, short half-life
Perhydroxyl radical	$HOO^{\bullet}$	Reacts with lipids and initiates chain reaction with formed lipid peroxyl radicals ($LOO^{\bullet}$)
Hydrogen peroxide	H_2O_2	No radical, moderately reactive, peroxisomes important source, signaling molecule
Hypochlorous acid	HOCl	No radical, generated in antibacterial defense and inflammatory reactions
Nitrogen monoxide	$NO^{\bullet}$	Signaling molecule
Peroxynitrite	$ONOO^-$	Highly reactive with DNA and proteins

(0.15–2%) "leak" to oxygen prematurely and give rise to the $O_2^{\bullet -}$ radical. This is generated at a rate that is largely determined by the mitochondrial membrane potential $\Delta\Psi$. Because of this regulation, the metabolic intensity and electron flux through the mitochondrial electron transport chain does not necessarily correlate with the rate of ROS generation in the mitochondria, and this has important biological implications. As will be seen later (Section 5.1.5), the concentration of the $O_2^{\bullet -}$ radical is of major importance in the regulation of cell proliferation, differentiation, and also in aging. Although this radical is mostly produced in the mitochondria it can also be generated in several enzymatic reactions (Figure 5.1). $O_2^{\bullet -}$ is a major source of other important ROS, in particular $HOO^{\bullet}$, $^{\bullet}OH$, and H_2O_2.

Not all members of the ROS family are radicals. Singlet oxygen ($^1O_2$) and hypochlorous acid (HOCl) and, most importantly, hydrogen peroxide (H_2O_2) are examples of such compounds. The peroxisomes are specialized organelles in all eukaryotic cells and constitute a major source for H_2O_2 (but not $O_2^{\bullet -}$), since oxidative reactions catalyzed by different oxidases are typical for these organelles. H_2O_2 is rather stable in cells but it may be converted to the much more aggressive hydroxyl radical ($^{\bullet}OH$) in the presence of metals such as Fe^{2+} or Cu^{2+} (Fenton reaction). The cellular iron concentration is tightly controlled, and in healthy individuals the free iron concentration is negligible; however, some ROS (in particular $O_2^{\bullet -}$ and H_2O_2) may interfere with the elaborate control mechanisms that regulate intracellular metal concentrations, and bound Fe^{2+} may then become available to react with H_2O_2 to produce the aggressive $^{\bullet}OH$ radical:

$$H_2O_2 + Fe^{2+} \rightarrow Fe^{3+} + {}^{\bullet}OH + OH^-$$

The required divalent Fe^{++} is generated when Fe^{3+} reacts with $O_2^{\bullet -}$. The entire Fe-catalyzed process is known as Haber–Weiss reaction, and can be summarized as follows:

$$O_2^{\bullet -} + H_2O_2 \rightarrow O_2 + {}^{\bullet}OH + OH^-$$

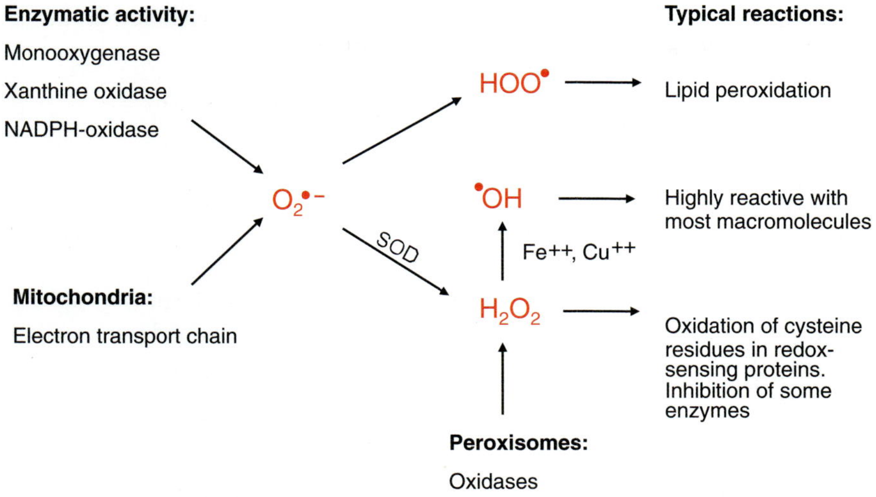

Figure 5.1 Cellular sources of ROS. Superoxide ($O_2^{\bullet -}$) is an important ROS species which is mainly produced by mitochondria and to a smaller extent also in the cytoplasm in several enzymatic reactions. The reactive $O_2^{\bullet -}$ may damage macromolecules and contributes to the initiation of the harmful lipid peroxidation. Normally, $O_2^{\bullet -}$ is quickly converted to H_2O_2 by the enzyme superoxide dismutase (SOD). This important ROS species is produced by several cellular peroxidases. Peroxisomes are rich in these enzymes. The most reactive hydroxyl radical ($^{\bullet}OH$) may be generated from H_2O_2 in the presence of metals such as Fe^{2+} and Cu^{2+}. The ROS are shown in red.

Obviously, the intact homeostatic control of iron and copper metabolism is essential to control this reaction. The $^{\bullet}OH$ radical is a most reactive and short-lived ROS, with a half-life of 10^{-9} s, and for this reason it reacts only with molecules in the vicinity of its site of origin. $^{\bullet}OH$, and in particular the perhydroxyl radical ($HOO^{\bullet}$), are thought to react with lipids and to initiate a chain reaction that may lead to massive lipid oxidation and severe damage of the cell membranes.

The nitrogen oxide radical ($NO^{\bullet}$) is generated in tissues by specific nitrogen oxide synthases (NOS) which metabolize the amino acid arginine to citrulline. $NO^{\bullet}$ is an important second messenger, and acts as a signal in several physiological and developmental processes (see Section 5.1.2). This RNS can diffuse through membranes and has a half-life of several seconds. When $NO^{\bullet}$ and $O_2^{\bullet -}$ react with each other, the most aggressive peroxynitrite ($ONOO^-$) is formed which may react with proteins, lipids, and DNA.

As some ROS act as important signaling molecules, a strict control of ROS generation is required in the tissues, as well as an efficient mechanism to prevent high ROS concentrations and oxidative stress (Section 5.1.3). Because of the vital importance of redox homeostasis, robust control mechanisms have evolved in all organisms – ranging from bacteria to humans – that ensure balanced intracellular concentrations of electron donor and electron acceptor molecules. The regulation is not only cell-type specific but each cellular compartment (mitochondrion, nucleus and cytoplasm) is controlled separately. Basically, three different mechanisms ensure ROS homeostasis:

1) Chemical compounds may act as radical scavengers.
2) Enzymes may react with ROS and convert them into less-reactive products.
3) Redox-active peptides that can be reversibly oxidized/reduced and act as redox-buffers.

Ascorbic acid (vitamin C) is water-soluble and able to scavenge major ROS species such as $O_2^{\bullet -}$, H_2O_2 and $^{\bullet}OH$. Other endogenous antioxidants include uric acid and taurine. Vitamin E (α-tocopherol) is more hydrophobic and acts as radical scavenger in fatty tissues. A particularly important role of α-tocopherol is its ability to interrupt the lipid peroxidation chain reaction.

Several enzymes can react efficiently with ROS, and this represents the second line of ROS control. Superoxide dismutase (SOD) is a phylogenetically old enzyme that converts the aggressive $O_2^{\bullet -}$ radical to the less-reactive H_2O_2:

$$2\,O_2^{\bullet -} + 2H^+ \rightarrow H_2O_2 + O_2$$

This reaction may occur spontaneously, but it is strongly accelerated by the activity of SOD. It has been seen that H_2O_2 may give rise to the damaging $^{\bullet}OH$ radical when free iron is available, but under normal physiological conditions H_2O_2 is an important ROS species, the intracellular concentration of which largely determines the redox-state of the cell. Because of this important role, the concentration of H_2O_2 is tightly controlled, mainly by action of the enzyme catalase, which converts excess H_2O_2 to water (H_2O) while oxygen (O_2) is liberated. H_2O_2 can also be reduced by the enzyme ascorbate peroxidase which employs ascorbate (vitamin C) as an electron donor (Figure 5.2).

Apart from catalase and ascorbate peroxidase, other cellular peroxidases are also able to reduce H_2O_2. For example, the enzymes glutathione peroxidase and peroxiredoxin can oxidize the important redox-sensitive tripeptide glutathione (GSH) and the protein thioredoxin (TRX), respectively, which are components of the third line of antioxidative defense and redox regulation. GSH is present at high concentrations in the nucleus (3–15 mM), the cytosol (1–11 mM), and in the mitochondria (5–11 mM). As GSH possesses a cysteine residue, two molecules can become covalently linked by a sulfur bridge and this represents the oxidized form (GSSG). The enzyme glutathione peroxidase reduces H_2O_2 and converts GSH to the oxidized form GSSG. The ratio of GSH/GSSG serves as an indicator for the redox status of cells, and a high GSSG concentration is taken as an indicator of oxidative stress. The two reaction partners that form the sulfur bridge under oxidative conditions need not necessarily be two GSH peptides. The oxidation of thiol groups may result in the formation of mixed disulfides between, for example, GSH and the cysteine residue of another protein. Some natural substances, such as isothiocyanates, may also react with the thiol group of the cysteine and form a covalent bond.

The antioxidative potential of a cell depends on its capacity to reverse the reaction and to catalyze the reduction of GSSG by the enzyme glutathione

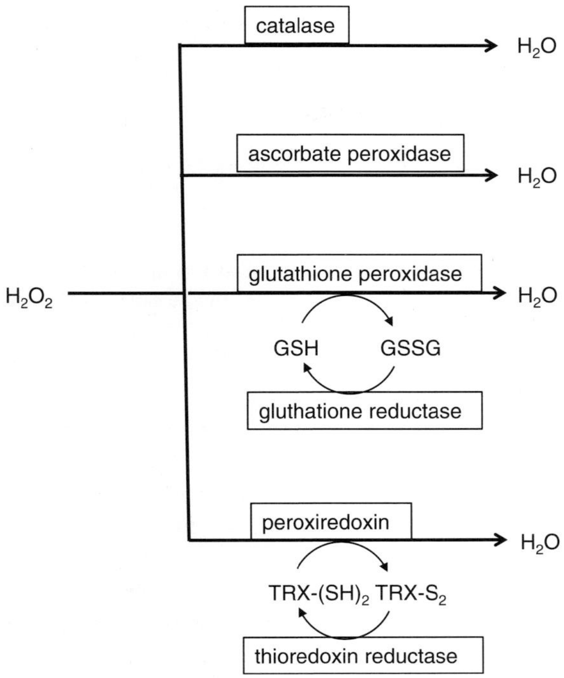

Figure 5.2 Regulation of the intracellular H_2O_2 concentration. The concentration of H_2O_2 is tightly controlled. Some major enzyme-catalyzed degradative reactions are shown. The biologically important glutathione (GSH) and the thioredoxin (TRX) systems can be regenerated. GSH and TRX-(SH)$_2$ represent the reduced states of these two redox-buffering systems. The oxidized states with formed sulfur bridges are abbreviated GSSG and TRX-S$_2$, respectively. The reduced forms can be regenerated by special reductases.

reductase (Figure 5.2). The oxidation of the cellular ROS scavengers vitamin C and E is reversible so that the reduced state can be regenerated. Importantly, the redox-active compounds can interact with each other so as to form a regulatory redox network. In an example illustrating this interaction, a lipid radical (LOO$^•$) reacts with the ROS scavenger vitamin E, which in turn is converted to a radical in this redox reaction (Figure 5.3). Ascorbate (vitamin C) may then act as an electron donor so that vitamin E is regenerated and the ascorbate is reduced to dihydroascorbate after passing through an intermediate radical stage (semiascorbyl-radical). In this example, the GSH/GSSG system may finally come into play and the regeneration of glutathione is provided by NADPH produced during the course of glucose metabolism.

These described reactions allow the maintenance of a regulated redox-state, even in the presence of a moderate amount of oxidants. In order to react to changing concentrations of oxidants, it is necessary to measure the amount of oxidants present in a cell by using a molecular redox-sensor. Cysteine residues are again used for this purpose (Figure 5.4) and may become oxidized to form a sulfur

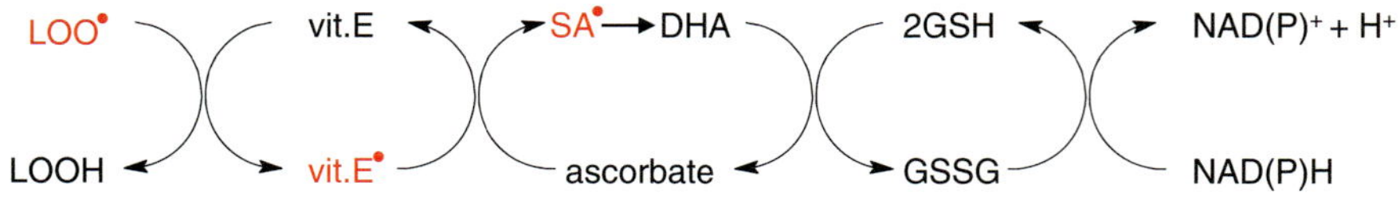

Figure 5.3 Example of a redox-controlling network. Redox-buffering systems form a network in which single components can be regenerated in consecutive redox reactions. In this example, a lipid radical (LOO•) reacts with the lipophilic α-tocopherol (vitamin E). The generated vitamin E radical can, in turn, react with the antioxidant ascorbic acid (vitamin C), which is converted to a semiascorbyl radical (SA•) and finally to dihydroascorbate (DHA). To regenerate the reduced form of vitamin C the glutathione (GSH) system may come into play. The oxidized form of glutathione (a dimer linked by a sulfide bridge) can be regenerated by NADPH which is produced during the course of glucose metabolism. The radicals are shown in red.

bridge. When this molecular switch signals an oxidative redox shift, a transcriptional response is triggered that results in an upregulation of stress-response genes. The expression of enzymes that are able to regenerate the reduced forms of the different redox-buffering components forms part of the stress reaction. The redox-dependent protein modification is a post-translational mechanism to regulate protein activity, similar to the phosphorylation of proteins by kinases. Both reactions control important physiological pathways and their dysregulation is implicated in many diseases.

The thioredoxin (TRX) system has numerous functions and regulates the activity of other proteins containing thiol groups. Members of the TRX gene family are present in all cellular compartments, including the nucleus, cytoplasm and mitochondria. Regulation of the human TRX protein, which consists of 105 amino acids, may serve as an example to illustrate the structural modifications in response to changing ROS concentrations. Two cysteine residues at positions 32 and 35 in the polypeptide chain represent the active redox site (Figure 5.5). Far away from this site there are other cysteines with regulatory functions that modify the activity of the protein. For example, ^{69}cysteine can be S-nitrosylated by covalent attachment of NO, a condition associated with active ROS scavenging function. Furthermore, ^{73}cysteine may be involved in dimerization of the TRX

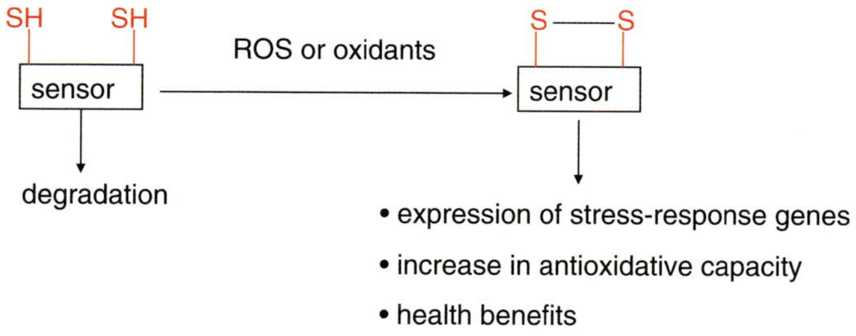

Figure 5.4 A redox-sensor for oxidants. The mechanistic principle of a redox sensor is shown which is able to respond to low concentrations of oxidants. The formation of a disulfide bond between two neighboring thiol groups results in a conformational change of the respective sensor protein and the gained activity triggers a biological reaction.

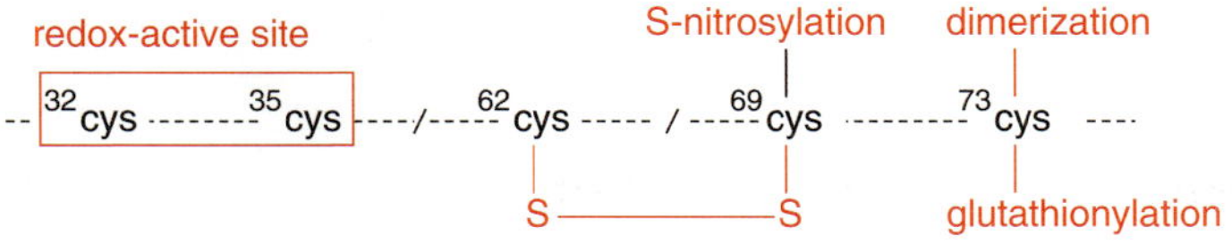

Figure 5.5 Regulation of thioredoxin (TRX) activity. The function of the redox-regulating protein TRX is controlled by redox-sensitive thiol groups of cysteine residues in the polypeptide chain. The positions of the cysteines in the protein are indicated by the numbers. The oxidation of the thiol groups occurs in different ways, and may involve the formation of disulfates, the reaction with nitric oxide (*S*-nitrosylation) or the reaction with a thiol group of another TRX molecule (dimerization) or with the thiol group of glutathione (glutathionylation). Redrawn from Kondo, N. *et al.* (2006) *Antioxid. Redox Signal.*, **8**, 1881–1890.

molecule; this is thought to inhibit TRX function, a condition that may be prevented by glutathionylation of the same cysteine residue. The formation of mixed disulfides can be reversed by the enzyme glutaredoxin, which is also member of the TRX protein family. These different options for modifying cysteine residues allow a fine-tuning in the control of protein function.

An oxidative shift of the cellular redox state will alter the activity of TRX and other redox-sensors, such that an antioxidative response of the cell is elicited. The main transcription factors involved in this response are AP-1, NFκB, and Nrf2. These and other transcription factors bind to an antioxidant response element (ARE), a specific nucleotide sequence of the respective promoters of redox-responsive genes. Such an ARE sequence controls the expression of TRX and of other genes which strengthen the antioxidant capacity, or which are involved in Phase I and II metabolism (see Section 4.1). The reaction to oxidative stress also involves the expression of heat shock proteins (HSPs) that help the cell to cope with a variety of different stress situations. The HSPs play a particularly important role in the repair of damaged proteins, so that they regain their active three-dimensional conformation.

The following example of an important regulatory protein illustrates the principle of a cellular reaction to the presence of electrophilic substances. The transcription factor Nrf2 is associated with the ROS-sensor protein keap, which binds to Nrf2 and regulates its function. In a current model (Figure 5.6), Nrf2 binds to a keap dimer with low- and a high-affinity binding sites (LBS and HBS, respectively). The protein complex is normally degraded rapidly by ubiquitination, but the keap dimer is able to act as a sensor as the cysteine residues of the protein are normally in the reduced state. If ROS or electrophilic natural substances are present, the SH-residues become alkylated and this reaction leads to the formation of a covalent bond with the substance. Oxidation of the thiol groups leads to the liberation of bound Zn^{2+} and a conformational change is induced which results in a loss of the keap/Nrf2 interaction at the LBS site. Only in this conformation can Nrf2 act as an active transcription factor by binding to ARE and inducing the transcription of ROS defense genes such as SOD and different peroxidases. In this case, the thiol groups of the redox-sensor are oxidized by alkylation, and this reaction may be particularly relevant for the antioxidative effect of flavonoids and

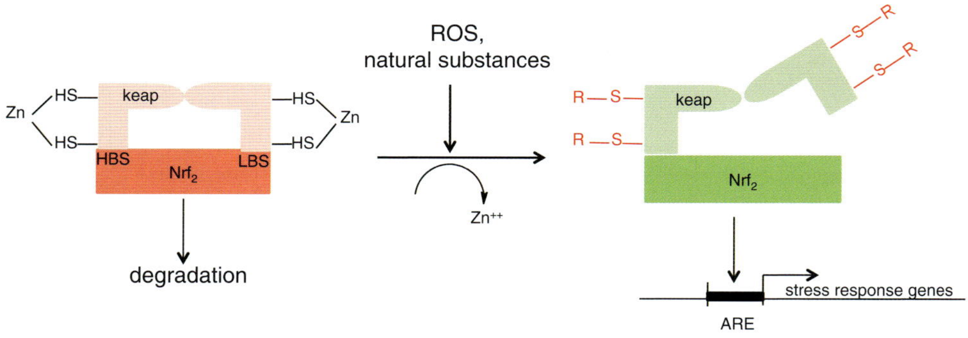

Figure 5.6 Regulation of the redox-sensor Nrf2. An important redox sensor is the transcription factor Nrf2 which is regulated by the activity of its inhibitor, the keap dimer, which binds to Nrf2 with high- and low-affinity binding sites (HBS and LBS, respectively). In the presence of oxidants thiol groups become oxidized by the formation of a covalent bond with an alkyl group (R). With this reaction, bound Zn becomes liberated and keap changes its conformation, thus liberating Nrf2 which is now able to act as transcription factory by binding to the antioxidant response element (ARE) and initiating the transcription of stress response genes. Redrawn from Hur, W. and Gray, N.S. (2011) *Curr. Opin. Chem. Biol.*, **15**, 162–173.

other natural compounds (Section 5.1.4). Apparently, there is an alternative way to liberate Nrf2 from its suppressor and to trigger the stress-response: the protein can be phosphorylated by, for example the extracellular signal-regulated protein kinase (ERK1/2), protein kinase C (PKC), or c-Jun N-terminal kinase (JKN). Whether the redox-sensitive cysteines in these kinases act as sensors is not known; however, the fine-tuning of the redox-status is organ-specific and even at the subcellular level the redox potential will differ between the mitochondria, cytoplasm, and nucleus. These highly complex homeostatic control mechanisms highlight the importance of the redox potential for cellular functions.

5.1.2
Function of ROS/RNS as Signaling Molecules for Cellular Functions

All organisms have developed highly sophisticated control mechanisms to cope with ROS and to ensure redox homeostasis. If the goal was only to reduce ROS and RNS to an insignificant concentration in a cell, then this elaborate control system would be unnecessary. ROS are not harmful substances *per se*, but are also important endogenous signaling molecules that control numerous biological processes. The necessary distinction between biologically meaningful ROS and detrimental ROS and oxidative stress is the reason why organisms have developed such elaborate mechanisms to control intracellular ROS concentrations.

It is unclear which ROS and RNS species are suitable for cellular signaling during normal development, and how the signal is transduced into a biological response. Some ROS species such as H_2O_2, $NO^{\bullet}$ and $O_2^{\bullet-}$ are believed to interact with specific sensor proteins and elicit the adaptive response. The aggressive $^{\bullet}OH$

does not seem to play a role in signaling as it reacts with essentially all macromolecules in the vicinity and hence lacks specificity. The chemical reaction of $O_2^{\bullet-}$ with thiol groups of target proteins is rather slow and might qualify as signaling molecule, although presumably H_2O_2 is the more important signaling molecule. Both, the formation and degradation of H_2O_2 are efficiently controlled by the enzymes SOD and several peroxidases, respectively (see Figures 5.1 and 5.2), and this precise regulation is a precondition to act as signaling molecule. The importance of a critical H_2O_2 concentration is nicely illustrated by the observation that the migration of leukocytes towards a wound is directed by a H_2O_2 gradient. With respect to RNS, only $NO^\bullet$ has been identified as a signaling molecule. Induced modifications include the covalent attachment of $NO^\bullet$ to thiol groups forming *S*-nitrosothiols, while tyrosine nitration is also a common reaction that affects the activity of the respective target proteins. The nitrosylation of a cysteine residue in TRX (Figure 5.5) is an example of such a reaction.

ROS/RNS signaling has been implicated in angiogenesis, meiosis, fertilization, and many other developmental processes. The importance and versatility of these molecules in rather different situations can be illustrated by three examples:

1) In the blood vessels, $NO^\bullet$-mediated signaling controls the function of the endothelium that lines the inner surface of the blood vessels. The $NO^\bullet$ radical is produced in these cells by the enzyme endothelial nitric oxide synthase (eNOS). The radical has a half-life of a few seconds, and diffuses through the cell membranes and relaxes the smooth muscle cells of the vessels; this vasodilatory effect causes a reduction in blood pressure. Additional positive effects are ascribed to $NO^\bullet$ due to its anti-inflammatory activity and the inhibition of platelet aggregation, which is implicated in the progression of atherosclerosis. The positive function of $NO^\bullet$ may be abolished in the presence of unphysiologically high ROS concentrations in the endothelium (Figure 5.7). Under these conditions, the redox-sensitive cofactors of the enzyme divert the electron transfer in the enzyme, and under such "uncoupling" conditions $O_2^{\bullet-}$ rather than $NO^\bullet$ may be produced. Recent experiments have suggested that eNOS may itself be able to sense the cellular redox state and respond by changing from an active to an inactive state. The downregulation of $NO^\bullet$ and upregulation of $O_2^{\bullet-}$ production leads to endothelial dysfunction that favors common diseases including hypertension, diabetes, atherosclerosis, and myocardial infarction. This example illustrates the double-faced character of radicals in cell biology: $NO^\bullet$ signaling is essential for a healthy cardiovascular system, but if this signaling is disturbed then damaging ROS species may reverse the effect and disturb the normal functions of the tissue.
2) Endogenous ROS production can be strongly upregulated under special conditions as part of a normal physiological response. Bacterial or fungal infections trigger a defense reaction in granulocytes (neutrophils) or phagocytes, and this response involves a massive increase in ROS levels that is referred to as oxidative burst due to the increased consumption of molecular oxygen. The activation of membrane-bound NADPH oxidase (NOX) produces large amounts of

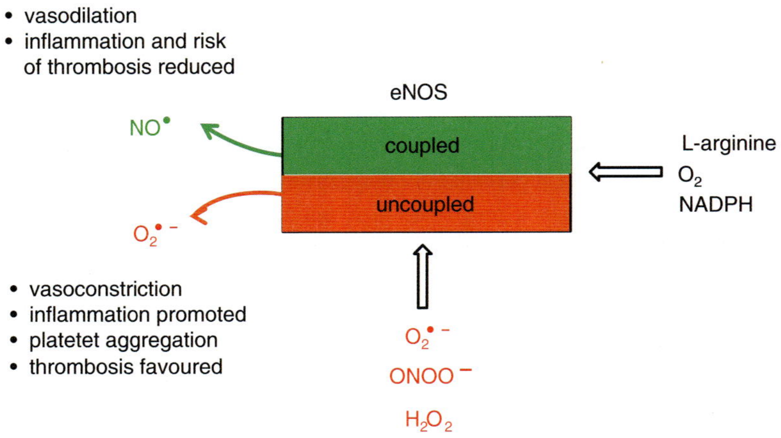

Figure 5.7 Uncoupling of the endothelial enzyme nitric oxide synthase (eNOS) under oxidative stress. The enzyme eNOS of the blood vessel endothelium produces NO• from L-arginine in the presence of O_2 and NADPH. NO• acts as a signaling molecule to regulate essential physiological functions. In case of oxidative stress the same enzyme may become uncoupled and harmful $O_2^{\bullet -}$ is produced which initiates pathological developments. Based on Montezano, A.C. and Touyz, R.M. (2011) *Basic Clin. Pharmacol. Toxicol.*, **110**, 87–94.

$O_2^{\bullet -}$ (Figure 5.8) and this, in turn, leads to a high concentration of H_2O_2 (10 to 100 μM) due to the activity of SOD. H_2O_2 may then be the source of other ROS and in the presence of Fe^{2+} the reactive and toxic •OH may form (Fenton reaction). Furthermore, the enzyme myeloperoxidase (MPO) catalyzes the oxidation of chloride anions (Cl^-) with the abundant ROS (H_2O_2 or •OH) to yield

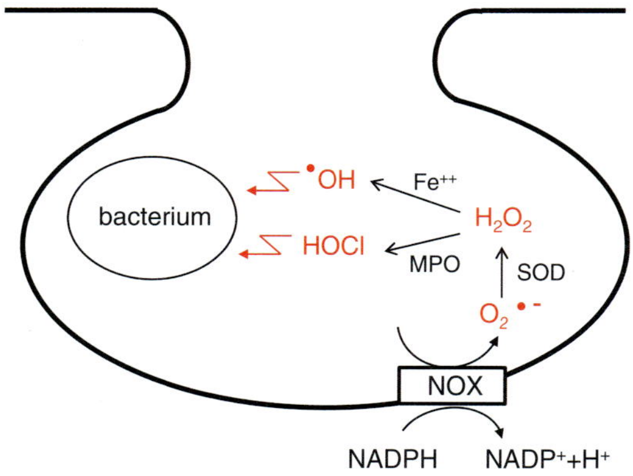

Figure 5.8 Generation of ROS in the defense against bacterial infections. A normal physiological reaction to bacterial or fungal infections involves a massive production of ROS (oxidative burst). The reaction is initiated by the activity of NADPH oxidase (NOX) which produces $O_2^{\bullet -}$. This generates other ROS species (see also Figure 5.1) and, in addition, hypochloric acid due to the activity of the enzyme myeloperoxidase (MPO). This bombardment with damaging ROS is an effective defense against infecting organisms.

hypochlorous acid. If the enzyme NOS is also activated, as is the case in inflammatory reactions, the highly reactive peroxynitrite ($ONOO^-$) is produced:

$$NO^\bullet + O_2^{\bullet -} \rightarrow ONOO^-$$

This second example illustrates a new physiological aspect of ROS generation: in this case, the toxic effect of ROS is used to kill infecting microorganisms by induced DNA strand breaks, lipid oxidation, and protein modifications.

Do respiratory bursts also occur under normal physiological reactions? Indeed, most cell types are able to respond to cytokines and growth factors or other signal molecules by eliciting a small oxidative burst. The reaction is generally well-controlled so that the oxidative damage is minimized, but the change in the redox-equilibrium in the respective cell serves as second messenger in several signal transduction pathways, including the important mitogen-activated protein kinase pathway (MAPK). The activation of this pathway stimulates growth and/or cell differentiation. An interesting example of a controlled oxidative burst is ovulation in mammals, when luteinizing hormone (LH) produced by the pituitary gland triggers an inflammatory reaction that results in the production of ROS, and this reaction is essential to initiate ovulation. Interestingly, H_2O_2 is able to act as a trigger in a similar way as does LH.

3) The importance of redox-signaling in the control of normal cellular functions is illustrated by recent findings concerning the maintenance and differentiation of stem cells. Quiescent adult stem cells seem to be characterized by low levels of ROS. Germline stem cells and hematopoietic stem cells (HSCs) must maintain their DNA integrity for many years, and it is obviously advantageous to minimize oxidative damage. When these adult stem cells are induced to divide and differentiate, ROS-mediated signals seem to be required in order to activate the stem cells in different tissues. Stem cells in the intestine and the brain, as well as HSCs share this property. Stem cells of the colon respond to NADPH oxidase signaling, and the generated $O_2^{\bullet -}$ triggers the proliferation of the stem cells and their subsequent differentiation of the progeny which is directed by the differentiation genes *wnt* and *Notch1*.

Regulation of the developmentally important wnt pathway by ROS has also been observed in other developing systems, and the redox-sensor that mediates the effect is nucleoredoxin (NRX), a member of the TRX family of redox-regulators. TRX is also implicated in the control of HSC regulation. These examples show that ROS signals control important developmental processes.

5.1.3 Oxidative Stress in Human Diseases and in Aging

When the normal mechanisms controlling the ROS production are compromised, or the ROS concentration exceeds the redox buffering capacity, excessive oxidative stress ensues. When RNS are in focus an imbalance is often referred to as "nitrosative stress." The balance can shift to the oxidative state under pathological conditions, or when environmental agents induce excessive ROS production or impair the ROS defense system. For example, in smokers a severe depletion of circulating antioxidants has been reported. Common environmental pollutants such as halogenated or polycyclic hydrocarbons have also been linked to oxidative stress, and an increased incidence of cancer and immune dysfunction. Furthermore, exposure to heavy metals or pesticides such as paraquat or the organophosphorous compound malathion has a prooxidant effect and may induce oxidative stress. Recently, small particulate matter ($<2.5\,\mu m$ in size) has been added to the list of environmental pollutants that may affect the redox-status of exposed individuals. Finally, the exposure of tissues to UV-light is known to produce ROS (in particular $O_2^{\bullet -}$), and ionizing radiations such as X-rays and γ-radiation will also increase the concentrations of ROS in these tissues.

Excessive ROS levels result in cellular damage, and since all classes of macromolecules are ROS target molecules, important cellular functions may be severely impaired. For example, the DNA may become modified by oxidation of the purine or pyrimidine bases, or of deoxyribose; these defects may lead to an incorrect pairing of the DNA strands and mutations or even strand breaks. Such changes are implicated in carcinogenesis and also in aging. Lipids, and in particular polyunsaturated fatty acids, are also a major target of ROS, and as cell membranes consist largely of phospholipids the oxidative damage in cell membranes may be especially severe. The major end products of lipid oxidation reactions are the toxic 4-hydroxy-2-nonenal and the mutagenic substance, malondialdehyde.

Proteins are also susceptible to oxidative damage. ROS-mediated oxidative damage can be quantified by the concentration of protein carbonyl groups. Most sensitive are the sulfur-containing amino acids cysteine and methionine, the oxidation of which may lead to an uncontrolled formation of mixed disulfides between the thiol group of, for example, GSH and the thiol of another protein. Further oxidation will generate sulfinic acid ($-SO_2^-$) or even sulfonic acid ($-SO_3^-$), and these oxidation states are irreversible and the affected protein will remain damaged. Nitrosative stress may also cause proteins to be damaged by nitrosylation, and when these reactions occur in an uncontrolled manner the affected proteins will become nonfunctional. Typically, such damaged proteins are efficiently degraded.

Oxidative stress is thought to promote common diseases such as diabetes, cardiovascular and neurogenererative diseases, and also cancer. In principle, excessive ROS production may cause pathological developments in two different ways:

1) The redox equilibrium is shifted towards a more oxidized state, so that the homeostatic control is challenged and the redox buffering capacity exhausted. Under these conditions the oxidative forms of the GSH and TRX systems (GSSG and $TRX-S_2$) may prevail in the stressed cell. The impaired cellular control processes will favor diseases such as diabetes and cancer.
2) Oxidative stress may lead to inflammatory reactions. In the case of vascular functions it has been shown that eNOS produces $O_2^{\bullet -}$ instead of $NO^{\bullet}$ (Figure 5.7). Along with this detrimental reaction other major ROS producers such as NADPH oxidase (NOX) and xanthine oxidase (XO) become stimulated and contribute to excessive $O_2^{\bullet -}$ levels. The development of cardiovascular diseases is promoted by an ROS-mediated inflammatory reaction of the endothelium.

The redox homeostasis can be challenged in exceptional situations. For example, some diseases may weaken the control mechanisms, while redox-control gradually becomes less efficient during aging; indeed, an accumulation of oxidative damage in the tissues is considered to be a hallmark of the aging process. Interestingly, in animal model systems ranging from worms to mammals a strong correlation has been found between stress resistance and longevity. In mutant strains of some organisms (e.g., *Drosophila*, *C. elegans*, mouse) an increase in stress-resistance was shown to prolong life expectancy.

The germline stem cells in the gonads are a special case, as any ROS-induced molecular damage will be transmitted to the next generation. A specific metabolic regulation and strong resistance to oxidative stress by high concentrations of zinc with antioxidative properties, and particularly active SOD, may help to minimize mutations and other detrimental damage.

5.1.4 Natural Substances Affecting Redox Homeostasis

Whilst secondary metabolites from plants have numerous biological activities, substances with antioxidant properties have attracted much attention as beneficial effects on human health have been ascribed to this property. Which natural substances possess this activity, and what are the biochemical reaction pathways? An antioxidative effect can be realized in different ways, and experimental evidence has been obtained for the following four different reaction pathways.

5.1.4.1 Reduction of ROS in a Redox Reaction

Flavonoids and other secondary plant metabolites which have electrophilic properties may act as ROS scavengers in a simple chemical reaction. Scavenging activity of numerous flavonoids for ROS and RNS has been reported, but the antioxidative properties have been demonstrated mostly *in vitro*. The phenolic hydroxyl groups of flavonoids are the preferred reaction partners, and quinines are the typical oxidized forms of flavonoids such as quercetin (Figure 5.9a). Flavonoids may react directly not only with ROS; rather, amino acid radicals in macromolecules may also be reaction partners of flavonoids. Quercetin was found to repair experimentally induced amino acid radicals in apolipoprotein B100 by intramolecular electron transfer. Oxidative damage of apolipoprotein B100 in low-density lipoprotein (LDL) is thought to be a hallmark in the genesis of atherosclerosis.

The analysis of a chemical reaction *in vivo* is hampered by the rapid metabolic conversion of flavonoids in the tissue (see Section 4.1), and it is often not clear if the measured effect is produced by the test substance itself, or by a metabolite. The inhibition of lipid peroxidation after the consumption of antioxidants illustrates the difficulty of a mechanistic explanation, as the effect may be produced by activation of the endogenous redox defense system and may not necessarily be the result of a direct chemical interaction with a natural compound with antioxidative properties.

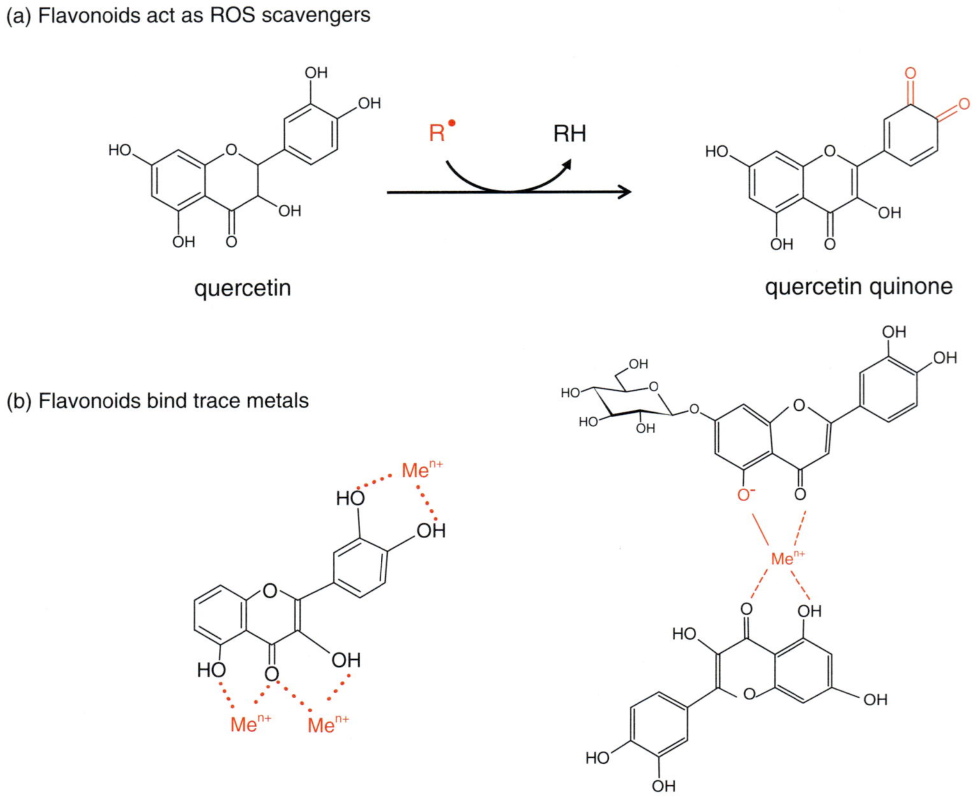

Figure 5.9 Flavonoids may react with ROS and chelate trace metals. (a) The flavonoid quercetin can act as a ROS scavenger. A possible reaction involves the oxidation of hydroxyl groups and the formation of quinones; (b) Metals (Me) can be bound by flavonoids and in this way the generation of ROS and in particular $^{\bullet}OH$ generated by the Fenton reaction may be prevented. Trace metals cations such as Cu, Mg, Co, Ni, or Fe may bind to the flavonoid or form complexes by binding with two or more molecules. In the example shown a flavonoid glycoside and an aglycone are connected by binding to a metal cation. (a) Redrawn from Pietta, P.G. (2000) *J. Nat. Prod.*, **63**, 1035–1042. (b) Drawing based on data from: Davis, B.D. and Brodbelt, J.S. (2004) *J. Am. Soc. Mass Spectrom.*, **15**, 1287–1299.

5.1.4.2 Formation of Chelates with Metals

Many flavonoids with hydroxyl-groups are able to bind metals and form chelates (Figure 5.9b). This reduces the concentration of free Fe^{2+} and Cu^{2+} in the cell, and might also inhibit the Fenton reaction which gives rise to the damaging $^{\bullet}OH$ radicals. Although the chelating properties of flavonoids have been demonstrated *in vitro*, their relevance *in vivo* is still unknown. It is also unclear if flavonoids can bind to other metals in the cell, such as Se or Zn. As many enzymes require trace metals for their activity it is conceivable that flavonoids (and possibly also other natural substances) inhibit metal-containing enzymes in this way. The important enzyme SOD, which scavenges $O_2^{\bullet-}$ and converts the radical into H_2O_2, contains metals at its catalytic center. Three isoforms of SOD have been identified in mammals: two isoforms require Cu and Zn for their activity, and these enzymes are located either in the cytoplasm or in the extracellular space. The third isoform, located in the mitochondria, uses Mn as a cofactor. Whether flavonoids affect the activity of these and many other metal-containing enzymes involved in ROS homeostasis remains to be seen.

5.1.4.3 Inhibition of Enzymes Involved in Stress Defense

A vast amount of information is available on the interaction of flavonoids and other natural compounds with mammalian enzymes. Several substances have been reported to inhibit enzymes that produce ROS or are involved in ROS-producing pathways, such as lipoxygenase and cyclooxygenase, both of which mediate inflammatory reactions. The inducible form of nitric oxide synthase (iNOS), which generates $NO^{\bullet}$ and also plays a role in inflammatory and autoimmune diseases is inhibited by silymarin, while xanthine oxidase, which generates $O_2^{\bullet-}$ (Figure 5.1) is strongly inhibited by quercetin and baicalein. Considering the important role of NADPH oxidase (NOX) in eliciting an oxidative burst (Figure 5.8), natural inhibitors of this enzyme are of special interest. Some natural substances have been identified that inhibit members of the NOX family of enzymes (Figure 5.10); examples of these inhibitors include norathyriol (a xanthone aglycone from *Tripterospermum lanceolatum*), abruquinone (an isoflavanquinone from the roots of *Abrus*

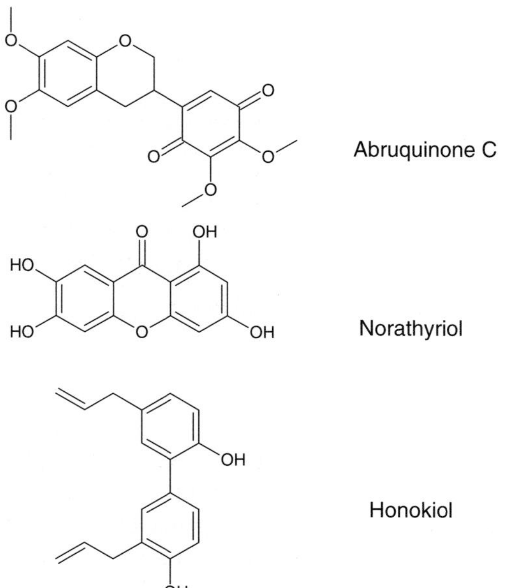

Figure 5.10 Natural inhibitors of the enzyme NADPH oxidase (NOX). The enzyme generates $O_2^{\bullet-}$ when, for example, a respiratory burst is initiated (see also Figure 5.8). The inhibition of the enzyme and the reduction of oxidative stress are desirable in some pathological situations. The compounds shown are examples of identified natural NOX inhibitors.

precatorius), or honokiol (from the medicinal plant *Magnolia officinalis*). Typically, such substances have additional pharmacological properties and interact with several target proteins in the cell, apart from NOX.

Because of the importance of GSH and TRX for redox homeostasis, any interference of natural compounds with the enzyme-controlled balance between the respective reduced and oxidized forms (Figure 5.2) will greatly influence the redox status of a cell. TRX can react with quinones and reduce some compounds that would act as prooxidants. On the other hand, the inhibition of TRX-reductase (TRX-R) will lower the concentration of the reduced form of TRX (TRX-$(SH)_2$) and hence weaken the antioxidance defense. Curcumin and the flavonoids myricetin and quercetin target the selenothiol-containing active site, and are potent inhibitors of TRX-R.

5.1.4.4 **Initiation of an Adaptive Response**

The fourth mechanism that needs to be considered when discussing the antioxidative effects of natural compounds is unexpected, but most interesting. Natural substances may assist the endogenous cellular ROS defense in an indirect manner by a temporary oxidative challenge of the redox controlling system. The induced defense reaction strengthens the antioxidative capacity of the cell, and this effect is measured as an antioxidative property of the respective natural substance. A careful examination of several natural compounds with different chemical scaffolds has revealed that the antioxidant mechanism is indirect and involves an initial prooxidative activity. Such a mechanism is, for example, indicated in well-known natural products with unsaturated carbonyl groups, such as caffeic acid phenylester (from honey), curcumin (turmeric), shogaol and zerumbone (ginger), xanthohumol (hops, beer) or cinnamaldehyde (cinnamon oil). Other "antioxidants" that act in the same indirect manner include isothiocyanates and disulfides such as lipoic acid (in meat, spinach and broccoli) and sulforaphane (also in broccoli). Some naphthoquinones in food provide further examples of this interesting biological activity. To date, more than 60 natural substances have been identified which act through the Nrf2 pathway (Figure 5.6). In fact, the ARE-mediated cellular defense response holds great promise for pharmacological intervention, and there is some evidence that ARE inducers might indeed possess therapeutic potential against the initiation and development of cancer and also protect against inflammation or neurodegenerative diseases.

Some of the above-mentioned natural substances will be discussed in greater detail in the following subsections in order to illustrate the practical relevance of the adaptive redox-response elicited by natural compounds. The best-studied example of an Nrf2-mediated stress response is provided by the isothiocyanate (ITC) sulforaphane, which is abundant in broccoli and other plants of the family Brassicaceae. Sulforaphane can react with the amino groups and thiol groups present in the three cysteine residues in keap, so that carbamodithioate is formed (Figure 5.11). The induced expression of the stress response genes has interesting pharmacological implications. A chemoprotective effect was reported in mice that had been treated topically with two different carcinogens whereby, if sulforaphane was applied before the treatment, the incidence of skin tumors was decreased

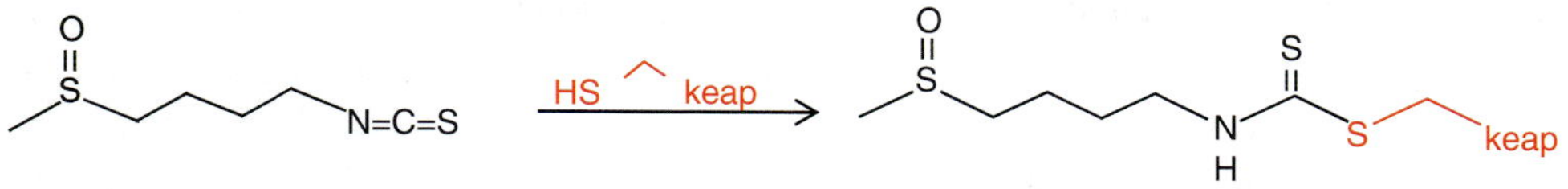

Figure 5.11 Initiation of an antioxidative cellular reaction. Some natural substances are able to strengthen the antioxidative cellular defense. The isothiocyanate sulforaphane, for example, reacts with keap cysteines and the alkylation activates the transcription factor Nrf2 (see Figure 5.6) and triggers an adaptive antioxidative response. Based on Hur, W. and Gray, N.S. (2011) *Curr. Opin. Chem. Biol.*, **15**, 162–173.

drastically while, at the same time, an increase in Nrf2 was observed. As this effect was observed in wild-type mice but not in knock-out mice (Nrf2$^{-/-}$), it indicated that Nrf2 was involved in the observed chemoprotective effect. Interestingly, ITC-containing foods were reported to reduce the risk in different types of cancer, and also to have a neuroprotective effect. Natural compounds typically interact with many target proteins, and this also seems to be the case for sulforaphane, which interacts with other important regulatory proteins; for example, it can interact with the transcription factor of NFκB by binding to a cysteine residue. Sulforaphane was also found to be an inducer of TRX.

The spice turmeric and its active component curcumin provide another example of an indirect antioxidative effect. Curcumin induces ROS in human hepatoma cells and this, in turn, was shown to induce Nrf2 activation. When curcumin was added at a concentration of 0.05% to the diet of mice for 15 days, Nrf2 was translocated from the cytoplasm to the nucleus and found to associate with the ARE sequence. Positive *in vivo* effects were reported by several groups, with curcumin reducing the experimentally induced oxidative damage of proteins, DNA and lipids in the kidneys of mice. In another study, a neuroprotective effect of curcumin was observed when using a murine model of Alzheimer's disease.

Resveratrol is a particularly interesting polyphenol in grapes, as various health benefits have been ascribed to this compound (see Section 6.1.2). Initially, resveratrol has a prooxidative effect as the ROS concentration increases and GSH levels are depleted. Again, the induced adaptive response is mediated by Nrf2, as confirmed experimentally in human cells that had been stressed by exposure to cigarette smoke. Resveratrol also had a positive effect in a mouse model of diabetes, causing a significant reduction in oxidative stress.

Flavonoids and other polyphenols are special cases, and the reactions may be more complex. Although ROS-scavenging activity has been reported for some flavonoids (as described above), it has also been shown that prooxidative reactions may occur both *in vitro* and *in vivo*. When a flavonoid such as quercetin or myricetin is added to cells cultured in normal tissue culture media, H_2O_2 is produced. The metal-catalyzed autoxidation of flavonoids is a possible source of $O_2^{\bullet -}$, which may then dismutate to H_2O_2. Alternatively, peroxidases which are abundant in cells may catalyze the oxidation of polyphenols; the phenoxyl radicals generated can then further react, leading to the production of $O_2^{\bullet -}$. This type of mechanism may apply not only for flavonoids but also for resveratrol and other natural substances.

Flavonoids may also act as prooxidants when catechols (1,2-diphenols) and hydroquinones (1,4-diphenols) are converted to quinones by autoxidation. It has been shown that, in the presence of oxygen and catalyzed by Cu^{2+}, hydroquinones are oxidized to quinones that react with cysteine residues in keap and activate the transcription factor Nrf2. Similarly, flavones and flavanones may first need to be metabolized to chalcones before they can elicit an adaptive antioxidant response that typically involves the activation of Nrf2 and the depletion of GSH.

The above-described reactions of different natural substances that are considered to be health-promoting converge on the activation of the Nrf2-mediated antioxidative response. Nonetheless, despite their initial prooxidative reactivity these substances will still be referred to as "antioxidants" in the following sections, in view of the stimulated antioxidant defense system.

5.1.5
Hormesis and the Benefits of Temporary Mild Oxidative Stress

Biological effects of flavonoids studied *in vitro* using cell cultures or isolated enzymes are typically recorded at high concentrations ($\geq 20\,\mu M$). However, the bioavailability of flavonoids *in vivo* is much lower, usually in the nanomolar range, when the natural substances are taken up in food. Yet a beneficial effect on health has been inferred for some food and beverages (Sections 5.2 and 6.1), and the health-promoting effects of broccoli and turmeric are examples of food with redox-active substances that are able to elicit a biological reaction even at a low concentration. One possible explanation for this effect is provided by the concept of "hormesis," defined as a biphasic dose–response effect with low-dose stimulation and high-dose inhibition (Figure 5.12). The ROS sensors are apparently able to induce an adaptive reaction of the redox-buffering system on the transcriptional and post-transcriptional level, even at very low concentrations.

It has been argued that a low ROS production is life-prolonging. This argument was based on the finding that a limited food intake clearly has a positive effect on life expectancy in invertebrates and vertebrates, including humans. This effect of caloric restriction (CR) was thought to be based on a lower metabolism and a reduced ROS production so that oxidative damage in the tissues would be minimized. However, this argument must be revisited in light of the adaptive response to oxidative stress. The mitochondria appear to play a pivotal role in the aging process, as there is a close correlation between age-related conditions such as cardiovascular diseases, diabetes and cancer and mitochondrial dysfunction. As the development of these diseases is positively affected by fasting (CR), what might be the connection between aging, ROS production, and mitochondrial function?

The mitochondria are major sites of intracellularly produced $O_2^{\bullet -}$ (Figure 5.1), and these organelles play an important role in the hormetic redox-response (also referred to as mitohormesis). Recent evidence has indicated that the respiratory activity of the mitochondria does not decrease but, surprisingly, is increased and ROS production is stimulated during fasting. Mitochondrial metabolism is also stimulated when insulin levels are low. Downregulation of the signal transduction

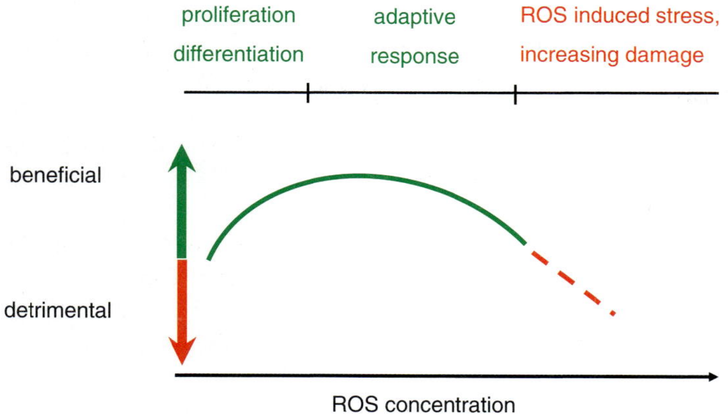

Figure 5.12 Oxidative stress and hormesis. The biological reaction to increasing ROS concentrations is biphasic according to the concept of hormesis. At low concentrations ROS act as signaling molecules to initiate cell proliferation and differentiation. An adaptive antioxidative response may be initiated at higher ROS concentration. At a certain critical ROS level oxidative stress is initiated and the effects are detrimental. Based on Siow, R.C. and Mann, G.E. (2010) *Mol. Aspects Med.*, **31**, 468–477.

pathways controlled by insulin (and insulin-like growth factor) correlates with longevity in several animal models. The relevant signaling pathways converge in the regulation of the key enzyme mTORC (mammalian target of rapamycin complex), which controls metabolic functions and plays an important role in regulating the lifespan (see Section 6.1.1). When the mitochondrial metabolism is stimulated by low mTORC activity, an adaptive response may be triggered similar to the health-promoting effects of prooxidative natural substances. The suggested pathways are illustrated in Figure 5.12, although details of the reaction pathways involved will need to be resolved in future research.

Although the dynamics of the aging process is species-specific, the basic mechanisms have been conserved during evolution and much of the present knowledge on human aging derives from studies in yeast and invertebrates (*Drosophila* and *C. elegans*). In yeast (*Saccharomyces cerevisiae*), the interesting observation has been made that not only CR but also the inactivation of catalase increases the lifespan. Under these conditions H_2O_2 was shown to increase, but the $O_2^{\bullet -}$ radical did not accumulate and was efficiently removed by induced SOD activity. Interestingly, in catalase-deficient mutants the extended lifespan was accompanied by increased ROS-inflicted damage. This finding cannot be explained by the long-held assumption that oxidative damage is the primary cause of aging. In this system H_2O_2 appears to be the important signaling molecule, while the $O_2^{\bullet -}$ concentration is kept low due to the activity of SOD. The results of these and other experiments in model organisms, including mammals, suggest that the extension of the lifespan involves a ROS signal (Figure 5.13).

Are antioxidants healthy? There is no straightforward answer to this question. Clearly, ROS are not *per se* harmful as they are used as important signaling

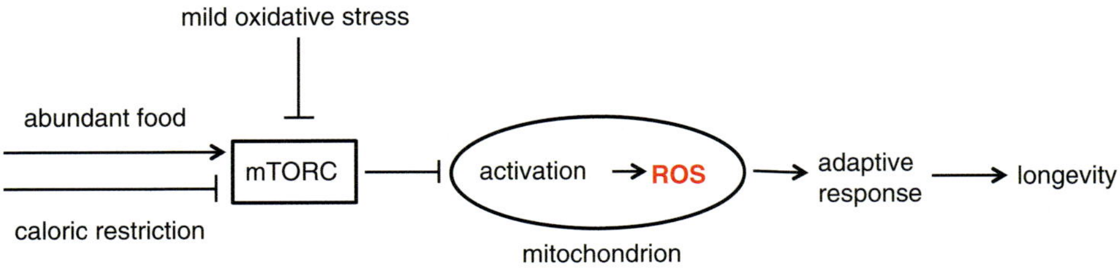

Figure 5.13 A mild oxidative stress may elicit a beneficial physiological reaction. The kinase TOR plays a central role for the control of metabolic activity. The downregulation of TOR leads to a prolongation of life in invertebrate and vertebrate model organisms. Mild oxidative stress also inhibits TOR and leads to longevity. The downregulation of TOR results in mitochondrial activation and the generation of ROS. The resulting adaptive stress response may be an essential step for longevity. Modified from Ristow, M. and Schmeisser, S. (2011) *Free Radic. Biol. Med.*, **51**, 327–336.

molecules. Oxidative stress, however, is clearly undesirable and is associated with the development of diseases. Numerous clinical studies have been carried out with high doses of antioxidants but the results have been disappointing, with no beneficial effects being observed in some studies while other studies showed even some negative effects. Physical exercise is health-promoting and associated with increased ROS production, but the positive effects are diminished by treatment with antioxidants. More dramatically, the treatment of mice with radical scavengers led to a significant reduction in the rate of ovulation. Apparently, the controlled oxidative burst elicited during LH-induced ovulation (Section 5.1.2) is an essential feature of this developmental process, and its suppression by a high dose of antioxidants resulted in a reduced fertility.

It has become clear during recent years that redox homeostasis and redox signaling are basic control mechanisms in cell differentiation, aging and, in case of dysfunction, in the development of diseases. The effect of high-dose antioxidants depends in part on the health status of the individual. Healthy persons do not benefit from antioxidants added to a normal diet, although under pathological conditions and in the elderly the situation may be different. There is, however, no disease that has been successfully treated with high concentrations of antioxidants. One likely reason for the failure of high-dose antioxidant treatment is the suppression of normal ROS signaling and also the inhibition of the beneficial hormetic response to polyphenols present in fruits and vegetables.

5.2 Endocrine Effects of Natural Substances

5.2.1 Are Endocrine Disruptors a Threat to Animal and Human Health?

The term "endocrine disruption" has been coined to describe the biological effect of compounds which affect the endocrine homeostasis of animals or humans alike, either directly or indirectly. The effect can be brought about by interference with hormone synthesis, transport, metabolism, or signal transduction. The list of hormones of which the biological function is affected by endocrine disrupting chemicals (EDCs) is long, and includes hormones of different chemical nature such as steroids, retinoic acid, and thyroid hormones. These hormones are typical representatives of the endocrine system, and are distributed throughout the body via the vascular system. In taking a broader view, locally acting hormones in a particular tissue will be briefly discussed, since such paracrine signaling may also be disturbed by natural compounds. In an extreme case the signaling may only affect two specialized cells, as is the case of synaptic transmission in the nervous system (Section 5.3).

Much attention has been paid to the endocrine disruption caused by synthetic chemicals used as industrial solvents (e.g., polychlorinated biphenyls and dioxins), plasticizers (phthalates), fungicides (vinclozolin), pesticides (DDT, methoxychlor), and many others. These substances are chemically very heterogeneous and the prediction of possible endocrine effects based on their structures is not possible. Typically, the biological degradation of these substances is slow, and if they are hydrophobic they may accumulate in fatty tissues. At this point, attention will be focused only on biologically active substances from natural sources. The metabolism of these compounds is comparatively rapid (see Section 4.1), but the amount taken up is often very large compared to the synthetic hormonal disruptors in the environment and in industrial products. As humankind is exposed to a "cocktail" of hormonally active substances from different sources, it is important to consider possible additive or synergistic effects of natural and synthetic substances in the affected tissues.

During the mid-1990s the threat of EDCs for human health became a matter of a heated debate in the scientific literature, as well as in public media. In particular, the fears that estrogenic chemicals in the environment would lead to a decline in sperm counts, and also to an increased risk of women developing breast cancer, fueled public discussions. Observations from the wildlife supported this fear; for example, the feminization of male fish and the formation of intersexes were observed in polluted rivers. The estrogenic properties of natural substances (phytoestrogens) were described as early as 1930, and in 1946 it was reported that sheep grazing on highly estrogenic clover pastures developed various abnormalities that were referred to as "clover disease." These sheep had low lambing rates, increased embryonic loss, a prolapse of the uterus, and several other abnormalities of the reproductive tract. Depending on the estrogenic activity in the pastures, ewes became temporarily or permanently infertile, but the rams were unaffected. In subterranean clover (*Trifolium subterraneum*), genistein, biochanin A and formononetin were identified as major estrogenic compounds. However, only formononetin was considered responsible for the symptoms of clover disease because it is metabolized in the rumen, first to the flavonoid daidzein and then further to the more potent estrogenic compound, equol. In contrast, genistein and biochanin A are efficiently degraded to nonestrogenic substances in the sheep's rumen.

The relevance of these findings to humans became apparent when numerous substances with estrogenic properties in food were identified. The resultant concern led to intensive research efforts to identify all substances that might act as phytoestrogens in humans, and to characterize their properties *in vitro* and *in vivo*. Some substances may act antagonistically (antiestrogenic) and attenuate or even abolish the normal activity of estradiol. As the interpretation of the obtained data depends in part on the test system used for the characterization of the estrogenic substances, relevant test systems will be briefly described first.

In 1996 the Environmental Protection Agency (EPA) in the USA implemented a screening program to identify substances which affect estrogen, androgen, or thyroid hormone signaling. To this end, a number of validated *in vitro* and *in vivo* tests were chosen to identify and characterize EDCs. *In vitro* assays with steroid receptors provide information as to whether the test compound binds to the estrogen receptor (ER) or to the androgen receptor (AR). A compound possesses estrogenic activity if it competes successfully with [^{3}H]-estradiol bound to ER present in cytosolic proteins from the rat uterus. A similar test using rat prostate cytosol allows the identification of androgenic AR-binding substances.

An assay with recombinant yeast cells has been developed to monitor estrogenic activity mediated by nuclear ER (Figure 5.14). As yeast cells do not normally contain an ER, transgenic strains were established which expressed a human ER (hER); the cells also contained an expression plasmid with an estradiol response element (ERE). In the presence of estradiol or a phytoestrogen, the formed ligand–receptor complex binds to the ERE sequence and a reporter gene on the plasmid is expressed. In the example shown in Figure 5.14, the reporter is an enzyme (β-galactosidase) which is secreted and converts the yellow substrate chlorophenol

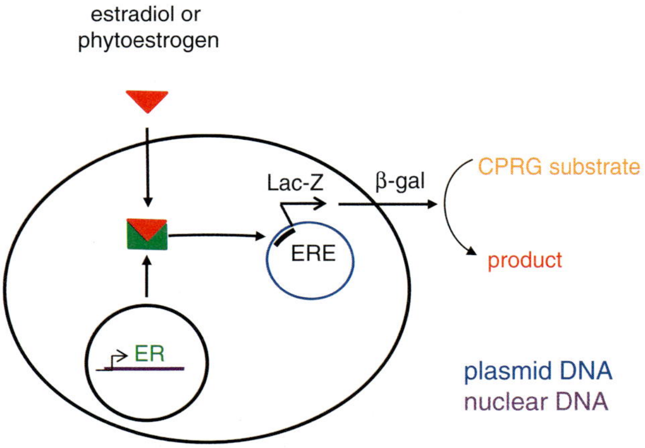

Figure 5.14 The estrogenic activity of substances can be quantified by transgenic yeast reporter strains (transactivation assay). Yeast cells are genetically engineered to express the human estradiol receptor (ER). If a ligand is present which is able to bind to the receptor, the ligand–receptor complex translocates to the nucleus and binds to an estradiol response element (ERE) which drives the expression of a lac-Z (β-galactosidase) reporter gene on a plasmid. The enzyme converts a yellow substrate to a red product, which can be quantified photometrically. CPRG: chlorophenol red β-D-galactopyranoside. Modified from Routledge, E.J. and Sumpter, J.P. (1996) *Environ. Toxicol. Chem.*, **15**, 241–248.

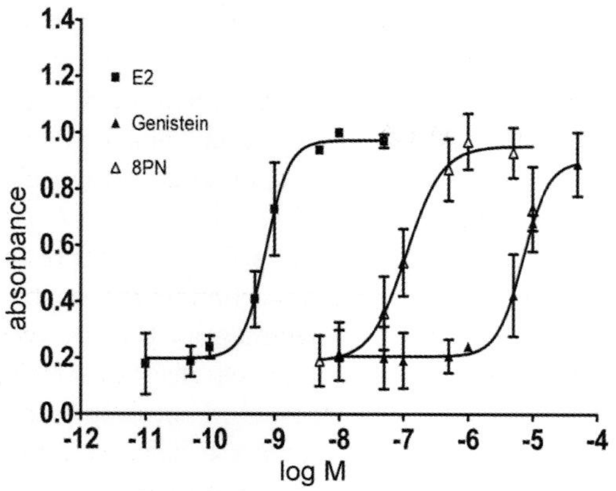

Figure 5.15 Quantification of the estrogenic activity of phytoestrogens by the transactivation assay. To evaluate the estrogenicity of genistein and 8-prenylaringenin (8-PN) the transgenic yeast assay (see Figure 5.14) was applied. These phytoestrogens show a concentration-dependent response and induce lac-Z synthesis. 17-β-estradiol (E_2) was assayed as a control. The hormone induces lac-Z synthesis at comparatively small concentrations. The EC_{50}-values reflect the different activities: $E_2 = 8.4 \times 10^{-10}$ M; genistein $= 9.3 \times 10^{-6}$ M; 8-PN $= 1.3 \times 10^{-7}$ M. Data from Zierau, O. *et al.* (2005) *Biochem. Biophys. Res. Commun.*, **326**, 900–916.

red-β-D-galactopyranoside (CPRG) to a red product, the concentration of which can be determined photometrically at 540 nm. Several variants of this test have been developed with different cell lines and reporter systems. The activity of phytoestrogens such as genistein and the highly estrogenic 8-prenylnaringenin (8-PN) present in hops were quantified using the described transactivation test (Figure 5.15). Compared to the endogenous hormone estradiol-17β, all phytoestrogens proved to be much less active.

The normal hormonal signaling may not only be disturbed by interaction with the respective receptor, but synthesis of the hormone itself may also be a target of EDCs. Hence the arsenal of *in-vitro* tests also includes assays that monitor steroidogenesis in a suitable cell line. Particular attention is paid to the enzyme aromatase, the activity of which is required to catalyze the conversion of testosterone to estradiol. Several natural substances are known to interfere with this process.

A number of validated *in vivo* assays are available to screen for EDCs and phytoestrogens. First, the uterotrophic assay should be mentioned as this has been subject to an extensive international validation program. The test measures increases in uterus weight after exposure to an estrogenic substance. The animals used (rats or mice) are ovariectomized in advance, so that the major source of endogenous estradiol is absent. A similar strategy uses the Hershberger assay to identify androgenic and antiandrogenic compounds in male rodents. Finally, female and male pubertal assays are available in which endpoints such as the onset of puberty and organ weights are determined. Substances that disturb thyroid function are identified using an amphibian metamorphosis test, which makes use of a thyroid hormone-dependent developmental process. Based on these standard assays, further tests have been developed that address more specific questions concerning interactions of EDCs with target proteins or the promotion (or prevention) of common diseases, using animal disease models. These test systems will be explained in the context of the respective experiments described below.

Despite the availability of validated test systems and advanced genetic techniques, it is often difficult to determine the mode of action of a particular substance. Typically, there are too many variables that cannot be controlled and may give rise to varying, and occasionally contradictory, results. For example, a phytoestrogen may act as either an agonist or antagonist, depending on the experimental circumstances and the particular tissue under investigation. Unexpected effects may also result from metabolites which differ in their properties from the original compound and interact with a different set of target proteins. Furthermore, the ligand-induced ER-mediated transcriptional response depends to a large extent on the activity of regulatory proteins which may also be targets of natural substances. Estrogenic substances often affect more than one of these regulatory levels. In males, testosterone-driven processes may be disturbed in similar ways, and natural substances may act as either androgens (mimicking the effect of testosterone) or as antiandrogens.

Natural substances with estrogenic activity have been discovered in many fruits, vegetables, and spices (Table 5.2). Some of these substances, such as the stilbene resveratrol and the isoflavone genistein, are of particular importance and their biological activities will be discussed in different contexts as estrogenic activity is only one aspect of their known biological properties. Other phytoestrogens were identified among flavonols, flavanones (including the highly estrogenic 8-PN), catechins, and coumestrans. The isoflavane equol (a metabolite of daidzein) is of special interest due to its strong estrogenicity. Lignans are components of the plant cell walls and

Table 5.2 Examples of polyphenolic natural compounds with estrogenic properties and the dietary sources. The chemical structures of selected compounds are shown in Figure 6.1. Examples taken from Patisaul, H.B. and Jefferson, W. (2010) *Front. Neuroendocrinol.*, **31**, 400–419.

Class of natural compounds	Examples of substances with estrogenic properties	Dietary source
Flavanones	Eriodictyol, Hesperetin, Homoeriodictyol, Naringenin	Citrus fruits and juices
Flavones	Apigenin, Luteolin, Tageritin	Parsley, celery, pepper, capsicum
Flavonols	Rhamnazin, Pachypodol, Myricetin, Quercetin, Fisetin, Kaepferol	Kale, broccoli, onions, tomatoes, apples, grapes, red wine
Catechins	Proanthocyanides	Chocolate, green tea, beans, apricots, cherries, berries
Isoflavones	Genistein, Daidzein, Formononetin, Biochanin A, Glycitein	Soy beans and other legumes
Isoflavanes	Equol	Metabolite of daidzein
Stilbenes	Resveratrol	Grape skin, red wine
Coumestans	Coumestrol	Clover, alfalfa, spinach
Lignans	Enterolactone	Flax seed, sesame seed

are typically present in berries, seeds, nuts, and fruits. The compound secoisolariciresinol diglycoside is highly concentrated in flax seed and is metabolized by bacteria of the gut to enterodiol and finally to the estrogenic enterolactone.

Phytoestrogens are marketed as health-promoting nutritional supplements; examples include the lignan-rich flax seed or red clover preparations with a variety of isoflavonoids such as daidzein, genistein, glycitein, formononetin, biochanin A, and irilon. The most interesting and relevant source of human phytoestrogens is soy, as it forms part of human food requirements, particularly in Asia where large amounts are consumed. For this reason, there is a strong interest in the biological effects of genistein, the major phytoestrogen in soy, and its possible effects on human health.

Nutritional differences between populations in Asia, Central Europe and the USA have formed the focus of research to explain the lower rates of cardiovascular diseases, diabetes, obesity – and in particular of breast cancer and other hormone-dependent cancers – in Asia. A major difference in the diet is the consumption of soy in Japan and other Asian countries, and this has led to the suggestion that biologically active components in soy may be responsible for the observed health benefits. The difference cannot be explained by genetic factors alone, since for persons who migrate early in life from one region to the other the risk of the respective diseases is essentially the same as in the new host region. Apart from the influence of nutrition, environmental factors and the lifestyle are additional factors that affect the incidence of specific diseases.

The relevant estrogenic compounds in soy are the isoflavones genistein and daidzein, and in lower concentrations glycitein. These compounds are present in soy beans, soy milk, in Tofu and other soy products in large amounts. In Asia, the daily isoflavone uptake with food is in the range of 15–50 mg, whereas in Europe and the USA the daily intake was estimated to be less than 2 mg. These differences are reflected in the plasma concentrations: in Japanese populations the mean serum concentration is in the order of 0.5 μM for genistein and 0.2 μM for daidzein, while in the British population the concentration is one order of magnitude lower. When high amounts of isoflavones were ingested by volunteers, a serum concentration of up to 5 μM was achieved, and about the same maximal level can be reached in infants fed with soy-based milk. However, the serum concentration may vary widely between individuals depending on their metabolic activity and their lifestyle. Vegetarians and persons consuming dietary supplements will obviously increase their isoflavonoid plasma level. Evaluation of the possible biological effects must take into account that, in the mammary gland, prostate or thyroid, the isoflavonoid concentration may differ significantly from the plasma concentration and may reach much higher values.

5.2.2
How Do Phytoestrogens Interfere with the Endocrine System?

The endocrine system is controlled in a hierarchical fashion (Figure 5.16). In the brain, the special centers that regulate endocrine activity are located in the hypothalamus. The pituitary in turn reacts to hypothalamic activity, and this constitutes

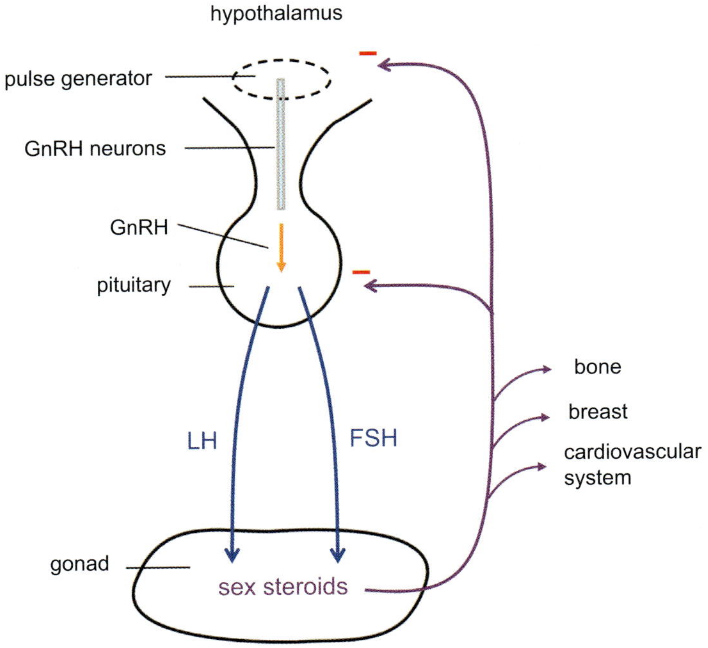

Figure 5.16 The endocrine system of vertebrates is hierarchically controlled. The hypothalamus in the central nervous system plays a dominant role and controls the activity of the subordinate endocrine glands, including the gonads. Neuroendocrine cells synthesize and secrete gonadotropic releasing hormone (GnRH) in a pulsatile fashion. The hormone is transported in a vascular portal system for a short distance in the pituitary, where it triggers release of the gonadotropins luteinizing hormone (LH) and follicle-stimulating hormone (FSH). These hormones are then distributed by the circulation system in the body and reach the gonads, where they control synthesis of the sex steroids estradiol and testosterone in males and females. These steroids are essential for the control of gametogenesis in the gonads; the hormones also control physiological processes in other organs. The indicated feedback loops to the pituitary and the hypothalamus limit steroid synthesis.

the second level of the regulatory hierarchy. Finally, the pituitary controls the activity of endocrine organs such as the gonads. This regulatory system, which is known as the hypothalamus-pituitary-gonad (HPG) axis, controls reproduction in mammals. In order to illustrate the principles of the endocrine control and the interference with natural substances, only the gonads as endocrine glands for sex steroids will be considered at this point, although other endocrine glands may also be affected by natural substances.

Neurons originating in the hypothalamus function simultaneously as nerve cells and endocrine cells. The neurons possess long axons that extend towards the pituitary gland, and they are also able to synthesize, transport and secrete hormones such as gonadotropin-releasing hormone (GnRH), which is relevant in this context. GnRH is released into the portal blood circulation of the pituitary gland and controls the release of the two gonadotropins from the pituitary gland, namely luteinizing hormone (LH) and follicle-stimulating hormone (FSH) (Figure 5.16). The neurons in the hypothalamus are characterized by a physiologically important periodicity. In postmenopausal woman this "GnRH pulse generator" has a periodicity of about 90 min. The periodicity is essential for release of the gonadotropins and their physiological function in the gonads. The principle of the hierarchical endocrine control system is conserved in all vertebrates.

Steroid biosynthesis in the gonads of both sexes is controlled by LH and FSH. Two types of somatic cell are relevant to this process, namely Leydig cells and Sertoli cells in the male, and theca and granulosa cells in the female. Estradiol and testosterone are synthesized in the gonads of both sexes, but the concentrations show pronounced gender-specific differences. Androgens and estrogens have a physiological role in both males and females, and this fact must be considered when the benefits and risks of phytoestrogens are investigated. For example, estradiol stimulates spermatogonial proliferation, while in females there is also some testosterone synthesis in the ovary. Testosterone is not only the substrate for estradiol synthesis, but also stimulates aromatase synthesis and promotes granulosa cell proliferation in concert with FSH.

The steroids synthesized in the gonads are truly endocrine hormones and are distributed through the body via the blood circulation and the lymph system. The hormones act and control physiological processes in numerous organs, and this feature is important when the biological effects of phytoestrogens are under consideration. The circulating hormones also reach the pituitary and the hypothalamus, and in this way a feedback loop of endocrine signaling of the HPG axis is established (Figure 5.16). Only a small fraction of the steroids is present in the serum as free hormone; rather, most steroids are transported by a special protein known as sex hormone-binding globulin (SHBG). This protein regulates the access of androgens and estrogens to the target tissues, as well as their metabolic clearance and their bioavailability.

5.2.2.1 **The Estradiol Receptor is a Target of Natural Substances**

How do the circulating steroids act within the respective target cells? Steroids are hydrophobic molecules that readily pass through cell membranes. The classical signal transduction pathway involves two intracellular estrogen receptors (ERα and ERβ). The ligand–receptor complex and additional regulatory proteins bind to a specific recognition sequence, the ERE, and initiate the transcription of a set of estradiol-responsive genes. This reaction is exploited in the described test system for estrogenicity (Figure 5.14). The tissue distribution of the ER subtypes differs: whereas ERα is predominant in the breast, uterus, cervix and vagina, the ERβ subtype prevails in the gonads of both sexes and in the lung, spleen, prostate, hypothalamus and thymus. During recent years it has become clear that the transduction pathways of estradiol signaling are even more complex. Apart from the classical nuclear receptor-mediated signaling, additional receptors with different signal transduction pathways play important roles (Figure 5.17). Some effects of estradiol occur so rapidly that regulation by nuclear receptors and induced transcription could not account for the effect. It emerged that some ERs are

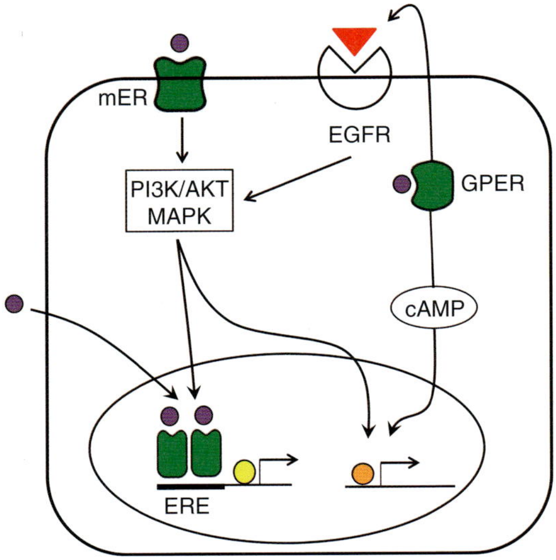

Figure 5.17 Genomic and nongenomic signaling of estradiol. The classical estradiol signaling involves the intracellular binding of the hormone (purple) to its receptor (green) which dimerizes and binds to a specific estradiol responsive element (ERE) and induces transcription of a specific set of estradiol-inducible genes. The reaction typically involves further transcription factors (yellow dot). A faster response to estradiol can be mediated by membrane receptors which signal via phosphatidylinositol-3 kinase (PI3K) and further kinases (AKT and MAPK). Such reactions may affect the genomic response to estrogens and, in addition, the activity of other transcription factors by phosphorylation (orange dot). Another estradiol-responsive pathway involves intracellular membrane-associated receptors which are coupled to G proteins (GPER). Signaling through this pathway involves cAMP and, again, the activity of kinases to activate transcription factors. In addition, GPER activates in a reaction cascade the important epidermal growth factor receptor (EGFR; the ligand EGF is shown as a red triangle). The signal transduction may then again proceed via the PI3K/AKT pathway. Modified from Prossnitz, E.R. and Barton, M. (2011) *Nat. Rev. Endocrinol.*, **7**, 715–726.

membrane-bound (mER) and use a different signal transduction pathway. The enzymes phosphatidylinositol-3-kinase (PI3K) and mitogen-activated protein kinase (MAPK) play a central role in controlling the physiological state of the cell. Their activation may ultimately result in the recruitment of transcription factors that control the expression of a set of genes which may or may not possess an ERE sequence (Figure 5.17).

Another important class of ERs has been discovered that is capable of triggering a rapid cellular response to estradiol; this receptor is G-protein-coupled and hence is referred to as GPER (also known as GPR30). The receptor is located intracellularly and its activation is mediated by the second messenger, cAMP. GPER is able to transactivate the epidermal growth factor receptor (EGFR), which is also an important receptor controlling a number of cellular processes. The rapid nongenomic signaling of mER or GPER has been reported in the hypothalamus and in GnRH neurons, as well as in the mammalian testis. Androgens may also use

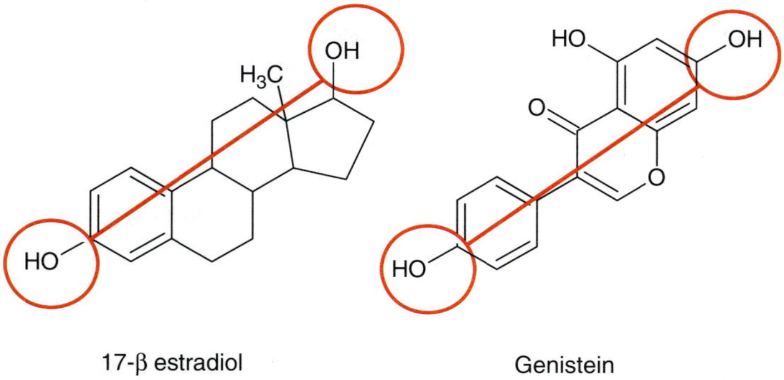

Figure 5.18 Genistein and 17β-estradiol share structural similarities. The estrogenic properties of genistein may result from structural properties of the phytoestrogen. The similar distance of OH-groups in genistein and 17β-estradiol is considered to be an important structural feature that allows the interaction with the ligand-binding domain of the receptor. From Wuttke, W. *et al.* (2007) *Ageing Res. Rev.*, **6**, 150–188 (see reference list). Reproduced with permission from Elsevier.

different signaling pathways, as the activation of kinases and a cAMP-mediated response is implicated in some androgen-mediated reactions.

Estradiol receptors are the classical target proteins of phytoestrogens, and this interaction has been studied intensively. The binding of phytoestrogens to the ligand-binding site of the ER is due to structural similarities of estradiol and ER ligands such as genistein (Figure 5.18). The dimensions of these molecules are similar, and in particular the distance between the two important hydroxyl groups is nearly identical. The distance between the hydroxyl groups is not the only important structural feature, however, as the estrogenic activities of flavonoids with similar structures differ widely and completely different chemical scaffolds may also possess estrogenic activity (Table 5.2). Typically, phytoestrogens bind with a higher affinity to ERβ than to ERα. For example, the estrogenic potency of genistein is 30-fold higher with ERβ than with ERα. The type of assay used in such an evaluation is critically important, as genistein (and other phytoestrogens) do not behave as straightforward agonists after binding to the ER. The observation that genistein interacts not only with ERβ but also with GPER, and hence interferes with the entire regulatory network (Figure 5.17), makes it clear that highly complex tissue-specific reactions are to be expected.

Compounds which bind to the ER may cause specific conformational changes of the ligand–receptor complex, so that different regulatory proteins can interact with the complex and alter its biological activity (Figure 5.19). The recruited regulatory proteins control the interaction with the ERE in the promoter of estrogen-responsive genes, and these interactions may result in therapeutically desirable tissue-specific effects. ER ligands with such interesting properties have been termed selective estrogen receptor modulators (SERMs). Pharmaceuticals acting as SERMs have been developed, although some phytoestrogens also possess such properties and small structural differences may produce different biological

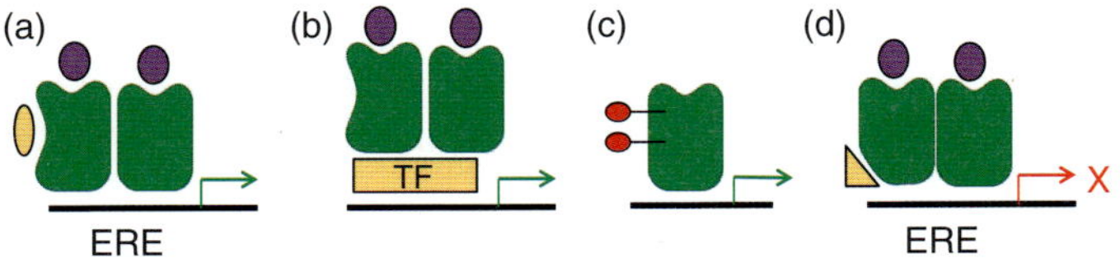

Figure 5.19 Regulatory proteins modulate the function of the estradiol receptor. Synthetic or natural products which bind to the ER (ER green, ligand purple) may induce slightly different protein conformations of the ER. Depending on the ligand-induced conformation, different regulatory proteins (orange) may be recruited which alter the biological activity of the protein complex. Ligands which act as selective estrogen modulators (SERMS) induce conformational changes of the ER that either promote (a) or inhibit (d) transcription (indicated by a green or red arrow, respectively); (b) Another possibility is that the ER dimer with its ligand binds to a tethered factor (TF) which may have different promoter specificity The recruitment of such ER-binding proteins is tissue-specific. Finally, the ER interaction with a promoter sequence may be ligand-independent and the ER activity (as monomer or dimer) may be modulated by phosphorylation (red dot). While the ER may act as transcription factor by binding to the estradiol response element (ERE) the TF may recruit other transcription factors like AP1.

effects; as a consequence, an estrogenic effect may turn into an antiestrogenic effect in a particular tissue or experimental set-up. For example, the prenylation and hydroxylation of genistein converts the substance from an agonist to an antagonist with antiestrogenic properties.

5.2.2.2 Natural Compounds Affecting Steroid Biosynthesis

An (anti)estrogenic effect may be elicited not only by the interaction of a phytoestrogen with the steroid receptors, but also by interfering with the enzymatic steps involved in steroid biosynthesis. Inhibitors of the steroidogenic pathway will also affect the synthesis of mineralocorticoid and glucocorticoid steroid hormones, but as research has focused largely on sex steroids the chosen examples will refer to these hormones only. In mammals, the sex hormones can be synthesized *de novo*, and cholesterol is a major precursor molecule (Figure 5.20). Phytoestrogens can interfere with almost every step in the synthetic pathway and either inhibit or stimulate sex steroid production in the endocrine glands (gonads and adrenal cortex). A critical and rate-limiting step is the transport of cholesterol from the outer to the inner mitochondrial membrane, where an enzymatic processing step takes place. The transport process is controlled by the steroidogenic acute regulatory protein (StAR). The transcriptional control and regulation of this protein is highly complex and involves several signal transduction pathways. Flavonoids have been identified which interfere with this process. Particular interest has been focused on stimulating compounds because of their possible therapeutic value when sex steroid levels decline during aging. A low testosterone level is also considered a risk factor for the development of Alzheimer's disease. Mouse Leydig cell cultures treated with the flavonoid apigenin were shown to upregulate StAR transcription in the presence of small amounts of cAMP, and as a result testosterone production in the cells was increased dramatically. This effect appeared due to the induced StAR protein and not to increased

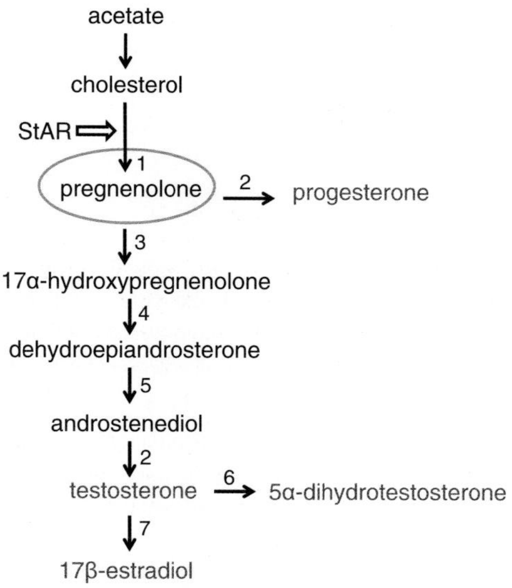

Figure 5.20 Biosynthesis of steroids. Some important steps in the biosynthesis of steroids are shown. The steroids 17β-estradiol and testosterone are synthesized from acetate and the basic precursor molecule cholesterol in a number of enzymatic reactions. An important rate-limiting step is the transport of cholesterol into the mitochondria, where it is converted to pregnenalone. The steroidogenic acute regulatory protein (StAR) controls the transport process. The remaining synthetic reactions are carried out outside the mitochondria, and several enzymes are targets of natural substances and can affect steroid production thus producing (anti)estrogenic or (anti)androgenic effects. Biologically active steroids are shown in purple. The numbers refer to the following enzymes: (1) P450 side-chain cleavage enzyme (P450scc, also referred to as 20,22-desmolase); (2) 3β-hydroxysteroid dehydrogenase; (3) 17α-hydroxylase; (4) 17,20-desmolase; (5) 17β-hydroxysteroid dehydrogenase; (6) 5α-reductase; (7) aromatase.

enzyme activities of the steroidogenic pathway. The flavonoid chrysin also had a stimulatory effect on testosterone production in Leydig cells in culture. When chrysin was given as dietary supplement to human volunteers, the results of two studies were inconclusive as the level of testosterone was found to increase in one study but remained unchanged in another. Injected catechin, epicatechin or epicatechin gallate increased the plasma testosterone level in rats, presumably by increasing the Leydig cell levels of cAMP, which is required for steroidogenesis. Flavonoids may attenuate steroidogenesis in many ways; for example, genistein was reported to inhibit 5α-reductase, while 3β- hydroxysteroid dehydrogenase activity was reduced by some isoflavones and coumestrol.

The final and rate-limiting enzymatic step in the synthesis of estradiol involves the conversion of androgens to estrogens (Figure 5.20). The required aromatization of a C19 androgenic steroid is catalyzed by the enzyme aromatase, whose common name reflects this activity. The enzyme belongs to the superfamily of cytochrome P450 enzymes and, more specifically, to family 19; hence it is also referred to as CYP19. There is a strong medical interest in this enzyme as a drug target, since most breast cancers are promoted by estradiol and, for this reason, a reduction in the estradiol level is therapeutically indicated. Some aromatase inhibitors are clinically available and are useful drugs for the treatment of breast cancer, which is the leading cause of death in women. Because synthetic aromatase inhibitors generally have adverse side effects, the search for a natural substance with inhibitory activity has been intense, and more than 300 substances with such inhibitory properties have been identified using *in-vitro* assays. Among the identified aromatase inhibitors are substances with different chemical structures, such as terpenoids, xanthones, fatty acids, alkaloids, and flavonoids; these comprise the largest group of identified inhibitors. Apigenin, chrysin, and myricetin are also examples of flavonoid inhibitors. IC_{50}-values in the low micromolar range have typically been reported but the most effective substances are active at nanomolar concentrations. *In-vitro* enzyme activity tests need first to be complemented by cellular assays, and finally by *in-vivo* tests to demonstrate possible useful medical applications. Occasionally, contradictory effects have been reported; for example, genistein was identified as an aromatase inhibitor *in vitro*, but in cultured human cells it induced aromatase mRNA expression. By using the zebrafish model, this finding was substantiated when it was observed that genistein induced the expression of an aromatase gene (*aroB*) in an estrogen-dependent pathway. Recent efforts have been aimed at downregulating the promoter activity of the aromatase gene by using suitable drugs. Medicinal chemistry also appears to be a promising approach for improving the specificity and bioavailability of an inhibitor by chemically modifying the structure of a natural substance with useful properties (see Section 6.2).

5.2.2.3 Other Activities of Phytoestrogens

Intensive studies of the estrogenic flavonoid genistein has revealed complex interactions with the endocrine system. For example, flavonoids may interfere with the important regulatory function of the SHBG which binds the bulk of steroids in the serum. Genistein was shown to stimulate the synthesis of this binding protein. Although low SHBG plasma levels are considered a risk factor for hyperglycemia and type 2 diabetes, elevated SHBG levels correlate with a higher risk of developing osteoporosis. The biological effect of the interaction of isoflavones with SHBG remains unclear.

The endocrine control of steroid synthesis is mediated by the hypothalamus and the pituitary, and steroids synthesized in the endocrine glands affect the activity of these neural centers in feedback loops (Figure 5.16). When the interference of phytoestrogens with these regulatory loops was studied in extensive *in vivo* experiments, the results were often difficult to interpret as genistein interacts with numerous target proteins that play a role in steroidogenesis, and both species- and tissue-specific effects must be taken into account.

5.2.3 Evaluation of Risks and Benefits of Phytoestrogens to Treat Common Diseases

Beneficial and toxic effects are typical properties of medicinal drugs, with the biological effect depending on the drug concentration (bioavailability), the health status and genetic constitution of the patient, and many other factors. The sex steroids estradiol and testosterone control numerous physiological functions, including sexual differentiation and the control of reproductive functions. Hence, natural substances that interfere with the respective hormonal signaling will have significant effects on human health. In the following sections possible health risks will first be discussed, followed by a brief review of the beneficial effects ascribed to phytoestrogens.

5.2.3.1 Phytoestrogens and Reproductive Functions

In lower vertebrates (bony fish), estrogenic substances produce dramatic developmental effects and are clearly reprotoxic. For example, when male Medaka (XY) are exposed to low concentrations of genistein (or estradiol) at a critical time of sexual differentiation they undergo sex reversal and develop into fertile females. The effect is less drastic in mammals but exposure to estrogens during susceptible periods during development may also lead to feminization. The brain is among the target organs that differentiate in a sex-specific manner under the influence of sex steroids. Numerous studies have addressed this issue, and it is clear that genistein and other isoflavones can affect neural differentiation as well as sex-specific behavior. Behavioral changes were also reported in monkeys; for example, when the animals were fed for 18 months with an isoflavone-rich soy protein diet, an increase in aggressive and submissive behavior was noted and the monkeys spent less time in physical contact with each other. The underlying cellular and molecular mechanisms for these effects are largely unknown.

The hypothalamic center controls the secretion of GnRH during critical phases around the time of birth, and genistein interferes with this sex-specific differentiation process. Surprisingly, in the male brain of rodents estradiol must be present during the neonatal period in order for masculinization to occur, since only in the presence of estradiol can the brain develop a male-type GnRH/LH secretion pattern. Conversely, estradiol must be absent from the female brain around the time of birth in order for the female-type GnRH/LH response (GnRH surge) to develop. Aromatase activity is required in the male brain to control the production of estradiol from the precursor testosterone, and, interestingly, neonatal genistein exposure has a masculinizing effect in this animal model. Unfortunately, these data provide no information on the effects of genistein in the developing human brain, as androgens rather than estrogens are required for masculinization in this case.

There is considerable evidence that isoflavone treatment affects reproductive physiology in rats or mice. The neonatal phase of development is a critical time and various anomalies were recorded. For example, animals treated with genistein show abnormal estrous cycles; this effect was also observed in animals treated with resveratrol or zearalenone (an estrogenic mycotoxin), and the abnormal cycles persisted in adults. Furthermore, neonatal treatment disturbs the normal differentiation of follicles in the ovaries, such that so-called multi-oocyte follicles are formed. This effect is apparently mediated through

ERβ, since the anomaly was not observed in mice lacking this receptor. Further evidence includes fewer pregnancies, an impaired function of the uterus to sustain pregnancies, and a smaller number of live pups. In mice, these reproductive anomalies resulted in a reduced fertility that was observed at a genistein blood level that can also be achieved in infants fed soy-based milk. These experimental findings send a clear warning with regards to the use of soy-milk for infants. However, as no significant long-term effects were reported in these children it remains unclear if rodents and humans differ in their response to isoflavones at that critical time of development. It is also possible that Asian and Caucasian infants differ in their response to high-dose isoflavones.

The estrogenicity of compounds can be quantified by studying the gain in uterus weight, as estradiol induces the cells of the myometrium (a layer of smooth musculature) and endometrium (the inner lining of the uterus) to proliferate. Phytoestrogens are also able to stimulate cell proliferation. When pregnant rats were fed with genistein-rich food and the newborns were raised with the same food, a proliferation of the endometrium was stimulated compared to control animals fed with genistein-depleted food (Figure 5.21) although, compared to estradiol, the effect was much smaller. Does the phytoestrogen-mediated stimulation of endometrial cell proliferation increase the risk of developing endometrial cancer? This type of cancer is common in women, and endometrial hyperplasia is an early indicator of an increased cancer risk. The long-term treatment of postmenopausal woman with soy phytoestrogens was shown to increase the incidence of endometrial hyperplasia, but this finding was not substantiated by other reports and similar studies performed in rodents did not resolve the issue. Based on the results of several studies, the picture emerges that genistein treatment (in particular neonatal exposure) has negative effects not only on ovarian differentiation but also on the estrous cycle and fertility.

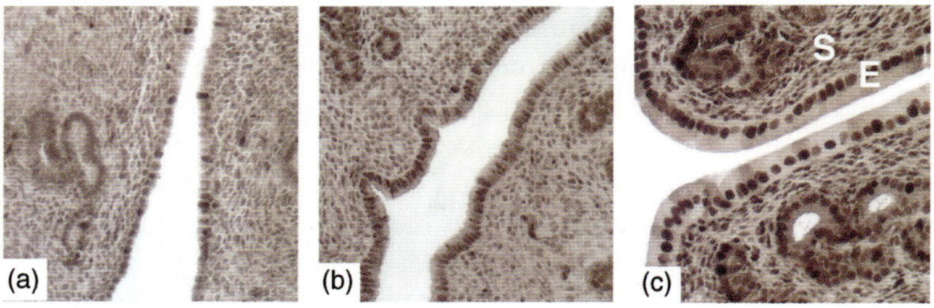

Figure 5.21 Estrogenic effect on the uterine epithelium in rats. Histological analysis of the uterine epithelium of rats fed with (a) phytoestrogen-free food, (b) genistein-rich food, and (c) 17β-estradiol-containing food. The exposure of animals to the different diets started *in utero* and was maintained after birth for a lifespan. The histological sections were stained with anti-PCNA antibodies which label the nuclei of proliferating cells. Estradiol strongly stimulates proliferation of the uterine epithelium (compare panels a and c). The cells of genistein-fed rats were also induced to proliferate, but to a lesser extent compared to estradiol. All images were recorded at the same magnification. Photographs courtesy of F. J. Möller, Institute of Zoology, TU Dresden.

What effects does isoflavone exposure have on the male reproductive system? The lower incidence of prostate cancer in Asian populations may suggest beneficial effects for men, though there is little evidence of a causal relationship. Hormone-dependent developmental disorders such as cryptorchidism (failure of the testes to descend into the scrotum), hypospadias (abnormally placed urethral opening at the penis) and testicular cancer doubled in North America between the 1970s and 1990s. Whether an increased exposure to phytoestrogens and synthetic EDCs contributed to these effects remains a matter of debate, but it was suggested that a "cocktail" of EDCs and phytoestrogens may have been responsible for the recognized declines in semen quantity and quality over the past decades. However, epidemiological data from various countries does not provide a coherent picture since, while declines in sperm count have been reported for some countries, there was no difference in other countries. The reason for these differences remains unresolved, and there is no apparent correlation with any nutritional variations. The few dietary intervention studies that have been carried out with phytoestrogens were also inconclusive with regards to semen quality and altered hormone levels.

The life-long exposure of male mice to phytoestrogens failed to provide any evidence that spermatogenesis is substantially affected by such treatment. However, during the final differentiation from round spermatids to the motile spermatozoa, some genes were downregulated compared to unexposed controls. Among affected genes were the androgen-responsive genes (*Pem*, *Testin*, *tPA*) and a glycolysis-regulation gene (*Gapd-s*) that is required for sperm motility and fertility. An interesting effect was observed when human spermatozoa were exposed to low concentrations (1–100 nM) of the estrogenic flavonoids genistein, daidzein, or 8-PN. In all cases the functional activation (capacitation) of the spermatozoa was accelerated and the acrosome (a vesicle filled with a cocktail of different enzymes that are essential for fertilization) reacted prematurely. The rapid flavonoid-induced reaction was apparently not mediated by nuclear ER but rather by a membrane-bound receptor; this led to the activation of a different signal transduction pathway that involved the upregulation of cytoplasmic cAMP levels.

It has already been pointed out that humans are exposed to a variety of substances that can modulate endocrine regulation. Although EDCs in the environment and phytoestrogens in food are the major sources of such compounds, little is known about possible synergistic interactions between synthetic and natural estrogenic substances. This question is undoubtedly highly relevant and a toxicological risk assessment is necessary. Because of their inherent complexity, epidemiological studies are unable to resolve this problem. Some experiments performed in rats, however, were of interest and have highlighted the significance of the problem. In this case the animals were exposed to low-dose mixtures of genistein in combination with the fungicide and a common food contaminant, vinclozolin, which acts as an antiandrogen. Interestingly, a low dose of this EDC mixture affected the fertility and reproductive functions of males to a greater degree than did a much higher dose. While the reason for this effect was not clear, such observations underscore the need to investigate low-dose effects as well as the effects of mixtures of phytoestrogens and EDCs.

5.2.3.2 Do Phytoestrogens Protect Against Breast Cancer?

Is there experimental evidence in support of the notion that phytoestrogens provide protection against breast cancer? The answer to this question emerged as a challenging research task. Epidemiological evidence has indicated that soy food consumption is protective for Asian populations, but some studies have suggested that this may not apply to other (Caucasian) populations. Dietary intervention studies have generally been disappointing and no clear picture has emerged. Moreover, even the suggested beneficial effect of isoflavones has been disputed as most breast cancers are estradiol-responsive, and there has also been concern that isoflavones might stimulate rather than inhibit tumor cell proliferation.

The different and contradictory results obtained could, in part, be due to different cellular responses to ERα and ERβ signaling. It has been claimed that ERα promotes cell proliferation and tumor progression in estrogen-sensitive tissues, whereas ERβ activity suppresses the ERα-mediated signaling and hence inhibits cell proliferation. The set of genes whose transcription is activated by each receptor differs in detail, and this may account for the receptor-specific biological reaction. Typically, the expression of ERα is higher in cancer cells than in healthy tissues, while ERβ prevails in many organs, including the steroid-responsive organs breast and prostate. Interestingly, the major phytoestrogens such as genistein, daidzein and equol bind with a higher affinity to ERβ. Whilst the implications of these observations are not fully understood, an additional dimension of complexity becomes apparent in view of the different tissue distributions of the ERs and the possibility to form homodimers and heterodimers. Furthermore, phytoestrogens act in some ways as selective estrogen modulators (SERMs; see Section 5.2.2) and may also show concentration-dependent and tissue-specific agonistic or antagonistic activities. For example, the isoflavone genistein can be estrogenic in some experiments but antiestrogenic in others, depending on the concentration and the physiological circumstances. Another factor which is difficult to control concerns the individual metabolism of phytoestrogens (see Section 4.1.2). An example of such variation is the conversion of isoflavones to the estrogenic equol, as a large proportion of the population is unable to produce equol.

Another interesting observation that will require future analysis is that the doses of phytoestrogens appear to be critically important, as a nonlinear or biphasic dose–response relationship was observed in some cases. For example, low doses of genistein stimulated the proliferation of tumor cells *in vitro*, while high concentrations ($>10\,\mu M$) inhibited tumor growth. This effect was observed with two different human breast cancer cell lines. Human cancer cells can be injected into an animal (xenograft) and further studied *in vivo*. In this case, human mammary cancer cells (MCF-4) were implanted into ovariectomized mice with a suppressed immune system (athymic). Subsequently, when the mice were fed sufficient genistein to create a plasma concentration similar to humans, the cancer cells were induced to proliferate. Treatment with genistein was shown to increase the number and size of the end buds of lactiferous ducts in the developing mammary gland (the buds are the sites where breast cancer is initiated), but the effect achieved was dependent on the genistein concentration and the developmental phase at the time of exposure. Conditions that stimulate cell proliferation and growth tend to increase the risk of breast cancer, whereas the induction of cell differentiation in the mammary gland reduces such risk. However, whether genistein can protect against breast cancer or will promote the disease depends largely on the developmental phase at the time of exposure, and on the concentration of the phytoestrogen.

Another experimental variable may have contributed to the incongruous experimental results obtained with phytoestrogens. In feeding experiments the composition of the isoflavone-rich food appears to be an important factor (e.g., the fraction of flavonoid glycosides, the mixture of isoflavones and other variables), as does the physiological state at the time of exposure. For example, there appears to be a difference if women consume isoflavones before or after the menopause. In premenopausal women the endogenous estradiol levels are high, whereas in menopausal women the phytoestrogens act in a low-estrogen environment. It is only in the latter situation that the phytoestrogens can stimulate the proliferation of estradiol-responsive cancer cells. Experiments with rodents have also shown that the reaction to isoflavones at an early developmental phase may differ from the typical reactions that might be expected at an adult stage. For example, when an isoflavone diet was given to newborns or pre-pubertal animals, the tumor incidence and growth later in development was significantly reduced. Exposure during the early phase of development may have longlasting effects up to the adult stage (this phenomenon will be discussed further in the context of epigenetic mechanisms; see Section 5.2.4). In experiments with cell cultures, the expression of relevant genes may differ from the respective target tissue *in vivo*. For example, experiments with MCF-7 breast cancer cells suffered from the disadvantage that only ERα is expressed but not the important protective ERβ. It must also be borne in mind that rodent model organisms (rats and mice) differ from humans in terms of flavonoid metabolism, estradiol plasma levels, and other physiological parameters of importance.

To summarize, the effect of phytoestrogens on the development of breast cancer is still unresolved. While animal models can provide only limited information with regards to the situation in humans, numerous epidemiological and dietary intervention studies have been performed to clarify the important issues for human health. However, as no clear picture has yet emerged, dietary recommendations cannot be given at the present time.

5.2.3.3 **Hormone Replacement Therapy**

Phytoestrogens may be useful for the treatment of diseases promoted by low estradiol levels. The prevention of osteoporosis and the relief of menopausal symptoms are typical applications, as each condition is caused by estradiol levels being decreased during the menopause. While these symptoms can be relieved by estradiol treatment, reports of severe side effects, such as increases in the incidence of breast cancer and cardiovascular disease, has spurred the search for a "natural" hormone replacement therapy.

Many neurons in the hypothalamus are estradiol-responsive, and GnRH release is controlled by estradiol levels. During the menopause, low estrogen levels provoke hyperactivation of the GnRH pulse generator, as the negative feedback loop of estradiol is inefficient for this purpose (see Figure 5.16). The resulting menopausal symptoms, such as hot flushes and tachycardic attacks (an abnormally high heart rate), can be relieved by treatment with low concentrations of estradiol, which will reduce the level of neurotransmitters that stimulate the GnRH pulse generator. Whether isoflavones may be suitable substitutes for estradiol treatment remains unclear, although the amounts of genistein required to reduce GnRH signaling are so high that adverse side effects (e.g., the stimulation of uterine endometrial growth) would surely occur. Alternative phytoestrogens have also been monitored for their effects on gonadotropin levels, as there is a strong correlation between LH pulses and hot flushes.

The intravenous infusion of coumestrol into ovariectomized rats caused a reduction in the activity of the GnRH pulse generator and also suppressed pulsatile LH secretion, but genistein administered in the same way had no effect. Similarly, administration of the flavanone 8-PN over a three-month period was found to inhibit pulsatile GnRH release in the rat brain and lowered the LH levels, but resveratrol tested in a similar fashion was ineffective. Further studies have indicated that some phytoestrogens affect both LH and FSH levels, although no clear picture has been obtained concerning the reaction pathways. Numerous clinical studies with phytoestrogens have provided conflicting results, and in most cases no relief from menopausal symptoms was obtained by phytoestrogen treatment.

5.2.3.4 Other Physiological Effects of Phytoestrogens

The claim that soy consumption reduces the risk of cardiovascular diseases was approved by the Food and Drug Administration (FDA) in 1999, and thereafter soy products have been marketed on a large scale. However, conflicting studies performed during the subsequent years failed to provide conclusive supporting evidence for these claims. The hope has always been that phytoestrogens may help to prevent atherosclerosis in a similar fashion as estradiol 17-β (which had been shown to have beneficial effects in monkeys), by lowering cholesterol levels and preventing the formation of atherosclerotic plaques. Other investigations in animals have been less convincing, however. In dietary intervention studies the reduction in serum (LDL) cholesterol levels was small at best; moreover, when isoflavones were removed from the soy diet the effect was the same, which indicated that the protein substitution rather than the isoflavonoids had produced the effect. In humans, data acquired from both clinical and diet intervention studies have been inconclusive, and no recommendations can be given based on present evidence.

The well-known vasodilatory effect of estradiol is a rapid process, and hence the involvement of nongenomic signal transduction has been inferred. Apparently, a subpopulation of ERs is functionally coupled to inducible eNOS. The generated $NO^{\bullet}$ has been identified as the essential signaling molecule that causes relaxation of the arterial myocytes (see Section 5.1.2); the subsequent lowering of the blood pressure reduces the risk factor for atherosclerosis. The NOS activity results in the S-nitrosylation of numerous endothelial proteins, causing their biological activity to be modulated. The nuclear ERs may also be S-nitrosylated, such that their activity as transcription factors would be inhibited. Genistein was found to activate both NOS and EGFRs, in the latter case via a G-protein-mediated reaction. The available data suggest that genistein affects both genomic and nongenomic pathways (Figure 5.17).

So far, the focus has been on estrogenic effects in organs with reproductive functions. Yet, other organs are also estrogen-responsive, and essentially the physiology of the entire organism is profoundly affected by phytoestrogens. For example, estradiol is known to control processes in the thyroid gland, the skeletal system, the gastrointestinal tract, and in the immune system. Furthermore, general physiological parameters are affected such as glucose homeostasis and adipositas (pathological obesity). Ideally, a risk assessment in clinical studies should take into account all possible effects of phytoestrogens, although because of the complexity this has not yet been attempted.

5.2.3.5 **Phytoestrogens Illustrate the Experimental Challenge of a Mechanistic Analysis**

Despite intensive research being conducted on the biological and pharmacological effects of phytoestrogens many unanswered questions remain. Notably, it is unclear why apparently straightforward questions relating to possible health benefits and hazards of phytoestrogens are difficult to answer. Currently, at the level of the receptors sophisticated techniques are available for the analysis of drug–target interactions. However, the existence of different signal transduction pathways (Figure 5.17) does not permit a straightforward biological interpretation of such molecular data. Furthermore, genistein and many other natural substances interact with numerous proteins, and consequently many different physiological processes may be affected. When the question is raised of whether a particular food is healthy (such as soy with estrogenic substances), the complex mixture of molecular constituents present thwarts a detailed molecular analysis of the underlying mechanisms. Moreover, other individual variables such as age and genetic constitution (see Chapter 4) add to these problems. The approach of "reverse pharmacology" (see Chapter 6) acknowledges the problem of complexity, and concentrates first on the induced biological effects of a particular substance *in vivo*; it is not the primary goal to provide a mechanistic explanation for the observed effects. In view of the enormous complexity of the physiological reactions involved it would be wise to view the health claims of phytoestrogens and other natural substances with caution, and to be aware of the limits of mechanistic explanations that are based on current analytical methods.

5.2.4
Transgeneration Effects

The early developmental stage is particularly sensitive, and natural substances that interfere with relevant signaling pathways are more likely to cause developmental or physiological defects. In mammals, exposure during the uterine and neonatal phases is particularly critical. Some diseases that become apparent as adults have their basis in the exposure to bioactive compounds during early development, although during recent years it has become clear that such an effect is not necessarily limited to the individual that has been exposed early in life but may also involve the next generations. Such persistent effects are not due to a mutation but rather to nongenomic processes which regulate genome activity by modulating chromatin structure. The term "epigenetics" has been coined to describe this phenomenon. In the case of transgenerational inheritance, exposure to a xenobiotic has a longlasting effect such that not only the exposed individual but also the next generation is affected.

Such surprising long-term effects are scientifically exciting but, at the same time, are disturbing in view of the consequences for health. Whether phytoestrogens and other natural compounds can produce transgeneration effects is unclear, but before considering that point it is necessary to examine epigenetic mechanisms more closely. The discovery of DNA methylation and the relevance of X-chromosome inactivation in females was a milestone in the research on the epigenetic control of transcriptional activity and the stable inheritance of this property during numerous mitotic divisions. Cytosine-guanine (CpG) dinucleotides which are typically clustered in gene promoter regions may become methylated, and about 50% of human genes possess such CpG islands. The chemical modification of histones is also an important epigenetic mechanism. The histone proteins form units called nucleosomes, around which the DNA is wound (Figure 5.22). The degree of chromatin condensation is regulated by DNA and histone modifications; histone acetylation is an important modification that has been identified. Genes in highly compacted chromatin are silenced and, therefore, the epigenetic mechanisms that regulate histone modifications also control transcriptional activity and either stimulate or inhibit gene expression. Epigenetic mechanisms also include the activity of noncoding RNA. Currently, research in this area is intense and the discovery of further epigenetic mechanisms is very likely.

The epigenetic silencing of genes is a normal developmental process. For example, the sex-specific silencing of genes in the male and female germline is a necessary process of normal development, and if the silencing of specific genes ("imprinting") does not occur then diseases might develop. Cancer and several other diseases have an epigenetic component; for example, hypermethylation of the ERβ promoter inhibits transcription and plays a role in the progression of cancer.

It is unclear in which way xenobiotics can interfere with the epigenetic regulation of gene activity. The effect of the fungicide vinclozolin, which has antiandrogenic activity, has been studied in detail; in some experiments a transgenerational epigenetic effect was documented, but the epigenetic effects of vinclozolin were variable depending on its concentration and the strain of animals used for the tests. When pregnant mice were exposed to vinclozolin, the sperm quality in male mice was reduced for two generations (F1 and F2), and dysfunctions in several organs (e.g., kidney, testis, immune system) were also observed. However, in the

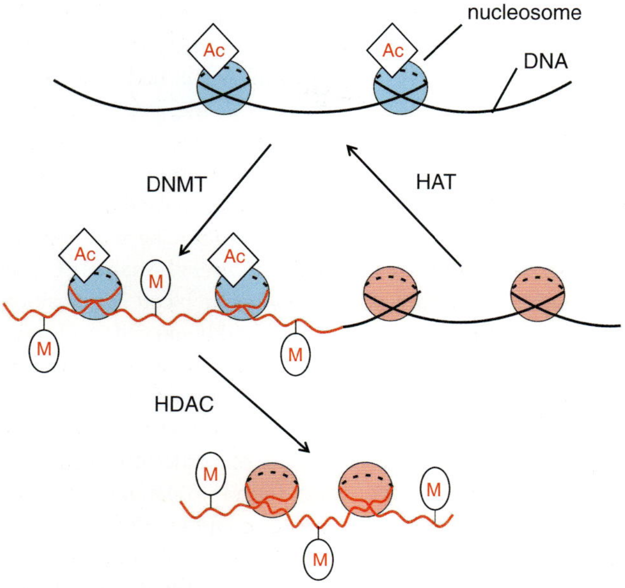

Figure 5.22 The chromatin structure controls the transcriptional activity. The chromatin (DNA with structural and regulatory proteins) may be in an open position that favors transcription, or in a condensed and silent condition. The DNA (black straight line) is wrapped around protein complexes (nucleosomes) which consist of histone proteins. The nucleosomes are spaced regularly along the DNA and give the chromatin its basic structure. In the open and active structure of the chromatin the nucleosomes (blue) are acetylated (Ac). Removal of the acetyl group from histones (nucleosomes in red) and methylation (M) of DNA (wavy line in red) are epigenetic changes which result in a more closed chromatin conformation. Under these conditions transcriptional activity is repressed. The epigenetic changes are catalyzed by, for example, the enzymes histone acetylase (HAT), DNA methyltransferase (DNMT), and histone deacetylase (HDAC). The activity of these enzymes may be affected by natural substances. Based on Link, A. *et al.* (2010) *Biochem. Pharmacol.*, **80**, 1771–1792.

third generation the animals were indistinguishable from controls. The reason for this transgenerational pattern was that not only was the mother exposed to the drug (F0) but also the embryo (F1) and, in the embryo, the germline cells of the next generation (F2). In some instances persistent epigenetic effects were also observed over more than two generations. It appears that, at least in some cases, the DNA/histone modifications are very stable. Indeed, molecular analyses confirmed that in vinclozolin-treated animals the methylation pattern in the germline was altered and expression of the genes that catalyze the methylation was reduced for three generations. Phytoestrogens such as genistein, coumestrol and equol were also found to alter the DNA methylation pattern. Although transgeneration effects of soy-fed rats have been reported, most studies have focused on epigenetic changes within the life-span of the animal.

The importance of the early phase of development has already been pointed out. Some of the effects of uterine or neonatal exposure to isoflavones manifest themselves much later in life at the adult stage; this phenomenon has been referred to as "late-onset-effect," for which epigenetic mechanisms offer a mechanistic explanation. The exposure of rat embryos in the uterus had longlasting effects on the endocrine systems, and long-term changes in the gene expression profile were also noted. In a recent study, a group of mice was fed food containing minimal concentrations of isoflavones, while a parallel group received food enriched with genistein. Exposure to both compounds was started three days before conception and continued until the end of pregnancy. The offspring of both groups was then fed on a low-isoflavone diet for 12 weeks. Analyses of the blood and bone marrow showed significant differences in the blood composition of mice exposed prenatally to genistein, with numbers of red blood cells and reticulocytes being elevated in the blood and bone marrow, and more granulocytes and pyrenocytes being detected compared to a control group. Genistein is known to stimulate bone marrow formation, and for this reason the gene expression profile of bone marrow cells was analyzed using whole-genome oligonucleotide microarrays. Among 21 742 annotated genes, 20% were found to be differentially expressed in the genistein-exposed animals compared to controls. The expression of the important signaling protein NFκB and of the NFκB-regulated genes was downregulated in the genistein-treated animals. Interestingly, this held true also for several estrogen-responsive genes. An analysis of the methylation pattern showed that the DNA of the genistein-treated animals was hypermethylated, and this presumably contributed to the observed downregulation of the genes. These data showed that genistein treatment before birth has a longlasting effect on cell differentiation and gene expression due to a stable change in the epigenetic methylation pattern. However, whether these epigenetic changes alter the risk of cancer is presently unknown.

Genistein may also affect adipocytes and induce obesity in mice. Adipocytes proliferate during early development and reach a plateau at adulthood. An excessive calorie intake during adult life would affect the adipocyte volume, but not their numbers. Epidemiological evidence has suggested that intrauterine exposure to EDCs predisposes the embryo to become obese later in life. In view of the important role of the adipose tissue as a central regulator of metabolism and as an endocrine organ, it is important to elucidate the effects of phytoestrogens and other natural substances on the differentiation of adipocytes early in life, and the inferred epigenetic changes that may result in diseases during adulthood.

One point to clarify would be whether these genistein-induced epigenetic changes can explain the different cancer incidences in populations that consume large amounts of soy, and those that do not. Neonatal exposure to equol led to a precocious differentiation of the mammary gland, and a similar effect was reported when prepubertal rats were fed with genistein. This anticancerogenic effect is thought to be due to an epigenetic mechanism. By using other cancer model systems, a protective effect of phytoestrogens to chemically induced carcinogenesis was also reported, though additional data are clearly required to resolve this issue.

Because of the known epigenetic effects in the development of cancer, many investigations have been conducted to identify polyphenols which affect DNA methylation and/or histone modifications (Figure 5.22). These epigenetic processes are controlled by a number of enzymes, although only DNA methylation and histone acetylation will be discussed in the following subsection. Three DNA methyltransferases (DNMTs) have been identified in humans, and with respect to histone acetylation both histone acetylase (HAT) and histone deacetylase (HDAC) have been investigated as possible targets for dietary polyphenols. Several well-known inhibitors have been identified that have already been discussed with regards to other biological properties. For example, epigallocatechin-3-gallate (EGCG; a component of green tea) is a potent inhibitor of DNMT, and in several tumor cell cultures the hypermethylation (Figure 5.22) of several tumor suppressor genes was reversed by EGCG. Tumor suppressor genes play an important role in controlling the cell cycle. The activity of their gene products prevents the uncontrolled proliferation of cells that is a hallmark of cancer. Some tumor suppressor genes are typically inactivated in cancer cells, and hence the epigenetic reactivation would be desirable. In contrast, oncogenes are characterized by an excessive activity in cancer, while the normal state is characterized by a tight regulatory control (then referred to as a proto-oncogene). In this case, and in contrast to tumor suppressor genes, epigenetic gene silencing is desirable (Figure 5.23). The epigenetic manipulation of cancer-promoting genes is a current therapeutic strategy for the treatment of cancer.

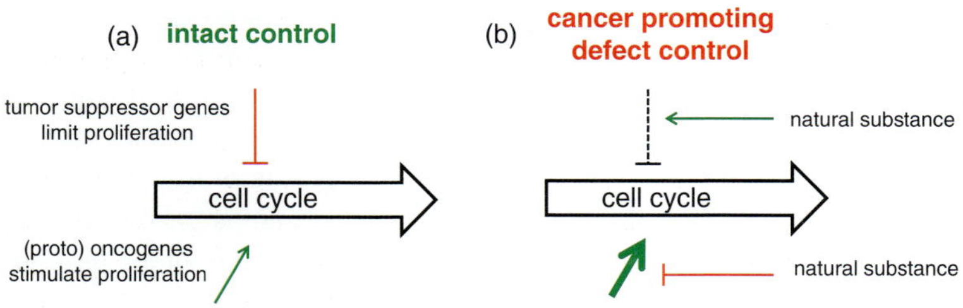

Figure 5.23 Some natural substances possess anticancerogenic activity by interfering with the control of the cell cycle. (a) Cell cycle control in normal cells. When cells proliferate they pass through defined stages of the cell cycle. The cycle is represented by an open arrow. Progression through the cell cycle, and hence the completion of a cell division, is controlled by numerous genes. Some genes have inhibitory activity (tumor suppressor genes) and some have a stimulatory influence (proto-oncogenes); (b) Defective cell cycle control. Mutations in cell cycle-controlling genes may lead to cancer. If a tumor suppressor gene is inactivated and hence the break on cell proliferation loosened (dashed line), or if a mutation causes hyperactivation of an oncogene (fat green arrow), the cell cycle control is compromised and this may result in excessive cell proliferation. Natural substances may exert an anticancerogenic effect if they either reactivate inactivated tumor suppressor genes or if they inhibit overexpressed oncogenes. The positive effect may also be mediated through epigenetic mechanisms (see Figure 5.22) when natural substances change the chromatin structure in such a way that the expression of mutated tumor suppressor genes is promoted or the expression of an oncogene repressed.

Although experimental data reported to date are promising, the system is complex and not understood in detail and, as a result, contradictory results have been reported. For example, the described effect on EGCG has not been observed by all groups; similarly, genistein has been described as a DNMT inhibitor *in vitro*, but the evidence for this activity *in vivo* is inconsistent. Other polyphenols have been studied and some DNMT inhibitory activity was ascribed to curcumin, rosmarinic acid or isothiocyanates. Curcumin was also identified as a potent HAT inhibitor, which is a desirable property for silencing oncogene activity.

Apparently, some of these compounds affect the activity of several target proteins. For example, EGCG and isothiocyanates are implicated in DNA methylation as inhibitors of DNMT, and at the same time appear to inhibit histone modifications as EGCG is also a HAT inhibitor and isothiocyanates inhibit HDAC. These data illustrate the complex molecular reactions that are to be expected in the search for a cancer therapy. It is also apparent that the promising *in vitro* data need to be confirmed in *in vivo* models, or by dietary intervention studies.

5.2.5
Natural Substances Interfering with Nonsteroidal Signaling

So far, attention has been focused on the endocrine and epigenetic activities of genistein and other natural substances. However, natural compounds typically interact with several target proteins, and many reports have been made of interactions between a natural substance such as genistein and specific target proteins. As a result, when a particular biological reaction is under investigation, side effects (also known as "off-target effects") must be expected. For example, genistein is not only a phytoestrogen but can also inhibit members of the protein tyrosine kinase (PTK) family of enzymes. These proteins play important regulatory roles in several signal transduction pathways, and control their own activity by phosphorylating their own tyrosine residues, as well as those of other proteins. PTK activity plays a role in the control of tumor growth and in various neural and cardiovascular functions. However, apart from PTKs numerous other enzyme activities are modulated by genistein and must be considered when the molecular mechanisms of isoflavonoid activity are being investigated. Unfortunately, off-target effects cannot be predicted and the mechanisms by which way a phytoestrogen functions are often unclear.

Of particular interest here is the thyroid as a possible target of EDC and of natural substances that interfere with thyroid function in various ways. Such goitrogenic substances have been identified in the environment. For example, essential iodine uptake into the thyroid gland is competitively inhibited by the universally present environmental contaminant perchlorate and, to a lesser extent, by thiocyanates which are present in cigarette smoke and some vegetables. Furthermore, some studies have provided evidence that flavonoids may have adverse effects on the thyroid and reduce thyroid hormone biosynthesis, decrease iodide uptake and, finally, cause the development of goiter. A possible molecular explanation for this finding was that thyroperoxidase is inhibited by flavonoids such as myricetin, kaempferol, morin, quercetin and naringenin. The enzyme is required for the iodination of tyrosine in the thyroglobulin molecule, and catalyzes the oxidation of iodine in the presence of H_2O_2. The major components of soy, genistein and daidzein were also shown to inhibit thyroperoxidase activity. Indeed, some evidence has suggested that infants fed with soy-based milk developed goiter that regressed after changing to an iodine-supplemented diet. However, recent clinical studies have shown that in postmenopausal woman genistein has no effect on thyroid function, though it remains to be seen if the patient age and duration of isoflavonoid exposure are critical factors. Based on the results of these studies, the picture emerges that in humans (though possibly not in rodents) the effects of flavonoids on thyroid function are relatively weak, but may be more severe under conditions of an insufficient iodine supply in the food. Given the antioxidative effects of some flavonoids (see Section 5.1.4), this might also enhance a goitrogenic effect as the reaction with H_2O_2 would inactivate the enzyme.

Compared to endocrine signaling, very little information is available concerning the effects of natural substances on paracrine signaling molecules which act locally in tissues, in contrast to endocrine hormones which are distributed throughout the body via the vascular system. Paracrine hormones are essential during development, and mediate intercellular communication that is essential for cell proliferation, cell migration, and cell differentiation. Given the enormous importance of such molecules it is not surprising that the dysregulation of paracrine signaling results in malformations during development and may also play a role in cancer development during adult life.

Retinoic acid, one such paracrine signaling molecule, is chemically closely related to vitamin A (retinol) and is involved in numerous processes of cell differentiation, both in normal development and in cancer. For example, retinoic acid plays a role in the formation of limbs during development and the formation of somites along the anterior–posterior body axis. Similar to the described steroid hormone receptors, retinoic acid is bound by nuclear receptors that bind to retinoic acid response elements and act as transcription factors. The Hox genes, which play essential roles in the axial organization of the vertebrate body, are among the retinoic acid-controlled genes; indeed, the inhibition of their expression leads to abnormalities of the axial skeleton. Screening to identify retinoic acid agonists and antagonists has been carried out using a similar genetic reporter gene system as described for the identification of estrogenic substances. EDCs in the aquatic environment were screened and several chemical contaminants were identified that would act as either agonists or antagonists of retinoic acid signaling. At present, very little information is available on endocrine disruptors of retinoic acid from plants, although the history of the discovery of estrogenic substances was similar. EDCs in the aquatic environment were studied first, before natural substances from plants were screened on a large scale.

The retinoic acid receptor β2 is a tumor suppressor gene that is repressed in some types of cancer. Reactivation in cultured cancer cells has been achieved with curcumin and genistein, albeit at rather high concentrations (20 μM).

Unfortunately, such high serum concentrations cannot be achieved for therapeutic cancer treatment, when toxic side effects would be very likely. However, a combination of bioactive molecules appears to be a more promising strategy. For example, the proliferation of B16 melanoma cells was significantly reduced when EGCG and retinoic acid were used in combination at very low concentrations (0.5 and 0.1 μM, respectively), but the substances administered alone showed no significant effects. The proliferation of B16 cells could also be inhibited when these cells were implanted into mice.

A second example of paracrine signaling and its disturbance by a natural compound relates to an important intracellular signaling pathway which was originally discovered in *Drosophila* and termed hedgehog (*hh*), because of its characteristic larval phenotype. Homologous genes were later also found in vertebrates, and in humans the relevant gene is referred to as sonic hedgehog (*shh*); gene activity is required for limb development, the differentiation of nerve cells, and in facial morphogenesis. The *shh* protein is attached to a cholesterol molecule, and this lipid is essential for the biological activity of *shh* and its diffusion properties. In the tissues the secreted molecule may diffuse over distances of a few hundred micrometers, where it can bind to the *shh* receptor, the integral membrane protein patched (*ptc*). The principal molecular reactions involved in the signaling are illustrated in a simplified way (Figure 5.24). In the absence of the ligand, *ptc* inhibits smoothened (*smo*) which is another membrane protein. The transcription factor *Gli*, which initiates the transcription of hh-inducible genes is blocked by the association with the protein suppressor of fused (*SuFu*). When the hh ligand binds to *ptc*, the inhibition of *smo* is relieved, after which the transcription factor *Gli* is released through the activation of *SuFu*, is translocated to the nucleus, and initiates transcription.

The discovery of a natural inhibitor of hh signaling was made during the 1950s, when herds of sheep in the state of Idaho were born with craniofacial defects. It emerged that these malformations were due to grazing on the flower *Veratrum californicum* (California False Hellebore), and a chemical analysis of the active compound led to the discovery of the alkaloid cyclopamine (Figure 5.24). The name reflects a severe craniofacial defect with a single median (Cyclops) eye formation. The critical time to induce this abnormal development is an exposure early during pregnancy.

Poison and medicine are often two sides of the same coin, and this also holds true for cyclopamine. hh pathways play a role in several types of cancer and for this reason there is an intensive search for effective inhibitors. Some cancers, such as basal cell carcinoma, originate from mutations of the *ptc* gene, while others are promoted by excessive hh signaling either from the cancer cells themselves (autocrine) or by the paracrine secretion of hh in the vicinity of the tumor. Cyclopamin treatment provided promising results in the treatment of pancreatic cancer. Today, by means of synthetic chemistry, new inhibitors are being synthesized which retain the activity of cyclopamine but possess improved pharmacokinetic properties. The use of natural lead compounds for the development of pharmaceuticals will be further discussed in Section 6.2.2.

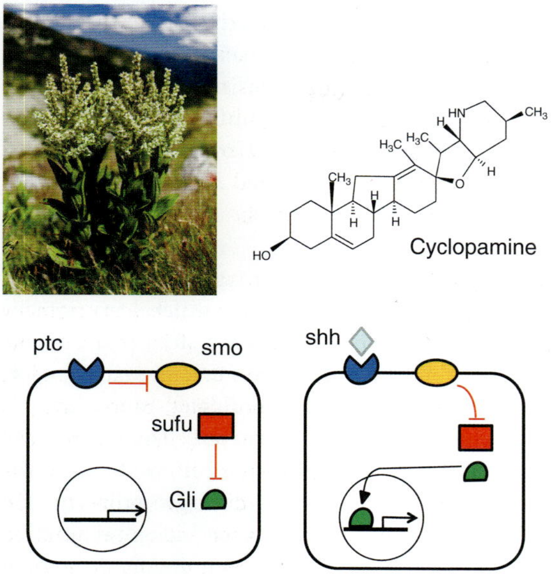

Figure 5.24 A natural inhibitor for a developmental signal causes malformations in vertebrates. An important local acting (paracrine) signal during development of vertebrates is the protein *sonic hedgehog* (shh). The two schematically drawn cells show the principle of the signal transduction pathway in the absence (left) and presence (right) of the shh ligand. When the shh protein with its attached cholesterol moiety binds to its receptor patched (ptch), the inhibition of the adapter protein smoothened (smo) is released. Smo can then regulate the protein *suppressor of fused* (sufu) which normally inhibits the transcription factor Gli. Through this interaction Gli is released from its inhibition by sufu and can act in the nucleus as transcription factor. The inhibition of shh during development leads to anomalies such as the formation of a median cyclops eye. A natural inhibitor is the alkaloid cyclopamine, which is a component of the false hellebore *Veratrum californicum*. Ewes feeding on that plant between the 10th and 15th weeks of gestation gave birth to malformed lambs. Cyclopamine inhibits smoothened and blocks the shh pathway. The photograph shows *Veratrum album* (© depositphotos/Xalanx), which is closely related to *Veratrum californicum*.

5.3 Interference with Neural Functions

5.3.1 Basic Features of the Nervous System and Synaptic Transmission

In the previous section it was seen that natural substances interfere with the hormonal signaling system, and can affect both endocrine and tissue-specific paracrine systems. The interference with neural cell function is a special case of paracrine signaling, and this subject will be covered here because of its medicinal and social importance. Some important biological reactions induced by natural substances of cannabis, opium and hallucinogenic mushrooms will be discussed, with attention focused not on the psychoactive effects and addiction but rather on the biological properties of the respective substances. Finally, the effects of natural substances on cholinergic signaling will be discussed. Their biology illustrates the Janus-faced nature of natural substances, in that they may be used as chemical weapons both in Nature and by humans, yet at the same time hold medical promise. In order to help understand the modes of action of these biologically active compounds, a brief description of the nervous system will first be provided.

The endocrine and nervous systems are two distinct signaling systems that control elementary body functions. The two systems have different properties and supplement each other perfectly; while the endocrine system reacts slowly and the physiological effects persist for some time, the nervous system reacts instantaneously. This velocity is a precondition for coordinated muscular contractions and the processing of acoustic or visual information. The specificities of the nervous system and the endocrine system have a different molecular basis. The endocrine signal must be interpreted in the target cells and the presence of receptors (and their regulators) determines the biological reaction to the endocrine signal, whereas in case of the nervous system the intercellular connections determine its specificity. The extremely complex pattern of neural connections is, in principle, established during development although there is some adaptive plasticity of neural signaling during the entire life.

Why has Nature invented so many strong poisons affecting the nervous system? Plants and animals alike have discovered the selective advantage that poisons may provide. The proteins α- and β-bungarotoxin of poisonous snakes (kraits) or the steroidal alkaloid batrachotoxin of poisonous frogs may serve as examples of bioactive molecules synthesized in animals to immobilize prey or to deter predators. From a biochemical point of view, such poisons may be considered as the animal equivalents to secondary plant metabolites. Many neural toxins are useful medicinal compounds, and they typically have a long history in human societies as hallucinogenic drugs or for the treatment of various diseases. Plant substances which affect neural activities have different chemical structures, and alkaloids and terpenoids are particularly prominent members of this class of substances; flavonoids, in contrast, are not represented.

The site of activity of the natural compounds chosen for this discussion is the neural synapse (Figure 5.25). The propagation of a neural action potential within the nervous system depends on transmission of the signal from the presynaptic neuron to the postsynaptic cell. Depolarization of the neuron involves a temporary massive influx of cations (mostly Na^+) into the cell through ion-selective channels in the membrane, formed by specialized proteins. The action potential also triggers an influx of Ca^{2+}-ions through gated Ca^{2+}-channels and induces the discharge of small signaling molecules known as neurotransmitters from the synaptic vesicles. The vesicles of the presynaptic neuron fuse with the cell membrane and release the stored neurotransmitter. These molecules bind to membrane

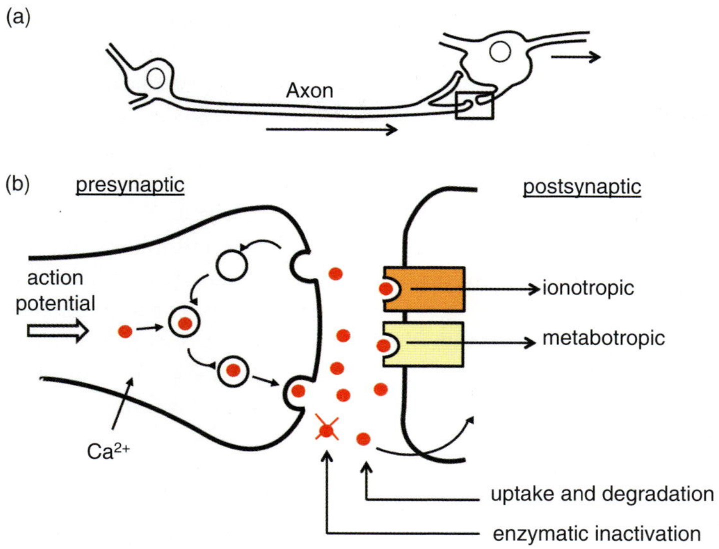

Figure 5.25 General features of a neural synapse. (a) Two neurons and their synapse are shown schematically. An action potential of the first neuron is transmitted along the axon, which branches out and makes contact with dendrites of the second neuron or contacts the cell body directly; (b) A synaptic contact is shown in close-up. At the terminal of the presynaptic cell the synaptic vesicles become loaded with the neurotransmitter (red dots). An action potential triggers the influx of Ca^{2+} into the cytoplasm, and this signal leads to fusion of the vesicles with the cell membrane and release of the neurotransmitter into the synaptic cleft. At the cell membrane of the postsynaptic cell the neurotransmitter acts as a ligand of specific membrane receptors and produces a biological response. Two types of membrane receptor may be distinguished: ionotropic receptors are specific ion channels which open in the presence of the ligand. In the case of a massive cation influx the cell will depolarize and an action potential is generated which is transmitted along the axon to the next neuron. If the ionotropic receptor is specific for anions the cell will hyperpolarize and the formation of an action potential is prevented. The second type of receptor is metabotropic and its activation involves complex signal transduction reactions which typically involve GTP-binding proteins (G proteins) which trigger a reaction cascade that modulates the response to the synaptic input. After release of the neurotransmitter and its interaction with the receptor, the transmitter substance must be removed quickly from the synaptic cleft so that the synapse is functional again.

receptors of the postsynaptic cell and trigger a biological response. Two types of signaling pathways can be distinguished:

- *Ionotropic receptors* open cation channels and, if a certain depolarization threshold is reached, a new action potential is triggered and a wave of depolarization travels along the axon of the neuron to the next synapse.
- *Metabotropic receptors* belong to the large family of G-protein-coupled receptors. Their effects are much slower (between seconds and hours) and they affect the membrane potential of the postsynaptic neuron indirectly through specific signal transduction pathways.

Numerous transmitters have been identified. Typically, they are simple low-molecular-weight substances, the enzymatic synthesis of which in the presynaptic cell is highly efficient and tightly controlled. The chemical structures of the transmitters are quite different and include, for example, peptides (e.g., encephalin or endomorphines), biogenic amines (e.g., acetylcholine, dopamine, and serotonin) and amino acids (e.g., glycine, glutamic acid, and γ-aminobutyric acid, abbreviated GABA). Important neurotransmitters are listed in Table 5.3, and in many cases natural substances are known to interfere with the signaling pathways. Some neurotransmitters, such as GABA and glycine, do not lead to depolarization of the postsynaptic cell but rather to a hyperpolarization, and in this way the generation of an action potential is inhibited. These inhibitory synapses are common in the central nervous system (CNS), where the inhibition is brought about by an influx of anions (in particular Cl^-) through special ion channels, and by the export of cations out of the neuron. Both mechanisms counteract a cellular depolarization and prevent the generation of an action potential.

A typical neuron has numerous inputs from axons of other neurons which form synapses at the dendrites of the neuron or at its cell body (two such synapses are shown in Figure 5.25a). The synapses can be either inhibitory or excitatory; the neuron integrates all synaptic inputs and, if a critical depolarization potential is reached, the neuron will trigger an action potential. If inhibitory inputs prevail, the reaction is suppressed, and in this way each neuron can integrate information from different sources and act as a processing unit. The electrical signal (if elicited) is propagated along the single axon. A special case here is the neuromuscular synapse that is formed between a neuron and a muscle cell, and where receipt of the action potential initiates a muscular contraction.

The biological processes described in synaptic signaling provide effective targets for natural substances. The synthesis or release of a synapse-specific neurotransmitter into the synaptic cleft can be inhibited at the presynaptic side. Alternatively, synaptic signaling can be inhibited at the postsynaptic cell by blocking neurotransmitter–receptor binding, or by inhibiting the signal transduction. The intensity and duration of synaptic signaling depends on the concentration of the transmitter in the synaptic cleft. The rapid removal of the transmitter or its enzymatic degradation is essential to maintain the function of the synapse (Figure 5.25). The involved enzymes also serve as targets for natural substances.

The CNS, comprising the brain and spinal cord, collects and processes information from all parts of the body. The afferent neural input is of very different quality and involves, for example, information concerning muscle posture, the perception of pain, and information generated by the sense organs (Figure 5.26). After processing all of the incoming information in the CNS, the neural efferent output controls voluntary and nonvoluntary activities: the somatic motor pathways innervate the somatic voluntary muscles, while the autonomic nervous system innervates organs without conscious (i.e., involuntary) control. Two different neural pathways – the sympathetic nervous system and the parasympathetic nervous system – control the physiological activity of all organs in the body. The sympathetic system typically activates functions that help to cope with stressful situations, such as increased heart rate and blood pressure, and dilation of the pupil in the eye. The

Table 5.3 List of selected neurotransmitters with relevance to the neural functions discussed in the text.

Major neurotransmitters and natural substances affecting synaptic activity				
Neurotransmitter	**Receptor**	**Distribution**	**Synaptic action**	**Natural substances affecting synaptic activity**
Amino acids				
Glycine	Glycine receptor (ionotropic)	CNS, brain stem, spinal cord, cerebellum	Inhibition	Strychnine
Aspartate	Aspartate receptor (ionotropic)	CNS	Excitation	
Glutamate	AMPA (ionotropic)	CNS, spinal ganglia	Excitation	
	NMDA (ionotropic)	CNS	Excitation	
	mGlu1-8 (metabotropic)	CNS	Excitation or inhibition	
GABA	GABA-A (ionotropic)	CNS	Inhibition	Muscimol
	GABA-B (metabotropic)	CNS	Inhibition	
Catecholamine				
Dopamine	D1 – D5 (metabotropic)	CNS, basal ganglia	Excitation or inhibition	
Noradrenaline (=norepinephrine)	α- or β-adrenergic (metabotropic)	CNS and peripheral NS, sympathetic ganglia, heart muscle, smooth muscles	Excitation or inhibition	Ephedrine
Adrenaline	α- or β-adrenergic (metabotropic)	CNS, heart muscle, smooth muscles	Excitation or inhibition	

(continued)

Table 5.3 *(Continued)*

Neurotransmitter	Receptor	Distribution	Synaptic action	Natural substances affecting synaptic activity
Peptide				
Opioids		CNS	Inhibition	Morphine and other opioids
Endorphin	μ (MOR, metabotropic)			
Endomorphins	μ (MOR, metabotropic)			
Morphiceptins	μ (MOR, metabotropic)			
Encephalins	δ (DOR, metabotropic)			
Dynorphins	κ (KOR, metabotropic)			
Nociceptin	NOP/ORL (metabotropic)			
Neuropeptide P	NK1 (metabotropic)	CNS, limbic system, peripheral pain perception	Excitation or inhibition	
Others				
Serotonin (5-HT)	5-HT_{1A} (metabotropic) 5-HT_2 (metabotropic)	CNS; 5-HT synthesizing neurons in the raphe nuclei	Excitation or inhibition	Psilocybin, psilocin, mescaline, ergotamine (5-HT_1)
	5-HT_3 (ionotropic)	CNS, peripheral NS	Excitation	
Acetylcholine	nAChR, (ionotropic)	CNS, skeletal muscles	Excitation	Nicotine, tubocurarine, AChE inhibitors
	mAChR (metabotropic)	CNS, parasympathicus: Smooth muscles, heart muscle, endocrine and exocrine glands	Excitation or inhibition	Pilocarpine, atropine, AChE inhibitors
Adenosine	Purin receptor (metabotropic)	CNS	Inhibition	
Morphine	μ_3 or M3R (metabotropic)	CNS, peripheral NS	Excitation or inhibition	Morphine
Cannabinoids AEA 2-AG	CB_1 (metabotropic) CB_2 (metabotropic) GP55 (metabotropic)	CNS, peripheral NS	Inhibition	Δ^9-THC and other cannabinoids

CNS: Central nervous system; NS: Nervous system; AEA: Arachidonylethanolamine, 2-AG: 2-Arachidonylglycerol. See text for further abbreviations.

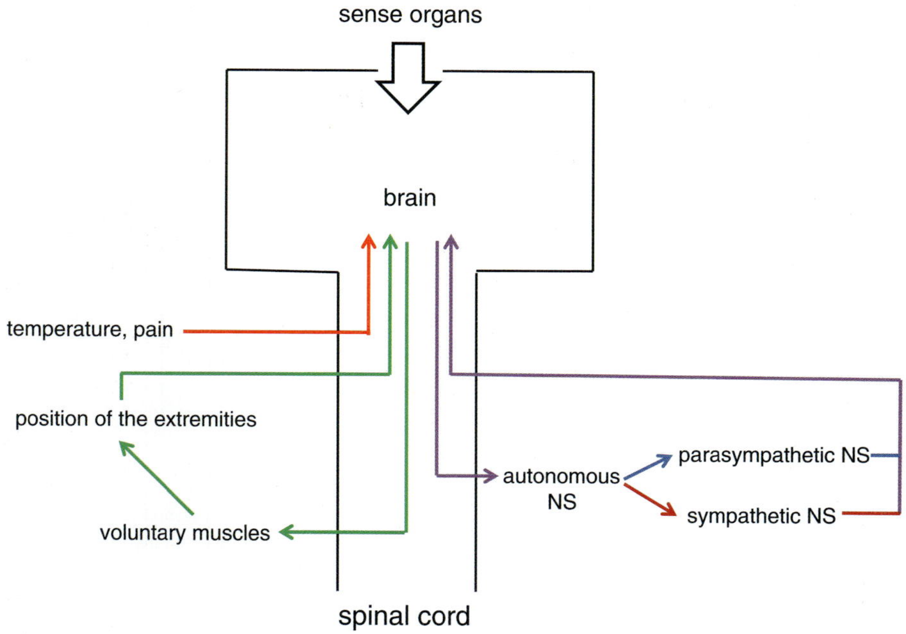

Figure 5.26 Major ascending and descending neural pathways. Important afferent (ascending) and efferent (descending) neural pathways are shown which are under the control of the brain. Both, the autonomous and the voluntary nervous system are regulated by feed-back loops. This also holds true for the pain pathways (see Figure 5.35 for details). The activity of the pathways shown can be affected by natural compounds.

parasympathetic system is important for maintenance functions, and controls physiological processes such as digestion (by increasing peristaltic movements of the gut) and the stimulation of secretions (e.g., tears, mucus). The parasympathetic nerves also stimulate reproductive functions.

The neurotransmitter acetylcholine (ACh) plays an important role in the control of both voluntary and involuntary processes (Figure 5.27). The voluntary contraction of skeletal muscles that is initiated at the neuromuscular synapse is controlled by ACh. The neurotransmitter is also an important signaling molecule in the autonomic nervous system; in particular, the parasympathetic system is entirely cholinergic and the neurons form synapses in ganglia which are located close to the innervated organs. In contrast, the neurons of the sympathetic system are relayed in ganglia that are close to the CNS (Figure 5.27), and only the preganglionic nerves are cholinergic while the postganglionic fibers are noradrenergic (the transmitter used is noradrenaline). The postsynaptic receptors of the sympathetic nerves can also be activated by the hormone adrenaline, which is released from the adrenal medulla upon cholinergic stimulation under stress conditions. Two types of ACh receptor can be distinguished in the postsynaptic membrane: (i) muscarinic receptors (mAChR); and (ii) nicotinic receptors (nAChR; see Section 5.3.6). The afferent fibers of the autonomic system feed-back the information concerning the activity status of the respective organ to the CNS.

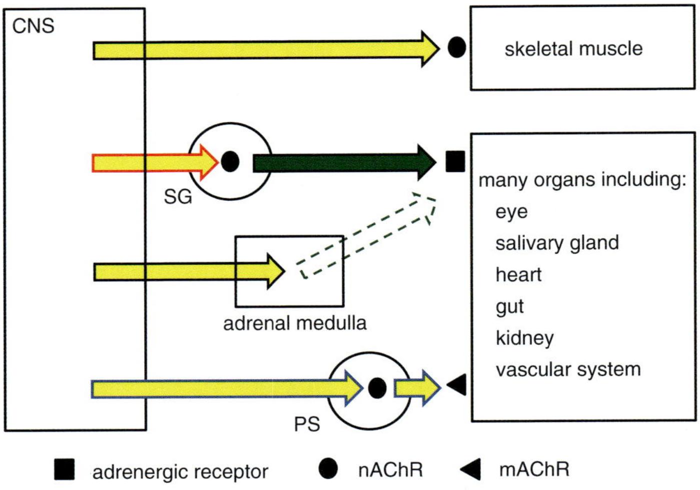

Figure 5.27 Cholinergic signaling is essential for the activity of the voluntary and autonomous nervous systems. Acetylcholine (ACh) is the transmitter of cholinergic neurons and their synapses (yellow arrows). Muscular contractions are initiated by cholinergic signaling at the neuromuscular synapse. Cholinergic nerves of the sympathetic system (red lining of the yellow arrows) form synapses at a ganglion (SG) with postganglionic fibers which are noradrenergic (transmitter noradrenaline, green arrows). The adrenergic receptors can also be stimulated with the hormone adrenaline (dashed green arrow) which is released from the adrenal medulla upon cholinergic stimulation. The parasympathetic system (blue lining) is entirely cholinergic, and also in this case synapses are present in a ganglion which is typically located in the vicinity of the innervated organs. ACh can be bound by two different receptors at the postsynaptic membrane, the nicotinic receptor (nAChR) or the muscarinic receptor (mAChR) in other neurons. The presence of the different receptors is indicated by different symbols (dot or triangle).

The complex neural networks of the CNS enable the performance of intellectual achievements that are the hallmark of human abilities. Indeed, the enormous enlargement of the telencephalon (Figure 5.28) during the course of vertebrate evolution represents the morphological manifestation of these gained abilities. The telencephalon has overgrown the older parts of the vertebrate brain during evolution, so that vital functions have become located in the central regions of the brain. Of particular interest with regards to the following discussions of hallucinogenic drugs is the limbic system; this term designates several brain structures such as the hippocampus, the amygdala, the fornix, and the limbic cortex, which are involved in the generation of emotions, specific behavioral reactions, and long-term memory (Figure 5.28).

5.3.2 Bioactivity of Cannabinoids

The plant *Cannabis indica* or *Cannabis sativa* (Figure 5.29) has been cultured for at least 7000 years. The plant has been known in traditional Chinese medicine and was mentioned in texts of early herbal medicine by Greek authors more than 2000 years ago. In particular, the plant's analgesic, anti-inflammatory and anticonvulsive

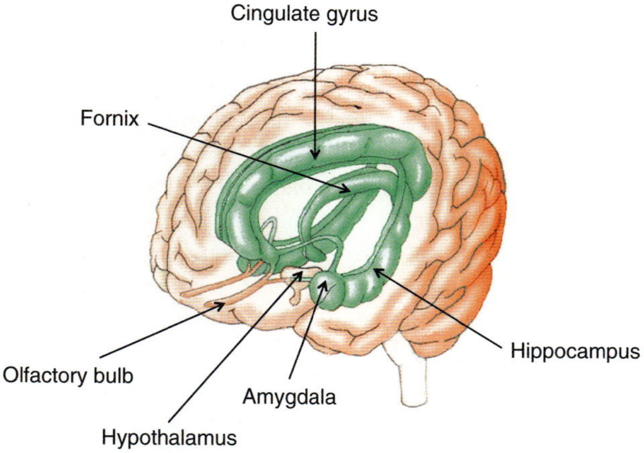

Figure 5.28 The human "emotional brain." The human brain is characterized by an enormous enlargement of the telencephalon (shaded brown) compared to vertebrate ancestors. As a consequence, the phylogenetically older parts become displaced to central areas of the human brain. The limbic system (green) encompasses different structures that are involved in the control of emotion, memory, and behavior.

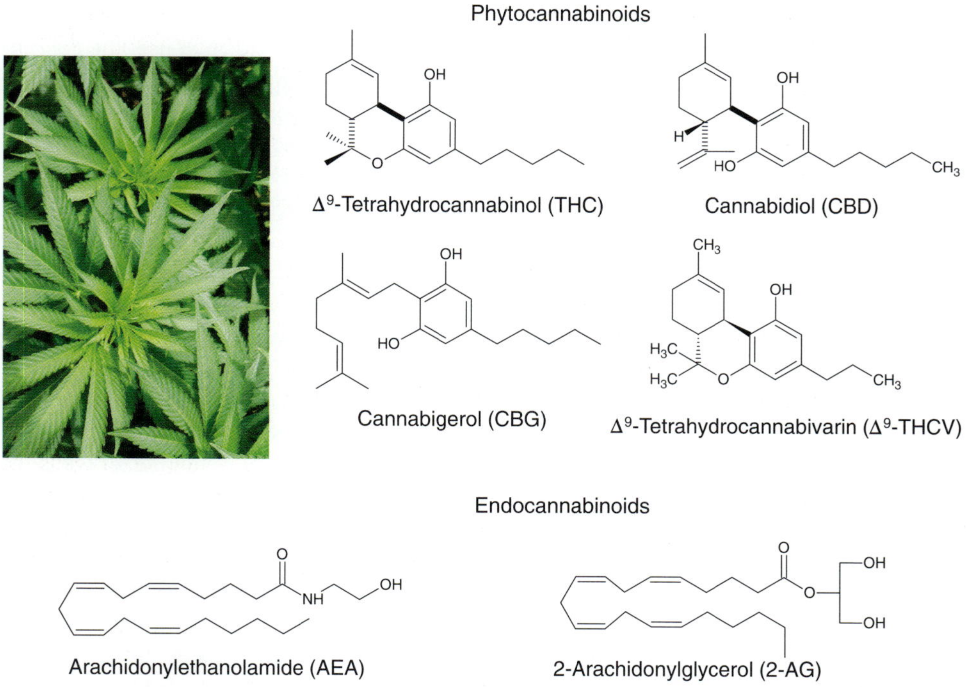

Figure 5.29 Examples of biologically active cannabinoids. A number of chemically related substances with different biological properties have been isolated from the plant *Cannabis sativa*. The structures of some representative phytocannabinoids are shown. Δ^9-THC is the main psychoactive compound. Note that the endocannabinoids are structurally distinct. Photograph © depositphotos/Stéphane Bidouze.

properties were appreciated in medical applications. The plant has also been used worldwide for the production of textiles, paper, ropes, and fuel. As the plant grows rapidly, the gain in biomass is enormous and favors such commercial applications. Different varieties of *Cannabis* are cultured, depending on their use: for industrial applications, long stems with little branching are favored, whereas varieties grown for hemp seed oil are rich in proteins and fatty acids. Plants used for medicinal and recreational purposes are enriched in bioactive cannabinoids. The substance Δ^9-tetrahydrocannabinol (Δ^9-THC; Figure 5.29) is particularly interesting as this substance has psychoactive properties. Between 1919 and 1930 it was common practice in the USA to smoke cannabis, but thereafter cannabis was banned, first in the USA and later in many other countries. It is only recently that some beneficial properties of Δ^9-THC and related compounds have been rediscovered and pharmaceuticals based on these compounds marketed for some medical indications.

The medicinal variety of *Cannabis sativa* produces more than 100 phytocannabinoids with similar chemical structures but different biological properties. Apart from Δ^9-THC, the cannabinoids cannabidiol (CBD), cannabigerol (CBG), and Δ^9-tetrahydrocannabivarin (Δ^9-THCV) represent major species of the cannabinoids present in the plant (Figure 5.29). Phytocannabinoids are hydrophobic molecules synthesized from fatty acid precursor molecules, and become highly concentrated in the resin secreted by trichomes which are abundant on female plants. When the enrichment of a particular component is desirable for medicinal use, the concentrations of the different components can be altered by applying genetic techniques.

When Δ^9-THC was identified as the psychoactive component of cannabis, two receptors for the compound were discovered, namely CB_1 and CB_2. The discovery of the endogenous ligands arachidonylethanolamine (AEA; also known as anandamide) and 2-arachidonylglycerol (2-AG; Figure 5.29) was a major achievement that allowed the receptor functions to be determined experimentally. CB_1 and CB_2 are transmembrane receptors of the rhodopsin G-protein-coupled receptor (GPCR) family. The CB_1 receptors are localized in the presynaptic terminals of synapses in the cerebral cortex, hippocampus and cerebellum; they are also present in the basal ganglia which gather and process cognitive and limbic information and also receive information from the motor system. CB_2 receptors are expressed mainly in non-neural (glial) cells of the CNS.

Endocannabinoid signaling has been identified as a key regulator of synaptic communication throughout the CNS. The CB_1 receptor is the most abundant GPCR in the brain and is present in many different types of synapse. The activity provides a synaptic feedback system that dampens excess presynaptic activity. Synaptic signaling by endocannabinoids is retrograde from the postsynaptic to the presynaptic neuron, and hence opposite to the normal anterograde signaling (Figure 5.25). Another difference is the absence of synaptic vesicles to store the neurotransmitters. Both, 2-AG and AEA (which are hydrophobic) are stored in the postsynaptic cell membrane in the form of inactive precursors that become enzymatically activated by phospholipases and released into the synaptic cleft upon regulatory stimuli. 2-AG and AEA each bind to the presynaptic CB_1 receptor and activate the GPCR. The G-protein regulates adenylate cyclase activity as well as the Ca^{2+} and K^+ ion channels.

The principal reaction cascade is illustrated by a glutamatergic synapse (Figure 5.30) which is common in the CNS. At the presynaptic end an action

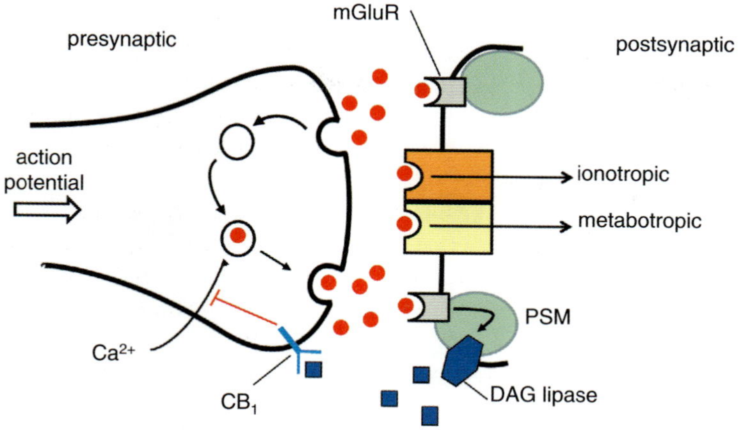

Figure 5.30 Cannabinoid signaling at the glutamatergic synapse. The activity of a glutamatergic synapse is modulated by endocannabinoid signaling. In this example of an excitatory synapse the transmitter glutamate is bound by ionotropic or metabotropic receptors and depolarization of the postsynaptic neuron is triggered. If excess glutamate is released into the synaptic cleft, the neurotransmitter will diffuse to a specialized area in lateral positions, the perisynaptic signaling machinery (PSM), where a specific glutamate receptor becomes activated. This triggers the enzymatically controlled release of the endocannabinoid 2-AG from the postsynaptic membrane. 2-AG acts retrogradely (from the postsynaptic to the presynaptic side) and through activation of the CB_1 receptor the Ca^{2+} influx in the presynaptic neuron is inhibited, and hence also release of the glutamate transmitter into the synaptic cleft.

potential triggers the influx of Ca^{2+} through voltage-gated channels, and this leads to the exocytosis of synaptic vesicles and release of glutamate into the synaptic cleft. At the postsynaptic side, the ligands bind to the glutamate receptors (AMPA or NMDA) and trigger a depolarization of the cell. These receptors can be either ionotropic or metabotropic. If excessive amounts of glutamate reach the peripheral area of the synapse, the perisynaptic signaling machinery (PSM) will be activated. The glutamate spill-over activates type 1 metabotropic glutamate receptors (mGluRs) and 2-AG is released from the postsynaptic membrane. The cannabinoid travels retrogradely and activates the presynaptic CB_1 receptor which, in turn, inhibits the voltage-gated Ca^{2+} channel and blocks the release of glutamate.

This synaptic regulation by endocannabinoids is not limited to excitatory glutamatergic neurons; indeed, other types of synapse can also be controlled in the same manner. The basal ganglia in the brain have important functions in motor control, and a network of neurons has been identified which forms part of the central processing circuits (Figure 5.31). Apart from the excitatory signaling by glutamate, the inhibitory GABAergic synapses and dopaminergic neurons with regulatory functions (not shown in Figure 5.31) form this network. The synapses in this network possess either CB_1 receptors or $TRPV_1$ receptors, which are also known as vanilloid receptors. This cation channel is activated by different ligands including the endocannabinoid AEA and *N*-arachidonyl dopamine. The receptor is

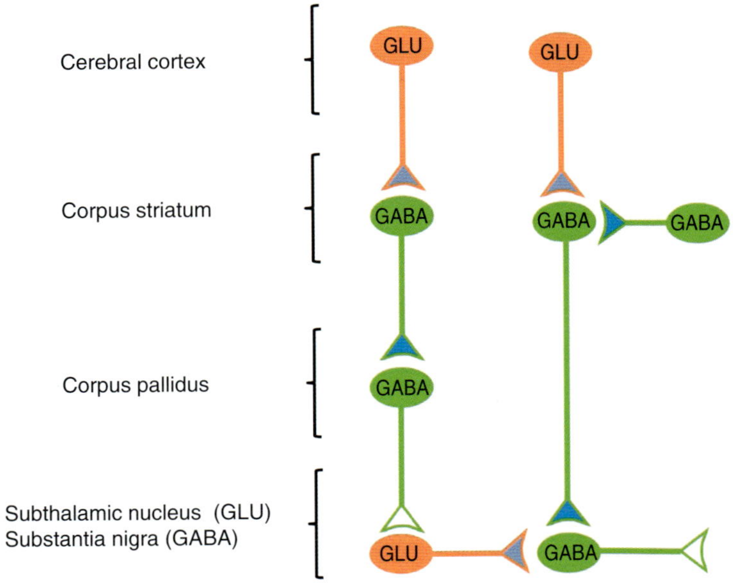

Figure 5.31 Example of a neural network whose activity is modulated by endocannabinoid signaling. The neural network shown in the basal ganglia of the brain consists of excitatory (glutamatergic) neurons and inhibitory (GABAergic) neurons. Most of the neurons express presynaptic CB_1 receptors (blue). In this simplified network, which spans different brain regions, modulation of the activity by endocannabinoids includes excitatory as well as inhibitory neurons. Based on Fernández-Ruiz, J. (2009) *Br. J. Pharmacol.*, **156**, 1029–1040.

present in nociceptive pathways in the peripheral nervous system as well as in the CNS (Figure 5.31). The $TRPV_1$ receptor is also activated by natural compounds with a pungent taste. such as capsaicin (in hot chili pepper) and allyl isothiocyanate (in mustard and wasabi). This example of a neural circuit illustrates the complexity of the control that involves stimulatory and inhibitory synapses with different neurotransmitters, and the modulation of their activity by endocannabinoids.

5.3.2.1 **Effects of Phytocannabinoids on Neural Functions**

The negative psychoactive effects of phytocannabinoids limit the medical application of these compounds, or of synthetic cannabinoids. The symptoms of cannabinoid intoxication may vary depending on the source and chemical composition of the preparation, and on the physiological state of the affected individual. One of the most serious side effects is an impairment of the ability to transiently hold and process the information that is essential for reasoning, learning, and active thinking. The molecular mechanisms of cannabinoids on this "working memory" have been elucidated recently in rats. The CB_1 receptor plays a major role in this process as the impairment of working memory by cannabinoids (or marijuana) is abolished in CB_1 receptor mutants. The cannabinoid ligands act via non-neural glial cells which possess the receptor and modulate the signaling of closely associated glutamatergic neurons. The signaling network of the working memory is located in the hippocampus.

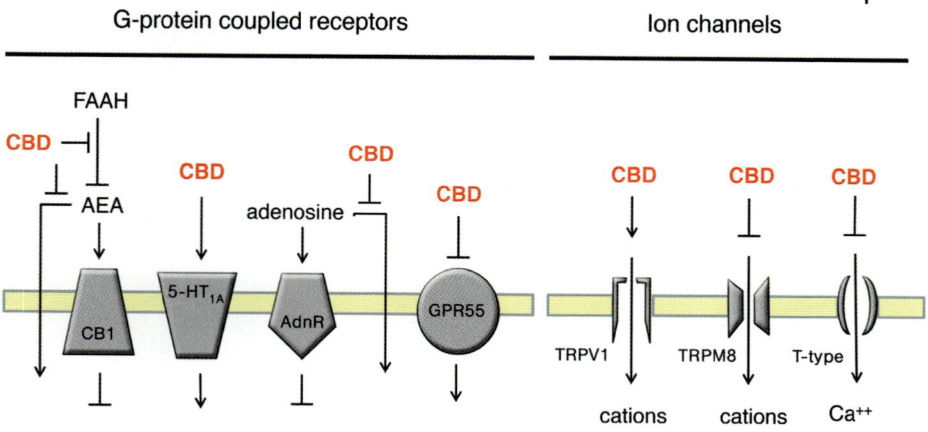

Figure 5.32 CBD interacts with several neural membrane receptors. The phytocannabinoid CBD interacts with several membrane-associated receptors and modulates different signal transduction pathways in the CNS. The figure summarizes experimental results using different mammalian test systems. CBD inhibits the degradation of AEA by the enzyme FAAH and, in addition, inhibits the uptake of AEA thus increasing the tissue level of the endocannabinoid. CBD acts as agonist at the human serotonin receptor 5-HT$_{1A}$. Uptake of the neurotransmitter adenosine is inhibited by CBD. The effect of cannabinoids acting as agonists at the GRP55 receptor is antagonized by CBD. The cannabinoid also affects the activity of various cation channels, including the vanilloid receptor (TRPV1). Stimulatory or inhibitory effects were reported using different test systems. The interactions of CBD with the shown target proteins are not specific for this substance. Other cannabinoids may react with different sets of target proteins with substance-specific affinity.

The molecular reactions of Δ^9-THC and other phytocannabinoids which result in observed neural effects have been analyzed, and several target proteins have been identified. The CB$_1$ receptor has attracted attention as Δ^9-THC has a high affinity to this receptor and acts as a (partial) agonist; in contrast, Δ^9-THCV has antagonistic activity, at least at low concentrations. Interestingly, CBD has a minimal affinity to the CB$_1$ receptor and there is also no psychoactive effect. The biological functions of phytocannabinoids cannot be described by an interference of the CB$_1$ signaling pathway alone; rather, the list of target proteins is long and complex interactions of different transduction pathways must be taken into account. The problem is exemplified by the phytocannabinoid CBD, which has been studied intensively because of its pharmaceutical promise.

Several membrane receptors have been identified among the CBD target proteins (Figure 5.32). CBD acts as an agonist for the 5-hydroxytryptamine receptor subtype 1A (5-HT$_{1A}$), which belongs to the family of serotonin receptors. Adenosine, a ubiquitous molecule in the CNS, activates adenosine receptors including the A$_{2A}$ receptor which is an important regulator of inflammation. CBD inhibits adenosine uptake, thus prolonging adenosine signaling at the receptor. The excitatory GPR55 receptor is also affected by CBD, which in this case acts as an antagonist. This receptor has little similarity to CB$_1$ and CB$_2$ receptors, but is also stimulated by Δ^9-THC and other cannabinoids. Although little is known of its function, the receptor is common in the CNS; the endogenous ligand appears to be the phospholipid lysophosphatidylinositol, which seems to induce the release of Ca^{2+} from intracellular stores. CBD also affects important cation channels and can act as an agonist or antagonist. The biological effects of the interaction are not known in detail but, interestingly, the vanilloid receptor (TRPV1) is at least in part responsible for the analgesic effect of CBD (see below). CBD has also been shown to lower the intracellular Ca^{2+} concentration by blocking intramitochondrial Ca^{2+} stores, thus reducing hyperexcitability. The effects of CBD on cation levels are to modulate neuron excitability, while the different target molecules allow a subtle and sophisticated regulation of excitability in specific neuron populations.

The list of targets is not yet complete. CBD is also known to interact with glycine receptors as well as μ- and δ-opioid receptors (see Section 5.3.3). Receptor–ligand interactions are not the only way in which neural signaling can be affected, as the enzymes involved in the synthesis or degradation of endocannabinoids may also be targets of phytocannabinoids. Such indirect effects mirror the situation described for phytoestrogens, which inhibit hormone synthesis very effectively (see Section 5.2.2). Several phytocannabinoids (except for Δ^9-THC) were found to inhibit the activity of enzymes involved in the synthesis and metabolism of endocannabinoids, so that the control of neural activity is strongly affected. Inhibitory activities were reported for the 2-AG-synthesizing enzyme diacylglycerol lipase α (DAGLα) and for the catabolic enzymes fatty acid amide hydrolase (FAAH) and monoacyl glycerol lipase (MAGL) which hydrolyze AEA and 2-AG, respectively. Other cannabinoids, including CBD, inhibit the uptake of AEA from the synapse and hence also modulate endocannabinoid activity (Figure 5.32). Several enzymes which play important roles in inflammation, such as phospholipase A2 and 15-lipoxygenase, are also affected by CBD. Presumably, these and other target proteins contribute to the biological effects of CBD which have been described as analgesic, anti-inflammatory, neuroprotective, immunosuppressive, and antioxidative.

Each cannabinoid has a specific set of target molecules. However, the comparatively long list of CBD targets is likely to reflect the intensity of the research efforts rather than the biological properties of the different phytocannabinoids. These data show that the molecular interactions are highly complex and affect not only the receptor and its signaling activity but also the synthesis and degradation of cannabinoids. The property of natural substances to interact with numerous target proteins is a recurrent theme and must always be considered in pharmaceutical applications.

5.3.2.2 **Medical Applications of Phytoestrogens**

Cannabis has been used in traditional medicine to treat epilepsy and other hyperexcitability disorders. Since 1970, therapeutic applications of cannabinoids for the treatment of seizures has been subject to intensive research using animal models, with additional information derived from case studies and clinical studies. Unfortunately, the inconsistent results of these studies and the well-known psychoactive properties of Δ^9-THC have limited its use. A promising pharmaceutical strategy is to use cannabinoid mixtures and to search for other bioactive cannabinoid substances that have therapeutic potential for different neural disorders. With respect to the control of hyperexcitability symptoms, Δ^9-THCV has been identified as a promising substance as this cannabinoid acts as a CB_1 receptor antagonist of inhibitory GABAergic transmission in cerebellar neurons. Furthermore, Δ^9-THCV was reported to dampen the excitatory activity of glutaminergic NMDA receptors in brain slices. The observed anticonvulsant property of Δ^9-THCV may be due to this inhibitory effect. This contrasts with the agonistic activity of Δ^9-THC at CB_1 receptors of cerebrellar Purkinje cells, and this effect may explain the induced deficits in motor coordination, which is a well-known side effect of cannabis intoxication. Since CB_1 receptors are common in the brain, and may affect inhibitory as well as excitatory synapses, it remains to be shown which types of neuron are affected preferentially. Similarly impressive as the number of target proteins is the range of indications for which cannabinoid treatment has been considered promising. CBD, for example, has been tested for the treatment of psychotic diseases, epilepsy, anxiety, sleep disorders, neurodegenerative diseases, cerebral and myocardial ischemia, inflammation, pain, cancer and type-1 diabetes. Particularly promising is the anticonvulsant activity of CBD, which has led to it being extensively tested as an antiepileptic drug. As CBD has no side effects on motor activity and lacks psychoactive properties, it appears to be a promising drug candidate to reduce neural hyperexcitability, despite the complex interaction with numerous receptors.

Cannabinoids have also shown promise to treat neurodegenerative disorders such as Alzheimer's disease, Parkinson's disease and multiple sclerosis (MS). The use of cannabis in the treatment of various neural disorders has a long history, and its muscle-relaxant properties help to relieve the symptoms of MS; phytocannabinoids also protect neurons and reduce inflammatory reactions in this autoimmune disease. Sativex is an approved pharmaceutical for the treatment of MS and consists of a mixture of Δ^9-THC and CBD.

Although these diseases have very different etiologies, they are all characterized by neurodegenerative processes and an inflammatory response. The enzymes acetylcholinesterase (AChE) and iNOS are both inhibited by cannabinoids, and both also play a role in mediating the symptoms of Alzheimer's disease. AChE inhibition is an acknowledged therapy to treat this disease and, therefore, competitive inhibitors such as Δ^9-THC have therapeutic potential. The enzyme iNOS has already been discussed in the context of oxidative stress and ROS generation. As neurodegenerative diseases are typically associated with inflammatory reactions and ROS production, it is not surprising that CBD, which inhibits iNOS activity, may help to attenuate the symptoms of Alzheimer's disease. Beneficial effects of phytocannabinoids such as Δ^9-THC, CBD and Δ^9-THCV have also been reported for patients with Parkinson's disease. The molecular reaction pathways that could explain the observed effects in these neurodegenerative diseases are not known in detail, however.

The therapeutic use of cannabinoids is not limited to signaling in the CNS. Cannabinoids play important roles in the control of the immune system and in the physiology of the skin; the cannabinoid system also appears to have important regulatory functions in non-neural cells. Both, CB_1 and CB_2 receptors are, for example, present on B and T cells, macrophages and neutrophils of the immune system, and on keratinocytes of the epidermis. In a mouse model cannabinoids attenuated an induced contact dermatitis and opened up new ways for the treatment of this common disease. Animals lacking CB_1 and CB_2 receptors displayed severe symptoms, while mice with elevated AEA levels (in FAAH knockout mice) showed a great reduction in allergic symptoms. In keeping with these data, treatment with Δ^9-THC and a CB_1 agonist caused a significant reduction in the allergic reaction. It is the balance between beneficial effects and undesirable side effects that determines the usefulness of the respective drugs in the treatment of diseases. To date, cannabinoids have provided many promising lead compounds for pharmacological developments to treat a wide range of diseases.

5.3.3
Discovery and Bioactivity of Opioids

Opium is the milky fluid harvested by cutting the green capsules (seed pods) of the poppy *Papaver somniferum*. The dried juice is known as raw opium and about 20–50 mg of this material can be obtained from one capsule. Morphine is the most prominent substance in raw opium (Figure 5.33), but many other alkaloids with interesting biological properties are also present. Codeine (3-*O*-methylmorphine), which is used as a remedy for cough, is a prominent example of these compounds (Section 4.1.2). The term "opioids" is used in the following sections to designate all substances that produce morphine-like effects by interacting with the opioid receptor, whereas the different alkaloids present in the poppy are referred to as "opiates." Morphine is the most potent herbal remedy against pain.

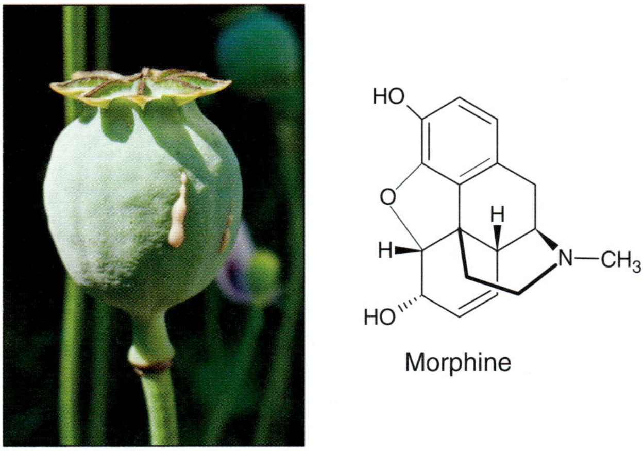

Figure 5.33 Morphine is a major alkaloid of the opium poppy. Here, the seed capsule of the opium poppy (*Papaver somniferum*) has been cut open. The milky exudation containing active substances such as morphine and codeine is collected and dried. Photograph © depositphotos/Andrey Koturanov.

Figure 5.34 The "Goddess of Poppies" symbolizes the long cultural history of Opium usage. The terracotta figure is exhibited in the Archeological Museum of Heraklion, Crete. This exhibit from Knossos (about 1400 BC) with the cut seed capsule (inset) documents the early usage of opioids in the Minoan civilization.

Poppy extracts have been used for thousands of years in different cultures as an analgesic, and presumably also for their psychoactive effects. A terracotta figure from Crete ("Goddess of Poppies") gives witness to the early use of opium in the Minoan civilization (Figure 5.34). Because of their interesting medical properties, opioids have been synthesized chemically. As early as 1898 the substance diacetylmorphine was introduced as a remedy against pain and cough and marketed under the name "heroin."

Phytocannabinoids and opiates have a similar history with respect to their early cultural usage. In both cases, the endogenous ligands of the respective receptors were discovered much later. In the case of opioid receptors, the ligands acting as neurotransmitters are peptides and the endorphins (endogenous morphine) are prominent members of the ligand family (Table 5.3). The active peptides are part of a much longer precursor protein which is cleaved proteolytically to produce, for example, the pentapeptides metencephalin and leuencephalin, which differ only in the one eponymous amino acid. The length of the peptide determines the half-life of the molecule; whereas the pentapeptides have a half-life of only 60 s, that of the longer endorphins is 15–20 min.

Different types of receptors can be distinguished with different ligand specificities. The best-studied receptor is the μ-receptor (MOR), although the δ (DOR) and κ (KOR) are physiologically important receptors. In addition, the nociceptin-receptor was discovered by different groups independently and abbreviated as either NOP or ORL (Table 5.1). Different subtypes of the receptors are likely to exist, but have not yet been characterized in detail. All receptors belong to the large family of GPCRs. Opioid receptors are common in the brain but are also found in all major organs of mammals. The wide distribution of the receptors is reflected by the numerous medical indications in which the opioid system plays a role. Opioid receptors are, therefore, interesting target proteins and are considered druggable targets for pharmacological developments. The relief of pain (analgesia) is the best known and medically exploited property of endogenous and natural opioids (see below for details). Endogenous opioids are also involved in the regulation of the immune system, in the control of obesity and, for example, in respiratory and cardiovascular control. In the brain, opioids regulate emotional responses and are involved in epileptic seizures. Opioids also play a role in male fertility.

Surprisingly, morphine also appears in the list of endogenous ligands (Table 5.3). The synthesis of this substance using dopamine as precursor molecule has been documented in nervous tissue and in white blood cells. Present evidence indicates that a splice variant of MOR (named μ3 or M3R) is activated by morphine but not by any endogenous opioid peptide. Interestingly, M3R is coupled to NOS and the production of NO is thought to be an important mediator of the biological effects that are induced by the receptor activation. Whether this observation is specific to morphine receptors is unclear, but it does not seem to be the case as NO was also found to play a role in DOR- and MOR-mediated signaling, and this also holds true for the cannabinoid receptor CB_{2R}. All of these receptors are implicated in the perception of chronic (neuropathic) pain. The inhibition of NOS by suitable inhibitors has an analgesic effect, thus confirming the role of NO in pain perception. The term "gasotransmitter" was coined to point out the importance of NO in neural signaling. These recent findings show the intimate connection between the signaling pathways. The role of NO has been discussed previously in the context of signaling by ROS and the effects of natural and endogenous antioxidants. The activity of evolutionarily conserved signaling pathways in different physiological and developmental contexts is a recurrent theme. Unfortunately, such complexity hampers the molecular analysis of opioid-mediated effects *in vivo*.

5.3.4
Analgesic Properties of Opioids and Cannabinoids

The analgesic properties of natural substances such as cannabinoids and morphine are discussed here because of their strong medical interest. The major neural pathways of pain perception (nociception) will be introduced first. Pain can result from exposure to extreme temperatures, to acids and other chemical triggers, or it may be caused by mechanical injury which typically is associated with an inflammatory reaction that modulates the neural response. Nociceptors in the skin, bone, muscle or viscera are free nerve endings. The vanilloid receptor (TRPV1), which also accepts cannabinoids as ligand, is present in the membrane of these neurons. Other receptors contribute to the generation of pain at the free nerve ending. The bradykinin receptor (NK1), which is activated by a neurotransmitter termed peptide P, is an example. If an action potential is elicited the peripheral nerve will conduct the signal to the spinal cord. The nerve enters at the dorsal root (Figure 5.35) and forms a

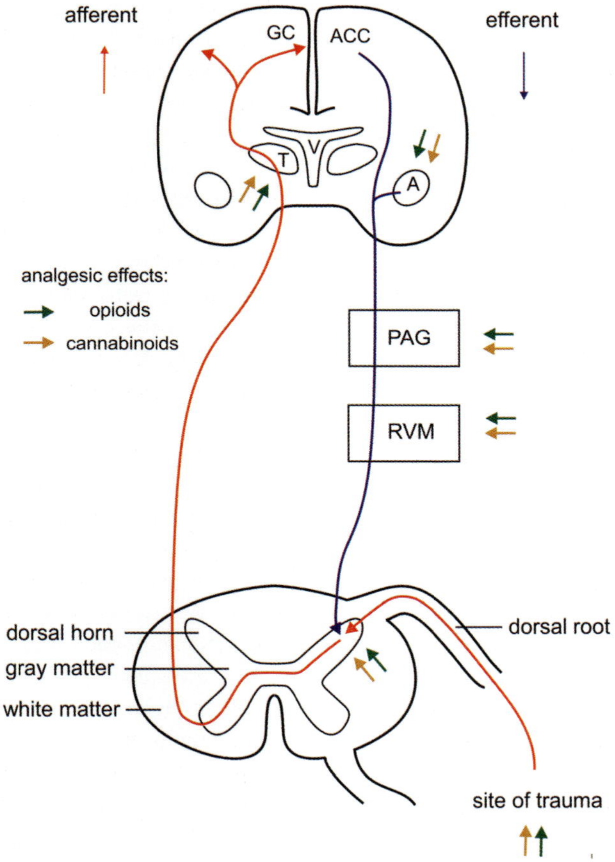

Figure 5.35 Important pain pathways. The major ascending (red) and descending (blue) nociceptive pathways are shown schematically, with the brain and the spinal cord drawn in cross-section. From the site of pain, free nerve endings are activated and the action potential travels along the ascending nerve to the dorsal horn of the spinal cord. The cell bodies of neurons and glial cells are located in the gray matter (substantia grisea), while the white matter (substantia alba) harbors the axons of the neurons. At the excitatory synapse in the dorsal horn the action potential is transferred to an ascending fiber which reaches the thalamus (T) and is relayed to the gyrus cinguli (GC) and other cortical areas. The information of descending pathways originates in the amygdala (A) and the anterior cingulate cortex (ACC). From two nuclei in the midbrain (PAG: periaqueductal gray) and in the posterior part of the brainstem (RVM: rostral ventromedial medulla) descending fibers project to the dorsal horn where the nociceptive information becomes modulated. Cannabinoids and opioids affect nociception at all levels of the respective pathways (brown and green arrows).

synapse with second-order neurons in the dorsal horn of the spinal cord. The action potential is transmitted by glutamate and neuropeptides (substance P) to the second-order neurons which ascend in the spinothalamic tract to the thalamus. The information is processed in the thalamus and then relayed to the sensomotory cortex of the telencephalon, where the strength of the pain is perceived, and also to parts of the limbic system (*Gyrus cinguli*, hippocampus) where the emotional component of pain is generated (Figures 5.28 and 5.35). Neural networks in the brain process the information, and the activity of the afferent nociceptive pathway can be modulated in an efferent feedback loop. The efferent nerves collect information from the *Gyrus cinguli anterior* and the amygdala, and pass two important relay stations (the periaqueductal gray and the rostral ventromedial medulla) on their way to the dorsal horn (see Figure 5.35). The strength of the signal generated by the injury can be modulated and either strengthened or attenuated at all levels of the afferent pathway. At the free nerve ending the vanilloid receptor may be inhibited, while in the dorsal horn synaptic transmission to the second-order neuron may be modulated by inhibitory or excitatory interneurons. In addition, serotonergic neurons originating in a special area of the brain stem (raphe nuclei) modulate the neural signaling in the dorsal horn. Of course, the processing of nociceptive information in the brain and the generated efferent information also play important roles in the control of pain perception.

5.3.4.1 **Analgesic Effects of Opioids**

The perception of pain is largely mediated by endogenous opioids. The peptides are present at the site of injury as well as in all processing centers of the nociceptive pathways in the CNS (Figure 5.35). The injury of a muscle results in a local increase in β-endorphin, while local stress induced by surgery may also lead to elevated concentrations of endogenous opioids. The importance of opioids in the perception of pain can be shown experimentally by inhibiting the opioid-mediated signaling; such treatment results in a loss of the analgesic effect. In contrast, stimulation of the opioid receptors by the uptake of morphine strongly attenuates the perception of pain.

The molecular mechanisms of opioid-mediated analgesic effects have been analyzed in detail. At the site of injury the free nerve endings function as pain receptors (nociceptors). The activation of the vanilloid receptor (TRPV1) by phosphorylation plays a central role in this process. Opioids inhibit the respective protein kinase, and in this way prevent the sensibilization of the receptor. Opioid receptors (MOR, DOR, KOR) are typically present at free nerve endings and DOR, for example, has been detected in dental pulps. The generation of pain may, therefore, become attenuated already at the site of origin. If an action potential is generated, the first synapse is in the dorsal horn of the spinal cord (Figures 5.35), which is another regulated site of the nociceptive pathway. Excitatory and inhibitory inputs from other neurons are present both at the presynaptic and the postsynaptic side. MOR agonists are known to inhibit the transmission of pain at the dorsal horn and also KOR and DOR seem to play a role. The latter receptors were identified at the presynaptic side and hyperpolarize the cell by inhibiting the Ca^{2+} influx and promoting outward-directed K^+ currents.

Opioid receptors are widely present in the brain, with the highest densities found in the thalamus, the basal ganglia, and some parts of the limbic system. MOR acting both presynaptically and postsynaptically have been localized also in the hypothalamus and the amygdala. The presence of opioid receptors in central areas of pain perception (thalamus, gyrus cinguli and cortical areas; see Figure 5.35) strongly suggests that opioids are essential mediators of pain perception. MOR agonists produce an analgesic effect by activating inhibitory neurons in the midbrain, which form part of the efferent pathways leading to the dorsal horns. The importance of this descending pathway has been shown experimentally by microinjecting morphine into the periaqueductal gray of mice, where a particularly high density of MOR is present. This treatment induced analgesia. Morphine was also shown to activate inhibitory cells in the rostral ventromedial medulla (Figure 5.35). When MOR signaling was prevented in knockout mice strains, the animals showed an increased sensitivity to heat.

5.3.4.2 **Analgesic Activity of Cannabinoids**

Although cannabis has been used for the treatment against pain for thousands of years, this analgesic effect has not been exploited during the past decades because of the compound's psychoactive properties. In fact, only recently have the useful pharmacological properties of cannabis come back into focus, with animal studies and clinical trials being carried out. Whilst cannabinoids have been shown to be particularly promising in the treatment of chronic pain, the evidence for analgesic effects in acute pain is inconclusive. In patients with opioid-resistant pain, cannabinoids may be a useful alternative; a combined application would also be conceivable as opioids and cannabinoids act synergistically in the relief of chronic pain. The considerable overlap with regards to the site of analgesic activity of opioids and cannabinoids (Figure 5.35) may offer an explanation for this effect.

The molecular reactions that result in pain relief are poorly understood. Since cannabinoids inhibit both excitatory glutaminergic transmission as well as inhibitory GABAergic signaling, the net effect is difficult to assess. The results of both electrophysiological and neurochemical studies have suggested that endocannabinoids and phytocannabinoids act primarily at the peripheral site where the pain originates, and at the synaptic transmission in the dorsal horn of the spinal cord (Figure 5.35). Apparently, it is not only the ascending pathways but also the descending pain pathways that are affected by cannabinoids, since it has been shown that the release of 2-AG and AEA in the periaqueductal gray (Figure 5.35) activates the descending pain pathways and has an analgesic effect. One promising strategy to employ the analgesic effect would be to increase the endocannabinoid concentration *in vivo*; this can be achieved by inhibiting the catabolic inactivation of AEA by the enzyme fatty acid amide hydrolase (FAAH). Subsequently, knockout mice for this enzyme have shown a reduced sensitivity to pain.

5.3.5
The Hallucinogenic Drugs Psilocin, Psilocybin, and Mescaline

5-Hydroxytryptamine (5-HT; better known as serotonin) is an important neurotransmitter in the CNS (Table 5.3). Attention here will be focused on two substances that belong to two classes of chemicals: the indolamines psilocin and psilocybin, and the phenylalkylamine mescaline (Figure 5.36). Preparations of psilocybin-containing mushrooms were used by the Aztecs to induce a state of inebriation, while in Europe a mushroom known as liberty cap (*Psilocybe similanceata*), as well as some other species, are known to contain this drug. Mescaline is a component of the peyote cactus (*Lophophora williamsii*), and also has a long history of human usage. The hallucinogenic drugs lead to a change of the mood, a sensitization of sensory perception to the point of illusions and hallucinations, and an altered perception of space and time. Typically, these effects are all drug concentration-dependent. The potent synthetic drug lysergic acid diethylamide (LSD) also acts on serotonergic synapses and has similar psychoactive effects as the natural compounds.

Serotonin is a common signaling molecule that is present in nearly all animals, including protozoa, and is also found in plants; the concentration is particularly high in walnuts. Serotonin may act as neurotransmitter as well as a neurohormone, and hence this interesting molecule can mediate both endocrine and

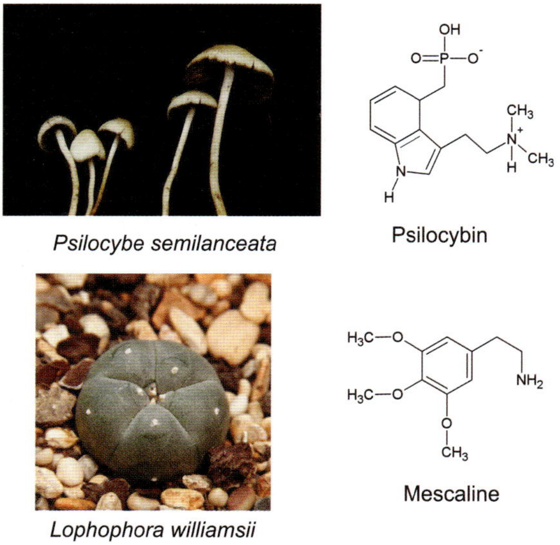

Figure 5.36 Hallucinogenic substances acting at serotonergic synapses. The mushroom *Psilocybe similanceata* is a source of the hallucinogenic drugs psilocin and psilocybin. Mescaline produces similar psychoactive effects, and is a product of the peyote cactos *Lophophora williamsii*. The natural substances target serotonin receptors and affect the signaling of serotonergic synapses. Photographs: *Psilocybe similanceata*, © depositphotos/yellow2j; *Lophophora williamsii*, © depositphotos/Edward Westmacott.

neural signaling. In its role as a neurohormone, serotonin can regulate the blood pressure by controlling the tonus of smooth vascular muscles. Serotonin is also involved in the control of blood clotting, and has regulatory functions in the digestive system. Most serotonin is synthesized in the enterochromaffin cells of the gut, where it is most concentrated.

Although the serotonin content of the CNS is comparatively small, some cells in the brain appear capable of its synthesis, as the compound is unable to cross the blood–brain barrier. Serotonin is synthesized by two consecutive enzymatic modifications of the amino acid L-tryptophan. A number of different serotonin receptors have been discovered and grouped into seven families. All of these receptors belong to the large group of GPCRs except for the 5-HT_3-receptors, which are ion channels. The molecular complexity of the different serotonin receptors and their tissue-specific distribution has made it difficult to develop receptor-specific inhibitors. Nonetheless, pharmacological research in this field has contributed greatly to the present knowledge of the biological functions of the respective receptors.

Serotonergic neurons are mainly located in the raphe nuclei located in the evolutionarily old brainstem from where the axons radiate to nearly all areas of the brain and also reach the synapse of the nociceptive neurons in the dorsal horn of the spinal cord. Some basic physiological functions are affected by serotonergic neurons, including circadian rhythm, sleep, temperature regulation and sexual behavior. Psilocin represses the activity of serotonergic neurons in the raphe nuclei in rats and cats by activating 5-HT_{1A} receptors. Such receptors are found on the cell bodies and dendrites of serotonergic neurons, and are hence termed "somatodendritic autoreceptors." The activation of these autoreceptors leads to an inhibition of serotonergic signaling at the serotonergic synapse. This curious situation leads to a biphasic dose–response relationship for serotonin, whereby a low dose of a 5-HT_{1A} receptor agonist results in an inactivation of the neurons by activating the autoreceptors, whereas a high dose has the opposite effect because the postsynaptic serotonin receptors are activated directly. Of particular interest here is the receptor 5-HT_{2A} to which the phenylalkylamine mescaline binds with great selectivity, while the indoleamines are more promiscuous. However, as the induced sensations are similar the psychoactivity is presumably mostly mediated by the 5-HT_{2A} receptor.

Interestingly, these hallucinogenic substances produce effects which show a striking similarity to the symptoms seen in schizophrenia. The effects of psilocybin in the brain can be visualized by using positron emission tomography (PET), which reflects a change in the cerebral blood flow (Figure 5.37). A marked prefrontal activation is noticeable in areas that are important for cognitive and affective processes, but there is an attenuated activity in other areas, such as regions of the thalamus which are important for gating and interpreting cortical information.

Studies of the biology of these hallucinogenic natural substances has greatly contributed to an understanding of neural function and has, furthermore, led to the identification of target proteins and useful drugs in manipulating serotonergic signaling. Whilst serotonin itself has no therapeutic value, drugs that alter serotonin levels have numerous applications and include antidepressants,

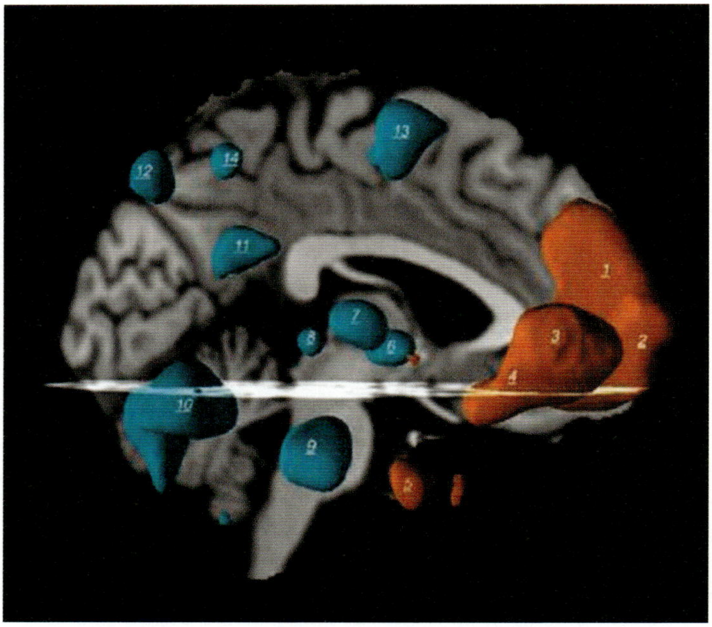

Figure 5.37 Changes in the cerebral blood flow caused by psilocybin can be visualized. Changes in brain activity are accompanied by changes in the cerebral blood flow which can be visualized using positron emission tomography (PET). The red areas show relative increases and blue areas relative decreases in regional brain activity induced by psilocybin in healthy volunteers. The drug induced a marked increase in anterior cortical areas (anterior to the right), while brain areas for processing cortical information (e.g., the thalamus) showed a reduced activity. This was also observed in the somatosensory areas, the cerebellum (involved in motor control), and areas of visuo-spatial processing. From Geyer, M.A. and Vollenweider, F.X. (2008) *Trends Pharmacol. Sci.*, **29**, 445–453 (see reference list). Reproduced with permission of Elsevier.

neuroleptics, tranquillizers, or antimimetics. A particularly promising application is to inhibit serotonin uptake at the synapse, as this would bypass any unwanted effects of the autoreceptors.

5.3.6
Natural Substances Affecting Cholinergic Signaling

5.3.6.1 The Cholinergic Synapse

The transmitter acetylcholine (ACh) and the respective cholinergic synapse are of enormous importance in the neural control of, for example, the muscular activity and the physiological control by the autonomous nervous system (Figure 5.27). So, why has Nature invented so many potent inhibitors of the cholinergic synapse? The toxicity of inhibitors is a powerful weapon in the fight for survival in a specific ecological niche. Plants may defend themselves against insects or other animals that may feed on them, and animals producing such poisons will be spared from

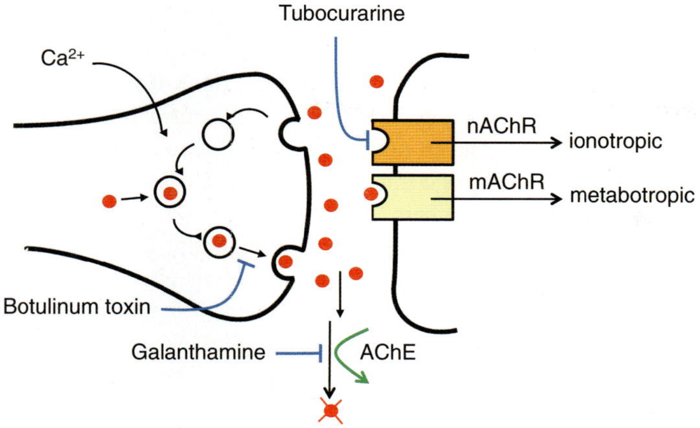

Figure 5.38 Cholinergic synapses are a target of natural substances. Natural compounds may affect cholinergic signaling by, for example, interfering with the exocytosis of synaptic vesicles (botulinum toxin). The nicotinic acetylcholine receptor (nAChR) can be blocked by tubocurarine, or the activity of acetylcholine esterase (AChE) may be inhibited by galanthamine and several other compounds (see Figure 5.39). The enzyme is present as a soluble protein in the synaptic cleft, or is membrane-bound. The generated choline is recycled and taken up by the presynaptic neuron. The two types of AChR, the metabotropic muscarinic acetylcholine receptor (mAChR) and the ionotropic nAChR, are present in the membranes of specific cholinergic neurons (see Figure 5.27).

predators; similarly, carnivorous species may use neurotoxins to immobilize their prey. Natural substances that affect cholinergic signaling have also opened up industrial and military applications that will be discussed in this section.

Acetylcholine is the ester of acetic acid and choline, and is a very simple molecule with a long evolutionary history. Intercellular communication with ACh is present in bacteria, protozoa and sponges and, of course, in all higher organisms. Although the molecule plays a role in many non-neural differentiation processes such as cell adhesion, hematopoiesis and cell migration, interest here is limited to its role as a neurotransmitter. The basic features of an excitatory cholinergic synapse are illustrated in Figure 5.38. At the presynaptic terminal, ACh is synthesized by the enzyme choline acetyltransferase (CHAT) from the simple and universally present substrates choline and acetyl-CoA. The transmitter is stored in synaptic vesicles until an action potential triggers a fusion of the vesicles with the membrane (exocytosis), when the transmitter is released into the synaptic cleft. At the postsynaptic side, two distinct ACh receptors are activated in specific neurons (for simplicity, both types are represented in the synapse shown in Figure 5.38). The receptors that must be distinguished are the metabotropic muscarinic acetylcholine receptor (mAChR) and the ionotropic nicotinic acetyl choline receptor (nAChR). Both, mAChR and nAChR owe their names to the natural substances nicotine and muscarine, which act as ligands for the respective receptors. In 1869, Schmiedeberg and Kopp discovered that extracts of the mushroom *Amanita muscaria* (fly agaric) would slow down the heartbeat of frogs, and they named the active substance muscarine. Nicotine was found to act as agonist for the second type of receptor. The mAChR is located mainly in the peripheral nervous system, the smooth musculature and in the heart muscle, whereas the nAChR has a different tissue distribution and is typically present in the CNS (controlling cognitive processes and memory functions) and in neuromuscular junctions. At this specialized junction the postsynaptic cell is a muscle cell which reacts to stimulation by ACh with contraction of the muscle fibers.

Removal of the transmitter from the synaptic cleft is important for all types of synapses, but in case of ACH the mechanisms are particularly efficient so that, after a short refractory period, the synapse is again functional. The enzyme acetylcholinesterase (AChE) is present in the synaptic cleft as a soluble enzyme, though some spliced forms of the enzyme are membrane-bound. The enzyme is characterized by extremely rapid reaction kinetics. In some tissues butyrylcholinesterase, an enzyme with a similar substrate specificity, is also capable of degrading ACh. The choline formed in the synaptic cleft is removed by a transporter that ensures recycling of the compound at the presynaptic nerve terminal.

5.3.6.2 Cholinergic Signaling is a Target for Natural Substances

Synaptic signaling by ACh can be inhibited in many ways: by transport of the transmitter into the vesicles being inhibited, by exocytosis of the vesicles, by binding of the ligand to the postsynaptic receptor, by degradation of the transmitter by AChE, or by the uptake of choline into the presynaptic cell. Natural or synthetic inhibitors are available for each of these processes (see Figure 5.38 for examples). One effective way to inhibit synaptic signaling is to prevent release of the neurotransmitter, and the poison botulinum toxin (which is synthesized by the bacterium *Clostridium botulinum*) is a prominent example of this. The extremely powerful neurotoxin inhibits the fusion of synaptic vesicles with the plasma membrane, and hence release of the neurotransmitter. The relaxation of muscles in the absence of cholinergic signaling is exploited in cosmetic applications (Botox).

One obvious target to prevent synaptic signaling would be to inhibit the postsynaptic receptor function. The biology of AChRs is highly complex, with five subtypes of mAChR having been identified (M1R to M5R), each with different tissue distributions and partly different functions. An example is the cholinergic control of the heart: whereas M2R activation leads to bradycardia (low heart rate), the activation of M1R has the opposite effect (tachycardia). The example illustrates that useful mAChR inhibitors should have a desired subtype specificity.

Another degree of complexity results from the distribution of the respective AChR. The nAChR consists of pentamers of different subunits, and the tissue-typical distribution must be taken into account in medicinal applications of nAChR inhibitors. At many junctions, receptors may also be present in the presynaptic cell membrane and can control the sensitivity of transmitter release. In case of the nicotinic synapse, the presynaptic nAChR facilitates exocytosis of the synaptic vesicles.

The compound tubocurarine, which is present in the climbing vine (*Chondrodendron tomentosum*), is a powerful competitive inhibitor of ACh at the nAChR, such that the voluntary contraction of skeletal muscles is inhibited. This results in a relaxation of the musculature, an effect which has been exploited during surgery, although an overdose of the inhibitor may lead to asphyxia and possibly to death. An antidote for tubocurarine is parathion, a strong AChE inhibitor that, based on its toxicity, has also been used as an insecticide. The resultant raised ACh concentration competes and mitigates the effects of tubocurarine at the cholinergic synapse.

Signaling through the nAChR is known to affect the synthesis and release of other transmitters, including serotonin and dopamine. Nicotine activates the dopaminergic "reward system" of the brain, which is an essential element of drug addiction. The dopaminergic system can be activated by tobacco smoke in another and surprising way; burning of the sugar present in tobacco leaves results in the formation of acetaldehyde, which in turn inhibits the enzyme monoamine oxidase B that degrades dopamine and serotonin.

5.3.6.3 Inhibitors of AChE

Numerous natural substances have been identified which inhibit AChE. The inhibition either delays or inhibits ACh degradation, such that cholinergic signaling at the postsynaptic cell is greatly enhanced and/or prolonged. At the neuromuscular junction, an excessive concentration of ACh will lead to involuntary muscle contractions, while at higher inhibitor concentrations muscle fibrillations or complete muscle paralysis will occur as the result of a lasting depolarization of the postsynaptic membrane potential. The abnormal stimulation of the cholinergic neurons of the autonomous nervous system (Figure 5.27) results in excessive secretions at parasympathetically controlled organs. Typical symptoms of acute toxicity – salivation, lacrimation, urination, defecation, gastrointestinal upset and emesis – are abbreviated as SLUDGE.

The toxic beans of the plant *Physostigma venenosum*, which is indigenous to Calabar in West Africa, were used as an ordeal poison by local natives to determine the guilt or innocence of accused criminals. If the person survived the acute toxicity of the beans this was interpreted as proof of innocence, but in the case of death this was considered proof of guilt. The first pharmacological studies were carried out more than 150 years ago, and in 1864 an alkaloid was isolated that was later named physostigmine (Figure 5.39). When the substance was ultimately found to be a potent AChE inhibitor, and its possible application in medicine and crop protection became apparent, the search for similar inhibitors and useful chemical derivatives was greatly intensified.

The natural AChE inhibitors detected in a variety of plants have quite different structures, and some examples may illustrate this point. One inhibitor with medical relevance is galanthamine (Figure 5.39), a tertiary alkaloid isolated from the Caucasian snowdrop (*Galanthus* spp.). Other plants belonging to the family Amaryllidaceae were also found to produce potent AChE inhibitors; for example, the alkaloid huperzine (Figure 5.39) from the clubmoss *Huperzia serrate* (Lycopodiaceae) has been used in Chinese medicine against memory loss and other indications. Benzylisoquinolines from *Corydalis* spp. (Papaveraceae) represent another class of natural inhibitors; the substance berberine (IC_{50} 0.23 μM; Figure 5.39)

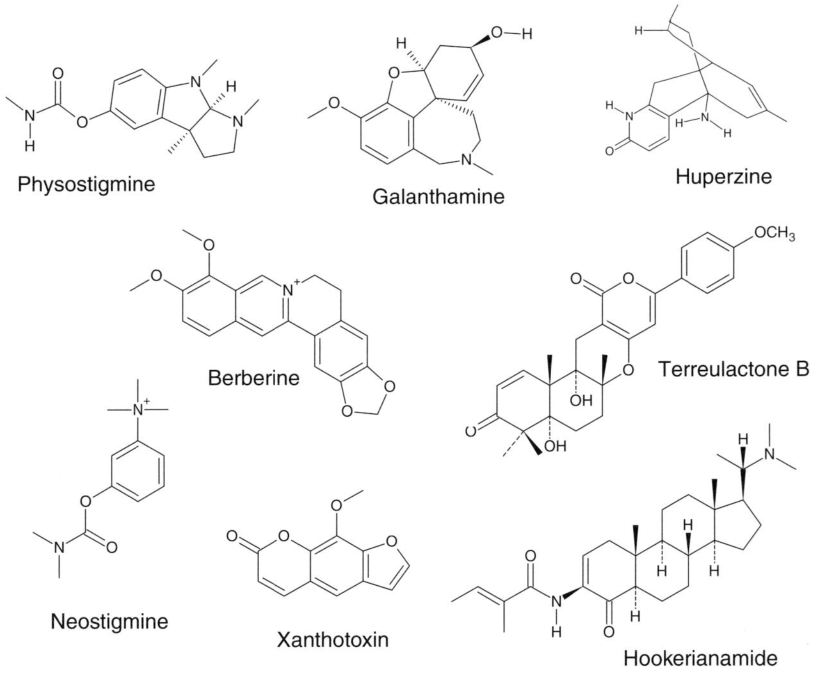

Figure 5.39 Natural inhibitors of acetylcholinesterase (AChE). Examples of natural compounds with different chemical structures are shown which inhibit AChE. From Houghton, P.J., Ren, Y. and Howes, M.J. (2006) *Nat. Prod. Rep.*, **23**, 181–199.

may serve as a representative example. The well-known arrow poison tubocurarine is not only a competitive inhibitor of ACh at the nAChR, but also has some AChE inhibitory activity. The plant family Buxaceae is known for the presence of toxic steroidal alkaloids, an example is hookerianamide-F from *Sarcococca hookeriana* (Figure 5.39). Several terpenoids have been shown to possess AChE inhibitory activity. Of particular interest are fungal meroterpenoids isolated from *Aspergillus terreus,* and terreulactone B (Figure 5.39) is an example of several structurally related compounds with similar inhibitory activity. Finally, some furanocoumarines (e.g., xanthotoxin; Figure 5.39) from the roots of *Angelica acutiloba* are highly active AChE inhibitors with an IC_{50}-value of 0.58 μM.

The toxicity of excessive cholinergic signaling has been exploited to fight insect pests; an example is the nAChR agonist nicotine, which was used as an insecticide until 1970 but then banned (see also Section 2.5.3). These powerful natural inhibitors inspired the development of chemical inhibitors, with natural carbamate inhibitors being replaced by synthetic organophosphate inhibitors that were highly effective AChE inhibitors. In this case, a critical serine at the enzyme's catalytic site is phosphorylated, which results in an irreversible inactivation of the enzyme (Figure 5.40). The serine forms part of an amino acid triad (serine-histidine-glutamate) that is essential for enzyme activity.

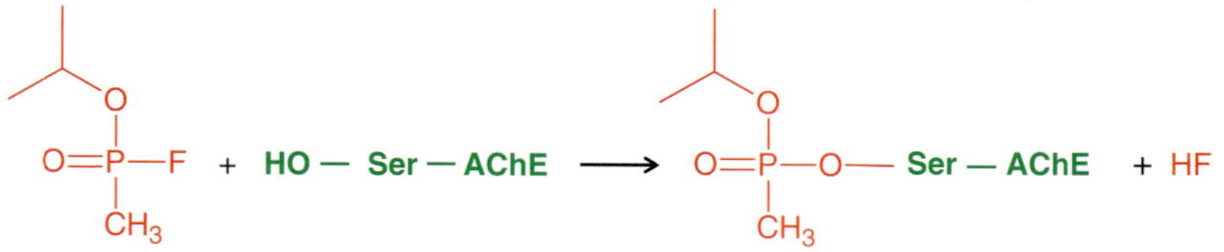

Figure 5.40 Irreversible inactivation of AChE. The organophosphate sarin, a chemical warfare agent, forms a stable ester bond with the serine in the catalytic center of AChE, thus inactivating the enzyme irreversibly.

Organophosphates have been used worldwide as insecticides, and today more than 100 different organophosphates are used by the agricultural industry. As a result, the contamination of food with such chemicals is of medical concern. It has been estimated by the World Health Organization that 200 000 deaths occur annually as a result of organophosphate poisoning, due to accidental and/or intentional suicidal intoxication. Unfortunately, these substances may also react with other target proteins; for example, pesticides such as profenofos and chlorpyrifos oxon (a metabolite of chlorpyrifos) are potent inhibitors of the catabolic enzyme FAAH, and hence affect the concentrations of endocannabinoids. The importance of these compounds in regulating the release of other transmitters, including ACh, has been highlighted previously. The exposure of mice to the insecticide methamidophos was shown to affect both GABA- and dopamine-signaling in the CNS.

Is it possible to develop insect-specific AChE inhibitors that do not affect vertebrates? Because the enzyme is highly conserved and the catalytic triad is present in all AChE enzymes, this is a difficult task. However, a careful examination of AChE of the greenbug aphid (*Schizaphis graminum*) indicated that, in principle, a discrimination is possible. A unique cysteine residue near the catalytic triad was identified and sulfhydryl inhibitors that reacted irreversibly with the cysteine were shown to be promising aphid-specific inhibitors. As many currently used organophosphate insecticides are chiral molecules, there is also the possibility that specific enantiomers can be used to achieve a degree of selectivity for the target organisms.

The toxicity of organophosphates has also been used to develop chemical weapons' which are known under the names of Sarin, Soman, and Tabun. These organophosphates are easy to synthesize chemically and several countries have produced these compounds for use as nerve poisons. In order to minimize the poisoning of soldiers who might be exposed during the course of chemical warfare, compounds have been sought that are able to reactivate the inhibited AChE. Paradoxically, the best antidote against organophosphate poisoning is to use other AChE inhibitors. For example, pyridostigmine binds to AChE reversibly and, when bound, will prevent the irreversible inactivation of the enzyme by organophosphates. The US military has used pyridostigmine in this way to protect against the potential exposure of soldiers to chemical weapons based on organophosphates. When this precaution was considered necessary the symptoms of SLUDGE were taken into account. Huperzine A and other AChE inhibitors were suggested as alternative antidotes, while various oximes were also reported to allow AChE reactivation.

5.3.6.4 Medical Applications of AChE Inhibitors

Medical applications of AChE inhibitors have a long history. During the late nineteenth century, physostigmine was used to treat the eye condition glaucoma, whereby the build-up of ocular liquid was relieved by cholinergic stimulation, due to an improved drainage. Formerly, the medical applications of natural AChE inhibitors were limited, but new pharmacological developments have led to changes in this scenario. For example, when physostigmine was modified chemically, the derivatives neostigmine (Figure 5.39) and pyridostigmine were shown to possess longer half-lives and to be more hydrophilic than physostigmine, so that crossing of the blood–brain barrier would be prevented. Neostigmine was also found to be useful for the treatment of myasthenia gravis, an autoimmune disease characterized by muscle weakness due to insufficient ACh at the neuromuscular synapse. Galanthamine has been used to treat myasthenia and other motor dysfunctions.

Although, during aging, a gradual loss of cholinergic functions is typical, in pathological situations such as Parkinson's disease, Down syndrome or Creutzfeld–Jakob disease a significant degeneration has typically been observed in the cholinergic neurons of some parts of the brain. The cortical loss of cholinergic neurons in Alzheimers's disease is particularly severe, and consequently it may be advantageous to increase the ACh concentration. In Alzheimer's disease the β-amyloid plaques present in the brain are diagnostic, and are held responsible for the neurodegenerative processes and diminished cholinergic function. Today, the treatment of Alzheimer's disease with AChE inhibitors is common practice, and several drugs have been approved for this purpose; it should be noted that only lipophilic compounds that cross the blood–brain barrier are applicable for this indication. In addition to such symptomatic treatment AChE inhibitors may also have disease-modifying functions. Notably, the enzyme colocalizes in β-amyloid peptides which are the main components of the senile plaques that are characteristic of Alzheimer's disease. The presence of AChE in the plaques seems to enhance their neurotoxicity, and AChE also stimulates amyloid fibril formation. As both negative effects are mitigated by AChE inhibitors the natural substance galantamine has shown therapeutic promise in this respect. In the USA, galantamine has even been used as a dietary supplement to improve memory functions.

Surprisingly, other possible therapeutic fields emerged when it was realized that both mAChR and nAChR mediate complex functions that go far beyond their role in neural signaling. For example, the activation of mAChR may trigger an activation of a metalloprotease which, in turn, activates the epidermal growth factor receptor (EGFR) that is involved in cancer-relevant signal transduction pathways. Crosstalk between the different signal transduction pathways involved in common diseases would allow new therapeutic approaches in which natural substances are likely to play an important role (see Section 6.3).

5.3.7 Sympathomimetic Substances

In the past, dramatic effects have resulted from the inhibition or stimulation of parasympathetic signaling. Of course, natural substances may also interfere with the sympathetic nervous system. Whilst a chemical stimulation of the noradrenergic sympathetic system (Figure 5.27) mimics a stress situation and increases activity and alertness, negative outcomes have also been reported including restlessness and insomnia, as well as psychoactive effects such as euphoria and hallucinations. Amphetamines (e.g., methamphetamine) are synthetic sympathomimetic substances that became "social drugs" consumed by millions of people, particularly in the USA and Japan. However, the dangers associated with amphetamine consumption became apparent towards the end of the twentieth century and, because of the risk of addiction and other health risks, the use of methamphetamine (also known as "crank," "speed," or "crystal") was prohibited and its medical use was discontinued.

Ephedrine is the main alkaloid of *Ephedra sinica* and other *Ephedra* species that are endemic in the Mediterranean, India and China, where extracts of the plant have been used for almost 4000 years to treat coughs. Although the chemical structure of ephedrine is similar to that of mescaline (Figures 5.36 and 5.41), the biological activities of these substances differ greatly. Ephedrine is taken up by an active transport mechanism into the presynaptic neuron of a noradrenergic synapse. In the cytoplasm, ephedrine inhibits the uptake of noradrenaline into the synaptic vesicles, so that the cytoplasmic concentration of the neurotransmitter is increased. In combination with the other synergistic effects of ephedrine, this results in an increased noradrenaline concentration in the synaptic cleft so that an indirect sympathomimetic effect is produced. Based on its β-adrenergic activity, ephedrine can be used pharmacologically to reduce swelling of the nasal mucosa in infectious diseases.

Several natural substances have a sympathomimetic component and also affect other types of synapses. For example, cocaine (Figure 5.41) from the coca plant *Erythroxylum coca* inhibits uptake of the catecholamines noradrenaline and dopamine, and also serotonin uptake. The consequent elevation of neurotransmitter levels leads not only to an increased alertness and euphoria but also to anxiety and paranoia and, at excessive concentrations, to tremors and convulsions. Unfortunately, the drug crosses the blood–brain barrier and is highly addictive. At high concentrations the Na^+ channels are blocked – an effect that led to cocaine being used as a local anesthetic in the late nineteenth century.

Ergotamine (Figure 5.41), an ergot alkaloid from the fungus *Claviceps purpurea*, is a phytopathogen that grows mainly on rye. As a result, rye bread may become contaminated with the alkaloid, and this proved historically to be a severe problem during the Middle Ages. Ergotamine affects the function of serotonin ($5\text{-}HT_1$) receptors, which makes it a useful drug for the treatment of migraine. Furthermore, because of its ability to activate α-adrenoreceptors and cause strong

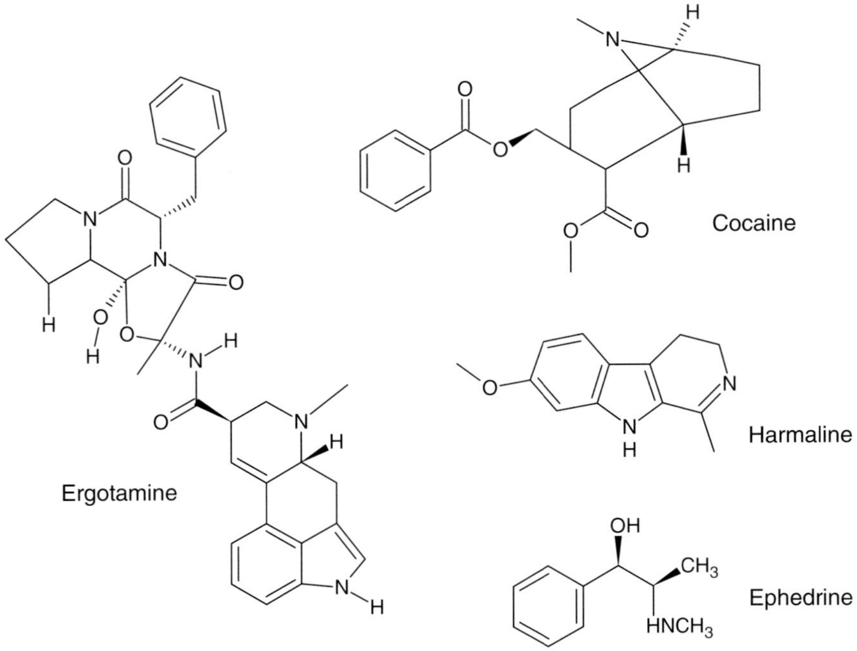

Figure 5.41 Natural substances with sympathomimetic activity. Note the structural similarity between ephedrine and mescaline (see Figure 5.36).

contractions of the uterine musculature, ergotamine has been used to stop maternal bleeding after childbirth. Unfortunately, the neurotropic activities of ergotamine lead to hallucinations and irrational behavior.

Finally, the hallucinogenic substances harmine and harmaline (Figure 5.41), which are synthesized in the jungle vine *Banisteriopsis caapi,* should be mentioned. These compounds are major components of the "Ayahuasca" preparation that was originally formulated by natives of the Amazons basin and consists of a mixture of different plants. The substances are natural inhibitors of monoamine oxidases (MAOs) which degrade catecholamines and serine. Inhibitors of these important mitochondrial enzymes (MAO-A and MAO-B) profoundly affect signaling in the respective neural pathways, and increase the levels of serotonin, noradrenaline and dopamine in the brain. The psychotropic effects of Ayahuasca resemble those of mescaline. Synthetic MAO inhibitors can be used to treat depression.

It has been seen in this chapter that numerous natural substances interfere with neural signaling and involve different neurotransmitter systems. Depending on the respective set of target molecules, and on the degree of crosstalk between different signal transduction pathways, essentially all neurally controlled processes may be affected by xenobiotics. Moreover, when the interference of natural substances with the endocrine system is also taken into account, the picture emerges that a broad spectrum of natural compounds is able to modulate all basic physiological control mechanisms in animals and humans alike. The question remains: Can advantage be taken of these biological reactions by selecting food with beneficial components, or perhaps by developing new pharmaceuticals? Some possible answers to these points are discussed in Chapter 6.

6
Nature's Drugstore for a Healthy Life

6.1
Biologically Active Food and Food Components

Today, humankind is swamped with information and recipes for a healthy life based on the consumption of health-promoting products marketed as "functional food" or "neutraceuticals." But, do the advertised additives in the food products hold their promise? Opinions about optimal nutrition differ widely and even the scientific debate is controversial on major issues. Moreover, public debates are often clouded by the economic interests of the agricultural sector, the food industry, and the "health industries" (drugstores and pharmaceutical companies). The possible benefits or perils of specific diets are not subject of this chapter; rather, an attempt will be made to explain some well-studied physiological reactions to food, or to specific substances in food. Natural substances in food not only determine its color and taste but also interfere with physiological process in healthy and sick persons. The reactions may differ widely in different individuals (see Section 4.1.5), and for this reason a general advice concerning the benefits of a particular food product must be viewed with caution. Some recent findings will be discussed below which shed some light on general physiological reactions elicited by food components.

All vegetarian food and plant-based beverages contain a complex mixture of low-molecular-weight secondary plant metabolites. Most of these substances are phenolic compounds with one or more phenol groups, but there are also some exceptions. Glucosinolates, for example, are derived from amino acids and their metabolic products (isothiocyanates) are also biologically active compounds. Different vegetables and fruits contain a specific set of natural substances which provide the specific color, taste, and nutritional value. Some relevant substances which are common components of food are compiled in Table 6.1. Most common secondary metabolites are phenolic components, and several of these have already been identified as estrogenic compounds (see Table 5.2).

What is meant by "healthy food?" Generally, food can be considered as healthy if its consumption prolongs life and postpones the development of age-related diseases. For persons who already suffer from a disease, however, the desirable

Plant Natural Products: Synthesis, Biological Functions and Practical Applications, First Edition.
Herwig O. Gutzeit and Jutta Ludwig-Müller.

Table 6.1 Common phenolic compounds in food.

Examples of phenolic compounds		Dietary source
Naringenin (Flavanone)	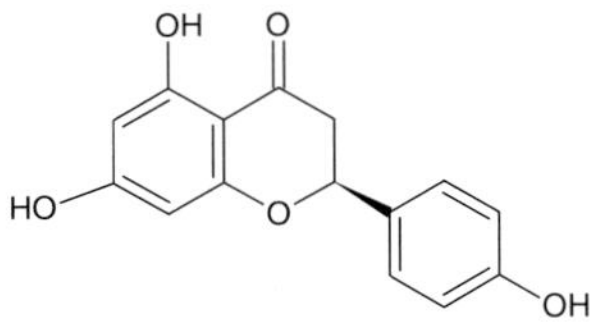	Citrus fruits and juices
Apigenin (Flavone)	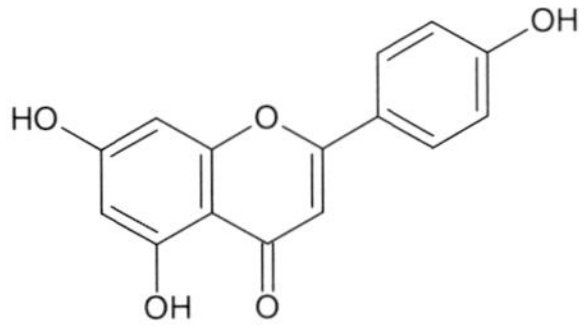	Parsley, celery, capsicum, pepper
Genistein (Isoflavone)	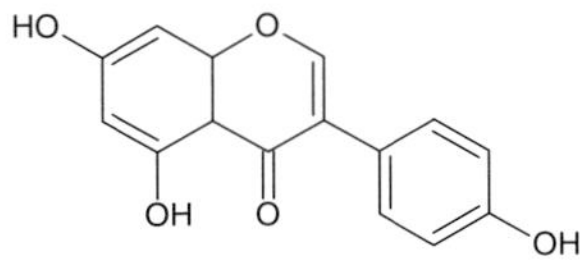	Soy beans and other legumes
Quercetin (Flavonol)	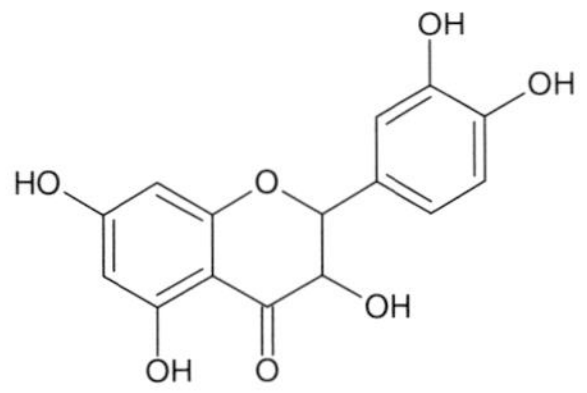	Broccoli, onions, tomatoes, lettuce, apples, grapes
Epigallocatechin (Catechin)	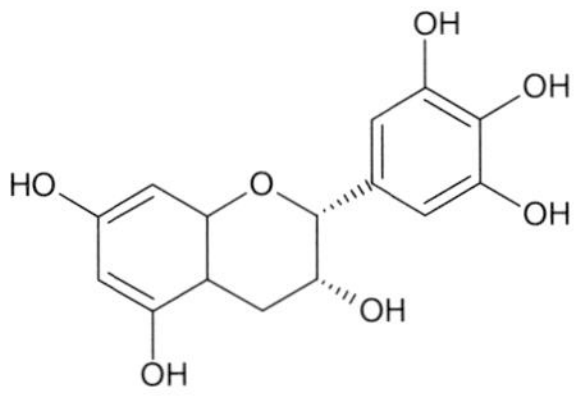	Green tea, cherries, berries, beans, chocolate
Cyanidin (Anthocyanidin)	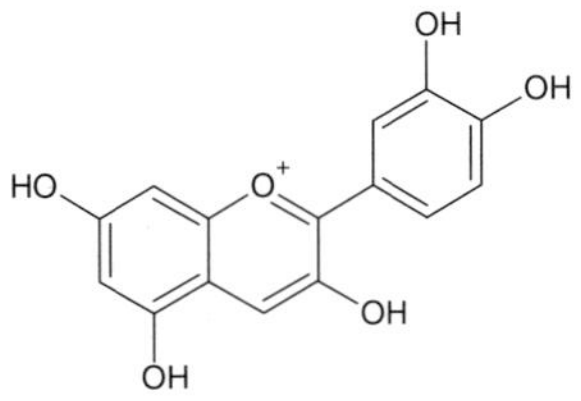	Berries, cherries, aubergine, red cabbage
Resveratrol (Stilbene)	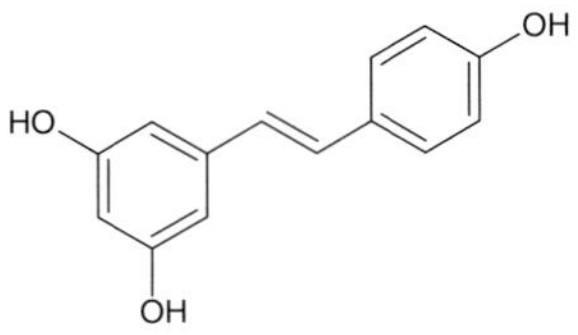	Grapes (particularly the skin), red wine

Coumestrol	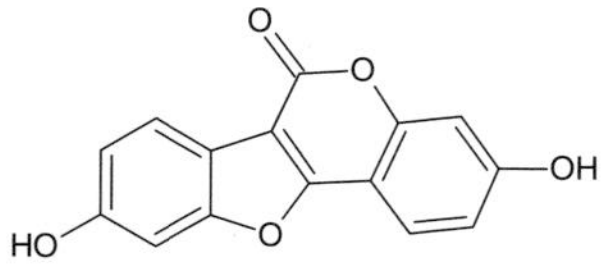	Clover, alfalfa, spinach
Enterolactone	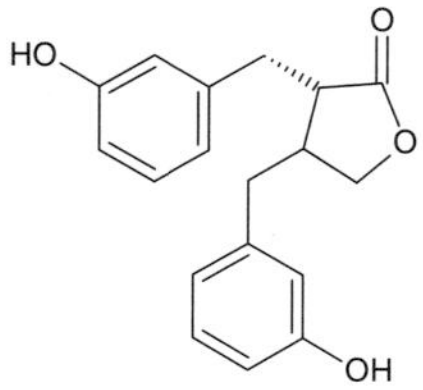	Flax and sesame seed
Cinnamaldehyde	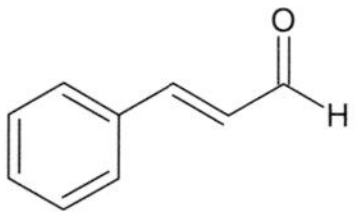	Cinnamon
Hydroxytyrosol	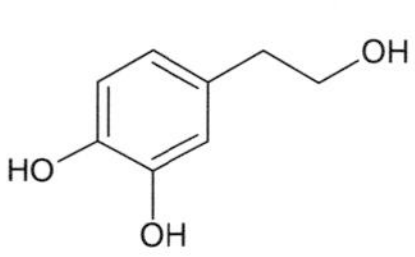	Olives
Rosmarinic acid	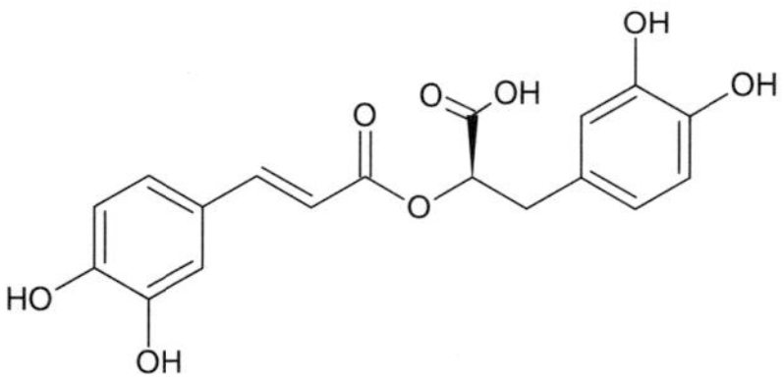	Rosemary, oregano

physiological response to the consumption of food may be significantly different and specific diets may be required. For healthy persons, the first question to be answered concerns the "healthy" amount of food that should be consumed. Interestingly, the quantity of the consumed food is as important as its quality.

6.1.1 The Quantity of Food Intake Affects Life Expectancy

In Western societies obesity is a major problem and presents a risk factor for common diseases such as diabetes and cardiovascular diseases. High-calorie diets result in physiological changes that are taken as indicators for a health risk. Abnormal cholesterol levels are typical indicators of a physiological disorder which is known as "Metabolic syndrome." A high concentration of low-density lipoprotein (LDL), which is typically associated with a low plasma level of high-density lipoprotein (HDL), is considered to favor the formation of arteriosclerotic plaques and so is regarded as a risk factor. Nicotinamide adenine dinucleotides

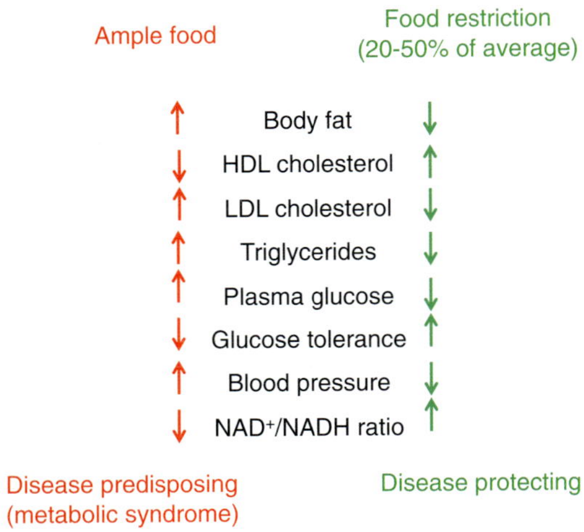

Figure 6.1 The quantity of food consumption affects the health status. The beneficial effects of caloric restriction are reflected by a positive influence on physiological parameters which are relevant for risk assessment. In contrast, calorie-rich diets affect these parameters negatively. Based on Guarente, L. (2006) *Nature*, **444**, 868–874.

(NAD^+ and NADH) play important roles in cellular energy metabolism, and a low NAD^+/NADH ratio is implicated in metabolic syndrome. An insufficient physiological reaction to increased glucose titers in the blood (glucose tolerance) and the reduced reaction to the hormone insulin (insulin resistance) are indicators for developing diabetes. Finally, high blood pressure is a risk factor for cardiovascular and other diseases (Figure 6.1).

Interestingly, a reduction in consumed food by 20–50% (with reference to average food consumption) leads to health benefits in invertebrate and mammalian animal models, and this also seems to hold true for humans. The indicators for health risks under conditions of metabolic syndrome improve with fasting (Figure 6.1). Benefits of caloric restriction (CR) concern common age-related diseases and reduce the risks for cancer, neurodegenerative disorders, autoimmune diseases, cardiovascular diseases and type II diabetes mellitus. Because of the impressive health benefits of CR an extension of the lifespan has been consistently observed in all animal studies, including mammals. But, what could be the biological reason for this astounding effect? A temporary shortage of food is a situation that animals have learned to cope with during the course of evolution. Energy-expensive reproductive functions are dispensable when survival of the individual is at stake under conditions of temporary food shortage, and reproduction can be resumed when environmental conditions improve. A postponed reproductive phase is apparently not compensated by accelerated aging during the post-reproductive phase, and hence the lifespan is prolonged. Although this interpretation is only supported by circumstantial evidence, it offers a plausible biological explanation for the unexpected effect of CR on life expectancy. Some mild stress

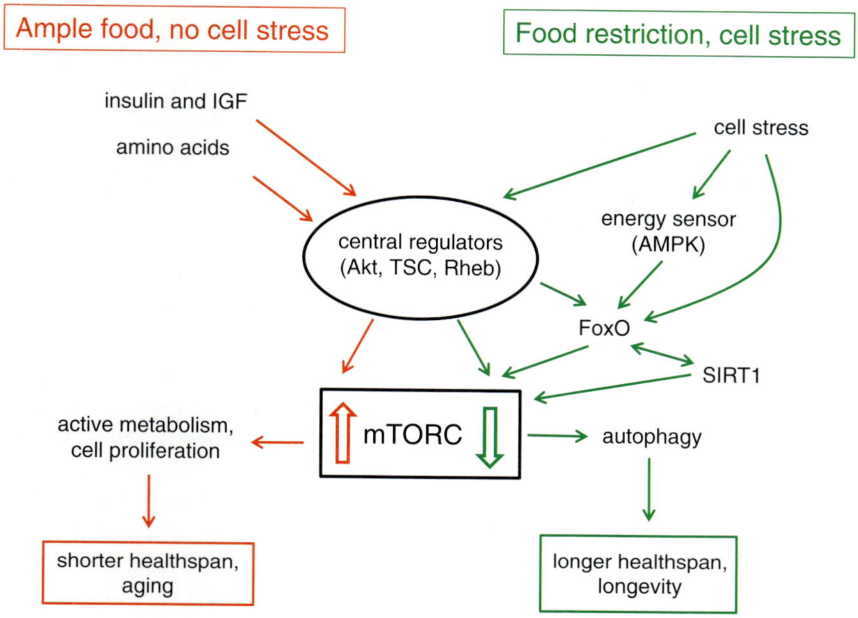

Figure 6.2 Major pathways which mediate the effect of food restriction and cell stress on health and longevity. The kinase mTORC plays a central role in the control of energy metabolism of cells. A low activity of the enzyme is life-prolonging, and at the same time cellular activities are downregulated. Some relevant positive and life-prolonging factors (green) and negative factors (red) are shown which affect the activity of mTORC. The quantity of consumed food and the degree of cell stress are measured by special molecular sensors that regulate important physiological pathways. Particularly important proteins include Akt (a regulatory serine/threonine kinase, also known as protein kinase B) and TSC (tuberous sclerosis complex) which regulates the GTPase Rheb. This enzyme controls the activity of mTORC. IGF: Insulin-like growth factor. See text for details of the stress-signal transduction.

situations such as hyperthermy or oxidative stress may also affect reproductive functions, and the observed life-prolonging effect in animals under such experimental conditions may be explained by the same argument. The increase in life-expectancy is tightly coupled to increased stress-resistance, and this correlation may be interpreted as a hormetic response to cellular stress (see Section 5.1.5).

The mechanisms which result in a longer lifespan after CR or mild cellular stress have been subjected to intensive research, and the molecular key players have been characterized (Figure 6.2). The atypical serine/threonine protein kinase mTORC (mechanistic target of rapamycin complex) is a key enzyme which integrates environmental and physiological cues and also regulates growth and cellular activities. The protein complex was named after the macrolide rapamycin, a fungal metabolite and an inhibitor of mTORC1. In mammals, two protein complexes (mTORC1 and mTORC2) cooperate in the control functions: mTORC2 regulates cell survival and cytoskeletal organization, while mTORC1 represents the complex which integrates metabolic information and has a central role in the regulation of the lifespan. An upregulation of mTORC is achieved by the presence of amino acids (in particular leucine and arginine), and in particular by the presence of carbohydrates so that the secretion of insulin and insulin-like growth factor (IGF-1) is stimulated. These hormones signal the presence of food, such that the respective membrane receptors trigger a signal transduction cascade that involves the release of the second messenger phosphatidylinositol 3′-phosphate (PIP3). This molecule activates the important serine/threonine kinase Akt, which is also known as protein kinase B (PKB). Downstream of Akt the tuberous sclerosis complex (TSC) acts as an inhibitor of Rheb, a small GTPase which is required to activate mTORC (Figure 6.2). When food is not limited, high serum concentrations of carbohydrates and amino acids stimulate the activity of mTORC1 and hence also metabolic functions, including cell proliferation, reproductive functions, and other energy-requiring processes. In contrast, a downregulation of mTORC1 brought about by food shortage and stress situations is known to prolong the lifespan. Because of the central role of mTORC in regulating cellular activity, it is not surprising that the (dys)regulation of this complex is critically involved in cancer and also in metabolic and neurodegenerative diseases.

Apart from the endocrine signaling pathway that reflects the availability of food (insulin and IGF-1), some additional regulatory proteins control mTORC1 activity. In order to attenuate energy-requiring processes under adverse conditions it is essential to have an energy-sensing molecule and, again, a kinase fulfils this important function (Figure 6.2). AMP-activated kinase (AMPK) is a highly conserved sensor which responds to low levels of ATP, the central energy-providing molecule in cells. A depletion of ATP results in an accumulation of the enzyme's cofactors AMP and ADP, and hence stimulates the enzyme activity. A low AMPK activity (reflecting an abundant food supply) favors insulin resistance and obesity and increases the risk for diabetes and cardiovascular diseases. AMPK also plays an important role in the regulation of an antioxidative response (see Section 5.1) by activating the transcription factor Nrf2 and the expression of thioredoxin. The forkhead transcription factors (FoxO) have been conserved in evolution from nematodes to mammals, and regulate essential cellular functions such as glucose and lipid metabolism, stress resistance, and apoptosis (Figure 6.2). The insulin/IGF-I pathway downregulates FoxO, while AMPK activates members of the family of FoxO transcription factors. FoxO efficiently prevents hyperactivation of mTORC and hence is a key regulator of the lifespan. The genes that are regulated by the AMPK/FoxO axis include stress resistance genes such as thioredoxin and DNA repair genes.

The Sirtuin family of enzymes – and in particular one member, the *Silent information regulator 1* (SIRT1) – is implicated in regulation of the stress response and the lifespan. The sirtuins deacetylate histones and transcription factors, and this post-translational modification regulates the function of the respective proteins (see Figure 5.22). The enzymes require NAD^+ for their activity, and in this way the sirtuins act as energy sensors. When food is not limiting the concentration of NADH is high; however, fasting and physical exercise will reduce the NADH level and, as a consequence, the NAD/NADH ratio is increased (Figure 6.1). The activation of SIRT as a result of energy depletion and raised NAD^+ levels may, therefore, be regarded as an adaptive response. Although SIRT and AMPK each act as energy sensors, the proteins respond to different metabolic parameters. Both sensors act in concert and are implicated in, for example, stress resistance, apoptosis, and inflammation. SIRT and AMPK also cooperate in the activation of FoxO transcription factors. When energy levels are low the upregulation of SIRT leads to a critical downregulation of mTORC (Figure 6.2). The stimulation of SIRT has been claimed as life-prolonging in animal models, and similar positive effects have been suggested for humans.

One important mTORC-regulated cellular function is the peculiar process of "self-digestion" (autophagy). During the course of autophagy, damaged proteins and defective organelles are digested within the cell in a tightly controlled process that involves the formation of vesicles (autophagosomes) for the degradation of selected cellular components. During aging, the capacity for autophagy declines and this is accompanied by the accumulation of waste products. mTORC-mediated autophagy is thought to be a crucial process for extending the lifespan (Figure 6.2). AMPK stimulates autophagy by attenuating the mTORC activity; FoxO was also shown to promote this process by activating inducers of autophagy.

AMPK is stimulated by mitochondrial reactive oxygen species (ROS), and this is presumably one reason for the positive effect of mitochondrial ROS signaling. The life-prolonging effect of moderate CR seems to trigger a similar hormetic response as the mild oxidative effects of natural substances (see Section 5.1.5). Excessive ROS, however, results in detrimental oxidative stress and, similarly, severe CR will at some point produce a negative physiological stress. Many questions concerning the regulation of longevity remain to be answered, and particularly puzzling is the observation that the experimental manipulation of one organ affects the lifespan of all organs in an animal. This suggests a hormonal regulation, but the nature of this signal is unknown. Another surprising finding is that some neurons seem to play particularly important roles in the regulation of the lifespan in flies and worms, while in mammals the role of the hypothalamus in CR has become a focal point of research.

It has been seen that CR and other stress situations which deplete the energy reserves of cells are life-prolonging. Since, as a general recipe for a healthy and long life this is unsatisfactory, might there be a less painful route to longevity? The search for CR-mimetic substances has spurred research into aging mechanisms, and bearing the described mechanisms in mind the question should now be asked if there is evidence for a health benefit connected with certain food or food products, and if certain specific substances interfere with the network regulating lifespan.

6.1.2
Resveratrol: Longevity and Health with Grapes?

Resveratrol-containing preparations have been used in traditional medicine in Asia for thousands of years. In 1997, the substance gained scientific interest when it was shown to inhibit cyclooxygenase (a mediator of inflammations) and also demonstrated a protective activity against cancer in a rodent model. Subsequently, interest in this natural substance increased even more dramatically when it was found that, in yeast, resveratrol could activate a histone deacetylase (in this species named sir2) and thus prolong the lifespan of the cells. This pathway has been conserved during evolution, and the critical role of sirtuins was confirmed in several animal models, including mammals. A systematic search for natural SIRT agonists revealed that the stilbene resveratrol (Table 6.1) was able to stimulate the enzyme activity, and that the chalcone butein and the flavonoids fisetin and quercetin were also active, although to a lesser degree. Since these discoveries, made in 2003, there have been numerous reports on resveratrol, notably because a natural anti-aging product would clearly have enormous medical (and economic!) potential. At this point the current knowledge of this topic will be reviewed in some detail, as the well-researched biological and pharmacological effects make resveratrol a "model substance" which provides important information about the biology of natural substances and the cumbersome road to a full understanding of the working mechanisms. With all the knowledge that has accumulated over the years the picture has become highly complex, but no less interesting!

In mammals, the sirtuin family of enzymes consists of seven members whose activity is NAD-dependent but the proteins differ in their enzymatic activity. They are involved in the regulation of different processes relevant to health, and also stimulate processes such as insulin sensitivity, improve genome stability, and reduce inflammatory reactions and tumor growth. In addition, the circadian rhythm is modulated by sirtuins, and they may protect against neurodegenerative diseases. Some of the effects resemble those observed under conditions of CR and, furthermore, the disease-protecting activities of SIRT1 and of resveratrol concern the same age-related diseases. In higher organisms it is not only SIRT1 that seems to mediate the positive physiological effects; SIRT3 is also expressed under CR conditions, while SIRT6 was found to play a role in the control of both genome stability and cell senescence.

Although the effect of resveratrol on sirtuins was detected first, other target proteins of resveratrol have more recently come into focus and now proteins in biochemical pathways upstream of the sirtuins have been identified as relevant targets (Figure 6.3). Under conditions of CR when the insulin and IGF-1 levels decline, catecholamine signaling increases and the hormone glucagon is upregulated to maintain vital physiological functions and to ensure an sufficient supply of glucose from stored glycogen. When the G-protein-coupled glucagon receptor is activated, cAMP levels rise due to the stimulated activity of adenylyl cyclase. Phosphodiesterase inhibits this rise in cAMP by converting it to adenosine 5′-monophosphate (AMP). Resveratrol is an efficient inhibitor of phosphodiesterase, and hence the cAMP levels are allowed to rise in its presence; this, in turn, activates AMPK, the energy sensor which has already been identified as a key enzyme in mediating the effects of CR (Figure 6.2). The AMPK-mediated increase in NAD^+ activates SIRT and this produces the described beneficial effects (Figure 6.3). AMPK activity was found to be stimulated by resveratrol, but in view of the upstream effect on

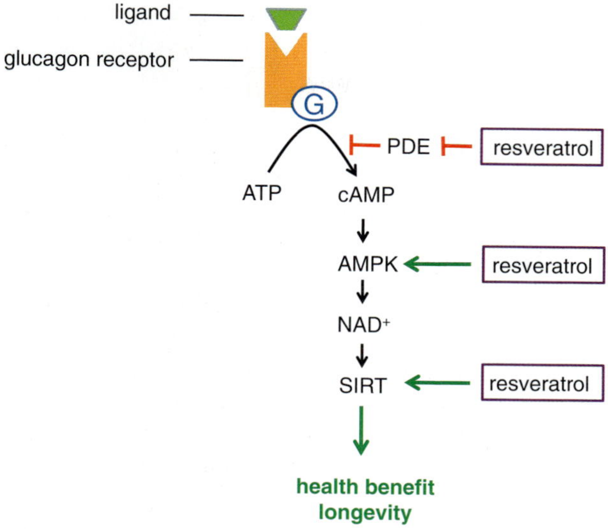

Figure 6.3 Target proteins of resveratrol. Resveratrol affects several proteins implicated in metabolic effects of caloric restriction. When food is limiting, glucagon is upregulated while insulin (and IGF) levels decline. The glucagon receptor recruits a G-protein (G) to activate adenylyl cyclase. This is prevented by the activity of phosphodiesterase (PDE) which converts cAMP to AMP. Resveratrol inhibits this enzyme and hence the cAMP concentration increases. The observed stimulation of AMPK and SIRT1 (see Figure 6.2) by resveratrol may be a secondary effect. Data from Park, S.-J. *et al.* (2012) *Cell*, **148**, 421–433.

phosphodiesterase it now seems that neither AMPK nor SIRT1 are the direct targets of resveratrol. The reaction cascade illustrated needs to be confirmed in future experiments, and it seems likely that the underlying network of reactions will turn out to be much more complex than presently envisaged. The interference with cAMP signaling prompts new questions concerning the role of resveratrol, since cAMP is a second messenger in several important signal transduction pathways and hence other and possibly unwanted reactions are conceivable.

Phosphodiesterase is clearly not the only target protein that is affected by resveratrol. Natural substances typically interact with numerous cellular proteins and resveratrol is no exception. For example, resveratrol has anti-inflammatory activity by inhibiting the important enzymes lipoxygenase and cyclooxygenase, and this activity is also thought to reduce the incidence of cancer. Furthermore, resveratrol was reported to inhibit DNA polymerase, aromatase and other enzymes with different and unrelated functions. The inhibition was determined *in vitro*, and the determined IC_{50}-values were typically higher than 20 μM. Whether these comparatively weak activities are relevant *in vivo* is not known. The interaction of resveratrol with glycerophospholipids of cell membranes is likely to affect the activities of a whole range of membrane-bound proteins.

Two major signal transduction pathways that are controlled by the peptides epidermal growth factor (EGF) and insulin/IGF-1 are particularly relevant to health

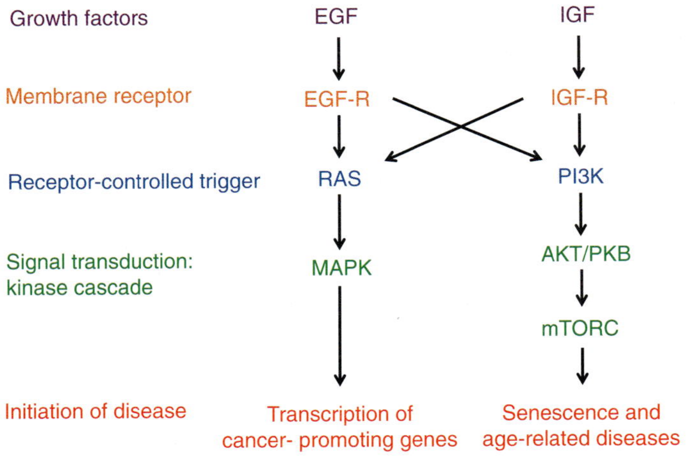

Figure 6.4 Two important signaling pathways in health and disease. Common diseases such as diabetes, neurodegenerative diseases and cancer may develop if the normal control mechanisms of major signaling pathways are compromised. Two important pathways are shown which are activated by the peptide hormones EGF and IGF. These extracellular signals are recognized by the respective membrane receptors (EGF-R and IGF-R) which initiate a cascade of further reactions. Two small molecules play important roles at the beginning of the signal transduction pathway: phosphatidylinositol-3-phosphate (PIP3), which is generated by the Pi3-kinase, and GTP, which is converted to GDP by the activity of the GTPase activity of the RAS. These two small molecules activate important kinases such as MAPK, Akt/PKB and mTORC. Both pathways may influence each other (crosstalk).

(Figure 6.4). Excessive signaling through these pathways promotes cancerogenesis. Resveratrol interferes with both pathways directly or indirectly, and in addition to the already discussed effect of resveratrol on mTORC, the natural compound also interferes with signaling by the EGF receptor (EGFR). The signaling pathways have some properties in common. In both cases the ligand–receptor complex triggers a signaling cascade by activating a tyrosine-specific protein kinase which, in turn, activates other downstream kinases. In the case of insulin/IGF-1-mediated signaling the phosphatidylinositol 3-kinase (PI3K) produces phosphatidylinositol (3,4,5)-triphosphate (PIP3) which activates an important downstream kinase in the pathway (Akt). In the case of EGF-R, the activation of Ras (a GTPase) is critical and this initiates another kinase signal transduction cascade. Both receptors belong to complex receptor protein families, and crosstalk between both pathways may occur (Figure 6.4). These signaling pathways are often hyperactivated in cancer, so that natural compounds which normalize the dysregulated pathways can be expected to be beneficial. The large number of known target proteins of resveratrol, and the complexity of the signal transduction pathways involved, do not permit a simple mechanistic explanation for the biological effect of resveratrol. Given the uncertainty about the relevant molecular mechanisms, how could resveratrol-containing food have a health-promoting effect?

6.1.3
Food with Health-Promoting Bioactive Substances

6.1.3.1 Grapes (Resveratrol)

The best natural sources of resveratrol are red grapes and, to a lesser extent, white grapes. The concentration of resveratrol in wine may vary from undetectable to 14.3 $mg\,l^{-1}$, depending on the grape variety, the geographic region of cultivation, possible stress conditions of the plants, the technology used to produce the wine, and many other factors. Small amounts of resveratrol are also found in some berries (e.g., cranberry, blueberry) and in peanuts and pistachios.

The metabolism of resveratrol has already been discussed (see Section 4.1.2), and it is important to remember that the bioavailability of resveratrol is poor and micromolar concentrations in the serum cannot be reached. As the toxicity of resveratrol is low, very high concentrations can be tested. For example, in clinical trials (Phase I) up to 7.5 g resveratrol per day were administered orally to the persons taking part. Numerous experiments with animal models have been carried out to demonstrate beneficial effects *in vivo*, and these have provided encouraging results.

Glucose tolerance was improved and other beneficial effects were also noted in animal experiments, and positive effects of resveratrol on some types of cancer have also been reported. Resveratrol suppresses the formation of atherosclerotic plaques, inhibits the oxidation of LDL, suppresses platelet aggregation, possesses anti-inflammatory activity, and demonstrates a vasodilatory effect by stimulating the production of NO by the endothelium. Activation of the antioxidative defense mechanisms, including downregulation of the ROS-producing NADPH oxidase, supports the cardioprotective effects. Of particular interest is resveratrol's neuroprotective activity, a finding which fosters hope of reducing the neurodegeneration associated with Alzheimer's and Parkinson's diseases. The risk of Alzheimer's disease occurring at a later age seems to depend on cardiovascular risk factors and metabolic syndrome earlier in life. However, the data acquired from experiments with rodent models need to be confirmed, and standardized clinical studies are required to translate these findings into medical treatments for these severe age-related diseases.

Although high doses of resveratrol are well tolerated in mammals, it is unclear how much resveratrol is really needed to obtain statistically significant effects. The small number of studies conducted with low concentrations of resveratrol have suggested that a few milligrams each day is sufficient, and a significant vasodilatory effect has even been observed with a daily dose of only 30 mg over a one-week period. Is it sufficient to drink grape juice or red wine to benefit from the positive effects of resveratrol? In France, red wine is a popular drink and the incidence of coronary heart diseases is indeed comparatively low. This effect is even more astounding considering the fact that the typical French diet is rather rich in saturated fats that are presumed to promote coronary heart diseases. This discrepancy is known as the "French paradox," and several explanations have been proposed to explain this effect. Particular attention has been paid to the possible protective effects of red wine on the blood vessels, and numerous studies have been conducted to demonstrate these. In most cases, nonalcoholic grape products were consumed in order to avoid a possible confounding influence of alcohol. However, no evidence has been acquired that a modest amount of alcohol either reduces or eliminates the beneficial effects of grape polyphenols. The cardioprotective effect of grape juice has been demonstrated in several studies, with critical parameters all being affected in a positive sense: blood pressure was decreased, platelet aggregation was lowered, NO production was stimulated (and hence vasodilation promoted), and LDL oxidation was reduced. Despite the different experimental conditions employed, there was a general agreement regarding the cardioprotective effect of grape products. The influence of grape products on other age-related diseases such as cancer, diabetes and Alzheimer's has also been studied and beneficial effects have been claimed.

It remains unclear as to whether these observed positive effects are due exclusively to resveratrol, however, as grapes also contain a variety of different polyphenols including phenolic acids, anthocyanins, hydroxycinnamates, monomeric flavonoids and oligomeric catechins (proanthocyanidins). The hormone melatonin and the biologically active substance hydroxytyrosol are typical components of grapes, albeit in low concentrations. As the content of these substances in the skin, flesh and seeds of grapes differs, the various grape products (juice, dealcoholized wine, extracts of seeds or skin) will differ in the concentrations of these components. White grapes typically contain fewer polyphenols than red grapes, and lack anthocyanins. It is generally believed that resveratrol is the main contributor with regards to the health benefits of grapes, although dietary intervention studies with grape products have failed to prove this point. In fact, proanthocyanidins and other polyphenols in grapes may also contribute to the observed beneficial effects. Of particular interest is the compound hydroxytyrosol (Table 6.1), a phenolic alcohol derived from oleuropein (Figure 6.5). Both oleuropein and hydroxytyrosol are very concentrated in olive oil, and are considered to be responsible for the positive effects of olive oil on the cardiovascular system, and the prevention of diabetes and cancer. Moreover, both are presumed to have neuroprotective functions. It is conceivable –

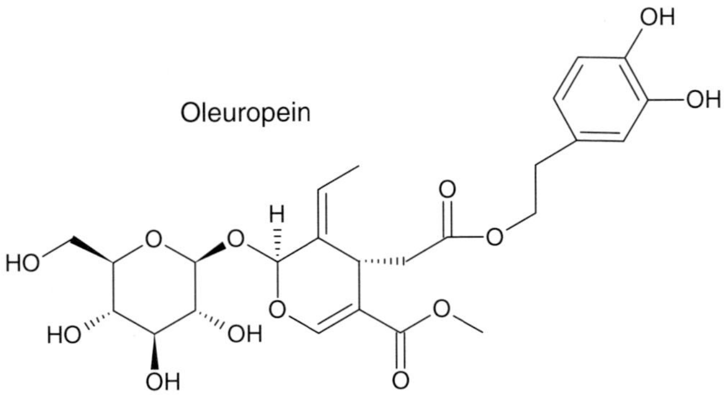

Figure 6.5 Chemical structure of oleuropein. This natural compound is present in high concentrations in olives and olive oil, and in lower concentration also in grapes.

but difficult to demonstrate experimentally – that different polyphenols in grapes have synergistic effects and contribute to the positive effects on health.

The antioxidative potential of grape polyphenols is high, and this property is shared by polyphenols with chemically different structures. Whilst resveratrol is a potent inhibitor of phosphodiesterase and also of the important kinase mTORC, its chemical properties as an electrophilic molecule are of interest and may, in part, explain the observed medicinal benefits. Resveratrol activates the Nrf2 transcription factor, and the elicited hormetic response (see Section 5.1.5) may contribute substantially to the observed health benefits. Similar prooxidative substances are typically present in the following examples of "healthy" food, and may elicit an adaptive beneficial response.

6.1.3.2 **Broccoli (Sulforaphane)**

Cruciferous vegetables such as broccoli contain vitamin C and E as well as several interesting secondary metabolites including carotenoids, the flavonoids quercetin and kaempferol, and several glucosinolates, which sets crucifers apart from other vegetables. Numerous beneficial effects have been ascribed to the consumption of broccoli. For example, it was suggested that broccoli lowers the risk for several types of cancer and for myocardial infarction (Figure 6.6). These beneficial effects may be due to an anti-inflammatory activity, the induction of apoptosis, and an increase in the antioxidative defense mechanisms.

The glucosinolates consist of modified amino acids such as tryptophan, methionine or phenylalanine (Figure 6.6). The glucosinolates are themselves biologically inactive, and need to be activated by enzymatic digestion (see also Figure 1.12). Sulforaphane is generated from glucosinolate due to the activity of plant myrosinases that are ingested with the vegetable. Although the cooking of broccoli denatures and inactivates the enzyme, the reaction can also be carried out by bacterial thioglucosidases of the gut microbiota (Figure 6.6). The available evidence suggests that most (but not all) of the biological activity of broccoli derives from the formation of sulforaphane, and for this reason various studies have been conducted with the pure substance or with water-soluble broccoli extracts. Several clinical intervention studies with cruciferous vegetables have also been carried out, in addition to epidemiological studies, to demonstrate the protective role of broccoli against cancer.

Sulforaphane is a reactive compound with prooxidative activity which activates the antioxidative defense system and elicits a hormetic response (see Section 5.1.5). The substance is taken up in the intestine and may form conjugates with glutathione, via a reaction catalyzed by glutathione-*S*-transferase (GST). Although human polymorphisms of these enzymes have been addressed in several studies, a clear picture with respect to the implication for cancer prevention has not yet emerged. The ability of sulforaphane to interact with biotransformation enzymes is presumably relevant to its cancer-protecting activity. In particular, Phase II biotransformation enzymes are upregulated, though this also applies to GST activity which was shown to increase after the daily ingestion of 200 g broccoli (and other crucifers) for six days. The activation of Phase II metabolism may help to remove carcinogens from the systemic circulation more efficiently. In contrast, Phase I

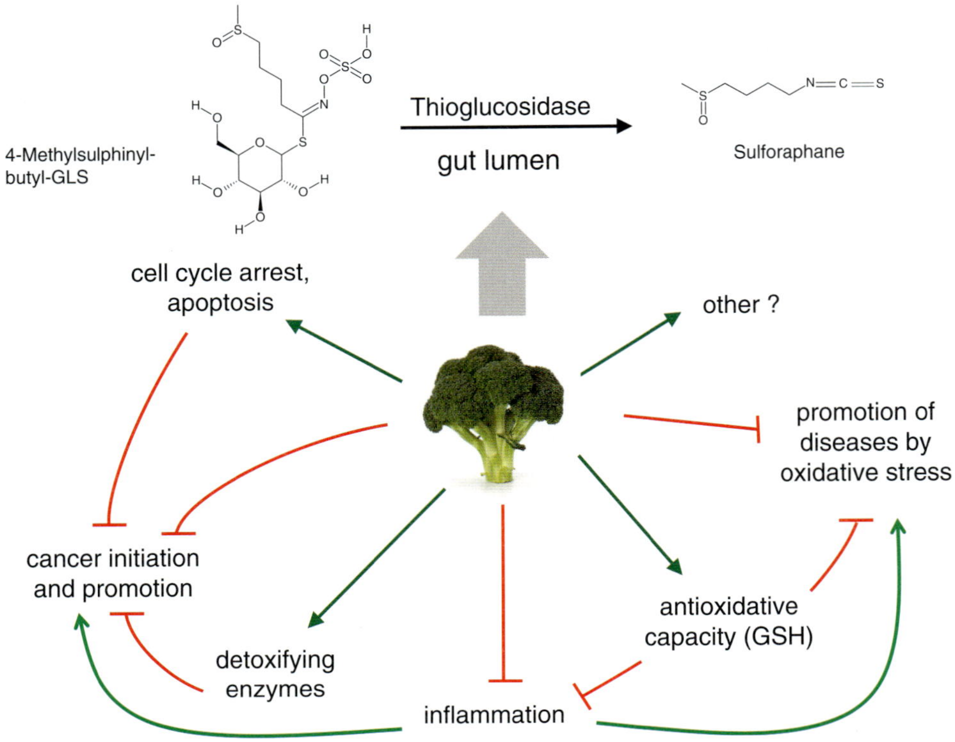

Figure 6.6 Health benefits ascribed to broccoli and the presumed molecular basis. Broccoli is thought to have positive effects in inflammatory diseases including cancer and in diseases resulting from oxidative stress. Broccoli contains glucosinolates (GLS), and an example for this class of compounds is shown. Plant enzymes (myrosinases) or bacterial enzymes in the gut (thioglucosidases) degrade the substance to yield sulforaphane, the biologically active compound. Photograph © fotolia/Natika.

biotransformation enzymes are generally inhibited, except that CYP1A1 is induced after broccoli ingestion. As this enzyme also inactivates estradiol, the possibility of adverse side effects should also be considered.

6.1.3.3 **Hops (Xanthohumol)**

Plants of the genus *Humulus* (Family Cannabinaceae) originate presumably from China. However, one species (*Humulus lupus*) has become very well known due to the hop cones having been used in the brewing of beer for many hundreds of years in Europe. Hops gives beer its characteristic flavor and also provide a preservative function due to its antimicrobial activity. Apart from being used for beer production, hops also has a medicinal history, and has been used for many indications in Chinese and Ayurvedic medicines. In recent times the properties of hops as a mild sedative and a remedy against sleeping disorders have been appreciated; moreover, hops may also activate gastric functions. In experimental studies the

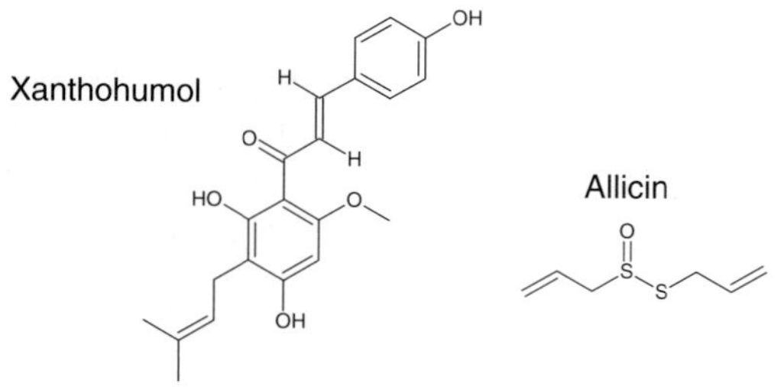

Figure 6.7 Xanthohumol and allicin. Xanthohumol is present in hops, while allicin is a natural substance in garlic. Beneficial effects are ascribed to both compounds.

biological effects of hops (or its bioactive components) have been analyzed both *in vitro* and *in vivo*; subsequently, anti-inflammatory, antioxidative and antiproliferative activities have been reported, and the anticancerogenic activity was explored.

Mature hop cones contain a variety of different polyphenols, including quercetin, kaempferol and rutin glycosides, (epi)catechin gallate and terpenes. Two flavonoids present in hops have attracted particular attention: the estrogenic substance 8-prenylnaringenin (see Section 5.2.4) and the prenylated chalcone xanthohumol (Figure 6.7). The above-mentioned beneficial effects are mostly ascribed to the activity of xanthohumol. In animal experiments, tumor-induced angiogenesis was inhibited, as was the proliferation of human breast cancer cells in a mouse xenograft model. These encouraging results have not been substantiated in clinical studies, however.

6.1.3.4 Green Tea (Epigallocatechin 3-Gallate)

Green tea is particularly popular in Asia, and amounts to about 20% of the tea consumed worldwide. The high content of polyphenols is held responsible for the anti-inflammatory and antioxidant effects. Epidemiological studies have suggested beneficial effects of green tea in reducing the incidence of some types of cancer, though the results of clinical trials have been inconclusive. Different flavonols such as quercetin, kaempferol and myricetin are present in tea, but the prevailing compounds belong to the class of catechins (Table 6.1), with epigallocatechin-3-gallate (EGCG) being the most abundant compound (50–75% of all catechins). The bioavailability of EGCG is poor and few data have been acquired concerning its pharmacokinetics. A large body of *in vitro* data has been produced, and numerous target proteins in different signal transduction pathways have been studied. For example, EGCG affects the activity of proteins regulating the cell cycle and apoptosis, the cancer-relevant Ras/MAPK pathway, and also the PI3K/Akt and the IGF-1 pathway (Figure 6.4). With regards to the anti-inflammatory effect of EGCG, an important target is the enzyme cyclooxygenase which controls the production of prostaglandins in inflammatory reactions. Clearly, EGCG has many target proteins and the reaction mechanisms *in vivo* are not known. Catechins also affect several important biotransformation enzymes and hence affect the metabolism of pharmaceuticals (see Section 4.1.1). These known interactions send a warning in view of the use of EGCG as a dietary supplement.

6.1.3.5 Garlic (Allicin)

Garlic is a well-known spice and herbal remedy, and has been used traditionally to treat infections, diarrhea, rheumatism, diabetes, and other disorders. Anticancerogenic and antibiotic activities are ascribed to garlic, but in recent years it is mainly the cardioprotective effects that have come into focus. It has been reported that garlic reduces platelet aggregation, lowers total blood cholesterol (and LDL concentrations), helps in the control of blood pressure (vasorelaxation), and also strengthens the antioxidant defense. These positive effects that protect against cardiovascular diseases are thought to be due to allicin, the major active compound in garlic (Figure 6.7). Allicin is a short-lived compound present in the volatile oil of garlic; it is also a component of onions and shallots. Because of its hydrophobic nature, allicin can penetrate cell membranes and is rapidly taken up in the intestine. The biologically important feature is the high reactivity of allicin with glutathione and the sulfhydryl groups of proteins. The prooxidant activity elicits an adaptive response (as discussed before; see Section 5.1.5), while the antioxidative capacity is increased by stimulating glutathione production. Several genes involved in the control of oxidative stress are also upregulated.

6.1.3.6 Other Polyphenol-Rich Foods

Various types of vegetables and fruits have been advertised as "healthy," but reliable and reproducible epidemiological or clinical studies with different diets are lacking, and hence the evidence is largely based on *in-vitro* studies and experimental evidence from animal models. Examples of food with high contents of different polyphenols (Table 6.1) may all be considered "healthy" in a broad sense. Cherries, for example, are rich in anthocyanidins, quercetin and carotenoids, and health benefits have been ascribed to their consumption. Similarly, some evidence suggests that flavonol-rich dark chocolate has anti-inflammatory activity and protects against myocardial infarction. The list of similar claims with different food and food products is long, and only spices as an especially rich source for polyphenols will be discussed briefly at this point.

Many spices have been reported to have anti-inflammatory, antiviral and anticarcinogenic activities. Furthermore, they may improve insulin-sensitivity and counteract weight gain and high blood pressure, thus reducing the risk of metabolic syndrome. More specifically, cinnamon has a long history as a spice, preservative and medicinal drug. Cinnamon oil obtained from the bark of the small cinnamon tree contains numerous secondary metabolites, including cinnamic acid and cinnamic alcohol and the prevailing compound cinnamaldehyde (Table 6.1; see also Section 3.2.2). Cinnamon has been claimed to be beneficial in cases of diabetes, in inflammatory and cardiovascular diseases, and also neurodegenerative diseases. The molecular pathways that mediate the effects are not known, but the expression of numerous genes involved in the insulin signaling pathway and in inflammatory reactions is known to be affected by cinnamon. In clinical trials, cinnamon had a positive effect on glycemic control and the potential of cinnamon for the treatment of diabetes was suggested.

Numerous other herbs and spices, including turmeric, rosemary and curcumin, have been studied for their beneficial effects on health. Based on the substantial literature on this subject, the picture is emerging that spices are much more than preservatives and flavors but rather should be viewed as valuable components which also add to the quality of the food from a health perspective.

6.1.3.7 Is There a General Mechanistic Principle for Health-Promoting Substances in Food?

Food contains a mixture of many different polyphenols and other low-molecular-weight secondary metabolites. Despite the different chemical contents, the beneficial effects of different food products are rather similar, and typically a positive influence on major diseases including cardiovascular and inflammatory diseases, cancer and diabetes have been reported. The risks for these diseases increases with obesity and the physiological factors associated with metabolic syndrome (see Figure 6.1). The broad beneficial effects of the respective natural substances seem to share a general "health-promoting" property, but what might this property be? A closer look at the molecules thought to be the responsible substances in food reveals that they are typically electrophilic molecules. For example, xanthohumol, curcumin and cinnamaldehyde (Table 6.1), and also the S-containing molecules sulforaphane and allicin (Figures 6.6 and 6.7), are all able to elicit a hormetic response (see Section 5.1.5). These and many more electrophilic molecules in different foods can generate thiol oxidation, a thiol/disulfide exchange, or S-alkylation. Induction of the Nrf2-pathway seems to be a hallmark of the induced hormetic response to the electrophilic substances. The stress-response proteins that are regulated in response to Nrf2 include glutathione peroxidase, superoxide dismutase, catalase, thioredoxin reductase, peroxiredoxin and thioredoxin glutathione reductase, as well as several biotransformation enzymes. Many dozens of electrophilic molecules in vegetarian foods have been detected which induce the Nrf2 network. The induced reaction strengthens the antioxidative capacity of cells and intensifies the detoxification of xenobiotics. The respective bioactive substances in food have, therefore, been referred to as xeno-hormetic phytochemicals.

It is unclear how flavonoids, as major polyphenolic constituents in food, fit into this picture. In fact, flavonoids such as quercetin and epigallocatechin were also found to stimulate the Nrf2 stress-response network. The mechanism involves, presumably, the generation of electrophiles by the conversion of diphenols to quinones or, in the case of flavones and flavanones, into chalcones (see Section 5.1.4 for details concerning the prooxidant property of flavonoids). Even the stilbene resveratrol was found to stimulate the Nrf2-mediated hermetic response. Based on the activation of this common reaction pathway, the picture emerges that the elicitation of a hormetic stress response seems to explain the major – but not all – benefits of "healthy" food.

There is an interesting evolutionary aspect to the observed hormetic reactions. Many organisms such as viruses and parasites use plants as hosts, while other organisms use plants as a food source. The plant reacts to these threats and other environmental stressors by synthesizing secondary metabolites (Figure 6.8). The chemical defense of the plant includes compounds with antiviral, antibiotic and

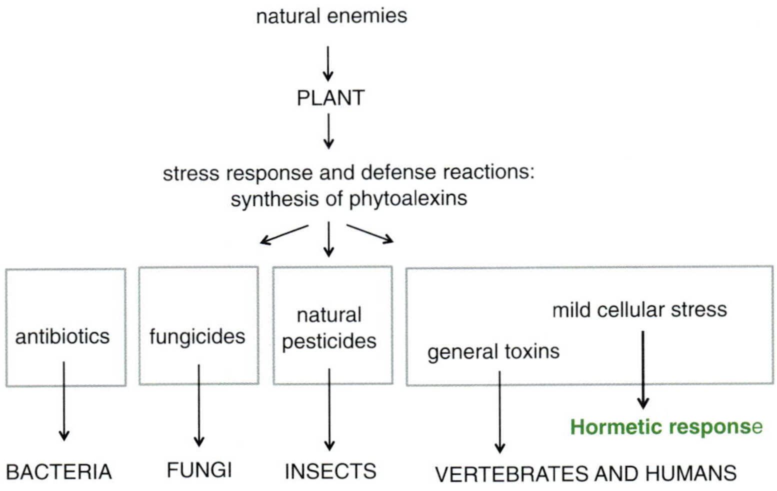

Figure 6.8 Toxicity of phytoalexins for other organisms. Plants produce various toxic substances to defend themselves against bacteria, fungi, or animals. Some of the substances are also toxic to humans, but vertebrates have apparently learned to use many of these substances to their benefit. The elicitation of an adaptive stress response is thought to be beneficial and to protect against common diseases.

antifungal activities; other substances deter insects or other animals will feed on the respective plant species. Although some of these natural compounds are toxic for humans (see Section 4.2.2 for examples), the vast majority of secondary metabolites seem to possess no toxicity for humans at low concentrations; indeed, many of these compounds actually possess health-promoting properties! Historically, human societies have always been dependent on vegetarian foods, and the selection and cultivation of suitable plant species has clearly played a major role in human lifestyles. The cultural tradition of food selection and preparation is also intimately associated with health aspects. Today, it is possible to profit from knowledge that has accumulated over millennia in different societies worldwide.

The low general toxicity of natural compounds for humans should also be viewed from an evolutionary perspective. Mammals have acquired a huge arsenal of enzymes involved in Phase I and II metabolism (see Section 4.1), and efficient mechanisms of detoxification have been developed to cope with the thousands of plant secondary metabolites. It is conceivable that adaptations to biologically active components in plants have gone a step further, however, and that xenobiotics have become signals to elicit a beneficial stress-response. The advantage of this reaction is reflected by the tight correlation between stress-resistance and longevity (Section 6.1.1). As bioactive components are often present in very small amounts in plants, the adaptive physiological reactions of humans have been optimized for these concentrations. This may explain the beneficial low-dose effect that is typical of the biphasic hormetic response (see Figures 5.12 and 6.8).

Is the food consumed by humans today balanced in view of their adaptation to vegetarian food during evolution, or have there been major shifts in the human diet during the past millennia to which humankind has not adapted? It may be illuminating to take a historical view on the human diet. Are the "modern" diseases in developed countries (e.g., obesity and metabolic syndrome) due to an inability to cope metabolically with the diet of modern human societies? About 10 000 years ago, at the end of the Paleolithic period and before agriculture emerged, the human diet consisted mainly of fruits, vegetables, nuts and honey (65%), while lean game, eggs and fish accounted for the remaining 35%. The present-day American diet differs significantly, in that about half of the calories are derived from carbohydrates and one-third from ingested fat. Some preliminary experiments with animals (piglets) have suggested that the Paleolithic diet would offer a better protection against cardiovascular diseases than would a modern cereal-based swine feed. A short-term dietary intervention study with human volunteers came to similar conclusions, but the data were still insufficient to draw firm conclusions. However, it seems clear that polyphenol-rich foods formed a greater fraction of the Paleolithic diet, and that the food was presumably richer in metabolically resistant fibers and n-3 fatty acids, both of which are considered health-promoting. Consequently, should a more Paleolithic type of diet be adopted because it would better reflect the diet to which humans have optimally adapted as a result of evolutionary processes? Whilst this issue has been discussed rather controversially, it would seem advisable to avoid the health-threatening components of modern diets (excessive carbohydrates and fat) and to reduce the calorie intake, and hence the risk for obesity and metabolic syndrome.

The question is: Can general guidelines for healthy food be defined based on the biological properties of natural substances? Some clues to this question have been provided in previous chapters:

- Food should not contain toxic substances (Section 4.2.2).
- Food should contain sufficient (but not excessive) amounts of trace elements (Chapters 4 and 5).
- The amount of food ingested should be reasonable, and occasional fasting might be beneficial (see above and Section 6.1.1).
- Finally, and importantly, some food is particularly rich in bioactive components which stimulate the body's own defense system and elicit a beneficial hormetic response (see Chapters 5 and 6 for examples).

Whilst these general rules apply to healthy persons, individual differences in the metabolism of xenobiotics must be taken into account; indeed, for persons with diseases specific diets may be required.

6.1.4
Functional Food: Does it Hold the Promise?

An increasing number of food products are marketed with the promise of promoting health due to the addition of one or more bioactive substances. Such modified foods have been referred to as "functional foods" or as "neutraceuticals" to indicate the claimed additional health benefit. This claim is a powerful marketing instrument and the commercial interest in functional foods is enormous.

The intensive investigations on resveratrol, combined with its low toxicity, makes this substance an excellent candidate as a food additive. However, the evidence for health benefits by high-dose resveratrol is inconclusive. In a recent study, Vang *et al.* (see Further Reading) summarized all of the available evidence and concluded:

> "Published evidence today is not sufficiently strong to justify recommendation for the chronic administration of resveratrol to human beings, beyond the dose which can be obtained from dietary sources."

The argument for other well-studied natural substances is similar. In the case of genistein, for example, epidemiological evidence suggested a cancer-protecting activity (see Chapter 5) but clinical studies with estrogenic isoflavones were inconclusive. In fact, substances that are essential and health-promoting at low doses may have serious side effects when administered at high doses. No natural compound administered at a high dose has yet shown exclusively beneficial effects, and this also holds true for vitamins and trace elements that are provided in sufficient amounts in the normal diet. Similar conclusions can be drawn from attempts to elicit beneficial effects with high-dose antioxidants such as vitamin C or polyphenols since, in clinical studies, the results of such treatments were generally disappointing. Apparently, it is impossible to obtain a linear dose–response relationship for antioxidants (see Section 5.1.5), and this also holds true for many natural substances that show low-dose beneficial health effects. In contrast, an excessive stimulation of the Nrf2 antioxidant network was shown to increase the cancer risk.

The apparent low-dose effects observed *in vivo* contrast with reported effects of bioactive substances *in vitro*. The concentration required to inhibit the activity of a particular enzyme *in vitro* frequently exceeds 10 μM. Whilst a low-dose hormetic effect offers a partial explanation for this discrepancy, the molecular promiscuity of natural substances is another important factor. As shown for quercetin, resveratrol and many other compounds, these substances can interact with numerous cellular target proteins and in this way affect the complex network of regulatory proteins (see Section 6.3).

6.2 Natural Compounds in Pharmacological Research

6.2.1 Mining for Bioactive Compounds in Marine and Terrestrial Organisms

Bioactive substances with interesting biological properties have been isolated from many organisms. Although microorganisms and plants have been the most important sources of useful natural compounds, the huge and largely untapped source of secondary metabolites in marine organisms has recently come into focus. To date, only about 22 000 natural substances have been identified from aquatic organisms, compared to the estimated 130 000–160 000 substances from terrestrial organisms. Some marine ecosystems such as coral reefs are characterized by an enormous biodiversity, and when the competition for an ecological niche is strong the development of useful secondary metabolites can confer a selective advantage to the respective organism. Hence, ecosystems with great biodiversity can be expected to serve as good sources for bioactive compounds. The following examples of natural substances from marine organisms were not selected for therapeutic relevance; rather, they illustrate the approach taken to identify substances with potentially useful pharmacological properties.

Next to cardiovascular diseases cancer is the second most frequent cause of death worldwide, and the development of new drugs to treat cancer is currently a central theme of biomedical research. The general importance of natural products in the fight against cancer is illustrated by the high percentage (50%) of approved antitumor drugs that are derived from natural sources or are synthetic analogs of natural compounds. With respect to secondary metabolites from marine sources, only three drugs have been approved for the treatment of different types of cancer; however, in screens for bioactive substances from marine sources numerous compounds have been identified that affect six critical processes in the progression of cancer (Figure 6.9). Examples of such candidate drugs affecting any of the six hallmarks of cancer are provided in the following subsections.

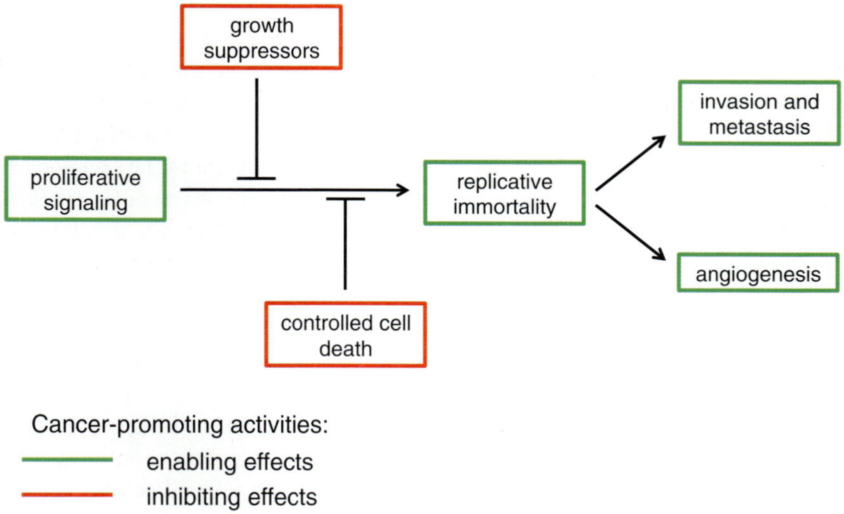

Figure 6.9 Dysregulation of critical regulatory processes leads to cancerogenesis. The latter process proceeds stepwise, and the failure of important activating (green) and inhibitory processes (red) are hallmarks of cancer. The underlying molecular control mechanisms are of particular interest for pharmacological intervention, and many natural substances with anticancerogenic activity have been identified (see Figure 6.10 for examples). Based on Hanahan, D. and Weinberg, R.A. (2011) *Cell*, **144**, 646–674.

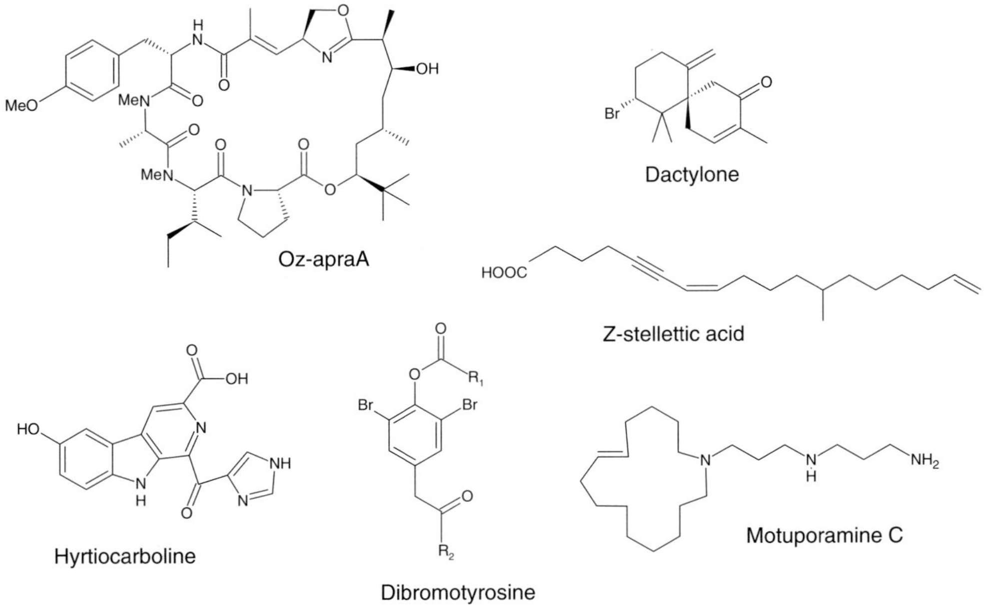

Figure 6.10 Anticancerogenic substances from marine organisms. The chosen compounds target the six hallmarks in the development of cancer (see Figure 6.9). Examples from Schumacher, M. *et al.* (2011) *Biotechnol. Adv.*, **29**, 531–547.

An important signal of cell proliferation is mediated by the EGFR and the downstream signaling cascade (the first hallmark of cancer; Figure 6.9). Excessive signaling can be inhibited by an induced endocytosis and degradation of the membrane receptor by the cyclodepsipeptide oz-apraA (an analog of apratoxin A; Figure 6.10), produced by the marine cyanobacterium *Lyngbya majuscula*. Cancer cells need to evade the controlling influences of cell cycle checkpoints (hallmark 2), and the sesquiterpene dactylone (Figure 6.10) was found to inhibit proteins that mediate the transition from G_1 to S phase of the cell cycle. Cancer cells are resistant to apoptosis (hallmark 3), and the induction of apoptosis-inducing pathways in cancer cells is a possible strategy for reducing tumor growth. A β-carboline alkaloid (hyrtiocarboline; Figure 6.10) was isolated from the sponge *Hyrtiosreticulatus* and shown to be cytotoxic to a number of human cancer cell lines at low (micromolar) concentrations. The replicative potential of tumor cells is increased and the normal senescence by telomere shortening delayed or abolished (hallmark 4). The enzyme telomerase restores the telomere length, and hence is a promising drug target. *Z*-stellettic acid (Figure 6.10) from the sponge *Stelletta* sp. was reported to inhibit telomerase, while secondary metabolites of the marine sponge *Aplysina* sp. are based on the scaffold of dibromotyrosine (Figure 6.10) and may serve as examples for inhibitors affecting angiogenesis (hallmark 5). Signaling by the vascular endothelial growth factor (VEGF) initiates the sprouting of new vessels that support the growth of solid tumors. Finally, invasion and metastasis (hallmark 6) can be slowed-down by a number of natural compounds. The molecular targets include membrane proteins such as E-cadherin or integrin, or actin and myosin as components of the cytoskeleton, which produce the force for the migration of metastatic tumor cells. The inhibitors of actin polymerization, latrunculin (Table 3.2) and motuporamine C (Figure 6.10), have been isolated from marine sponges.

The examples of natural substances with biological activities are structurally very different, and in most cases the substances never become approved anticancer drugs. A low bioavailability, rapid metabolism or undesirable biological effects and high toxicity due to interactions with other targets lead to most identified inhibitors being unsuitable for clinical applications. However, the identified chemical structure may inspire organic chemists to (partially) synthesize derivatives with improved structures (Section 6.2.2).

Tests which are carried out to identify the anticancer activity of natural substances often involve the use of human cancer cell lines. Induced cytotoxicity is a poor criterion unless some selectivity for transformed cancer cells is shown with reference to wild-type control cultures. Human primary cell cultures or nontransformed (but immortalized) cell cultures can be used for this purpose. An example of this critical experimental approach to demonstrate selective cytotoxicity for cancer cells involved the substance piperlongumine (Figure 6.11), which is not of marine origin but rather a secondary metabolite of the Long pepper (*Piper longum*). Piperlongumine affects the redox balance in cancer cells, and the increased oxidative stress in tested cancer cells is thought to be the cause of the cytotoxicity. The inhibition of mammary tumors by piperlongumine in a mouse model was also demonstrated. It is important to realize that such rigorous tests have been carried out very rarely with isolated natural substances, and many of the claims for antitumor activity must be viewed with caution. Clearly, many of the identified marine compounds still need to be tested rigorously, and few (if any) will gain final approval as anticancer drugs without further chemical modification.

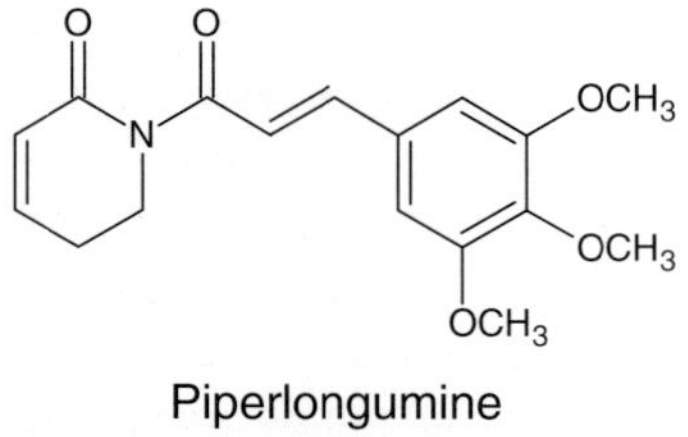

Figure 6.11 Piperlongumine has anticancerogenic activity. The anticancerogenic activity of perolongumine, a compound from long pepper (*Piper longum*), has been demonstrated in cell cultures as well as in breast cancer animal models.

6.2.2
Medicinal Chemistry Inspired by Natural Products

At the end of the nineteenth century, it was the rapidly developing synthetic chemistry that stimulated pharmacological applications, but it was during the twentieth century that the pharmaceutical companies really "boomed." The invention of Aspirin in 1899 laid the foundation of Bayer's dominance until the start of the First World War. Subsequently, the successful development of drugs against devastating infectious diseases were the highlights of pharmaceutical development during the twentieth century. For example, Paul Ehrlich developed a granular arsenic-containing efficient drug against syphilis ("Salvarsan") which became known as the "magic bullet," but this was later replaced by penicillin, discovered by Alexander Fleming.

The development of pharmaceutical drugs was driven by the belief that a cure for a specific disease such as an infection could be achieved without side effects by identifying a specific drug against the disease-causing agent. This mechanistic view of a disease based on a faulty component of the physiological machinery (much like a gearwheel in a machine) still prevails today. In modern terminology, a relevant target for a disease must be identified and validated, and a specific drug against this target then developed. In fact, until very recently this approach was highly successful and led to the development of numerous drugs. The most common targets which have been identified in pharmaceutical research are enzymes (47%) and G-protein-coupled receptors (30%). The activity of such proteins can be described as a chemical reaction that is initiated by the availability of an enzyme substrate or a ligand for the receptor. The chemical reaction can be interrupted by specific inhibitors (drugs) of the respective target protein. The concept of drug–target interaction, with the aim of affecting a specific disease-promoting pathway with a specific "magic bullet," has dominated pharmacological research and development to the present day. However, some doubt has recently been voiced about the sustained success of this concept for drug development (see Section 6.3).

How many pharmacologically relevant targets should be considered? The number is in the order of a few hundred, although published data differ depending on the exact definition of a target and whether receptor (sub)types, isoenzymes, individual differences and various combinations of multimeric proteins have been counted. This rather small number contrasts with a much higher number of possible drug–target interaction sites with human proteins, and the enormous number of identified natural compounds (on the order of 160 000). Because natural compounds typically interact with many proteins, it seems likely that natural substances can interact with all pharmacological targets. As a result, when natural substances are screened *in vitro* for the inhibition of a specific enzyme there are typically many "hits" and different flavonoids bind to and affect the activity of almost every class of enzyme. Investigations of the biology of natural substances have contributed substantially to the present knowledge of Medical Chemistry and also in many areas of Molecular Biology. The inspiration of Nature for pharmacological developments concerns not only biologically active compounds but also the respective target proteins involved in the genesis of diseases.

6.2.2.1 Lessons from Natural Substances About Relevant Targets

Taking a historical perspective first, many plant preparations that have been used for centuries in the treatment of diseases contained bioactive compounds which were later identified and, in the intensive research which followed, the target proteins were identified. The cannabinoid Δ^9-THC may serve as an example in this case: the discovery of the cannabinoid receptor and the endogenous ligand anandamide opened up a wide field of research with the discovery of new target proteins for drug development (see Section 5.3). Aspirin may serve as another example: the bark of the willow tree and other sources of salicylate-rich plants have been used to treat fever, pain, and inflammations, and at the end of the nineteenth century Bayer began to experiment with acetylsalicylic acid (marketed as Aspirin since 1899) to replace the common salicylate medicines. Much later, relevant targets were identified and the inhibition of cyclooxygenase activity and of prostaglandin production by Aspirin was reported.

In some cases a natural compound with surprising biological activity (toxic or beneficial) has spurred interest in the respective target protein, and this had an enormous impact on pharmacological research. With regards to a toxic natural product, developmental defects caused by cyclopamine stimulated research into the hedgehog pathway (see Section 5.2.5 and Figure 5.24). In patients with basal cell carcinoma (a form of skin cancer), this signaling pathway is reactivated through the smoothened protein. A synthetic inhibitor of this protein (Vismodegib) has been approved recently for the treatment of this type of cancer.

Resveratrol and its agonistic effect on SIRT1 (see above) is another example of a natural compound which greatly stimulated research in the field of aging and the prevention of age-related diseases. SIRT1 agonists which are potent, more specific, and show improved bioavailability have been sought by several companies, and some active substances have now been identified that are structurally unrelated to resveratrol. Studies on the AMPK and TOR-controlled network (Figure 6.2) led to attention being focused on other target proteins, and screens for CR mimetic substances are presently being carried out. To acquire the beneficial effects of CR with a potent pill would be an attractive and highly profitable achievement!

6.2.2.2 Natural Chemical Scaffolds and their Pharmacological Optimization

Natural substances often show low target specificity and low bioavailability, and this unfortunately limits their therapeutic use. For this reason, natural substances are rarely used as pharmaceutical drugs without modification (about 5% of newly introduced drugs in the period between 1981 and 2002). In most cases a natural substance which has been identified as a lead compound may require chemical modification to improve the pharmacological properties. The pharmacophore of a substance is defined as the structure that mediates the interaction with the target protein; consequently, a chemical modification of the pharmacophore is usually not possible without a loss of biological activity. However, it is possible to substitute or add side groups with the aim to alter the three-dimensional structure of the substance in order to increase the affinity of the drug to the target, and hence its activity as an inhibitor (or activator) of the respective target protein. Such optimization would typically include an improved specificity for the chosen target(s) and an improved bioavailability. The pharmacokinetic properties deserve special attention, and are determined by the lipophilic properties of the compound and its ability to bind to serum transport proteins (albumin). The lipophilicity of a compound is also important in order for it to cross the blood–brain barrier. Perhaps the most important information to be acquired when optimizing a lead compound is the exact spatial orientation of the substance within the target protein. The acquisition of such information generally involves extremely high resolution, X-ray crystallography as the method of choice (see Figure 3.17). Unfortunately, in many cases the optimal chemical modification cannot be predicted reliably, and the success is partly a matter of trial and error. Nonetheless, the possibility of accelerating the phase of drug development based on the identification of a lead compound and the knowledge about structural properties is of major economic importance for pharmaceutical companies.

Attempts to improve natural substances by chemical modification date back to the nineteenth century, and the synthesis of acetylsalicylic acid may serve as an example. Similarly, C.R. Alder Wright, an English chemist, prepared an acetylated form of morphine (diacetylmorphine) in an attempt to prepare a more potent and fast-acting drug that, during the early twentieth century, was marketed under the trademark Heroin. Modern chemistry has many more options to modify known chemical structures. Many polyphenols (including flavonoids) can be synthesized chemically, and this allows the replacement or introduction of new functional groups, and also the testing of biological consequences. This strategy was pursued in the case of epigallocatechin gallate (EGCG), a natural substance that is present in green tea in considerable amounts and has been ascribed beneficial effects on health (see Section 6.1.3). Unfortunately, EGCG is instable, its intestinal uptake and bioavailability are poor, and it is rapidly metabolized. This prompted a search for substances with improved properties, and as a consequence numerous derivatives based on the EGCG chemical scaffold were prepared. In the case of flavonoids and other polyphenols, the biological activity and specificity may often be changed if only a single hydroxyl group is removed or is substituted by another group. However, this approach had its shortcomings, since retaining the basic flavonoid structure did not resolve the principal problem of poor bioavailability and rapid metabolism.

An extension of this strategy is to make use of a given chemical scaffold and to produce a library of related compounds by varying the functional groups of the molecule. The strategy has been referred to as "diversity-oriented synthesis" (DOS), and the resultant library of synthetic compounds may consist of thousands of different substances that resemble natural products due to the presence of the same scaffold. For example, when galanthamine (an AChE inhibitor; see Section

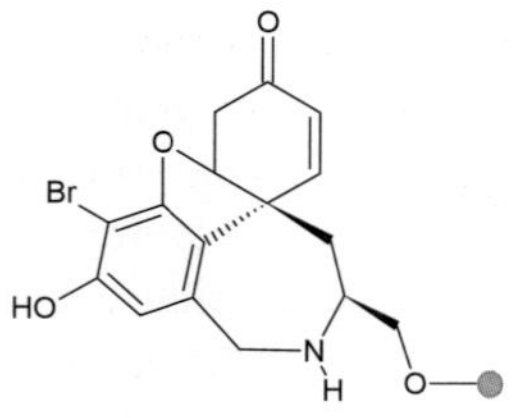

Galanthamine scaffold prepared for chemical modifications

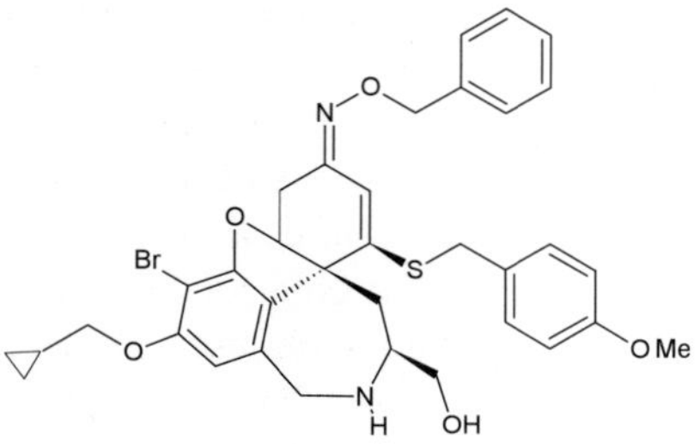

Secramine

Figure 6.12 Chemical modification of natural substances. New biologically active compounds have been generated by chemical modification of the galanthamine scaffold (see Figure 5.39). This strategy of diversity oriented synthesis has been successful and the drug secramine has been generated which has different biological activity compared to the parent compound. Data from Marcaurelle, L.A. and Johannes, C.W. (2008) *Prog. Drug Res.*, **66**, 189–216.

5.3.6) was used as the parent compound, different chemical procedures were applied to replace and modify different groups (see Figure 6.12), such that new compounds could be generated based on the galanthamine scaffold. As the newly generated compounds were structurally related to the natural substance, it could be expected that many biologically active substances would be generated. However, the derivatives may possess quite different activities compared to the parent compound, and this was the case for the galanthamine-derivative secramine (Figure 6.12). The newly generated compound did not inhibit AChE (as does galanthamine) but rather inhibited a Rho-GTPase and blocked dynamic vesicle transport in the cells. The DOS has been successful in several other cases and new drugs have been discovered. The main advantage of using this approach lies in the quality of the chemical library that can be used to screen for a desirable biological activity. A scaffold chosen for the introduction of various substituents need not necessarily be identical to the scaffold of the natural product, but can be closely related to it. In the latter case a *de-novo* synthesis of the structure is usually required, whereas in the former case the scaffold of the natural product can be used to construct a collection of derivatives.

6.2.2.3 Systematic Computational Analysis of Natural Scaffolds and Peptide Motifs

Combinatorial chemistry is a powerful chemical technique to generate a huge number of different substances; however, when such libraries were screened for compounds with a particular biological activity the results were disappointing. Such an approach, when based on synthetic compounds alone, ignores the evolutionary history of proteins, their structural motifs and the evolution of secondary plant metabolites that are selected by Nature to meet the specific structural properties. The chemical structures of natural substances represent only a very small subset of all possible low-molecular-weight organic substances that can be generated chemically. Hence, a distinction is made between the "biological space" in which the organic natural compounds can be positioned and the much larger "chemical space" with all possible structures. A library of "biological" compounds with a high degree of complexity and the prevalence of stereogenic centers will be most useful in screens, as the number of hits which can be expected is high (on the order of 1%). As a result, the library of such Nature-inspired compounds may be much smaller than a library generated by random combinatorial chemistry. The Biology-Oriented Synthesis (BIOS) of Nature-inspired substances has been suggested as a conceptual means for the generation of useful drugs.

Is it possible to identify typical structural features of natural compounds and to order them according to the family relationships? Databases of natural substances are available which can be subjected to computational analysis (cheminformatics). Glycosylation is a common feature of natural compounds (see Chapter 4) and contributes substantially to the diversity of products. Whilst this feature may influence the pharmacokinetic properties, it is not of interest with respect to the chemical scaffold of the substances. If a database (e.g., the CRC Dictionary of Natural Products) is corrected *in silico* for glycosylation diversity, more than 113 000 entries remain. On the basis of common structural features a tree of related substances can be established, starting with simple heterocycles or carboncycles and branching out to more complex ring systems (Figure 6.13). The analysis illustrates nicely the limited number of "families" into which the natural substances can be grouped.

As proteins and natural substances have a long common evolutionary history, it is not surprising that the limited variability in prevailing chemical scaffolds is reflected by a limited set of peptide motifs that are able to interact with the small molecules. Are there common structural motifs in proteins which permit or favor the interaction? The number of possible amino acid sequences is enormous but the structural features of proteins are less diverse. Secondary structural elements such as α-helices and β-sheets are common in proteins. The number of folds was estimated to be on the order of 1000–8000, although if only the number of folds of major protein families is considered then the figure will be much lower. A comparative analysis of protein structures and their similarity (Protein Structure Similarity Clustering) has been carried out with the aim of predicting suitable chemical scaffolds for small molecules that "fit" into the respective protein folds of interest. As protein

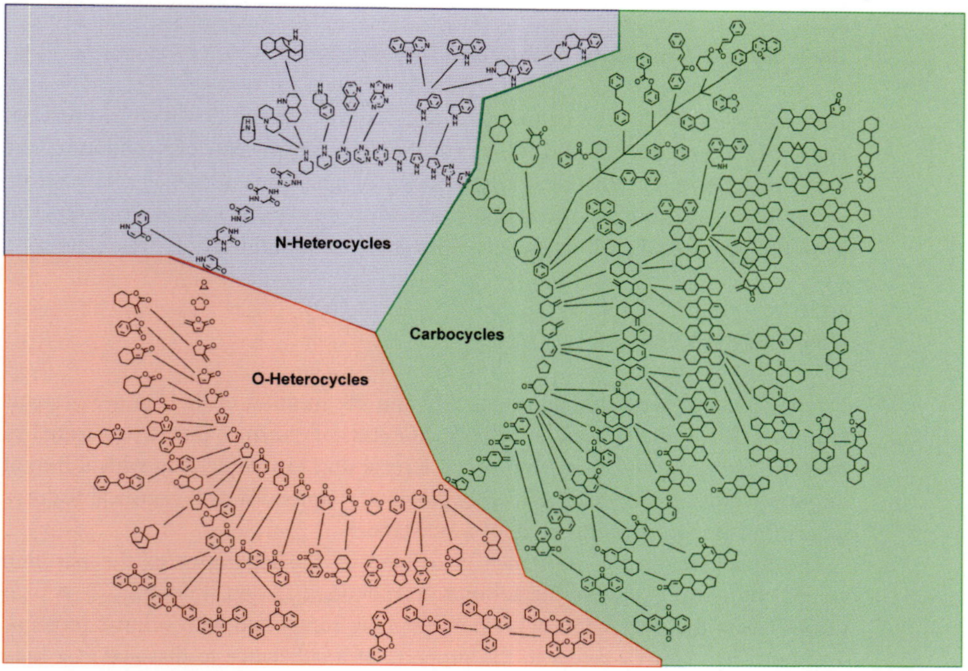

Figure 6.13 Natural substances can be grouped into families with similar structural properties. Only those chemical structures were considered which are listed in the CRC Dictionary of Natural Products and which represent cumulatively at least 0.2% of the natural products. Reproduced with permission from Koch, M.A. *et al.* (2005) *Proc. Natl Acad. Sci. USA*, **102**, 17272–17277 (see Further Reading); © 2005, National Academy of Sciences, USA.

folds are more conserved than the amino acid sequence, it is not possible to analyze structural similarities based on the amino acid sequence alone.

The functional coevolution of natural substances and proteins is most obvious for those enzymes which catalyze reactions in the biological synthesis of the substances by plants. It is, therefore, not surprising that the metabolic enzymes of biotransformation reactions are common targets of natural compounds. Many other proteins such as transcription factors, DNA-binding proteins or cytoskeletal proteins are highly conserved, and can be grouped into large families in which functional protein domains are structurally highly related, or even identical. The evolutionary conservation of protein domains has important implications for drug development: the pharmacophores of many natural substances tend to bind to more than one protein. The dimension of this problem has not yet been analyzed systematically and this is a deplorable gap in the present knowledge. Often, unexpected interactions are learned about only when drugs selected for inhibiting a specific function show unexpected and undesirable side effects due to their interaction with other proteins.

6.3
The Problem of Complexity: Analysis and Consequences

The interaction of natural compounds and Nature-inspired drugs with several or many target proteins has practical implications. A mechanistic explanation for the health effects of specific food components or pharmaceuticals is challenging. Some theoretical considerations and practical approaches to exploit the promiscuity of the compounds will be subject of these final subsections.

6.3.1
Many Targets: A Problem?

Flavonoids have been studied intensively and the number of their target proteins is impressive. There appears to be no major class of enzymes that is unaffected by these compounds and many biological effects due to interactions with different target proteins have been described in previous chapters. Which chemical features determine the target promiscuity of a substance?

The lipophilicity of a substance is defined as the logarithm of the partition coefficient (cLog*P*) between the solvents octanol ($C_{octanol}$) and water (C_{water}). A high cLog*P* favors the interaction of such a lipophilic compound with cellular proteins. The term lipophilicity refers to the property of a substance to dissolve in a lipid phase, and hence hydrophobicity is not a synonymous expression. An increase in lipophilicity correlates strongly with low-affinity promiscuity with different target and nontarget proteins. According to Leeson and Springthorpe (see Further Reading), "Promiscuity is defined as the number of hits for a compound when >30% inhibition is found at a concentration of 10 μM." When testing 2133 compounds in an *in-vitro* panel of 200 different proteins the number of recorded hits was on the order of 40 at a $\text{cLog}P = 4$. As this degree of lipophilicity is representative for drugs listed in medicinal libraries, a high promiscuity must be expected in most cases. A high cLog*P* would tend to increase the toxicity of the substance as well as its tendency to integrate into biomembranes (e.g., curcumin; see Figure 3.19). The molecular weight and the complexity of a substance will also affect its promiscuity.

Most polyphenols such as flavonoids (as aglycones) are lipophilic substances that can readily pass the cell membrane. As polyphenols are present in food as complex mixtures, and given the fact that minor differences in structure (such as the presence or absence of a hydroxyl group) changes the specificity of the compound, the situation arises that a large fraction of the cellular receptors enzymes and proteins in signal transduction cascades is (to different degrees) affected by the polyphenol content of the diet. As a rule, the interactions will be weak and there is no apparent harm, as has been seen with the beneficial effects of selected vegetarian food. In the case of herbal medicine, the promiscuity of the content is expected to be high; moreover, for those substances which are thought to confer health benefits, such as resveratrol and sulforaphane (Section 6.1.3), there is also considerable promiscuity. The drawbacks of target promiscuity call for new approaches for the medical treatment of diseases.

6.3.2 From Pharmacology to Polypharmacology

Drug discovery by the pharmaceutical industry was dominated by the belief that there is a monocausal origin of disease: one dysfunctional gene product gives rise to a disease, and a suitable drug will interact with the target protein and remedy the defect. The screens were designed to select for highly specific compounds for the respective target protein. Much research has been devoted to detect the "disease-causing genes" for cancer and other common diseases, with the aim of identifying key target proteins. This approach was quite successful in that many tumor suppressor genes and oncogenes have been identified which play a role in any of the hallmarks of cancer development. Screens have been devised to identify small molecules which act on the respective gene products and inhibit cancer progression. An example of this approach is the successful search for bioactive compounds from marine organisms (see Figure 6.10; Section 6.2.1).

While the search for bioactive compounds with desirable properties has met with much success over the past decade, this did not result in many drugs receiving approval for the effective combat against cancer. Today, despite enormous financial investments the productivity of the pharmaceutical companies in introducing new drugs has slowed down, and many new drugs fail in Phase II and Phase III clinical trials. The cell-based assays in the preclinical tests have a poor predictive value for the efficacy of the drug *in vivo*. Since the promiscuity of a candidate drug cannot be predicted, side effects are common and even if the drug has been approved and introduced to the market, adverse drug reactions (see Section 4.1.5) may occur that will result in the pharmaceutical being withdrawn from the market. Indeed, during recent years it has become increasingly clear that the concept of "one drug, one target and one disease" is insufficient and may, in part, be the reason for the disappointing results of recent drug developments.

The physiological reactions cannot be viewed as a reaction cascade in which one factor affects another downstream factor, which again affects the activity of another factor in the reaction sequence. Rather, the metabolic processes and numerous intercellular signaling pathways form a highly complex network in each tissue and the entire organism. An effect of a single factor (target) in the network will result in a disturbance of the network such that numerous processes will be affected. One constructive feature of such a biological network is stability and plasticity, and the ability to cope with adverse conditions. Apparently, the network has an enormous self-regulatory potential, and only during severe diseases such as diabetes, cancer or neurodegenerative diseases will the network be permanently disturbed and the homeostatic regulation seriously challenged. How, then, can the disturbance be compensated by pharmacological intervention? Is the search for the "magic bullet" a promising approach, or is it advisable to target more than one protein in the regulatory network?

In many animal models extensive collections of mutants are available, and the majority of single gene knockouts have little or no phenotype, with only about 19% of the genes being of vital importance. When attempting to delete each of the druggable target genes in the mouse genome, useful and pharmacologically interpretable phenotypes were obtained in only 10% of the cases. Such experimental data reflect the robustness of the signaling network and illustrate that to disable a single factor by a drug or by mutation can be compensated in many cases. However, the efficiency can be increased if two (or more) factors are targeted at the same time. In this situation, the compensatory reactions may fail or become increasingly ineffective. The simultaneous deletion of two genes results in an increased percentage of lethal phenotypes (synthetic lethality). A similar experience occurred with yeast, when single genes were mutated and the strains used to test a library of compounds under different environmental conditions. In this situation many phenotypes were obtained as the compensatory mechanisms were weakened by the combined effects.

How can a "multihit" strategy be implemented in the treatment of diseases? One possibility is to deliberately create a stress situation in combination with a biologically active substance. It has been seen that a mild oxidative stress can be beneficial (see Section 5.1.5), and it is conceivable that drugs will be more effective if they are administered at the same time. However, an experimentally simpler approach would be to use heat stress as a factor that sensitizes the organism to treatment with a therapeutic drug. The induced stress situation may be viewed as a physiological response that weakens the capacity of a cell or organism to maintain homeostatic control, as some cellular resources are devoted to combat the stress situation. The effect of hyperthermia on the expression of a specific set of gene products ("heat-shock proteins") and the downregulation of other proteins has been well studied. Interestingly, a mild heat shock is life-prolonging in *Drosophila*, and this effect may be viewed as a hormetic response as discussed in the context of oxidative stress (see Section 5.1.5). The effect of hyperthermia on the regulatory network has been exploited in the treatment of cancer, and the whole body or single affected organs may be sensitized by hyperthermia to subsequent drug treatment. Natural substances can be useful for this purpose. Typically, the treatment of cultured cancer cells with quercetin (or other flavonoids) makes the cells more susceptible to hyperthermia and consequently the cytotoxicity is enhanced. The effects of drugs in a particular cellular stress situation are difficult to predict, and must be studied experimentally in each particular combination. The sensitization of cells to a particular drug treatment illustrates the potential of a "two-hit approach" in rectifying a dysregulated network of regulatory factors for the treatment of cancer and other diseases.

Some traditional herbal preparations which have been used in the treatment of diseases were analyzed using modern analytical methods in the hope of identifying a single active substance. Unfortunately, such attempts were often unsuccessful since only a combination of bioactive substances in the respective preparation proved to possess the activity. These observations have contributed to a new approach in drug development. Drugs with multiple targets are no longer viewed as unsuitable, but the beneficial combinations of drugs that more effectively modulate the regulatory network in a disease are considered promising; the term "polypharmacology" has been coined to designate this new approach. An example of the success of this therapeutic approach is the treatment of cancer by a combination of paclitaxel (a tubulin inhibitor; see Section 3.2.2) and salicylihamide A (a cytotoxic ATPase inhibitor from a marine sponge). The approach is not limited to two factors; in a breast cancer mouse model the combined genetic and pharmacological inhibition of four genes (*COX2, MMP1, MMP2, epiregulin*) proved to be effective against metastatic progression of the tumor.

A multihit strategy is by no means a new pharmacological discovery. Nature has been practicing this approach in many variations, and a mixture of different biologically active compounds, as well as the characteristic promiscuity of single natural compounds, can produce the desired effect. An example of the former strategy is the venom of snakes, which typically consists of many substances that interact at several nodes of the network to achieve an instantaneous effect (see Section 4.2.2). The useful promiscuity of a single substance is illustrated by some antibiotics. The antibiotic fosfomycin (originally isolated from *Streptococcus*) inhibits two enzymes involved in the biosynthesis of murein, an essential component of the bacterial cell wall. Targeting two or more proteins limits the risk that the bacteria become resistant against the antibiotic. Interestingly, a synergistic effect was observed when an additional antibiotic (β-lactam) was administered. Antibiotics of this type block the synthesis of the bacterial cell wall and also interact with more than one target protein. Oregano essential oil is known to possess antimicrobial activity and has, therefore, been used as preservative for food, although the flavor of the oil has hampered its widespread use. Two substances in the essential oil, carvacrol and thymol, have been shown to possess antibiotic activity, but it is only a mixture of both substances that accounts for the antimicrobial effect.

A final example for the power of two drugs acting in synergy concerns the treatment of Alzheimer's disease. The development of drugs to treat prion disorders and amyloid-forming proteins requires suitable test systems. In yeast, infectious amyloids can be induced by the protein Sup35 which terminates translation, and the effect of small molecules on the formation, folding and stability of the formed prions can be conveniently studied using this system. Interestingly, the catechin EGCG blocks Sup35 prion formation and disrupts contacts between the already-formed prions. Unfortunately, this treatment facilitates the formation of resistant prion strains, although this undesirable effect can be antagonized by another small molecule (4,5-bis-(4-methoxyanilino)phthalimide). By combining both drugs, each of which act via distinctly separate mechanisms, prion formation is efficiently prevented.

A network of target proteins can be constructed on the basis of known drug–target interactions. Such a network analysis has been carried out with current pharmaceutical drugs, when the average number of known targets per drug is about 2.5. A simple network of drug targets associated with asthma is shown in Figure 6.14, where the nodes represent different targets and the edges represent drugs that bind to two (or more) targets. Such information is useful for the development of lead compounds with multitarget profiles. The chemical scaffolds of promiscuous drugs can help to develop new drugs with the desired

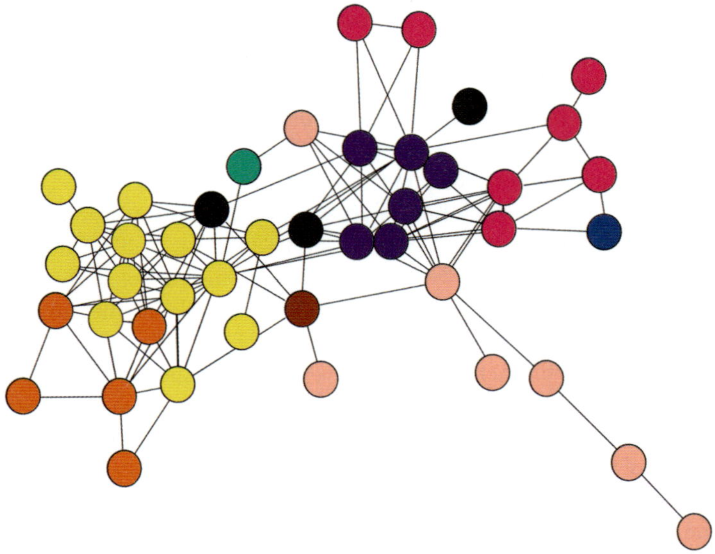

Figure 6.14 Example of a network of molecular interactions between drugs and different target proteins. The figure shows a network of interactions between drugs against asthma and different target proteins. Those proteins which are known to play a role in the development of the disease are represented as nodes (colored circles), while the edges represent different drugs that interact with the respective proteins. Many substances interact with two or more proteins. Different families of proteins are represented by different colors. From Hopkins, A.L. (2008) *Nat. Chem. Biol.*, **4**, 682–690 (see Further Reading). Reprinted (with minor modifications) with permission from Macmillan Publishers Ltd.

polypharmacological profile. In a similar way, drug–drug networks can be constructed in which the nodes are represented by the drugs and the edges by the number of affected targets.

A pharmacological analysis of multitarget drugs or of combinations of drugs calls for new testing strategies and new approaches in pharmacological research. First, the relevant network of interacting proteins and its properties need to be characterized, and for this new "omic" technologies have become available that allow the large-scale screening of genes and their polymorphisms, gene expression patterns (genomics) and the protein patterns (proteomics) in a tissue. Metabolomics provides information on the metabolic pathways and their modulation by drugs, environmental factors or diseases. On the molecular level, structural information of the proteome is very useful, and in future it may be possible to predict possible side effects of a drug based on the similarity of binding pockets in different proteins. For many proteins (enzymes, receptors), activity tests can be carried out to identify drug promiscuity with respect to members of protein families with related activities. Yet, despite of all these powerful techniques it is impossible at present to predict the biological effect of a drug, even when it was developed via the classic single-hit approach. The integration and interpretation of the enormously complex information that can be obtained with modern techniques is a challenge for future years, and this field of "systems biology" will continue to gain in importance. In the case of pharmaceutical applications, the terms "systems pharmacology" or "network pharmacology" have also been used.

6.3.3
Reverse Pharmacology Inspired by Traditional Medicine

The realization that drug promiscuity is not *per se* a negative feature, and that drug mixtures may provide a more robust and effective treatment against common diseases, has led to the concept of "polypharmacology." The enormous amount of data produced when applying modern techniques in biophysics, biochemistry and all of the "omics" technologies needs to be brought together, structured, and processed in a meaningful way. Apart from the technological problems, the acceptance of new pharmacological concepts will take time and the regulatory procedures for the admission of drug mixtures also need to be reviewed.

The knowledge in traditional medicine that has accumulated over the centuries is a source of inspiration for modern drug development. Much of the herbal prescriptions specified in the pharmacopoeias of different cultures are likely to be evidence-based but, of course, the efficacy needs to be verified and tested in clinical trials or in animal model systems. If ethnopharmacological prescriptions possess desirable activities, this may be viewed as equivalent to preclinical tests and a sufficient bioavailability of the active substance(s) is implicated by the positive results. The traditional approach differs from modern pharmacological developments in the order of the successive steps. Whereas, in the modern approach a specific target protein is the initial focus, the alternative is to concentrate on the biological effect *in vivo* at an early stage of drug development (Figure 6.15). The latter approach has been referred to as "reverse pharmacology" to emphasize the reversal of the discovery process which proceeds from "clinic to laboratory" instead of the conventional progression from "laboratory to clinic." Finally, the elucidation of molecular mechanisms which bring about observed biological effects is the subject of research in the modern interpretation of "reverse pharmacology." Network pharmacology and the application of modern techniques such as metabolomics will be the methods of choice to achieve this goal.

This route of discovery was first followed in Ayurvedic medicine in India, as well as in traditional Chinese medicine. It is often difficult to assess historical recipes: Are the decoctions evidence-based or is the evidence for the efficacy only anecdotal? In some cases historical recipes have been analyzed using modern analytical methods. For example, in traditional Chinese medicine *Angelica* herbal decoctions are used to treat menopausal problems in woman. In a popular decoction dating back to 1247 AD, two herbs are used: *Angelicae sinensis radix* (AR) and *Astragali radix* (ASR) in a ratio of 1 : 5. A biochemical analysis revealed the presence of the flavonoids formononetin, ononin, and calycosin, and some other secondary metabolites; the main compounds of ASR are ferulic acid and ligustilide. The estrogenic effects of some of these compounds are well known (e.g.,

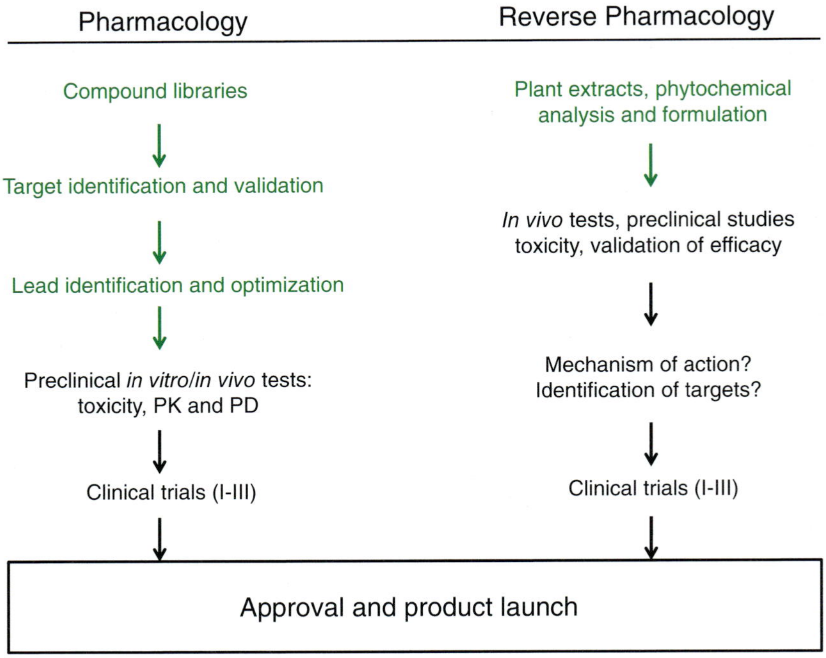

Figure 6.15 Comparison of different strategies in the development of pharmaceuticals. In classical pharmacological development the identification of an active substance and the chemical optimization (green) are the first steps, followed by the preclinical and clinical phase (blue). This strategy focuses on the identification of an active compound that acts on a single target molecule. In contrast, reverse pharmacology concentrates first on the activity of the substance (or mixture of substances) using suitable test systems. The underlying biological mechanisms may then be unraveled in a second step.

formononetin; see Section 5.2), and this presumably constitutes the mechanistic reason for the observed beneficial effects on menopausal disturbances. Interestingly, the exact procedure concerning ratio and extract preparation (ASR and AR are boiled together) maximizes the yield of the suspected beneficial substances, and suggests that the formulation was evidence-based. Interactions of the compounds are likely, however, and possible synergistic effects must be considered. This example shows that traditional medicine is a valuable source of inspiration, and that reverse pharmacology is a promising approach for the development of new herbal-based medicines.

When plant extracts are tested for their medicinal use, an important first step is that of quality control. Medicinal plant species from different locations may vary in their concentration of the secondary metabolites of interest, and this heterogeneity will be further increased by different growth conditions, genetic strains, and possible stress situations in particular locations that can strongly affect the pattern of synthesized metabolites (see Sections 2.3 and 2.5). A biochemical quality control rests on the identification of the relevant compounds, but this is not an easy task since often the bioactivity of all components has not been studied and chemical interactions may complicate matters. Once the quality control of the specific formulation is satisfactory, a combination of *in-vitro* and *in-vivo* tests is usually carried out. Since *in-vivo* tests at an early stage are an important part of the reverse pharmacology approach, the development of informative tests using a minimum number of animals requires special consideration. Depending on the disease that is in focus, animal models for diabetes, Alzheimer or cancer are particularly useful for these tests. Disease models in fish have shown promise for assessing the effects of drugs or drug mixtures. In particular, small fish (e.g., zebrafish or medaka) with their rapid development, transparency and easy handling, can lead to a considerable reduction in the number of mammals required for such screening experiments, and speed up drug development at the same time.

The numerous interactions of natural substances with cellular components (see Figure 3.14) make the molecular analysis of biological effects very difficult, and it is impossible to accurately predict such effects *in vivo*. The promiscuity of natural substances should, however, be viewed not only as a problem but also as a chance. The multihit approach appears to be a promising concept for pharmacological developments. Natural substances are essential components of human nutrition and hence are important for health. Nature has not only provided humankind with a drugstore containing many useful compounds but has also provided an invitation to make use of these treasures. Clearly, the biology of natural compounds will remain a fascinating – and controversial – area of research.

Further Reading

1
Biosynthesis and Chemical Properties of Natural Substances in Plants

1.1
Selected Classes of Secondary Metabolites

Wink, M. (ed.) (1999) *Functions of Plant Secondary Metabolites and Their Exploitation in Biotechnology. Annual Reviews*, vol. **3**, CRC Press, Boca Raton, Florida USA.

1.1.1
Occurrence and Compartmentation

Christensen, L.P. and Brandt, K. (2006) Bioactive polyacetylenes in food plants of the Apiaceae family: occurrence, bioactivity and analysis. *J. Pharm. Biomed. Anal.*, **41**, 683–693.

Hermann, F., Hamoud, R., Sporer, F., Tahrani, A., and Wink, M. (2011) Carlina oxide – a natural polyacetylene from *Carlina acaulis* (Asteraceae) with potent antitrypanosomal and antimicrobial properties. *Planta Med.*, **77**, 1905–1911.

Lange, B.M. and Turner, G.W. (2012) Terpenoid biosynthesis in trichomes – current status and future opportunities. *Plant Biotechnol. J.*, **11**, 2–22.

Steglich, W. and Strack, D. (1990) Betalains, in *The Alkaloids Chemistry and Pharmacology*, vol. **2** (ed. A.A. Brossi), Academic Press, San Diego, pp. 1–62.

Wink, M. (ed.) (1999) *Biochemistry of Plant Secondary Metabolism. Annual Reviews*, CRC Press, Boca Raton, Florida USA.

1.1.2
Biosynthesis

Andréasson, E., Jørgensen, L.B., Höglund, A.-S., Rask, L., and Meijer, J. (2001) Different myrosinase and idioblast distribution in *Arabidopsis* and *Brassica napus*. *Plant Physiol.*, **127**, 1750–1763.

Brader, G., Mikkelsen, M.D., Halkier, B.A., and Palva, E.T. (2006) Altering glucosinolate profiles modulates disease resistance in plants. *Plant J.*, **46**, 758–767.

Dong, X., Braun, E.L., and Grotewold, E. (2001) Functional conservation of plant secondary metabolic enzymes revealed by complementation of *Arabidopsis* flavonoid mutants with maize genes. *Plant Physiol.*, **127**, 46–57.

Fellermeier, M., Eisenreich, W., Bacher, A., and Zenk, M.H. (2001) Biosynthesis of cannabinoids. Incorporation experiments with ^{13}C-labeled glucoses. *Eur. J. Biochem.*, **268**, 1596–1604.

Glawischnig, E. (2007) Camalexin. *Phytochemistry*, **68**, 401–406.

Hagel, J.M., Krizevski, R., Marsolais, F., Lewinson, E., and Facchini, P.J. (2012) Biosynthesis of amphetamine analogs in plants. *Trends Plant Sci.*, **17**, 404–412.

Halkier, B.A. and Gershenzon, J. (2006) Biology and biochemistry of glucosinolates. *Annu. Rev. Plant Biol.*, **57**, 303–333.

Plant Natural Products: Synthesis, Biological Functions and Practical Applications, First Edition.
Herwig O. Gutzeit and Jutta Ludwig-Müller.

Han, X.-H., Gao, Z.-J., and Xiao, X.-G. (2009) Enzymes and genes involved in the betalain biosynthesis in higher plants. *Afr. J. Biotechnol.*, **8**, 6735–6744.

Harborne, J.B. (1980) Plant phenolics, in *Secondary Plant Products*, Encyclopedia of Plant Physiology vol. **8** (eds E.A. Bell and B.V. Charlwood), Springer-Verlag, Berlin, Heidelberg, New York.

Hatlestad, G.J., Sunnadeniya, R.M., Akhavan, N.A., Gonzalez, A., Goldman, I.L., McGrath, J.M., and Lloyd, A.M. (2012) The beet *R* locus encodes a new cytochrome P450 required for red betalain production. *Nat. Genet.*, **44**, 816–820.

Kissen, R., Rossiter, J.T., and Bones, A.M. (2009) The "mustard oil bomb": not so easy to assemble?! Localization, expression and distribution of the components of the myrosinase enzyme system. *Phytochem. Rev.*, **8**, 69–86.

Koroleva, O.A., Davies, A., Deeken, R., Thorpe, M.R., Tomos, A.D., and Hedrich, R. (2000) Identification of a new glucosinolate-rich cell type in *Arabidopsis* flower stalk. *Plant Physiol.*, **124**, 599–608.

Larcher, W. (1995) *Physiological Plant Ecology*, 3rd edn, Springer-Verlag, Berlin, Heidelberg, New York.

Lee, S., Kaminaga, Y., Cooper, B., Pichersky, E., Dudareva, N., and Chapple, C. (2012) Benzoylation and sinapoylation of glucosinolate R-groups in *Arabidopsis*. *Plant J.*, **72**, 411–422.

Ludwig-Müller, J., Ihmig, S., Bennett, R., Kiddle, G., Ruppel, M., and Hilgenberg, W. (1999) The host range of *Plasmodiophora brassicae* and its relationship to endogenous glucosinolate content. *New Phytol.*, **141**, 443–458.

Matile, P.H. (1980) 'Die Senfölbombe': Zur Kompartimentierung des Myrosinasesystems. *Biochem. Physiol. Pflanzen*, **175**, 722–731.

McGarvey, D.J. and Croteau, R. (1995) Terpenoid metabolism. *Plant Cell*, **7**, 1015–1026.

Müller, C. (2009) Role of glucosinolates in plant invasiveness. *Phytochem. Rev.*, **8**, 227–242.

Pedras, M.S.C., Jha, M., and Ahiahonu, P.W.K. (2003) The synthesis and biosynthesis of phytoalexins produced by cruciferous plants. *Curr. Org. Chem.*, **7**, 1635–1647.

Poulton, J.E. (1990) Cyanogenesis in plants. *Plant Physiol.*, **94**, 401–405.

Rohmer, M. (1999) The discovery of a mevalonate-independent pathway for isoprenoid biosynthesis in bacteria, algae and higher plants. *Nat. Prod. Rep.*, **16**, 565–574.

Verpoorte, R. (2000) Secondary metabolism, in *Metabolic Engineering of Plant Secondary Metabolism* (eds R. Verpoorte and A.W. Alfermann), Kluwer Academic Publisher, Dordrecht, Netherlands, pp. 1–29.

Vogt, T. (2010) Phenylpropanoid biosynthesis. *Mol. Plant*, **3**, 2–20.

Winkel, B.S. (2004) Metabolic channeling in plants. *Annu. Rev. Plant Biol.*, **55**, 85–107.

1.1.3
Transcriptional Control

Ahuja, I., Kissen, R., and Bones, A.M. (2012) Phytoalexins in defense against pathogens. *Trends Plant Sci.*, **17**, 73–90.

Bekaert, M., Edger, P.P., Hudson, C.M., Pires, J.C., and Conant, G.C. (2012) Metabolic and evolutionary costs of herbivory defense: systems biology of glucosinolate synthesis. *New Phytol.*, **196**, 596–605.

De Geyter, N., Gholami, A., Goormachtig, S., and Goossens, A. (2012) Transcriptional machineries in jasmonate-elicited plant secondary metabolism. *Trends Plant Sci.*, **17**, 349–359.

Gigolashvili, T., Berger, B., and Flügge, U.-I. (2009) Specific and coordinated control of indolic and aliphatic glucosinolate biosynthesis by R2R3-MYB transcription factors in *Arabidopsis thaliana*. *Phytochem Rev.*, **8**, 3–13.

Jung, J.-H., Seo, P.J., and Park, C.-M. (2009) MicroRNA biogenesis and function in higher plants. *Plant Biotechnol. Rep.*, **3**, 111–126.

Levy, M., Wang, Q., Kaspi, R., Parrella, M.P., and Abel, S. (2005) *Arabidopsis* IQD1, a novel calmodulin-binding nuclear protein, stimulates glucosinolate accumulation and plant defense. *Plant J.*, **43**, 79–96.

Yang, C.Q., Fang, X., Wu, X.M., Mao, Y.B., Wang, L.J., and Chen, X.Y. (2012) Transcriptional regulation of plant secondary metabolism. *J. Integr. Plant Biol.*, **54**, 703–712.

1.1.4
Transport of Secondary Metabolites

Alfenito, M.R., Souer, E., Goodman, C.D., Buell, R., Mol, J., Koes, R., and Walbot, V. (1998) Functional complementation of anthocyanin sequestration in the vacuole by widely divergent glutathione S-transferases. *Plant Cell*, **10**, 1135–1149.

Buer, C.S., Muday, G.K., and Djordjevic, M. A. (2007) Flavonoids are differentially taken up and transported long distances in *Arabidopsis*. *Plant Physiol.*, **145**, 478–490.

Grotewold, E. (2001) Subcellular trafficking of phytochemicals. *Recent Res. Devel. Plant Physiol.*, **2**, 31–48.

Nour-Eldin, H.H. and Halkier, B.A. (2009) Piecing together the transport pathway of aliphatic glucosinolates. *Phytochem. Rev.*, **8**, 53–67.

Nour-Eldin, H.H. and Halkier, B.A. (2013) The emerging field of transport engineering of plant specialized metabolites. *Curr. Opin. Biotechnol.*, **24**, 263–270.

Pakdeechanuan, P., Shoji, T., and Hashimoto, T. (2012) Root-to-shoot translocation of alkaloids is dominantly suppressed in *Nicotiana alata*. *Plant Cell Physiol.*, **53**, 1247–1254.

1.2
Evolution of Natural Compounds in Plants

Bittrich, B. and Amaral, M. (1991) Proanthocyanidins in the testa of centrospermous seeds. *Biochem. Syst. Ecol.*, **19**, 319–321.

Fahey, J.W., Zalcmann, A.T., and Talalay, P. (2001) The chemical diversity and distribution of glucosinolates and isothiocyanates among plants. *Phytochemistry*, **56**, 5–51.

Kimler, L., Mears, J., Mabry, T.J., and Roesler, H. (1970) On the question of the mutual exclusiveness of betalains and anthocyanins. *Taxon*, **19**, 875–878.

Rodman, J.E., Soltis, P.S., Soltis, D.E., Sytsma, K.J., and Karol, K.G. (1998) Parallel evolution of glucosinolate biosynthesis inferred from congruent nuclear and plastid gene phylogenies. *Am. J. Bot.*, **85**, 997–1006.

1.2.1
Parallel Evolution of Similar Pathways

Bak, S., Nielsen, H.L., and Halkier, B.A. (1998) The presence of CYP79 homologues in glucosinolate-producing plants shows evolutionary conservation of the enzymes in the conversion of amino acid to aldoxime in the biosynthesis of cyanogenic glucosides and glucosinolates. *Plant Mol. Biol.*, **38**, 725–734.

Frisch, T. and Møller, B.L. (2012) Possible evolution of alliarinoside biosynthesis from the glucosinolate pathway in *Alliaria petiolata*. *FEBS J.*, **279**, 1545–1562.

Jirschitzka, J., Schmidt, G.W., Reichelt, M., Schneider, B., Gershenzon, J., and D'Auria, J.C. (2012) Plant tropane alkaloid biosynthesis evolved independently in the Solanaceae and Erythroxylaceae. *Proc. Natl Acad. Sci. USA*, **109**, 10304–10309.

Ludwig-Müller, J., Ihmig, S., Bennett, R., Kiddle, G., Ruppel, M., and Hilgenberg, W. (1999) The host range of *Plasmodiophora brassicae* and its relationship to endogenous glucosinolate content. *New Phytol.*, **141**, 443–458.

1.2.2
Gene Duplication

Agerbirk, N. and Olsen, C.E. (2012) Glucosinolate structures in evolution. *Phytochemistry*, **77**, 16–45.

Chen, F., Tholl, D., Bohlmann, J., and Pichersky, E. (2011) The family of terpene synthases in plants: a mid-size family of genes for specialized metabolism that is highly diversified throughout the kingdom. *Plant J.*, **66**, 212–229.

Dick, R., Rattei, T., Haslbeck, M., Schwab, W., Gierl, A., and Frey, M. (2012) Comparative analysis of benzoxazinoid biosynthesis in monocots and dicots: independent recruitment of stabilization and activation functions. *Plant Cell*, **24**, 915–928.

Jenrich, R., Trompetter, I., Bak, S., Olsen, C. E., Møller, B.L., and Piotrowski, M. (2007) Evolution of heteromeric nitrilase complexes in Poaceae with new functions in nitrile metabolism. *Proc. Natl Acad. Sci. USA*, **104**, 18848–18853.

Piotrowski, M. (2008) Primary or secondary? Versatile nitrilases in plant metabolism. *Phytochemistry*, **69**, 2655–2667.

Ober, D. and Kaltenegger, E. (2009) Pyrrolizidine alkaloid biosynthesis, evolution of a pathway in plant secondary metabolism. *Phytochemistry*, **70**, 1687–1695.

1.2.3 Gene Clusters

Chu, H.Y., Wegel, E., and Osbourn, A. (2011) From hormones to secondary metabolism: the emergence of metabolic gene clusters in plants. *Plant J.*, **66**, 66–79.

DellaPenna, D. and O'Connor, S.E. (2012) Plant gene clusters and opiates. *Science*, **336**, 1648–1649.

Osbourn, A.E. and Field, B. (2009) Operons. *Cell Mol. Life Sci.*, **66**, 3755–3775.

Osbourn, A.E. (2010a) Secondary metabolic gene clusters: evolutionary toolkits for chemical innovation. *Trends Genet.*, **26**, 449–457.

Osbourn, A.E. (2010b) Gene clusters for secondary metabolic pathways: An emerging theme in plant biology. *Plant Physiol.*, **154**, 531–535.

Takos, A.M. and Rook, F. (2012) Why biosynthetic genes for chemical defense compounds cluster. *Trends Plant Sci.*, **17**, 383–388.

Wada, M., Takahashi, H., Altaf-Ul-Amin, M., Nakamura, K., Hirai, M.Y., Ohta, D., and Kanaya, S. (2012) Prediction of operon-like gene clusters in the *Arabidopsis thaliana* genome based on co-expression analysis of neighboring genes. *Gene*, **503**, 56–64.

Winzer, T., Gazda, V., He, Z., Kaminski, F., Kern, M., Larson, T.R., Li, Y., Meade, F., Teodor, R., Vaistij, F.E., Walker, C., Bowser, T.A., and Graham, I.A. (2012) A *Papaver somniferum* 10-gene cluster for synthesis of the anticancer alkaloid noscapine. *Science*, **336**, 1704–1708.

1.2.4 Natural Variation

Benderoth, M., Pfalz, M., and Kroymann, J. (2008) Methylthioalkylmalate synthases: genetics, ecology and evolution. *Phytochem. Rev.*, **8**, 255–268.

Burow, M. and Wittstock, U. (2009) Regulation and function of specifier proteins in plants. *Phytochem. Rev.*, **8**, 87–99.

Lambrix, V., Reichelt, M., Mitchell-Olds, T., Kliebenstein, D.J., and Gershenzon, J. (2001) The *Arabidopsis* epithiospecifier protein promotes the hydrolysis of glucosinolates to nitriles and influences *Trichoplusia ni* herbivory. *Plant Cell*, **13**, 2793–2807.

Routaboul, J.-M., Dubos, C., Beck, G., Marquis, C., Bidzinski, P., Loudet, O., and Lepiniec, L. (2012) Metabolite profiling and quantitative genetics of natural variation for flavonoids in *Arabidopsis*. *J. Exp. Bot.*, **63**, 3749–3764.

1.2.5 Modulation of Different Molecules with Similar Functions

Friml, J. (2010) Subcellular trafficking of PIN auxin efflux carriers in auxin transport. *Eur. J. Cell Biol.*, **89**, 231–235.

Mravec, J., Skupa, P., Bailly, A., Hoyerova, K., Krecek, P., Bielach, A., Petrasek, J., Zhang, J., Gaykova, V., Stierhof, Y.-D., Dobrev, P., Schwarzerova, K., Rolcik, J., Seifertova, D., Luschnig, C., Benkova, E., Zazimalova, E., Geisler, M., and Friml, J. (2009) Subcellular homeostasis of phytohormone auxin is mediated by the ER-localized PIN5 transporter. *Nature*, **459**, 1136–1140.

Peer, W.A., Blakeslee, J.J., Yang, H., and Murphy, A.S. (2011) Seven things we think we know about auxin transport. *Mol. Plant*, 4, 487–504.

1.2.6 Evolution by Stress Factors

Pollastri, S. and Tattini, M. (2011) Flavonols: old compounds for old roles. *Ann. Bot.*, **108**, 1225–1233.

1.3 Biotechnological Applications

Dewitt, D. and Bosland, P.W. (1996) *Peppers of the World: An Identification Guide*, Ten Speed Press, Berkeley, California.

Lewis, W.H. and Elvin-Lewis, M.P.F. (1977) *Medical Botany*, John Wiley & Sons, Inc., New York.

Schultes, R.E. (1976) *Hallucinogenic Plants*, Golden Press, New York.

Stamets, P. (1996) *Psilocybin Mushrooms of the World*, Ten Speed Press, Berkeley, California.

Verpoorte, R. and Alfermann, A.W. (eds) (2000) *Metabolic Engineering of Plant Secondary Metabolism*, Kluwer Academic Publishers, Dordrecht, Netherlands.

1.3.1 Identification and Quantification of Natural Products

Cannell, R.J.P. (1998) (ed.) *Natural Products Isolation. Methods in Biotechnology 4*, Humana Press, Totowa, New Jersey, USA.

Cohen, J.D., Baldi, B.G., and Slovin, J.P. (1986) ($^{13}C_6$-[Benzene ring)-indole-3-acetic acid: a new internal standard for quantitative mass spectral analysis of indole-3-acetic acid in plants. *Plant Physiol.*, **80**, 14–19.

Fang, J., Reichelt, M., Hidalgo, W., Agnolet, S., and Schneider, B. (2012) Tissue-specific distribution of secondary metabolites in rapeseed (*Brassica napus* L.). *PLoS One*, **7** (10), e48006.

Kim, H.K., Choi, Y.H., and Verpoorte, R. (2011) NMR-based plant metabolomics: where do we stand, where do we go? *Trends Biotechnol.*, **29**, 267–275.

Okazaki, Y. and Saito, K. (2012) Recent advances of metabolomics in plant biotechnology. *Plant Biotechnol. Rep.*, **6**, 1–15.

Sarsby, J., Towers, M.W., Stain, C., Cramer, R., and Koroleva, O.A. (2012) Mass spectrometry imaging of glucosinolates in *Arabidopsis* flowers and siliques. *Phytochemistry*, **77**, 110–118.

1.3.2 Biological Activity

Stracke, R., De Vos, R.C.H., Bartelniewoehner, L., Ishihara, H., Sagasser, M., Martens, S., and Weisshaar, B. (2009) Metabolomic and genetic analyses of flavonol synthesis in *Arabidopsis thaliana* support the in vivo involvement of leucoanthocyanidin dioxygenase. *Planta*, **229**, 427–445.

1.3.3 Biotechnological Production of Natural Compounds

Chandra, S. and Chandra, R. (2011) Engineering secondary metabolite production in hairy roots. *Phytochem. Rev.*, **10**, 371–395.

Chiu, L.-W., Zhou, X., Burke, S., Wu, X., Prior, R.L., and Li, L. (2010) The purple cauliflower arises from activation of a MYB transcription factor. *Plant Physiol.*, **154**, 1470–1480.

Georgiev, M.I., Ludwig-Müller, J., and Bley, T. (2010) Hairy root culture: copying nature in new bioprocesses, in *Medicinal Plant Biotechnology* (ed. R. Arora), CAB International, Oxford, United Kingdom, pp. 156–175.

Georgiev, M., Agostini, E., Ludwig-Müller, J., and Xu, J. (2012) Genetically transformed roots: from plant disease to biotechnology. *Trends Biotechnol.*, **10**, 528–537.

Matzke, M.A. and Matzke, A.J.M. (1995) How and why do plants inactivate homologous (trans)genes? *Plant Physiol.*, **107**, 679–685.

Napoli, C., Lemieux, C., and Jorgensen, R. (1990) Introduction of a chalcone synthase gene into *Petunia* results in reversible co-suppression of homologous genes in trans. *Plant Cell*, **2**, 279–289.

Runguphan, W., Qu, X., and O'Connor, S.E. (2010) Integrating carbon–halogen bond formation into medicinal plant metabolism. *Nature*, **468**, 461–464.

Runguphan, W., Glenn, W.S., and O'Connor, S.E. (2012) Redesign of a dioxygenase in morphine biosynthesis. *Chem. Biol.*, **19**, 674–678.

2
Function of Natural Substances in Plants

2.1
Secondary Compounds in Primary Metabolism

Buchanan, B.B., Gruissem, W., and Jones, R. L. (eds.) (2004) *Biochemistry and Molecular biology of Plants*, American Society of Plant Physiologists.

2.2
Development

Davies, P.J. (ed.) (2010) Plant hormones, in *Biosynthesis, Signal Transduction, Action!* Springer-Verlag, Heidelberg, Berlin, New York.

Howell, S.H. (1998) *Molecular Genetics of Plant Development*, Cambridge University Press.

Falcone Ferreyra, M.L., Rius, S.P., and Casati, P. (2012) Flavonoids: Biosynthesis, biological functions, and biotechnological applications. *Front. Plant Sci.*, **28** (3), 222.

2.2.1
Vegetative Growth

Arnao, M.B. and Hernandez-Ruiz, J. (2006) The physiological function of melatonin in plants. *Plant Signal Behavior*, **1**, 89–95.

Bender, J. and Celenza, J.L. (2009) Indolic glucosinolates at the crossroads of tryptophan metabolism. *Phytochem. Rev.*, **8**, 25–37.

Bennici, A., Schiff, S., and Mori, B. (2006) Morphogenic effect of colchicine in *Cichorium intybus* L. root explants cultured *in vitro*. *Caryologia*, **59**, 284–290.

Boerjan, W., Cervera, M.T., Delarue, M., Beeckman, T., Dewitte, W., Bellini, C., Caboche, M., Van Onckelen, H., Van Montagu, M., and Inze, D. (1995) *superroot*, a recessive mutation in *Arabidopsis*, confers auxin overproduction. *Plant Cell*, **7**, 1405–1419.

Boerjan, W., Ralph, J., and Baucher, M. (2003) Lignin biosynthesis. *Annu. Rev. Plant Biol.*, **54**, 519–546.

Brown, D.E., Rashotte, A.M., Murphy, A.S., Normanly, J., Tague, B.W., Peer, W.A., Taiz, L., and Muday, G.K. (2001) Flavonoids act as negative regulators of auxin transport *in vivo* in *Arabidopsis*. *Plant Physiol.*, **126**, 524–535.

Buer, C.S. and Djordjevic, M.A. (2009) Architectural phenotypes in the *transparent testa* mutants of *Arabidopsis thaliana*. *J. Exp. Bot.*, **60**, 751–763.

Close, D.C. and Beadle, C.L. (2003) The ecophysiology of foliar anthocyanin. *Bot. Rev.*, **69**, 149–161.

Ditengou, F.A., Raudaskoski, M., and Lapeyrie, F. (2003) Hypaphorine, an indole-3-acetic acid antagonist delivered by the ectomycorrhizal fungus *Pisolithus tinctorius*, induces reorganisation of actin and the microtubule cytoskeleton in *Eucalyptus globulus* ssp. *bicostata* root hairs. *Planta*, **218**, 217–225.

Facchini, P.J., Bird, D.A., and St. Pierre, B. (2004) Can *Arabidopsis* make complex alkaloids? *Trends Plant Sci.*, **9**, 116–122.

Kehr, A.E. (1971) A tetraploid *Rhododendrum carolinianum*. *J. Am. Rhododendron Soc.*, *25n1*. Available at: http://scholar.lib.vt.edu/ejournals/JARS/

Muday, G.K. and Murphy, A.S. (2002) An emerging model of auxin transport regulation. *Plant Cell*, **14**, 293–299.

Ntuli, N.R. and Zobolo, A.M. (2008) Effect of water stress on growth of colchicine induced polyploid *Coccinia palmata* and *Lagenaria sphaerica* plants. *Afr. J. Biotechnol.*, **7**, 3548–3652.

Peer, W.A., Bandyopadhyay, A., Blakeslee, J. J., Makam, S.N., Chen, R.J., Masson, P.H., and Murphy, A.S. (2004) Variation in expression and protein localization of the PIN family of auxin efflux facilitator proteins in flavonoid mutants with altered auxin transport in *Arabidopsis thaliana*. *Plant Cell*, **16**, 1898–1911.

Ramakrishna, A., Giridhar, P., and Ravishankar, G.A. (2011) Phytoserotonin: a review. *Plant Signal. Behavior*, **6**, 800–809.

Rusak, G., Cerni, S., Stupin Polancec, D., and Ludwig-Müller, J. (2010) The responsiveness of the *IAA2* promoter to IAA and IBA in *Arabidopsis thaliana* is differentially affected in roots and shoots by flavonoids. *Biol. Plant.*, **54**, 403–414.

Taylor, L.P. and Grotewold, E. (2005) Flavonoids as developmental regulators. *Curr. Opin. Plant Biol.*, **8**, 317–323.

Thimann, K.V. (1937) On the nature of inhibition caused by auxin. *Am. J. Bot.*, **24**, 407–412.

Waller, G.R. and Burström, H. (1969) Diterpenoid alkaloids as plant growth inhibitors. *Nature*, **222**, 576–578.

Woo, H.-H., Jeong, B.R., and Hawes, M.C. (2005) Flavonoids: from cell cycle regulation to biotechnology. *Biotechnol. Lett.*, **27**, 365–374.

2.2.2
Reproduction: Male Fertility

Tang, L.K., Chu, H., Yip, W.K., Yeung, E.C., and Lockwell, C. (2009) An anther-specific dihydroflavonol 4-reductase-like gene (*DRL1*) is essential for male fertility in *Arabidopsis*. *New Phytol.*, **181**, 576–587.

Taylor, L.P. and Jorgensen, R. (1992) Conditional male fertility in chalcone synthase-deficient petunia. *J. Hered.*, **83**, 11–17.

Vogt, T. and Taylor, L.P. (1995) Flavonol 3-O-glycosyltransferases associated with *Petunia* pollen produce gametophyte-specific flavonol diglycosides. *Plant Physiol.*, **108**, 903–911.

Vogt, T., Wollenweber, E., and Taylor, L.P. (1995) The structural requirements of flavonols that induce pollen germination of conditionally male fertile *Petunia*. *Phytochemistry*, **38**, 589–592.

2.2.3
Flower Biology: Pollination

Behmer, S.T. and Nes, W.D. (2003) Insect sterol nutrition and physiology: a global overview. *Adv. Insect Physiol.*, **31**, 1–72.

Bender, R., Klinkenberg, P., Jiang, Z., Bauer, B., Karypis, G., Nguyen, N., Perera, M.A.D.N., Nikolau, B.J., and Carter, C.J. (2012) Functional genomics of nectar production in the Brassicaceae. *Flora*, **207**, 491–496.

Cunningham, F.X. Jr and Gantt, E. (1998) Genes and enzymes of carotenoid biosynthesis in plants. *Annu. Rev. Plant Physiol. Plant Mol. Biol.*, **49**, 557–583.

Danforth, B.N., Sipes, S., Fang, J., and Brady, S.G. (2006) The history of early bee diversification based on five genes plus morphology. *Proc. Natl Acad. Sci. USA*, **103**, 15118–15123.

Dudareva, N., Pichersky, E., and Gershenzon, J. (2004) Biochemistry of plant volatiles. *Plant Physiol.*, **135**, 1893–1902.

Ito, K. and Seymour, R.S. (2005) Expression of uncoupling protein and alternative oxidase depends on lipid or carbohydrate substrates in thermogenic plants. *Biol. Lett.*, **1**, 427–430.

Klahre, U., Gurba, A., Hermann, K., Saxenhofer, M., Bossolini, E., Guerin, P. M., and Kuhlemeier, C. (2011) Pollinator choice in *Petunia* depends on two major genetic loci for floral scent production. *Curr. Biol.*, **21**, 730–739.

Müller, A., Diener, S., Schnyder, S., Stutz, K., Sedivy, C., and Dorn, S. (2006) Quantitative pollen requirements of solitary bees: Implications for bee conservation and the evolution of bee-flower relationships. *Biol. Conserv.*, **130**, 604–615.

Ollerton, J., Winfree, R., and Tarrant, S. (2001) How many flowering plants are pollinated by animals? *Nordic Society Oikos*, **120**, 321–326.

Raskin, I. (1992) Salicylate, a new plant hormone. *Plant Physiol.*, **99**, 799–803.
Raskin, I., Turner, I.M., and Melander, W.R. (1989) Regulation of heat production in the inflorescences of an Arum lily by endogenous salicylic acid. *Proc. Natl Acad. Sci. USA*, **86**, 2214–2218.
Rasmont, P., Regali, A., Ings, T.C., Lognay, G., Baudart, E., Marlier, M., Delcarte, E., Viville, P., Marot, C., Falmagne, P., Verhaeghe, J.-C., and Chittka, L. (2005) Analysis of pollen and nectar of *Arbutus unedo* as a food source for *Bombus terrestris* (Hymenoptera: Apidae). *J. Econ. Entomol.*, **98**, 656–663.
Rhoads, D.M. and McIntosh, L. (1992) Salicylic acid regulation of respiration in higher plants: Alternative oxidase expression. *Plant Cell*, **4**, 1131–1139.
Roulston, T.H. and Cane, J.H. (2000) Pollen nutritional content and digestibility for animals. *Plant Syst. Evol.*, **222**, 187–209.
Steenhuisen, S.-L., Raguso, R.A., and Johnson, S.D. (2012) Floral scent in bird- and beetle-pollinated *Protea* species (Proteaceae): Chemistry, emission rates and function. *Phytochemistry*, **84**, 78–87.
Svoboda, J.A., Feldlaufer, M.F., and Weirich, G.F. (1994) Evolutionary aspects of steroid utilization in insects, in *Isopentenoids and Other Natural Products*, vol. **562**, ACS Press, Washington, USA, pp. 126–139.
Tanaka, Y., Sasaki, N., and Ohmiya, A. (2008) Biosynthesis of plant pigments: anthocyanins, betalains and carotenoids. *Plant J.*, **54**, 733–749.
Tollsten, L. and Knudsen, J.T. (1992) Floral scent in dioecious *Salix* (Salicaceae) – a cue determining the pollination system? *Plant Syst. Evol.*, **182**, 229–237.

2.2.4 Fruit and Seeds

Giovannoni, J.J. (2004) Genetic regulation of fruit development and ripening. *Plant Cell*, **16**, S170–S180.
Goodwin, T.W. (1988) *Plant Pigments*, Academic Press, Oxford, UK.
Goulet, C., Mageroy, M.H., Lam, N.B., Floystad, A., Tieman, D.M., and Klee, H.J. (2012) Role of an esterase in flavor volatile variation within the tomato clade. *Proc. Natl Acad. Sci. USA*, **109**, 19009–19014.
Jaakola, L., Määttä, K., Pirttilä, A.M., Törrönen, R., Kärenlampi, S., and Hohtola, A. (2002) Expression of genes involved in anthocyanin biosynthesis in relation to anthocyanin, proanthocyanidin, and flavonol levels during bilberry fruit development. *Plant Physiol.*, **130**, 729–739.
Klee, H.J. and Giovannoni, J.J. (2011) Genetics and control of tomato fruit ripening and quality attributes. *Annu. Rev. Genet.*, **45**, 41–59.
Rodriguez, A., Alquezar, B., and Pena, L. (2013) Fruit aromas in mature fleshy fruits as signals of readiness for predation and seed dispersal. *New Phytol.*, **197**, 36–48.

2.3 Abiotic Stress

Nitzsche, A., Tokalov, S.V., Gutzeit, H.O., and Ludwig-Müller, J. (2004) Chemical and biological characterization of cinnamic acid derivatives from cell cultures of lavender (*Lavandula officinalis*) induced by stress and jasmonic acid. *J. Agric. Food Chem.*, **52**, 2915–2923.
Pollastri, S. and Tattini, M. (2011) Flavonols: old compounds for old roles. *Ann. Bot.*, **108**, 1225–1233.
Ramakrishna, A. and Ravishankar, G.A. (2011) Influence of abiotic stress signals on secondary metabolites in plants. *Plant Signal. Behavior*, **6**, 1720–1731.

2.3.1 Antioxidative Properties in Plants

Agati, G., Azzarello, E., Pollastri, S., and Tattini, M. (2012) Flavonoids as antioxidants in plants: Location and functional significance. *Plant Sci.*, **196**, 67–76.
Mahajan, M., Joshi, R., Gulati, A., and Yadav, S.K. (2012) Increase in flavan-3-ols by silencing flavonol synthase mRNA affects the transcript expression and activity levels of antioxidant enzymes in tobacco. *Plant Biol.*, **14**, 725–733.
Mittler, R. (2002) Oxidative stress, antioxidants and stress tolerance. *Trends Plant Sci.*, **7**, 405–410.

Petrov, V.D. and Van Breusegem, F. (2012) Hydrogen peroxide – a central hub for information flow in plant cells. *AoB Plants*, **pls014**. doi: 10.1093/aobpla/pls014

Suzuki, N., Koussevitzky, S., Mittler, R., and Miller, G. (2012) ROS and redox signalling in the response of plants to abiotic stress. *Plant Cell Environ.*, **35**, 259–270.

2.3.2 Light and UV Irradiation

Brown, B.A., Cloix, C., Jiang, G.H., Kaiserli, E., Herzyk, P., Kliebenstein, D.J., and Jenkins, G.I. (2005) A UV-B-specific signaling component orchestrates plant UV protection. *Proc. Natl Acad. Sci. USA*, **102**, 18225–18230.

Ouwerkerk, P.B.F., Hallard, D., Verpoorte, R., and Memelink, J. (1999) Identification of UV-B light-responsive regions in the promoter of the tryptophan decarboxylase gene from *Catharanthus roseus*. *Plant Mol. Biol.*, **41**, 491–503.

Mewis, I., Schreiner, M., Nguyen, C.N., Krumbein, A., Ulrichs, C., Lohse, M., and Zrenner, R. (2012) UV-B irradiation changes specifically the secondary metabolite profile in broccoli sprouts: induced signaling overlaps with defense response to biotic stressors. *Plant Cell Physiol.*, **53**, 1546–1560.

Tossi, V., Lombardo, C., Cassia, R., and Lamattina, L. (2012) Nitric oxide and flavonoids are systemically induced by UV-B in maize leaves. *Plant Sci.*, **193–194**, 103–109.

Wright, A.H., DeLong, J.M., Gunawardena, A.H.L.A.N., and Prange, R.K. (2011) The interrelationship between the lower oxygen limit, chlorophyll fluorescence and the xanthophyll cycle in plants. *Photosynthesis Res.*, **107**, 223–235.

2.3.3 High Temperatures

Bennett, R.N., Wenke, T., Freudenberg, B., Mellon, F.A., and Ludwig-Müller, J. (2005) The *tu8* mutation of *Arabidopsis thaliana* encoding a heterochromatin protein 1 causes defects in the induction of secondary metabolite biosynthesis. *Plant Biol.*, **7**, 348–357.

Gu, Z.-X., Guo, Q.-H., and Gu, Y.-J. (2012) Factors influencing glucoraphanin and sulforaphane formation in Brassica plants: a review. *J. Integr. Agric.*, **11**, 1804–1816.

Wahid, A., Gelani, S., Ashraf, M., and Foolad, M.R. (2007) Heat tolerance in plants: An overview. *Environ. Exp. Bot.*, **61**, 199–223.

2.3.4 Cold, Salt, and Drought Stress

Deng, W.-W., Wang, S., Chen, Q., Zhang, Z.-Z., and Hu, X.-Y. (2012) Effect of salt treatment on theanine biosynthesis in *Camellia sinensis* seedlings. *Plant Physiol. Biochem.*, **56**, 35–40.

Yancey, P.H. (2005) Organic osmolytes as compatible, metabolic and counteracting cytoprotectants in high osmolarity and other stresses. *J. Exp. Biol.*, **208**, 2819–2830.

Zhang, Q.-Y., Wang, L.-Y., Kong, F.-Y., Deng, Y.-S., Li, B., and Meng, Q.-W. (2012) Constitutive accumulation of zeaxanthin in tomato alleviates salt stress-induced photoinhibition and photooxidation. *Physiol. Planta*, **146**, 363–373.

2.3.5 Nutrient Deficiency

Larcher, W. (1995) *Physiological Plant Ecology*, 3rd edn, Springer Verlag, Heidelberg, Berlin, New York.

2.3.6 Environmental Pollution

Jüttner, F. (1988) Changes of monoterpene concentrations in needles of pollution-injured *Picea abies* exhibiting montane yellowing. *Physiol. Planta*, **72**, 48–56.

Keilig, K. and Ludwig-Müller, J. (2009) Effect of flavonoids on heavy metal tolerance in *Arabidopsis thaliana* seedlings. *Bot. Stud.*, **50**, 311–318.

Loreto, F., Mannozzi, M., Maris, C., Nascetti, P., Ferranti, F., and Pasqualini, S. (2001) Ozone quenching properties of isoprene and its antioxidant role in leaves. *Plant Physiol.*, **126**, 993–1000.

2.4 Symbioses

2.4.1 Rhizobia

Long, S.R. (2001) Genes and signals in the *Rhizobium*-legume symbiosis. *Plant Physiol.*, **125**, 69–72.

Wasson, A.P., Pellerone, F.I., and Mathesius, U. (2006) Silencing the flavonoid pathway in *Medicago truncatula* inhibits root nodule formation and prevents auxin transport regulation by rhizobia. *Plant Cell*, **18**, 1617–1629.

2.4.2 Mycorrhiza

Asensio, D., Rapparini, F., and Penuelas, J. (2012) AM fungi root colonization increases the production of essential isoprenoids vs. nonessential isoprenoids especially under drought stress conditions or after jasmonic acid application. *Phytochemistry*, **77**, 149–161.

Fester, T., Hause, B., Schmidt, D., Halfmann, K., Schmidt, J., Wray, V., Hause, G., and Strack, D. (2002) Occurrence and localization of apocarotenoids in arbuscular mycorrhizal plant roots. *Plant Cell Physiol.*, **43**, 256–265.

Fester, T., Lohse, S., and Halfmann, K. (2007) "Chromoplast" development in arbuscular mycorrhizal roots. *Phytochemistry*, **68**, 92–100.

Klingner, A., Hundeshagen, B., Kernebeek, H., and Bothe, H. (1995) Localization of the yellow pigment formed in roots of gramineous plants colonized by arbuscular fungi. *Protoplasma*, **185**, 50–57.

Meixner, C., Ludwig-Müller, J., Miersch, O., Gresshoff, P., Staehelin, C., and Vierheilig, H. (2005) Lack of mycorrhizal autoregulation and phytohormonal changes in the supernodulating soybean mutant *nts1007*. *Planta*, **222**, 709–715.

Vierheilig, H. (2004) Further root colonization by arbuscular mycorrhizal fungi in already mycorrhizal plants is suppressed after a critical level of root colonization. *J. Plant Physiol.*, **161**, 339–341.

Vierheilig, H., Bennett, R., Kiddle, G., Kaldorf, M., and Ludwig-Müller, J. (2000) Differences in glucosinolate patterns and arbuscular mycorrhizal status of glucosinolate-containing plant species. *New Phytol.*, **146**, 343–352.

Walter, M.H., Floß, D.S., Hans, J., Fester, T., and Strack, D. (2007) Apocarotenoid biosynthesis in arbuscular mycorrhizal roots: contributions from methylerythritol phosphate pathway isogenes and tools for its manipulation. *Phytochemistry*, **68**, 130–138.

Walter, M.H., Floss, D.S., and Strack, D. (2010) Apocarotenoids: hormones, mycorrhizal metabolites and aroma volatiles. *Planta*, **232**, 1–17.

2.4.3 Endophytic Fungi

Nongbri, P.L., Johnson, J.M., Sherameti, I., Glawischnig, E., Halkier, B.A., and Oelmüller, R. (2012) Indole-3-acetaldoxime-derived compounds restrict root colonization in the beneficial interaction between *Arabidopsis* roots and the endophyte *Piriformospora indica*. *Mol. Plant-Microbe. Interact.*, **25**, 1186–1197.

Rodriguez, R.J., White, J.F., Arnold, A.E., and Redman, R.S. (2009) Fungal endophytes: diversity and functional roles. *New Phytol.*, **182**, 314–330.

2.4.4 Lichens

Paracer, S. and Ahmadjian, V. (2000) *Symbiosis*, 2nd edn, Oxford University Press, Oxford, UK.

2.5 Defense

Agrios, G.N. (1997) *Plant Pathology*, 4th edn, Academic Press, Oxford, UK.

Bennett, R.N. and Wallsgrove, R. (1994) Secondary metabolites in plant defence mechanisms. *New Phytol.*, **127**, 617–633.

2.5.1 Pathogens

Bednarek, P., Piślewska-Bednarek, M., Ver Loren van Themaat, E., Maddula, R.K., Svatoš, A., and Schulze-Lefert, P. (2011) Conservation and clade-specific diversification of pathogen-inducible tryptophan and indole glucosinolate metabolism in *Arabidopsis thaliana* relatives. *New Phytol.*, **192**, 713–726.

Carle, R. (1992) Phytoalexine. *Pharm. Unserer Zeit*, **21**, 99–104.

Chaturvedi, R., Venables, B., Petros, R.A., Nalam, V., Li, M., Wang, X., Takemoto, L. J., and Shah, J. (2012) An abietane diterpenoid is a potent activator of systemic acquired resistance. *Plant J.*, **71**, 161–172.

Dempsey, D'M.A. and Klessig, D.F. (2012) SOS – too many signals for systemic acquired resistance? *Trends Plant Sci.*, **17**, 538–544.

Dixon, D.P., Sellars, J.D., Kenwright, A.M., and Steel, P.G. (2012) The maize benzoxazinone DIMBOA reacts with glutathione and other thiols to form spirocyclic adducts. *Phytochemistry*, **77**, 171–178.

Elstner, E.F., Oßwald, W., and Schneider, I. (1996) *Phytopathologie*, Spektrum Akademischer Verlag, Heidelberg, Berlin, Oxford.

Halbwirth, H., Fischer, T., Roemmelt, S., Spinelli, F., Schlangen, K., Peterek, S., Sabatini, E., Messina, C., Speakman, J.-B., Andreotti, C., Rademacher, W., Bazzi, C., Costa, G., Treutter, D., Forkmann, G., and Stich, K. (2003) Induction of antimicrobial 3-deoxyflavonoids in pome fruit trees controls fire blight. *Z. Naturforsch.*, **58c**, 765–770.

Hartmann, T. (1996) Diversity and variability of plant secondary metabolism: a mechanistic view. *Entomol. Exp. Appl.*, **80**, 177–188.

Hipskind, J.D., Hanau, R., Leite, B., and Nicholson, R.L. (1990) Phytoalexin accumulation in sorghum: identification of an apigeninidin acyl ester. *Physiol. Mol. Plant Pathol.*, **36**, 381–396.

Iven, T., König, S., Singh, S., Braus-Stromeyer, S.A., Bischoff, M., Tietze, L.F., Braus, G.H., Lipka, V., Feussner, I., and Dröge-Laser, W. (2012) Transcriptional activation and production of tryptophan-derived secondary metabolites in *Arabidopsis* roots contributes to the defense against the fungal vascular pathogen *Verticillium longisporum*. *Mol. Plant*, **5**, 1389–1402.

Kröner, A., Marnet, N., Andrivon, D., and Val, F. (2012) Nicotiflorin, rutin and chlorogenic acid: phenylpropanoids involved differently in quantitative resistance of potato tubers to biotrophic and necrotrophic pathogens. *Plant Physiol. Biochem.*, **57**, 23–31.

Ludwig-Müller, J. (2009) Glucosinolates and the clubroot disease: Defense compounds or auxin precursors? *Phytochem. Rev.*, **8**, 135–148.

Marinic, M., Piantanida, I., Rusak, G., and Zinic, M. (2006) Interactions of quercetin and its lanthane complex with double stranded DNA/RNA and single stranded RNA: Spectrophotometric sensing of poly G. *J. Inorg. Biochem.*, **100**, 288–298.

Mert-Türk, F. (2006) Saponins versus plant fungal pathogens. *J. Cell Mol. Biol.*, **5**, 13–17.

Nicholson, R.L. and Wood, K.V. (2001) Phytoalexins and secondary products, where are they and how can we measure them? *Physiol. Mol. Plant Pathol.*, **59**, 63–69.

Niemeyer, H.M. (2009) Hydroxamic acids derived from 2-hydroxy-2*H*-1,4-benzoxazin-3(4*H*)-one: Key defense chemicals of cereals. *J. Agric. Food Chem.*, **57**, 1677–1696.

Osbourn, A.E., Clarke, B.R., Lunness, P., Scott, P.R., and Daniels, M.J. (1994) An oat species lacking avenacin is susceptible to infection by *Gaeumannomyces graminis* var. *tritici*. *Physiol. Mol. Plant Pathol.*, **45**, 457–467.

Pedras, M.S.C. and Hossain, S. (2011) Interaction of cruciferous phytoanticipins with plant fungal pathogens: Indole glucosinolates are not metabolized but the corresponding desulfo-derivatives and nitriles are. *Phytochemistry*, **72**, 2308–2316.

Polster, J., Dithmar, H., Burgemeister, R., Friedemann, G., and Feucht, W. (2006) Flavonoids in plant nuclei: detection by laser microdissection and pressure catapulting (LMPC), in vivo staining, and UV-visible spectroscopic titration. *Physiol. Planta*, **128**, 163–174.

Rademacher, W., Spinelli, F., and Costa, G. (2006) Prohexanedione-Ca: Modes of action of a multifunctional plant bioregulator for fruit trees. *Acta Hort.*, **727**, 97–106.

Rouxel, T., Kolimann, A., Boulidard, L., and Mithen, R. (1991) Abiotic elicitation of indole phytoalexins and resistance to *Leptosphaeria maculans* within Brassiceae. *Planta*, **184**, 271–278.

Rusak, G., Krajačić, M., and Pleše, N. (1997) Inhibition of tomato bushy stunt virus infection using a quercetagetin flavonoid isolated from *Centaurea rupestris* L. *Antiviral Res.*, **36**, 125–129.

Rusak, G., Piantanida, I., Bretschneider, S., and Ludwig-Müller, J. (2009) Complex formation of quercetin with lanthanum enhances binding to plant viral satellite double stranded RNA. *J. Inorg. Biochem.*, **103**, 1597–1601.

Schlaeppi, K., Abou-Mansour, E., Buchala, A., and Mauch, F. (2010) Disease resistance of *Arabidopsis* to *Phytophthora brassicae* is established by the sequential action of indole glucosinolates and camalexin. *Plant J.*, **62**, 840–851.

Snyder, B.A. and Nicholson, R.L. (1990) Synthesis of phytoalexins in sorghum as a site-specific response to fungal ingress. *Science*, **29**, 1637–1639.

Yuan, L., Wang, L., Han, Z., Jiang, Y., Zhao, L., Liu, H., Yang, L., and Luo, K. (2012) Molecular cloning and characterization of *PtrLAR3*, a gene encoding leucoanthocyanidin reductase from *Populus trichocarpa*, and its constitutive expression enhances fungal resistance in transgenic plants. *J. Exp. Bot.*, **63**, 2513–2524.

2.5.2
Antimicrobial Potential in Plant Protection (Biofumigation)

Gimsing, A.L. and Kirkegaard, J.A. (2009) Glucosinolates and biofumigation: fate of glucosinolates and their hydrolysis products in soil. *Phytochem. Rev.*, **8**, 299–310.

Reardon, C.L., Strauss, S.L., and Mazzola, M. (2013) Changes in available nitrogen and nematode abundance in response to *Brassica* seed meal amendment of orchard soil. *Soil Biol. Biochem.*, **57**, 22–29.

2.5.3
Invertebrate Herbivores

Ali, J.G. and Agrawal, A.A. (2012) Specialist versus generalist insect herbivores and plant defense. *Trends Plant Sci.*, **17**, 293–302.

Baldwin, I.T., Halitschke, R., Kessler, A., and Schittko, U. (2001) Merging molecular and ecological approaches in plant–insect interactions. *Curr. Opin. Plant Biol.*, **4**, 351–358.

Barone, M. and Frank, T. (1999) Effect of plant extracts on feeding behaviour of the slug *Arion lusitanicus*. *Ann. Appl. Biol.*, **134**, 341–345.

Berger, J., Walczysko, S., Pávková, J., and Gutzeit, H.O. (2003) Effects of genistein on insect haemocytes. *J. Appl. Biomed.*, **1**, 161–168.

Chitwood, D.J. (2002) Phytochemical based strategies for nematode control. *Annu. Rev. Phytopathol.*, **40**, 221–249.

Clark, S.J., Henderson, I.F., Hill, D.J., and Martin, A.P. (1999) Use of lichen secondary metabolites as antifeedants to protect higher plants from damage caused by slug feeding. *Ann. Appl. Biol.*, **134**, 101–108.

Crespo, E., Hordijk, C.A., deGraaf, R.M., Samudrala, D., Cristescu, S.M., Harren, F. J.M., and van Dam, N.M. (2012) On-line detection of root-induced volatiles in *Brassica nigra* plants infested with *Delia radicum* L. root fly larvae. *Phytochemistry*, **84**, 68–77.

Frydman, A., Liberman, R., Huhman, D.V., Carmeli-Weissberg, M., Sapir-Mir, M., Ophir, R., Sumner, L.W., and Eyal, Y. (2013) The molecular and enzymatic basis of bitter/non-bitter flavor of citrus fruit: evolution of branch-forming rhamnosyltransferases under domestication. *Plant J.*, **73**, 166–178.

Gahukar, R.T. (2012) Evaluation of plant-derived products against pests and diseases of medicinal plants: A review. *Crop Protect.*, **42**, 202–209.
Halitschke, R. and Baldwin, I.T. (2005) Jasmonates and related compounds in plant-insect interactions. *J. Plant Growth Regul.*, **23**, 238–245.
Kuzina, V., Ekstrøm, C.T., Andersen, S.B., Nielsen, J.K., Olsen, C.E., and Bak, S. (2009) Identification of defense compounds in *Barbarea vulgaris* against the herbivore *Phyllotreta nemorum* by an ecometabolomic approach. *Plant Physiol.*, **151**, 1977–1990.
Lattanzio, V., Arpaia, S., Cardinali, A., Di Venere, D., and Linsalata, V. (2000) Role of endogenous flavonoids in resistance mechanism of *Vigna* to aphids. *J. Agric. Food Chem.*, **48**, 5316–5320.
Paré, P.W., Farag, M.A., Krishnamachari, V., Zhang, H., Ryu, C.-M., and Kloepper, J.W. (2005) Elicitors and priming agents initiate plant defense responses. *Photosynth. Res.*, **85**, 149–159.
Riddiford, L.M. (1994) Cellular and molecular actions of juvenile hormone. I. General considerations and premetamorphic actions. *Adv. Insect Physiol.*, **24**, 213–227.
Rostás, M., Simon, M., and Hilker, M. (2003) Ecological cross-effects of induced plant responses towards herbivores and phytopathogenic fungi. *Basic Appl. Ecol.*, **4**, 43–62.
Schmid, O. and Henggeler, S. (1989) *Biologischer Pflanzenschutz im Garten*, Ulmer, Stuttgart.
Shroff, R., Vergara, F., Muck, A., Svatos, A., and Gershenzon, J. (2008) Nonuniform distribution of glucosinolates in *Arabidopsis thaliana* leaves has important consequences for plant defense. *Proc. Natl Acad. Sci. USA*, **105**, 6196–6201.
Veromann, E., Toome, M., Kännaste, A., Kaasik, R., Copolovici, L., Flink, J., Kovács, G., Narits, L., Luik, A., and Niinemets, Ü. (2013) Effects of nitrogen fertilization on insect pests, their parasitoids, plant diseases and volatile organic compounds in *Brassica napus*. *Crop Protect.*, **43**, 79–88.
Vogt, H. and Händel, U. (1996) Zur Wirksamkeit und Selektivität von Neem-Produkten im Obstbau. *Mitt. Biol. Bundesanst.*, **321**, 420.

2.5.4 Parasitic Plants

Jamison, D.S. and Yoder, J.I. (2001) Heritable variation in quinone-induced haustorium development in the parasitic plant *Triphysaria*. *Plant Physiol.*, **125**, 1870–1880.
Press, M. and Graves, J. (eds.) (1995) *Parasitic Plants*, Springer Verlag, Berlin, Heidelberg.
Rubiales, D. and Heide-Jørgensen, H.S. (2011) *Parasitic Plants*, eLS. (Published Online: 15 February 2011). doi: 10.1002/9780470015902.a0021271
Yoshida, S. and Shirasu, K. (2012) Plants that attack plants: molecular elucidation of plant parasitism. *Curr. Opin. Plant Biol.*, **15**, 708–713.

2.6 Allelopathy

Rice, E.L. (1984) *Allelopathy*, 2nd edn, Academic Press, San Diego, California, USA.
Rizvi, S.J.H., Haque, H., Singh, V.K., and Rizvi, V. (1992) A discipline called allelopathy, in *Allelopathy. Basic and Applied Aspects* (eds S.J.H. Rizvi and V. Rizvi), Chapman & Hall, London, UK, pp. 1–8.

3 Biological Effects on Microorganisms and Animals

3.1 Coevolution

Berendsen, R.L., Pieterse, C.M.J., and Bakker, P.A.H.M. (2012) The rhizosphere microbiome and plant health. *Trends Plant Sci.*, **17**, 478–486.
Kallenbach, M., Bonaventure, G., Gilardoni, P.A., Wissgott, A., and Baldwin, I.T. (2012)

Empoasca leafhoppers attack wild tobacco plants in a jasmonate dependent manner and identify jasmonate mutants in natural populations. *Proc. Natl Acad. Sci. USA*, **109**, E1548–E1557.

Jones, J.D.G. and Dangl, J.L. (2006) The plant immune system. *Nature*, **444**, 323–329.

Labandeira, C.C., Dilcher, D.L., Davis, D.R., and Wagner, D.L. (1994) Ninety-seven million years of angiosperm-insect association: paleobiological insights into the meaning of coevolution. *Proc. Natl Acad. Sci. USA*, **91**, 12278–12282.

Labandeira, C.C. (1997) Insect mouthparts: ascertaining the paleobiology of insect feeding strategies. *Annu. Rev. Ecol. Syst.*, **28**, 153–193.

Schoonhoven, L.M. (2005) *Insect-Plant Biology*, Oxford University Press, New York, USA.

Steppuhn, A., Gase, K., Krock, B., Halitschke, R., and Baldwin, I.T. (2004) Nicotine's defensive function in nature. *PLoS Biol.*, **2**, 1074–1080.

Tiffney, B.H. (1992) The role of vertebrate herbivory in the evolution of land plants. *Palaeobotanist*, **41**, 87–97.

Wu, J., Hettenhausen, C., Schuman, M.C., and Baldwin, I.T. (2008) A comparison of two *Nicotiana attenuata* accessions reveals large differences in signalling induced by oral secretions of the specialist herbivore *Manduca sexta*. *Plant Physiol.*, **146**, 927–939.

3.1.1
Adaptation Mechanisms of Microbes and Herbivores

Boppré, M. (1986) Insects pharmacophagously utilizing defensive plant chemicals (pyrrolizidine alkaloids). *Naturwissenschaften*, **73**, 17–26.

Cardoso, M.Z. (2008) Herbivore handling of a plant's trichome: the case of *Heliconius charithonia* (L.) (Lepidoptera: Nymphalidae) and *Passiflora lobata* (Killip) Hutch. (Passifloraceae). *Neotrop. Entomol.*, **37**, 247–252.

Dussourd, D. and Eisner, T. (1987) Vein-cutting behavior: insect counterploy to the latex defense of plants. *Science*, **237**, 898–901.

Eisner, T., Eisner, M., and Hoebeke, E.R. (1998) When defense backfires: Detrimental effect of a plant's protective trichomes on an insect beneficial to the plant. *Proc. Natl Acad. Sci. USA*, **95**, 4410–4414.

Galletti, S., Sala, E., Leoni, O., Cinti, S., and Cerato, C. (2008) *Aspergillus flavus* transformation of glucosinolates to nitriles by an arylsulfatase and a β-thio-glucosidase. *Soil Biol. Biochem.*, **40**, 2170–2173.

Gibson, R.W. and Pickett, J.A. (1983) Wild potato repels aphids by release of aphid alarm pheromone. *Nature*, **302**, 608–609.

Hartmann, T. and Ober, D. (2000) Biosynthesis and metabolism of pyrrolizidine alkaloids in plants and specialized insect herbivores. *Topics Curr. Chem.*, **209**, 207–243.

Ishii, H., Fraaije, B.A., Sugiyama, T., Noguchi, K., Nishimura, K., Takeda, T., Amano, T., and Hollomon, D.W. (2001) Occurrence and molecular characterization of strobilurin resistance in cucumber powdery mildew and downy mildew. *Phytopathology*, **91**, 1166–1171.

Kuhn, J., Pettersson, E.M., Feld, B.K., Burse, A., Termonia, A., Pasteels, J.M., and Boland, W. (2004) Selective transport systems mediate sequestration of plant glucosides in leaf beetles: A molecular basis for adaptation and evolution. *Proc. Natl Acad. Sci. USA*, **101**, 13808–13813.

Pedras, M.S.C. and Khan, A.Q. (1996) Biotransformation of the Brassica phytoalexin Brassicanal A by the blackleg fungus. *J. Agric. Food Chem.*, **44**, 3403–3407.

Randlkofer, B., Obermaier, E., Hilker, M., and Meiners, T. (2010) Vegetation complexity: The influence of plant species diversity and plant structures on plant chemical complexity and arthropods. *Basic Appl. Ecol.*, **11**, 383–395.

Samuni-Blank, M., Izhaki, I., Dearing, M.D., Gerchman, Y., Trabelcy, B., Lotan, A., Karasov, W.H., and Arad, Z. (2012) Intraspecific directed deterrence by the

mustard oil bomb in a desert plant. *Curr. Biol.*, **22**, 1218–1220.
Sosa, T., Chaves, N., Alias, J.C., Escudero, J. C., Henao, F., and Gutiérrez-Merino, C. (2004) Inhibition of mouth skeletal muscle relaxation by flavonoids of *Cistus ladanifer* L.: A plant defense mechanism against herbivores. *J. Chem. Ecol.*, **30**, 1087–1101.
Stauber, E.J., Kuczka, P., van Ohlen, M., Vogt, B., Janowitz, T., Piotrowski, M., Bäuerle, T., and Wittstock, U. (2012) Turning the 'Mustard Oil Bomb' into a 'Cyanide Bomb': aromatic glucosinolate metabolism in a specialist insect herbivore. *PLoS One*, **7**(4), e35545.
Stefanato, F.L., Abou-Mansour, E., Buchala, A., Kretschmer, M., Mosbach, A., Hahn, M., Bochet, C.G., Métraux, J.-P., and Schoonbeek, H.-j. (2009) The ABC transporter BcatrB from *Botrytis cinerea* exports camalexin and is a virulence factor on *Arabidopsis thaliana*. *Plant J.*, **58**, 499–510.
Voigt, D. and Gorb, S. (2010) Locomotion in a sticky terrain. *Arthropod-Plant Interact.*, **4**, 69–79.

3.1.2
Attraction of Insects

Büchel, K., Malskies, S., Mayer, M., Fenning, T.M., Gershenzon, J., Hilker, M., and Meiners, T. (2011) How plants give early herbivore alert: Volatile terpenoids attract parasitoids to egg-infested elms. *Basic Appl. Ecol.*, **12**, 403–412.
Ghirardo, A., Heller, W., Fladung, M., Schnitzler, J.-P., and Schroeder, H. (2012) Function of defensive volatiles in pedunculate oak (*Quercus robur*) is tricked by the moth *Tortrix viridana*. *Plant Cell Environ.*, **35**, 2192–2207.
Kessler, D., Bhattacharya, S., Diezel, C., Rothe, E., Gase, K., Schöttner, M., and Baldwin, I.T. (2012) Unpredictability of nectar nicotine promotes outcrossing by hummingbirds in *Nicotiana attenuata*. *Plant J.*, **71**, 529–538.
Kim, S.-G., Yon, F., Gaquerel, E., Gulati, J., and Baldwin, I.T. (2011) Tissue specific diurnal rhythms of metabolites and their regulation during herbivore attack in a native tobacco, *Nicotiana attenuata*. *PLoS One*, **6** (10), e26214.
McCormick, A.C., Unsicker, S.B., and Gershenzon, J. (2012) The specificity of herbivore-induced plant volatiles in attracting herbivore enemies. *Trends Plant Sci.*, **17**, 303–310.
Neal, A.L., Ahmad, S., Gordon-Weeks, R., and Ton, J. (2012) Benzoxazinoids in root exudates of maize attract *Pseudomonas putida* to the rhizosphere. *PLoS One*, **7** (4), e35498.
Riffell, J.A., Lei, H., Abrell, L., and Hildebrand, J.G. (2013) Neural basis of a pollinator's buffet: olfactory specialization and learning in *Manduca sexta*. *Science*, **339**, 200–204.
Rosenstiel, T.N., Shortlidge, E.E., Melnychenko, A.N., Pankow, J.F., and Eppley, S.M. (2012) Sex-specific volatile compounds influence microarthropod-mediated fertilization of moss. *Nature*, **489**, 431–433.
Städler, E. and Reifenrath, K. (2009) Glucosinolates on the leaf surface perceived by insect herbivores: review of ambiguous results and new investigations. *Phytochem. Rev.*, **8**, 207–225.
Weinhold, A. and Baldwin, I.T. (2011) Trichome-derived O-acyl sugars are a first meal for caterpillars that tags them for predation. *Proc. Natl Acad. Sci. USA*, **108**, 7855–7859.

3.2
Effects on Vertebrates: An Overview

3.2.2
Molecular Interactions with Cellular Components

Barry, J., Fritz, M., Brender, J.R., Smith, P.E., Lee, D.K., and Ramamoorthy, A. (2009) Determining the effects of lipophilic drugs on membrane structure by solid-state NMR spectroscopy: the case of the antioxidant curcumin. *J. Am. Chem. Soc.*, **131**, 4490–4498.
Barthel, A., Ostrakhovitch, E.A., Walter, P.L., Kampkötter, A., and Klotz, L.O. (2007)

Stimulation of phosphoinositide 3-kinase/Akt signaling by copper and zinc ions: mechanisms and consequences. *Arch. Biochem. Biophys.*, **463**, 175–182.

Bernhardt, M.L., Kim, A.M., O'Halloran, T. V., and Woodruff, T.K. (2011) Zinc requirement during meiosis I-meiosis II transition in mouse oocytes is independent of the MOS-MAPK pathway. *Biol. Reprod.*, **84**, 526–536.

Celino, F.T., Yamaguchi, S., Miura, C., Ohta, T., Tozawa, Y., Iwai, T., and Miura, T. (2011) Tolerance of spermatogonia to oxidative stress is due to high levels of Zn and Cu/Zn superoxide dismutase. *PloS One*, **6**, e16938.

Le Bourvellec, C. and Renard, C.M. (2012) Interactions between polyphenols and macromolecules: quantification methods and mechanisms. *Crit. Rev. Food Sci. Nutr.*, **52**, 213–248.

Macpherson, L.J., Dubin, A.E., Evans, M.J., Marr, F., Schultz, P.G., Cravatt, B.F., and Patapoutian, A. (2007) Noxious compounds activate TRPA1 ion channels through covalent modification of cysteines. *Nature*, **445**, 541–545.

Quesada, I.M., Bustos, M., Blay, M., Pujadas, G., Ardèvol, A., Salvadó, M.J., Bladé, C., Arola, L., and Fernández-Larrea, J. (2011) Dietary catechins and procyanidins modulate zinc homeostasis in human HepG2 cells. *J. Nutr. Biochem.*, **22**, 153–163.

Wenzel, E. and Somoza, V. (2005) Metabolism and bioavailability of trans-resveratrol. *Mol. Nutr. Food Res.*, **49**, 472–481.

Wolford, J.L., Chishti, Y., Jin, Q., Ward, J., Chen, L., Vogt, S., and Finney, L. (2010) Loss of pluripotency in human embryonic stem cells directly correlates with an increase in nuclear zinc. *PLoS One*, **5**, e12308.

3.2.3 Visualization of Natural Substances in Living Cells

Asakawa, T., Hiza, A., Nakayama, M., Inai, M., Oyama, D., Koide, H., Shimizu, K., Wakimoto, T., Harada, N., Tsukada, H., Oku, N., and Kan, T. (2011) PET imaging of nobiletin based on a practical total synthesis. *Chem. Commun. (Camb.)*, **47**, 2868–2870.

Gutzeit, H.O., Henker, Y., Kind, B., and Franz, A. (2004) Specific interactions of quercetin and other flavonoids with target proteins are revealed by elicited fluorescence. *Biochem. Biophys. Res. Commun.*, **318**, 490–495.

Li, Z.H., Alex, D., Siu, S.O., Chu, I.K., Renn, J., Winkler, C., Lou, S., Tsui, S.K., Zhao, H. Y., Yan, W.R., Mahady, G.B., Li, G.H., Kwan, Y.W., Wang, Y.T., and Lee, S.M. (2011) Combined in vivo imaging and omics approaches reveal metabolism of icaritin and its glycosides in zebrafish larvae. *Mol. Biosyst.*, **7**, 2128–2138.

Moskaug, J., Carlsen, H., Myhrstad, M., and Blomhoff, R. (2004) Molecular imaging of the biological effects of quercetin and quercetin-rich foods. *Mech. Ageing Dev.*, **125**, 315–324.

Mueller-Harvey, I., Feucht, W., Polster, J., Trnková, L., Burgos, P., Parker, A.W., and Botchway, S.W. (2012) Two-photon excitation with pico-second fluorescence lifetime imaging to detect nuclear association of flavanols. *Anal. Chim. Acta*, **719**, 68–75.

4 Metabolism and Toxicity of Natural Substances in Mammals

Aktories, K.F., Förstermann, U., Hoffmann, F., and Starke, K. (2009) *Allgemeine und spezielle Pharmakologie und Toxikologie*, 10th edn, Elsevier, Urban & Fischer, Munich.

Efferth, T. (2006) *Molekulare Pharmakologie und Toxikologie*, Springer, Berlin.

Newsholme, E.A.L. and Anthony, R. (2010) *Functional Biochemistry in Health and Disease*, 1st edn, Wiley-Blackwell, Chichester.

4.1 Metabolism of Natural Substances in Mammals

Scheline, R.R. (1991) *Handbook of Mammalian Metabolism of Plant*

Compounds, CRC Press Inc., Boca Raton, Florida, USA.

4.1.1 Principles of Pharmacokinetics

Atkinson, C., Frankenfeld, C.L., and Lampe, J.W. (2005) Gut bacterial metabolism of the soy isoflavone daidzein: exploring the relevance to human health. *Exp. Biol. Med. (Maywood)*, **230**, 155–170.

Clemente, J.C., Ursell, L.K., Parfrey, L.W., and Knight, R. (2012) The impact of the gut microbiota on human health: an integrative view. *Cell*, **148**, 1258–1270.

Cuyckens, F. and Claeys, M. (2005) Determination of the glycosylation site in flavonoid mono-O-glycosides by collision-induced dissociation of electrospray-generated deprotonated and sodiated molecules. *J. Mass Spectrom.*, **40**, 364–372.

Day, A.J., Bao, Y., Morgan, M.R., and Williamson, G. (2000) Conjugation position of quercetin glucuronides and effect on biological activity. *Free Radical Biol. Med.*, **29**, 1234–1243.

Galindo, P., Rodriguez-Gómez, I., González-Manzano, S., Dueñas, M., Jiménez, R., Menéndez, C., Vargas, F., Tamargo, J., Santos-Buelga, C., Pérez-Vizcaíno, F., and Duarte, J. (2012) Glucuronidated quercetin lowers blood pressure in spontaneously hypertensive rats via deconjugation. *PloS One*, **7**, e32673.

Kohalmy, K. and Vrzal, R. (2011) Regulation of phase II biotransformation enzymes by steroid hormones. *Curr. Drug Metab.*, **12**, 104–123.

Morris, M.E. and Zhang, S. (2006) Flavonoid–drug interactions: effects of flavonoids on ABC transporters. *Life Sci.*, **78**, 2116–2130.

Muller, F. and Fromm, M.F. (2011) Transporter-mediated drug–drug interactions. *Pharmacogenomics*, **12**, 1017–1037.

Spencer, J.P., Abd-el-Mohsen, M.M., and Rice-Evans, C. (2004) Cellular uptake and metabolism of flavonoids and their metabolites: implications for their bioactivity. *Arch. Biochem. Biophys*, **423**, 148–161.

Xiao, J. and Kai, G. (2012) A review of dietary polyphenol-plasma protein interactions: characterization, influence on the bioactivity, and structure-affinity relationship. *Crit. Rev. Food Sci. Nutr.*, **52**, 85–101.

Zhang, L., Zuo, Z., and Lin, G. (2007) Intestinal and hepatic glucuronidation of flavonoids. *Mol. Pharm.*, **4**, 833–845.

4.1.2 Case Studies of Selected Natural Compounds

Andersen, G., Christrup, L., and Sjøgren, P. (2003) Relationships among morphine metabolism, pain and side effects during long-term treatment: an update. *J. Pain Symptom. Manage.*, **25**, 74–91.

Cottart, C.H., Nivet-Antoine, V., Laguillier-Morizot, C., and Beaudeux, J.L. (2010) Resveratrol bioavailability and toxicity in humans. *Mol. Nutr. Food Res.*, **54**, 7–16.

Daly, A.K. (2010) Pharmacogenetics and human genetic polymorphisms. *Biochem. J.*, **429**, 435–449.

Gutzeit, H.O., Tokalov, S.V., Ludwig-Müller, J., and Rusak, G. (2005) Monitoring flavonoid metabolism in human cells by exploiting fluorescence elicited by quercetin/protein interactions. *Croatica Chemica Acta*, **78**, 337–342.

Hukkanen, J., Jacob, P., and Benowitz, N.L. (2005) Metabolism and disposition kinetics of nicotine. *Pharmacol. Rev.*, **57**, 79–115.

Ludwig-Müller, J., Tokalov, S.V., Franz, A., and Gutzeit, H.O. (2005) Quercetin metabolism in vital and apoptotic human leukaemia cells. *Biol. Chem.*, **386**, 279–283.

Meyer, M.R. and Maurer, H.H. (2011) Absorption, distribution, metabolism and excretion pharmacogenomics of drugs of abuse. *Pharmacogenomics*, **12**, 215–233.

Monagas, M., Urpi-Sarda, M., Sánchez-Patán, F., Llorach, R., Garrido, I., Gómez-Cordovés, C., Andres-Lacueva, C., and Bartolomé, B. (2010) Insights into the metabolism and microbial biotransformation of dietary flavan-3-ols and the bioactivity of their metabolites. *Food Funct.*, **1**, 233–253.

Russo, M., Spagnuolo, C., Tedesco, I., Bilotto, S., and Russo, G.L. (2012) The flavonoid quercetin in disease prevention and therapy: Facts and fancies. *Biochem. Pharmacol.*, **83** (1), 6–15.

4.1.3 Metabolic Bioactivation

Chen, C.H., Dickman, K.G., Moriya, M., Zavadil, J., Sidorenko, V.S., Edwards, K.L., Gnatenko, D.V., Wu, L., Turesky, R.J., Wu, X.R., Pu, Y.S., and Grollman, A.P. (2012) Aristolochic acid-associated urothelial cancer in Taiwan. *Proc. Natl Acad. Sci. USA*, **109**, 8241–8246.

Chen, X.W., Serag, E.S., Sneed, K.B., and Zhou, S.F. (2011) Herbal bioactivation, molecular targets and the toxicity relevance. *Chem. Biol. Interact.*, **192**, 161–176.

Stiborová, M., Frei, E., Arlt, V.M., and Schmeiser, H.H. (2008) Metabolic activation of carcinogenic aristolochic acid, a risk factor for Balkan endemic nephropathy. *Mutat. Res.*, **658**, 55–67.

4.1.4 Effects on Enzymes Controlling Biotransformation and Bioavailability

Alvarez, A.I., Real, R., Pérez, M., Mendoza, G., Prieto, J.G., and Merino, G. (2010) Modulation of the activity of ABC transporters (P-glycoprotein, MRP2, BCRP) by flavonoids and drug response. *J. Pharm. Sci.*, **99**, 598–617.

Kimura, Y., Ito, H., Ohnishi, R., and Hatano, T. (2010) Inhibitory effects of polyphenols on human cytochrome P450 3A4 and 2C9 activity. *Food Chem. Toxicol.*, **48**, 429–435.

Lin, H.L., Kenaan, C., and Hollenberg, P.F. (2012) Identification of the residue in human CYP3A4 that is covalently modified by bergamottin and the reactive intermediate that contributes to the grapefruit juice effect. *Drug Metab. Dispos.*, **40**, 998–1006.

Rodríguez-Fragoso, L., Martínez-Arismendi, J.L., Orozco-Bustos, D., Reyes-Esparza, J., Torres, E., and Burchiel, S.W. (2011) Potential risks resulting from fruit/vegetable-drug interactions: effects on drug-metabolizing enzymes and drug transporters. *J. Food Sci.*, **76**, R112–R124.

4.1.5 Genetic Polymorphisms

Daly, A.K. (2010) Pharmacogenetics and human genetic polymorphisms. *Biochem. J.*, **429**, 435–449.

Somogyi, A.A., Barratt, D.T., and Coller, J.K. (2007) Pharmacogenetics of opioids. *Clin. Pharmacol. Ther.*, **81**, 429–444.

Zhou, S.F. (2009) Polymorphism of human cytochrome P450 2D6 and its clinical significance: Part I. *Clin. Pharmacokinet.*, **48**, 689–723.

Zhou, S.F., Liu, J.P., and Chowbay, B. (2009) Polymorphism of human cytochrome P450 enzymes and its clinical impact. *Drug Metab. Rev.*, **41**, 89–295.

4.2 Toxicity

4.2.1 Basic Toxicology

Seibert, H., Maser, E., Schweda, K., Seibert, S., and Gülden, M. (2011) Cytoprotective activity against peroxide-induced oxidative damage and cytotoxicity of flavonoids in C6 rat glioma cells. *Food Chem. Toxicol.*, **49**, 2398–2407.

Timbrell, J.A. (2002) *Introduction to Toxicology*, 3rd edn, Taylor & Francis, London, UK.

4.2.2 Toxicity of Selected Natural Substances

Moldenhauer, G., Salnikov, A.V., Lüttgau, S., Herr, I., Anderl, J., and Faulstich, H. (2012) Therapeutic potential of amanitin-conjugated anti-epithelial cell adhesion molecule monoclonal antibody against pancreatic carcinoma. *J. Natl Cancer Inst.*, **104**, 622–634.

4.2.3
Strategies and Difficulties in Toxicological Evaluations

Boverhof, D.R., Chamberlain, M.P., Elcombe, C.R., Gonzalez, F.J., Heflich, R. H., Hernández, L.G., Jacobs, A.C., Jacobson-Kram, D., Luijten, M., Maggi, A., Manjanatha, M.G., Benthem, J., and Gollapudi, B.B. (2011) Transgenic animal models in toxicology: historical perspectives and future outlook. *Toxicol. Sci.*, **121**, 207–233.

Doktorova, T.Y., Pauwels, M., Vinken, M., Vanhaecke, T., and Rogiers, V. (2012) Opportunities for an alternative integrating testing strategy for carcinogen hazard assessment? *Crit. Rev. Toxicol.*, **42**, 91–106.

Harwood, M., Danielewska-Nikiel, B., Borzelleca, J.F., Flamm, G.W., Williams, G.M., and Lines, T.C. (2007) A critical review of the data related to the safety of quercetin and lack of evidence of in vivo toxicity, including lack of genotoxic/carcinogenic properties. *Food Chem. Toxicol.*, **45**, 2179–2205.

Spielmann, H. (2005) Predicting the risk of developmental toxicity from in vitro assays. *Toxicol. Appl. Pharmacol.*, **207**, 375–380.

Tralau, T. and Luch, A. (2012) Drug-mediated toxicity: illuminating the 'bad' in the test tube by means of cellular assays? *Trends Pharmacol. Sci.*, **33**, 353–364.

5
Examples of Physiological Effects of Natural Substances Present in Food and in Medicinal Plants

Polya, G.M. (2003) *Biochemical Targets of Plant Bioactive Compounds a Pharmacological Reference Guide to Sites of Action and Biological Effects*, Taylor & Francis, London, UK.

5.1
Are Antioxidants Good for Health?

Birringer, M. (2011) Hormetics: dietary triggers of an adaptive stress response. *Pharm. Res.*, **28**, 2680–2694.

Valko, M., Leibfritz, D., Moncol, J., Cronin, M.T., Mazur, M., and Telser, J. (2007) Free radicals and antioxidants in normal physiological functions and human disease. *Int. J. Biochem. Cell Biol.*, **39**, 44–84.

5.1.1
Generation and Homeostatic Control of Reactive Oxygen and Nitrogen Species

Kondo, N., Nakamura, H., Masutani, H., and Yodoi, J. (2006) Redox regulation of human thioredoxin network. *Antioxid. Redox Signal.*, **8**, 1881–1890.

Limón-Pacheco, J. and Gonsebatt, M.E. (2009) The role of antioxidants and antioxidant-related enzymes in protective responses to environmentally induced oxidative stress. *Mutat. Res.*, **674**, 137–147.

5.1.2
Function of ROS/RNS as Signaling Molecules for Cellular Functions

Coant, N., Ben Mkaddem, S., Pedruzzi, E., Guichard, C., Tréton, X., Ducroc, R., Freund, J.N., Cazals-Hatem, D., Bouhnik, Y., Woerther, P.L., Skurnik, D., Grodet, A., Fay, M., Biard, D., Lesuffleur, T., Deffert, C., Moreau, R., Groyer, A., Krause, K.H., Daniel, F., and Ogier-Denis, E. (2010) NADPH oxidase 1 modulates WNT and NOTCH1 signaling to control the fate of proliferative progenitor cells in the colon. *Mol. Cell Biol.*, **30**, 2636–2650.

Dröge, W. (2002) Free radicals in the physiological control of cell function. *Physiol. Rev.*, **82**, 47–95.

Jeong, M., Piao, Z.H., Kim, M.S., Lee, S.H., Yun, S., Sun, H.N., Yoon, S.R., Chung, J. W., Kim, T.D., Jeon, J.H., Lee, J., Kim, H. N., Choi, J.Y., and Choi, I. (2009) Thioredoxin-interacting protein regulates hematopoietic stem cell quiescence and mobilization under stress conditions. *J. Immunol.*, **183**, 2495–2505.

Montezano, A.C. and Touyz, R.M. (2012) Reactive oxygen species and endothelial function–role of nitric oxide synthase uncoupling and Nox family nicotinamide

adenine dinucleotide phosphate oxidases. *Basic Clin. Pharmacol. Toxicol.*, **110**, 87–94.
Niethammer, P., Grabher, C., Look, A.T., and Mitchison, T.J. (2009) A tissue-scale gradient of hydrogen peroxide mediates rapid wound detection in zebrafish. *Nature*, **459**, 996–999.
Noble, M., Pröschel, C., and Mayer-Pröschel, M. (2011) Oxidative-reductionist approaches to stem and progenitor cell function. *Cell Stem Cell*, **8**, 1–2.
Shkolnik, K., Tadmor, A., Ben-Dor, S., Nevo, N., Galiani, D., and Dekel, N. (2011) Reactive oxygen species are indispensable in ovulation. *Proc. Natl Acad. Sci. USA*, **108**, 1462–1467.

5.1.3
Oxidative Stress in Human Diseases and in Aging

Celino, F.T., Yamaguchi, S., Miura, C., Ohta, T., Tozawa, Y., Iwai, T., and Miura, T. (2011) Tolerance of spermatogonia to oxidative stress is due to high levels of Zn and Cu/Zn superoxide dismutase. *PloS One*, **6**, e16938.
Weseler, A.R. and Bast, A. (2010) Oxidative stress and vascular function: implications for pharmacologic treatments. *Curr. Hypertens. Rep.*, **12**, 154–161.

5.1.4
Natural Substances Affecting Redox Homeostasis

Arnér, E.S. (2009) Focus on mammalian thioredoxin reductases–important selenoproteins with versatile functions. *Biochim. Biophys. Acta*, **1790**, 495–526.
Davis, B.D. and Brodbelt, J.S. (2004) Determination of the glycosylation site of flavonoid monoglucosides by metal complexation and tandem mass spectrometry. *J. Am. Soc. Mass Spectrom.*, **15**, 1287–1299.
Filipe, P., Morlière, P., Patterson, L.K., Hug, G.L., Mazière, J.C., Mazière, C., Freitas, J. P., Fernandes, A., and Santus, R. (2002) Mechanisms of flavonoid repair reactions with amino acid radicals in models of biological systems: a pulse radiolysis study in micelles and human serum albumin. *Biochim. Biophys. Acta*, **1571**, 102–114.
Galati, G., Sabzevari, O., Wilson, J.X., and O'Brien, P.J. (2002) Prooxidant activity and cellular effects of the phenoxyl radicals of dietary flavonoids and other polyphenolics. *Toxicology*, **177**, 91–104.
Hatcher, H., Planalp, R., Cho, J., Torti, F.M., and Torti, S.V. (2008) Curcumin: from ancient medicine to current clinical trials. *Cell Mol. Life Sci.*, **65**, 1631–1652.
Hur, W. and Gray, N.S. (2011) Small molecule modulators of antioxidant response pathway. *Curr. Opin. Chem. Biol.*, **15**, 162–173.
Jaquet, V., Scapozza, L., Clark, R.A., Krause, K.H., and Lambeth, J.D. (2009) Small-molecule NOX inhibitors: ROS-generating NADPH oxidases as therapeutic targets. *Antioxid. Redox Signal.*, **11**, 2535–2552.
Mesquita, A., Weinberger, M., Silva, A., Sampaio-Marques, B., Almeida, B., Leão, C., Costa, V., Rodrigues, F., Burhans, W.C., and Ludovico, P. (2010) Caloric restriction or catalase inactivation extends yeast chronological lifespan by inducing H_2O_2 and superoxide dismutase activity. *Proc. Natl Acad. Sci. USA*, **24** (107), 15123–15128.
Middleton, E., Kandaswami, C., and Theoharides, T.C. (2000) The effects of plant flavonoids on mammalian cells: implications for inflammation, heart disease, and cancer. *Pharmacol. Rev.*, **52**, 673–751.
Pietta, P.G. (2000) Flavonoids as antioxidants. *J. Nat. Prod.*, **63**, 1035–1042.
Wang, X.J., Hayes, J.D., Higgins, L.G., Wolf, C.R., and Dinkova-Kostova, A.T. (2010) Activation of the NRF2 signaling pathway by copper-mediated redox cycling of *para*- and *ortho*-hydroquinones. *Chem. Biol.*, **17**, 75–85.

5.1.5
Hormesis and the Benefits of Temporary Mild Oxidative Stress

Calabrese, V., Cornelius, C., Cuzzocrea, S., Iavicoli, I., Rizzarelli, E., and Calabrese, E. J. (2011) Hormesis, cellular stress response and vitagenes as critical

determinants in aging and longevity. *Mol. Aspects Med.*, **32**, 279–304.
Ristow, M. and Schmeisser, S. (2011) Extending life span by increasing oxidative stress. *Free Radic. Biol. Med.*, **51**, 327–336.
Siow, R.C. and Mann, G.E. (2010) Dietary isoflavones and vascular protection: activation of cellular antioxidant defenses by SERMs or hormesis? *Mol. Aspects Med.*, **31**, 468–477.

5.2 Endocrine Effects of Natural Substances

5.2.1 Are Endocrine Disruptors a Threat to Animal and Human Health?

Andres, S., Abraham, K., Appel, K.E., and Lampen, A. (2011) Risks and benefits of dietary isoflavones for cancer. *Crit. Rev. Toxicol.*, **41**, 463–506.
Diamanti-Kandarakis, E., Bourguignon, J.P., Giudice, L.C., Hauser, R., Prins, G.S., Soto, A.M., Zoeller, R.T., and Gore, A.C. (2009) Endocrine-disrupting chemicals: an Endocrine Society scientific statement. *Endocr. Rev.*, **30**, 293–342.
Patisaul, H.B. and Jefferson, W. (2010) The pros and cons of phytoestrogens. *Front. Neuroendocrinol.*, **31**, 400–419.
Routledge, E.J. and Sumpter, J.P. (1996) Estrogenic activity of surfactants and some of their degradation products assessed using a recombinant yeast screen. *Environ. Toxicol. Chem.*, **15**, 241–248.
Wuttke, W., Jarry, H., and Seidlová-Wuttke, D. (2007) Isoflavones–safe food additives or dangerous drugs? *Ageing Res. Rev.*, **6**, 150–188.
Zierau, O., Hamann, J., Tischer, S., Schwab, P., Metz, P., Vollmer, G., Gutzeit, H.O., and Scholz, S. (2005) Naringenin-type flavonoids show different estrogenic effects in mammalian and teleost test systems. *Biochem. Biophys. Res. Commun.*, **326**, 909–916.

5.2.2 How do Phytoestrogens Interfere with the Endocrine System?

Balunas, M.J., Su, B., Brueggemeier, R.W., and Kinghorn, A.D. (2008) Natural products as aromatase inhibitors. *Anticancer Agents Med. Chem.*, **8**, 646–682.
Li, W., Pandey, A.K., Yin, X., Chen, J.J., Stocco, D.M., Grammas, P., and Wang, X. (2011) Effects of apigenin on steroidogenesis and steroidogenic acute regulatory gene expression in mouse Leydig cells. *J. Nutr. Biochem.*, **22**, 212–218.
Prossnitz, E.R. and Barton, M. (2011) The G-protein-coupled estrogen receptor GPER in health and disease. *Nat. Rev. Endocrinol.*, **7**, 715–726.

5.2.3 Evaluation of Risks and Benefits of Phytoestrogens to Treat Common Diseases

Eustache, F., Mondon, F., Canivenc-Lavier, M.C., Lesaffre, C., Fulla, Y., Berges, R., Cravedi, J.P., Vaiman, D., and Auger, J. (2009) Chronic dietary exposure to a low-dose mixture of genistein and vinclozolin modifies the reproductive axis, testis transcriptome, and fertility. *Environ. Health Perspect.*, **117**, 1272–1279.
Zhang, H.H., Feng, L., Livnat, I., Hoh, J.K., Shim, J.Y., Liao, W.X., and Chen, D.B. (2010) Estradiol-17beta stimulates specific receptor and endogenous nitric oxide-dependent dynamic endothelial protein S-nitrosylation: analysis of endothelial nitrosyl-proteome. *Endocrinology*, **151**, 3874–3887.

5.2.4 Transgeneration Effects

Cederroth, C.R., Auger, J., Zimmermann, C., Eustache, F., and Nef, S. (2010) Soy, phyto-oestrogens and male reproductive function: a review. *Int. J. Androl.*, **33**, 304–316.
Chalabi, N., Coxam, V., Satih, S., Davicco, M. J., Lebecque, P., Fontana, L., Bignon, Y.J., and Bernard-Gallon, D.J. (2010) Gene signature of rat mammary glands: Influence of lifelong soy isoflavones consumption. *Mol. Med. Rep.*, **3**, 75–81.
Janesick, A. and Blumberg, B. (2011) Endocrine disrupting chemicals and the

developmental programming of adipogenesis and obesity. *Birth Defects Res. Part C, Embryo Today: Reviews*, **93**, 34–50.
Jefferson, W.N. and Williams, C.J. (2011) Circulating levels of genistein in the neonate, apart from dose and route, predict future adverse female reproductive outcomes. *Reprod. Toxicol. (Elmsford, NY)*, **31**, 272–279.
Li, Y. and Tollefsbol, T.O. (2010) Impact on DNA methylation in cancer prevention and therapy by bioactive dietary components. *Curr. Med. Chem.*, **17**, 2141–2151.
Link, A., Balaguer, F., and Goel, A. (2010) Cancer chemoprevention by dietary polyphenols: promising role for epigenetics. *Biochem. Pharmacol.*, **80**, 1771–1792.
Skinner, M.K., Manikkam, M., and Guerrero-Bosagna, C. (2010) Epigenetic transgenerational actions of environmental factors in disease etiology. *Trends Endocrinol. Metab.*, **21**, 214–222.
Vanhees, K., Coort, S., Ruijters, E.J., Godschalk, R.W., vanSchooten, F.J., and Barjesteh van Waalwijk van Doorn-Khosrovani, S. (2011) Epigenetics: prenatal exposure to genistein leaves a permanent signature on the hematopoietic lineage. *FASEB J.*, **25**, 797–807.
Walker, D.M. and Gore, A.C. (2011) Transgenerational neuroendocrine disruption of reproduction. *Nat. Rev. Endocrinol.*, **7**, 197–207.

5.2.5 Natural Substances Interfering with Nonsteroidal Signaling

de Souza Dos Santos, M.C., Gonçalves, C.F., Vaisman, M., Ferreira, A.C., and de Carvalho, D.P. (2011) Impact of flavonoids on thyroid function. *Food Chem. Toxicol.*, **49**, 2495–2502.
Heretsch, P., Tzagkaroulaki, L., and Giannis, A. (2010) Cyclopamine and hedgehog signaling: chemistry, biology, medical perspectives. *Angew. Chem. Int. Ed.*, **49**, 3418–3427.
Lee, J.H., Kishikawa, M., Kumazoe, M., Yamada, K., and Tachibana, H. (2010) Vitamin A enhances antitumor effect of a green tea polyphenol on melanoma by upregulating the polyphenol sensing molecule 67-kDa laminin receptor. *PloS One*, **5**, e11051.

5.3 Interference with Neural Functions

5.3.1 Basic Features of the Nervous System and Synaptic Transmission

Brodal, P. (2010) *The Central Nervous System: Structure and Function*, Oxford University Press, Oxford, UK.

5.3.2 Bioactivity of Cannabinoids

Di Marzo, V. (2008) Targeting the endocannabinoid system: to enhance or reduce? *Nat. Rev. Drug Discov.*, **7**, 438–455.
Fernández-Ruiz, J. (2009) The endocannabinoid system as a target for the treatment of motor dysfunction. *Br. J. Pharmacol.*, **156**, 1029–1040.
Han, J., Kesner, P., Metna-Laurent, M., Duan, T., Xu, L., Georges, F., Koehl, M., Abrous, D.N., Mendizabal-Zubiaga, J., Grandes, P., Liu, Q., Bai, G., Wang, W., Xiong, L., Ren, W., Marsicano, G., and Zhang, X. (2012) Acute cannabinoids impair working memory through astroglial CB1 receptor modulation of hippocampal LTD. *Cell*, **148**, 1039–1050.
Hill, A.J., Williams, C.M., Whalley, B.J., and Stephens, G.J. (2012) Phytocannabinoids as novel therapeutic agents in CNS disorders. *Pharmacol. Ther.*, **133**, 79–97.
Izzo, A.A., Borrelli, F., Capasso, R., Di Marzo, V., and Mechoulam, R. (2009) Non-psychotropic plant cannabinoids: new therapeutic opportunities from an ancient herb. *Trends Pharmacol. Sci.*, **30**, 515–527.
Katona, I. and Freund, T.F. (2008) Endocannabinoid signaling as a synaptic circuit breaker in neurological disease. *Nat. Med.*, **14**, 923–930.

Karsak, M., Gaffal, E., Date, R., Wang-Eckhardt, L., Rehnelt, J., Petrosino, S., Starowicz, K., Steuder, R., Schlicker, E., Cravatt, B., Mechoulam, R., Buettner, R., Werner, S., Di Marzo, V., Tüting, T., and Zimmer, A. (2007) Attenuation of allergic contact dermatitis through the endocannabinoid system. *Science (New York, NY)*, **316**, 1494–1497.

Turu, G. and Hunyady, L. (2010) Signal transduction of the CB1 cannabinoid receptor. *J. Mol. Endocrinol.*, **44**, 75–85.

5.3.3
Discovery and Bioactivity of Opioids

Feng, Y., He, X., Yang, Y., Chao, D., Lazarus, L.H., and Xia, Y. (2012) Current research on opioid receptor function. *Curr. Drug Targets*, **13**, 230–246.

Karst, M., Wippermann, S., and Ahrens, J. (2010) Role of cannabinoids in the treatment of pain and (painful) spasticity. *Drugs*, **70**, 2409–2438.

Kream, R.M., Stefano, G.B., and Ptáček, R. (2010) Psychiatric implications of endogenous morphine: up-to-date review. *Folia Biol. (Praha)*, **56**, 231–241.

Subirán, N., Casis, L., and Irazusta, J. (2011) Regulation of male fertility by the opioid system. *Mol. Med.*, **17**, 846–853.

5.3.4
Analgesic Properties of Opioids and Cannabinoids

Fields, H. (2004) State-dependent opioid control of pain. *Nat. Rev. Neurosci.*, **5**, 565–575.

5.3.5
The Hallucinogenic Drugs Psilocin, Psilocybin, and Mescaline

Geyer, M.A. and Vollenweider, F.X. (2008) Serotonin research: contributions to understanding psychoses. *Trends Pharmacol. Sci.*, **29**, 445–453.

Halberstadt, A.L. and Geyer, M.A. (2011) Multiple receptors contribute to the behavioral effects of indoleamine hallucinogens. *Neuropharmacology*, **61**, 364–381.

5.3.6
Natural Substances Affecting Cholinergic Signaling

Houghton, P.J., Ren, Y., and Howes, M.J. (2006) Acetylcholinesterase inhibitors from plants and fungi. *Nat. Prod. Rep.*, **23**, 181–199.

Muñoz-Torrero, D. (2008) Acetylcholinesterase inhibitors as disease-modifying therapies for Alzheimer's disease. *Curr. Med. Chem.*, **15**, 2433–2455.

Pope, C., Karanth, S., and Liu, J. (2005) Pharmacology and toxicology of cholinesterase inhibitors: uses and misuses of a common mechanism of action. *Environ. Toxicol. Pharmacol.*, **19**, 433–446.

Shah, N., Khurana, S., Cheng, K., and Raufman, J.P. (2009) Muscarinic receptors and ligands in cancer. *Am. J. Physiol Cell Physiol.*, **296**, C221–C232.

Soreq, H. and Seidman, S. (2001) Acetylcholinesterase – new roles for an old actor. *Nat. Rev. Neurosci.*, **2**, 294–302.

6
Nature's Drugstore for a Healthy Life

Ramawat, K.G. and Mérillon, J.-M. (2010) *Bioactive Molecules and Medicinal Plants*, Springer-Verlag, Berlin, Heidelberg.

Rehner, G.D. and Hannelore, D. (2002) *Biochemie der Ernährung*, 2nd edn, Spektrum, Heidelberg.

Widhalm, K. (2005) *Ernährungsmedizin*, 2nd edn, Verlagshaus d. Ärzte, Vienna.

6.1
Biologically Active Food and Food Components

Watson, R.R. (2010) *Bioactive Food in Promoting Health: Fruits and Vegetables*, Elsevier, London, UK.

Webb, G.P. (2011) *Dietary Supplements and Functional Food*, 2nd edn, Wiley & Sons Ltd, Chichester, West Sussex, UK.

6.1.1
The Quantity of Food Intake Affects Life Expectancy

Bishop, N.A. and Guarente, L. (2007) Genetic links between diet and lifespan: shared mechanisms from yeast to humans. *Nat. Rev. Genet.*, **8**, 835–844.

Guarente, L. (2006) Sirtuins as potential targets for metabolic syndrome. *Nature*, **444**, 868–874.

Houtkooper, R.H., Pirinen, E., and Auwerx, J. (2012) Sirtuins as regulators of metabolism and healthspan. *Nat. Rev. Mol. Cell Biol.*, **13**, 225–238.

Salminen, A. and Kaarniranta, K. (2012) AMP-activated protein kinase (AMPK) controls the aging process via an integrated signaling network. *Ageing Res. Rev.*, **11**, 230–241.

6.1.2
Resveratrol: Longevity and Health with Grapes?

Baur, J.A., Ungvari, Z., Minor, R.K., Le Couteur, D.G., and de Cabo, R. (2012) Are sirtuins viable targets for improving healthspan and lifespan? *Nat. Rev. Drug Discov.*, **11**, 443–461.

Park, S.-J., Ahmad, F., Philp, A., Baar, K., Williams, T., Luo, H., Ke, H., Rehmann, H., Taussig, R., Brown, A.L., Kim, M.K., Beaven, M.A., Burgin, A.B., Manganiello, V. and Chung, J.H. (2012) Resveratrol ameliorates aging-related metabolic phenotypes by inhibiting cAMP phosphodiesterases. *Cell*, **148**, 421–433.

Vang, O., Ahmad, N., Baile, C.A., Baur, J.A., Brown, K., Csiszar, A., Das, D.K., Delmas, D., Gottfried, C., Lin, H.Y., Ma, Q.Y., Mukhopadhyay, P., Nalini, N., Pezzuto, J. M., Richard, T., Shukla, Y., Surh, Y.J., Szekeres, T., Szkudelski, T., Walle, T., and Wu, J.M. (2011) What is new for an old molecule? Systematic review and recommendations on the use of resveratrol. *PloS One*, **6**, e19881.

Vislocky, L.M. and Fernandez, M.L. (2010) Biomedical effects of grape products. *Nutr. Rev.*, **68**, 656–670.

6.1.3
Food with Health-Promoting Bioactive Substances

Birringer, M. (2011) Hormetics: dietary triggers of an adaptive stress response. *Pharm. Res.*, **28**, 2680–2694.

Jeffery, E. and Araya, M. (2009) Physiological effects of broccoli consumption. *Phytochem. Rev.*, **8**, 283–298.

Kanwar, J., Taskeen, M., Mohammad, I., Huo, C., Chan, T.H., and Dou, Q.P. (2012) Recent advances on tea polyphenols. *Front. Biosci. (Elite Ed).*, **4**, 111–131.

Russo, M., Spagnuolo, C., Tedesco, I., Bilotto, S., and Russo, G.L. (2012) The flavonoid quercetin in disease prevention and therapy: facts and fancies. *Biochem. Pharmacol.*, **83**, 6–15.

Surh, Y.J. (2011) Xenohormesis mechanisms underlying chemopreventive effects of some dietary phytochemicals. *Ann. N. Y. Acad. Sci.*, **1229**, 1–6.

Zanoli, P. and Zavatti, M. (2008) Pharmacognostic and pharmacological profile of *Humulus lupulus* L. *J. Ethnopharmacol.*, **116**, 383–396.

6.2
Natural Compounds in Pharmacological Research

Imming, P. (2010) Molecular targets of natural drug substances: idiosyncrasies and preferences. *Planta Med.*, **76**, 1794–1801.

6.2.1
Mining for Bioactive Compounds in Marine and Terrestrial Organisms

Hanahan, D. and Weinberg, R.A. (2011) Hallmarks of cancer: the next generation. *Cell*, **144**, 646–674.

Raj, L., Ide, T., Gurkar, A.U., Foley, M., Schenone, M., Li, X., Tolliday, N.J., Golub, T.R., Carr, S.A., Shamji, A.F., Stern, A.M., Mandinova, A., Schreiber, S.L., and Lee, S. W. (2011) Selective killing of cancer cells by a small molecule targeting the stress response to ROS. *Nature*, **475**, 231–234.

Schumacher, M., Kelkel, M., Dicato, M., and Diederich, M. (2011) Gold from the sea: marine compounds as inhibitors of the hallmarks of cancer. *Biotechnol. Adv.*, **29**, 531–547.

6.2.2
Medicinal Chemistry Inspired by Natural Products

Koch, M.A., Schuffenhauer, A., Scheck, M., Wetzel, S., Casaulta, M., Odermatt, A., Ertl, P., and Waldmann, H. (2005) Charting biologically relevant chemical space: A structural classification of natural products (SCONP). *Proc. Natl Acad. Sci. USA*, **102**, 17272–17277.

Marcaurelle, L.A. and Johannes, C.W. (2008) Application of natural product-inspired diversity-oriented synthesis to drug discovery. *Prog. Drug Res.*, **66** (187), 189–216.

Wetzel, S., Bon, R.S., Kumar, K., and Waldmann, H. (2011) Biology-oriented synthesis. *Angew. Chem. Int. Ed.*, **50**, 10800–10826.

6.3
The Problem of Complexity: Analysis and Consequences

Leeson, P.D. and Springthorpe, B. (2007) The influence of drug-like concepts on decision-making in medicinal chemistry. *Nat. Rev. Drug Discov.*, **6**, 881–890.

6.3.2
From Pharmacology to Polypharmacology

Hopkins, A.L. (2008) Network pharmacology: the next paradigm in drug discovery. *Nat. Chem. Biol.*, **4**, 682–690.

Roberts, B.E., Duennwald, M.L., Wang, H., Chung, C., Lopreiato, N.P., Sweeny, E.A., Knight, M.N., and Shorter, J. (2009) A synergistic small-molecule combination directly eradicates diverse prion strain structures. *Nat. Chem. Biol.*, **5**, 936–946.

Zhao, S. and Iyengar, R. (2012) Systems pharmacology: network analysis to identify multiscale mechanisms of drug action. *Annu. Rev. Pharmacol. Toxicol.*, **52**, 505–521.

6.3.3
Reverse Pharmacology Inspired by Traditional Medicine

Patwardhan, B. and Vaidya, A.D. (2010) Natural products drug discovery: accelerating the clinical candidate development using reverse pharmacology approaches. *Indian J. Exp. Biol.*, **48**, 220–227.

Index

Plant Natural Products: Synthesis, Biological Functions and Practical Applications, First Edition.
Herwig O. Gutzeit and Jutta Ludwig-Müller.

n